HUMAN
AVI

HUMAN FACTORS IN AVIATION

2ND EDITION

Eduardo Salas and Dan Maurino

AMSTERDAM • BOSTON • HEIDELBERG • LONDON
NEW YORK • OXFORD • PARIS • SAN DIEGO
SAN FRANCISCO • SINGAPORE • SYDNEY • TOKYO

Academic Press is an imprint of Elsevier

Academic Press is an imprint of Elsevier

30 Corporate Drive, Suite 400, Burlington, MA 01803, USA
525 B Street, Suite 1800, San Diego, California 92101-4495, USA
84 Theobald's Road, London WC1X 8RR, UK

Library of Congress Cataloging-in-Publication Data
Application Submitted

British Library Cataloguing-in-Publication Data
A catalogue record for this book is available from the British Library

ISBN: 978-0-12-374518-7

For information on all Academic Press publications
visit our web site at www.book.elsevier.com

Transferred to Digital Printing in 2013

Working together to grow
libraries in developing countries

www.elsevier.com | www.bookaid.org | www.sabre.org

ELSEVIER   BOOK AID
International   Sabre Foundation

Table of Contents

I

INTRODUCTION

II

ORGANIZATIONAL PERSPECTIVES

III

PILOT AND CREW PERFORMANCE ISSUES

IV

HUMAN FACTORS IN AIRCRAFT DESIGN

V

VEHICLES AND SYSTEMS

VI

NEXTGEN COMMENTARY

For more material please visit our companion website at: www.elsevierdirect.
com/companions/9780123745187

Foreword

A lot of air has passed over the foil since the first edition of *Human Factors in Aviation* appeared in 1988—mergers, bankruptcies, security concerns, new generation "glass cockpit" aircraft. This revised edition of the classic volume edited originally by Earl Wiener, David Nagel, and Edward Carterette is a much needed addition to the aviation literature.

What is human factors? A common definition that it is the discipline that deals with human-machine interface. It deals with the psychological, social, physical, biological, and safety characteristics of individuals and groups at the sharp end of organizations and the environmental context in which they perform. I certainly didn't know this when I began to pursue my Ph.D. My doctoral training at Yale University was in social psychology. Yale had no courses or program in human factors. Indeed, the zeitgeist there and in much of academic psychology was that the only valid approach to research was through carefully controlled laboratory experiments using as data sources two- (or four-)legged subjects—usually bored undergraduates fulfilling a course requirement.

Before entering graduate school, I spent four years as a destroyer officer in the U.S. Navy. My naval experience included searching for Russian submarines off Cuba during the missile crisis. When I arrived at graduate school fresh from sea I was directed toward laboratory studies, trying to change student attitudes about mundane issues. As a president was assassinated, riots broke out in American cities, and the country continued in a frenzied race to land humans on the moon, I became less and less enamored with my research path.

Because of my military background, I was given the opportunity for my doctoral dissertation to study the behavior of aquanauts living on the ocean floor during Project Sealab. Using in-habitat video cameras and open microphones, the crew's behavior in an extremely dangerous and stressful environment could be captured and analyzed—a precursor of today's confidential observations of flightdeck behavior. I certainly did not realize that my study could best be classified as research into the human factors of an extreme environment.

This completely revised and updated version of *Human Factors in Aviation* is edited by two distinguished human factors experts. Eduardo Salas is an eminent scientist whose research in aviation psychology has contributed greatly to aviation safety, training, and flight operations.

Daniel Maurino, through his role at the International Civil Aviation Organization, has communicated the importance of human factors and system safety throughout the world.

This edition covers human performance from its system, organizational, and cultural context to relevant kinesthetic and cognitive components. Especially important, given the growing number of extremely long flights crossing many time zones is the summary of research into fatigue effects and circadian rhythms.

Section IV, discussing design issues from cockpit displays to unmanned [sic] aircraft will be of particular interest to aeronautical engineers but also provides important insights for the rest of the aviation community.

Section V provides perspectives from the central components of the aviation domain. This section gives the audience—whether students, flight crews, managers, or regulators—a nuanced view of aviation, civil and airline. It should be required reading for all to gain a needed perspective on all aspects of aviation.

Finally, I must comment on safety. Aviation has an extraordinary record that has become even better in two decades since publication of the first edition. Advancing from 1.9 hull losses per million flights to less than 1.0 per million is a remarkable achievement. Despite this progress, aircraft continue to crash. In 2009, a regional jet crashed in the United States, raising questions about the training and continuing qualification of flight crews. In the same year, a modern, widebody jet plunged into the ocean off the coast of Brazil. Although the wreckage remains on the Atlantic seabed, attention has focused on speed sensors and their reliability. The clear lesson from these tragedies is that human factors remains a critical component of aviation safety. It is an exciting field that can save lives and increase operational efficiency and economy.

Robert L Helmreich, Ph.D., FRAes
Professor Emeritus of Psychology
The University of Texas at Austin
Granite Shoals, Texas

Contributors

Amy L. Alexander Ph.D.   Human Factors Scientist, Aptima, Inc., USA

Siobhan Banks, Ph.D.   Division of Sleep and Chronobiology, Department of Psychiatry, University of Pennsylvania School of Medicine, Philadelphia, PA, USA

John Bent   Manager Flight Training Centre, Hong Kong International Airport, Hong Kong, PEOPLE'S REPUBLIC OF CHINA

Kwok Chan Ph.D.   Head of Corporate Safety & Quality, Hong Kong Dragon Airlines, Hong Kong, PEOPLE'S REPUBLIC OF CHINA

Stephen M. Casner Ph.D.   NASA Ames Research Center, Moffett Field, CA, USA

Michael Curtis M.A.   Institute for Simulation & Training, University of Central Florida, Orlando, FL, USA

Sidney W.A. Dekker   Lund University School of Aviation, Ljungbyhed, Lund, SWEDEN

David F. Dinges Ph.D.   Division of Sleep and Chronobiology, Department of Psychiatry, University of Pennsylvania School of Medicine, Philadelphia, PA, USA

R. Key Dismukes   Chief Scientist for Human Factors, NASA Ames Research Center, Moffett Field, CA, USA

Francis T. Durso Ph.D.   School of Psychology, Georgia Institute of Technology, Atlanta, GA, USA

Thomas Ferris   Center for Ergonomics, Department of Industrial and Operations Engineering, University of Michigan, Ann Arbor, MI, USA

John M. Flach   Department of Psychology, Wright State University, Daton, OH, USA

Kevin Gluck Ph.D.   Air Force Research Laboratory – RHAC, Mesa, AZ, USA

David A. Graeber Ph.D.   Boeing Research & Technology, Seattle, WA, USA

Deborah DiazGranados   Institute for Simulation & Training, University of Central Florida, Orlando, FL, USA

Captain William Hamman   United Airlines Center of Excellence in Simulation Education and Research, USA

Robert L. Helmreich Ph.D.   FRAes, Professor Emeritus of Psychology, The University of Texas at Austin, Granite Shoals, Texas, USA

Alan Hobbs Ph.D. San Jose State University, Human Systems Integration Division, NASA Ames Research Center, Moffett Field, CA, USA

Alan R. Jacobsen Ph.D. Boeing Commercial Airplanes, Seattle, WA, USA

Florian Jentsch Ph.D. University of Central Florida, Orlando, FL, USA

Barbara G. Kanki Ph.D. NASA Ames Research Center, Moffett Field, CA, USA

Paul Krois Federal Aviation Administration, Washington, DC, USA

Ann-Elise Lindeis Ph.D. NAV CANADA, Ottowa, Ontario, CANADA

Tom McCloy Federal Aviation Administration, Washington, DC, USA

Melissa M. Mallis Ph.D. Institutes for Behavior Resources, Inc., Baltimore, MD, USA

Dan Maurino Capt. International Civil Aviation Organization, Montreal, Quebec, CANADA

Kathleen L. Mosier Dept. of Psychology, San Francisco State University, San Francisco, CA, USA

Manoj S. Patankar Ph.D. National Center for Aviation Safety Research, St. Louis University, St. Louis, MO, USA

Dino Piccione Federal Aviation Administration, Washington, DC, USA

Amy R. Pritchett School of Aerospace Engineering, Georgia Tech, Atlanta, GA USA

William L Rutherford MD, Capt. (ret.) United Airlines, USA

Edward J. Sabin Ph.D. National Center for Aviation Safety Research, St. Louis University, St. Louis, MO, USA

Eduardo Salas Ph.D. Institute for Simulation & Training, University of Central Florida, Orlando, FL, USA

Nadine Sarter Ph.D. Center for Ergonomics, Department of Industrial and Operations Engineering, University of Michigan, Ann Arbor, MI, USA

Thomas B. Sheridan Ph.D. DoT Volpe National Transportation Systems Center, Washington, DC, USA

Marissa L. Shuffler Institute for Simulation & Training, University of Central Florida, Orlando, FL, USA

Pamela S. Tsang Department of Psychology, Wright State University, Daton, OH, USA

Michael A. Vidulich Ph.D. 11th Human Performance Wing, Wright-Patterson Air Force Base, OH, USA

Christopher D. Wickens Alion Science Corporation, Boulder, CO, USA

John Wiedemann Boeing Commercial Airplanes, Seattle, WA, USA

John A. Wise Ph.D. C.P.E. Honeywell AES, Phoenix, AZ, USA

PART I

INTRODUCTION

Human Factors in Aviation: An Overview

Eduardo Salas
University of Central Florida
Dan Maurino
International Civil Aviation Organization
Michael Curtis
University of Central Florida

In 1988, Earl Wiener and David Nagel's *Human Factors in Aviation* was released. At a time when the stealth bomber, Hubble telescope, and *perestroika* were fresh ideas, this important book signified a symbolic shift in the role of human factors within the aviation industry. "Human factors" was not a new concept, and human factors research, which traces its origins to aviation, had slowly established its place in improving safety in aviation already. At that point in the intertwined history of aviation and human factors, though, human factors researchers were just beginning to find themselves prominently involved in the design of aviation systems. This was in stark contrast to previous decades when human factors was not emphasized in aircraft design and aviation operations, but instead, was generally a corrective science. This evolved role helped the expansion of human factors research in the field. Whereas the origin and early years of study had predominantly been in cockpit and cabin technology design, the industry was beginning to address other important topics like cockpit organization, crew interaction, crew fitness, judgment, and automation. In all, their book should be considered a seminal contribution to aviation research as a whole. It represents one of the first books to present human factors topics relevant to aviation in a manner accessible not just for human factors professionals but also to pilots, aviation industry personnel, and others casually interested in the topic area.

In March of the same year, Avianca Flight 410 crashed into mountains shortly after taking off from a Colombian airport, killing all passengers and crew aboard (Aviation Safety Network, n.d.). The official cause of the accident was determined as controlled flight into terrain. This was precipitated by poor crew teamwork and cockpit distractions, including nonflying personnel present in the cockpit. Mentioning the tragedy of this accident following praise on the impact of the initial edition of this book serves to illustrate an important point in aviation research: Despite what is accomplished in terms of safety and operational improvements in aviation, the work is never done. Like any other form of transportation, the prospect of having a 100% safety record in aviation is improbable. As humans, our propensity to propel technology design into futuristic states will always drive the need for the research community to address evolving demands of aviation.

At any point in the history of human factors in aviation, one could characterize its current state at that time by the progress that had been made and by the opportunities that presented themselves for the future. Keeping this in mind, in this chapter we aim to provide a brief snapshot of aviation human factors since the first edition of this book. We will first highlight a few brief themes that show the progress that has been made in aviation research since the first edition. Following this, we will talk about the opportunities that we as human factors professionals, pilots, instructors, maintenance personnel, air traffic controllers, and interested parties in general are presented with currently to drive future generations of aviation. To conclude, we will provide a short overview of the chapters that follow in this edition. In all, we hope this chapter orients the reader to the causes and effects that guide the cutting edge of our field now, and whets the appetite of inquisitive minds to explore the subsequent chapters of this edition further.

PROGRESS SINCE THE FIRST EDITION

A little more than two decades has passed since the first edition of this book. Since that time, the industry has continued to change, technology has improved, and demands on crews have shifted. In part, the progress that scientists and developers have made in the topic areas covered in the initial edition have perpetuated the maturation and evolution of aviation. There have been a number of advances to classic human factors topic areas such as cockpit design in addition to the further development of several topics that were just burgeoning at the time of the first edition. The progress that has been made is too far-reaching to sufficiently cover in this chapter—in fact, we really don't do that here; other forums will have to document that. Despite this, we feel that there are several general areas in which progress has helped shape the evolving industry. We next briefly touch on two topic areas and how these have influenced training, technology, and flight procedure in general.

Training Crews Instead of Individuals—the Importance of Teamwork

In the mid-1980s the focus of attention for most pilot training programs still relied heavily on development of technical skill. Little emphasis was placed on developing teamwork skills within the flight crew. Evidence in the form of accident and incident report data suggested that suboptimal crew interactions were contributing to a noticeable number of human errors in the skies. Around this time, a number of important publications began surfacing with a call for increased emphasis on developing crew team behaviors in addition to the already ingrained technical skills required in flight (Foushee, 1984; Foushee & Helmreich, 1988; Orlady & Foushee, 1987).

Since the genesis and initial investigations into how crew interaction influences the flight process, the concept of crew resource management (CRM) has been developed into an important aspect of aviation. The complexity of the interacting variables within the concept of CRM have spawned research and subsequent progress into many aspects of the inputs, processes, and outputs of CRM as a whole. This includes investigation of inputs such as how team composition, cohesiveness, and organizational structure influence crew behavior (Guzzo & Dickson, 1996). In addition, researchers have aimed to improve our understanding of team processes such as communication and leadership, among others (Salas, Burke, Bowers, & Wilson, 2001). This has led to the widespread incorporation of CRM training in the curriculum. This includes focusing on crew behavior in full flight training simulations such as Line Oriented Flight Training (LOFT) and development of evaluation methods to accurately assess the pilot skill not only for technical skill but also in the less concretely observed behaviors that make up CRM. What has resulted is progress from a well-thought-out, yet unproven, concept for improving cockpit performance to a tested, successful method that is widely used and accepted in the aviation community for flight crew training. The development of CRM has resulted in a shift from the individualistic focus of training technical skill to the complexity of teamwork that affects standard flight operations.

Technological Advances—Shifting Performance Requirements

As one would expect, technological advance has served both as a product of progress made in aviation human factors and as a catalyst for new directions of research in the field. Technological advance has influence in all aspects of aviation from operations to training. Although it is difficult to get a snapshot of all advances in this area due to continual and rapid growth, there are some general topics that represent many of the more specific areas of research that are constantly at the cutting edge.

In the 1980s the technological and operational realization of automated systems in aviation was just starting to take shape. There was little question as to whether or not automation would have an increased role in aviation, but the impact of that automation on crew performance was largely unsupported by research (Wiener, 1989; Wiener & Curry 1980). Automated systems became more sophisticated, and research into operational concerns such as mode awareness (Sarter & Woods, 1995) and automation surprise

(Sarter, Woods, & Billings, 1997), as well as theoretical works on utilizing automation (Parasuraman & Riley, 1997; Parasuraman, Sheridan, & Wickens, 2000) tempered the "rush to implement" automation mentality. Instead, these works encouraged thoughtful inclusion of automation and consideration of issues associated with automation in parallel to the obvious benefits. The modern aircraft now features many exclusively automated features not possible in previous generations of aircraft. Although there are still challenges with optimizing human-automation interaction, progress in the last two decades has afforded automated features that are designed with human operational needs in mind.

Similar to these developments in automation, the cockpit instrumentation itself has continually evolved in line with the technological advances in computer and display technology. Display development, one of the more heavily studied areas of aviation human factors, originally focused on optimal display of information in a visually crowded cockpit. In that case, physical location and display size were among the most important features. Now, with increasingly computerized displays, it is possible to house much more information in a single display. Although the principles of display used in classic cockpits such as spatial proximity (Wickens & Carswell, 1995) are still applicable to modern instrumentation, the newer "glass cockpit" has led to a need for investigations into how the organization of information in virtual space can impact flight (Podczerwinski, Wickens, & Alexander, 2002; Woods & Watts, 1997). The glass cockpit has, in effect, decluttered the instrument panel, reducing the potential for visual overload, but heightening the need to address the different yet equally challenging prospect of cognitive overload. Improved display and computing capabilities have even spawned new conceptions for multimodal (Van Erp, 2005) and increasingly integrated displays using enhanced and synthetic imagery (Bailey, Kramer, & Prinzel, 2007). As the evolution of cockpit instrumentation continues toward more computerized display, the amount of information and way in which it is integrated will become increasingly important issues for cockpit design.

Technology has also shaped the development of crew training. Although full motion simulators have been in use for several decades, the simulated imagery and available functions associated with simulation training have improved. With that, more realistic recreations of actual flight are available for the implementation of simulation-based training methods like LOFT. In addition, findings suggest the use of low-cost, more accessible platforms deployed on personal computers or laptops may provide training benefits (Taylor, Lintern, Hulin, Talleur, Emanuel & Phillips, 1999; Prince & Jentsch, 2001; Salas, Bowers, & Rhodenizer, 1998). These low-cost solutions can help reduce the burden of time and cost that already dictates the availability of full-motion simulations.

In sum, progress in aviation human factors, since the publication of the last edition of this book, has both helped shape the current state of aviation and also uncovered new areas of focus. Fortunately, in line with the progress in research, industry-wide interest and implementation has driven improved safety. Developments like CRM training and collection of safety report data from sources such as Line Operations Safety Audit (LOSA) and Aviation Safety Action Plan (ASAP) have helped guide development of more effective training. Technological advance in the cockpit have produced more improved ways of keeping aircraft in safe operational states. Devices such as the Traffic alert and Collision Avoidance System (TCAS) have improved safety in an increasingly congested airspace. In addition, new theoretical concepts like threat and error management (Klinect, Wilhelm, & Helmreich, 1999) and situation awareness (Endsley, 1995) added to discussions on the factors affecting flight performance.

In all, the examples in this chapter just provide a small sampling of the progress that has been made in aviation through human factors research in the past two decades. Whereas most of the focus previously involved pilot performance and instrument design, today human factors researchers are investigating all aspects of the aviation industry. This includes but is certainly not limited to maintenance, air traffic, and even organizational structure, all with the goal of sending aviation into the future with improved safety, efficiency, and cost savings. Now that we have briefly touched on the progress in aviation human factors since the original edition of this book, we will now look toward the future and the opportunities that await.

A LOOK AHEAD ...

For several decades, aviation slowly evolved. Improvements in technology, operations, and organizational structure came about at a gradual pace, slowly improving safety and the efficiency of operation to levels far beyond previous generations of flight. As each improvement took effect, aviation slowly became more accessible, and thus, began to grow. The proverbial snowball began to roll, picking up additional demands for improvement in a growing field. Growth spawns research, which leads to technological advance, which leads to operational adjustments, which leads to organizational updates, which leads to growth to the point that there is no identifiable trigger for the beginning of this process, but each advance in the industry now has a ripple effect for change throughout. All of this has led to the current state in which change permeates every corner of aviation. The demand for flight is greater than ever before, leading to industry growth at unprecedented rates worldwide. Industry growth leads to an expanding workforce, which puts pressure on the organization to maximize training time and quickly produce capable pilots. Meanwhile, technological advance is facilitating the implementation of increasingly complex, automated systems that alter the way that pilots and personnel interact with aircraft systems. At the same time, a call for industry-wide improvement to the traffic management system is pressing researchers, industry professionals, and airlines to keep up. All of these influences, therefore, have an impact and human factors implications.

The growth and change that is occurring in the industry signifies not just a shift in the structure of the industry but also an opportunity to revolutionize the systems and methods in which all facets of the industry operate on. In this section we will briefly describe the predominant push for change in the aviation industry and the opportunity that is upon us.

Aviation in the NextGen

Although the Next Generation Air Traffic System (NextGen) program is only one of the catalysts for change in aviation currently, it represents the most dramatic call for change in the industry.

NextGen (FAA, 2009), which has a similar European counter-part (Single European Sky ATM Research [SESAR]; European Commission, 2005), is an initiative brought forth by the Federal Aviation Administration (FAA) to change the nature, necessity, and frequency of interaction between pilots and air traffic control. Ultimately, NextGen represents a shift from primarily a ground-based system of air traffic control to one that utilizes satellite-based technology. Current weather, data communication, and positioning technology now provide pilots the fortune of real-time updates in the cockpit. Using this technology, the NextGen program is geared toward shifting some of the flight path and traffic separation burden into the cockpit.

Ultimately, the NextGen program is intended to help address issues associated with the enormous burden of growth that air traffic controllers are already feeling worldwide. Before this goal can truly be realized, the industry has to prepare itself for the technological, operational, and organizational impact that NextGen will have. In the past, human factors was largely an afterthought in the design process of aviation systems. The job of the human factors professional would be to assess the state of a currently existing system and make recommendations for improvement. For NextGen, this may no longer be the case. In fact, at a recent panel discussion at an international aviation conference, Dino Piccione, a representative for the FAA on the panel, suggested that instead of being an afterthought, NextGen presents the opportunity for human factors to guide the design of systems that will comprise the future of aviation (Boehm-Davis, 2009). The balance of optimism and skepticism from the audience suggested that the entire aviation community may not be ready to accept this challenge, but Piccione drove home an important point. The dramatic change called for, and the timeframe in which it is proposed to take effect, will make it necessary for us collectively to take what we know about human factors and project that onto the future state of the industry. To design not just for the present but for the future of aviation.

The challenge of redesigning the air traffic management system, as we know it, is no small task. NextGen requires input from all phases of the aviation system. It will require new technology, role responsibility, crew interactions, training, organizational structure, and a host of other changes to fully transition to the new system. Whereas previous introduction of technology caused a spike in incident while everyone adjusted, perhaps we can use our knowledge of human factors to reduce the adjustment period for NextGen and related programs.

HUMAN FACTORS IN AVIATION: THIS EDITION

Although the message of this book is to discuss where we are, the sweeping changes briefly described earlier promise to alter the face of the industry in coming years. In addition to knowing where we have come from, and where we are in aviation, looking to and preparing for what the future holds is more important than ever before. With that in mind, we look to the following chapters to orient readers to human factors aviation and arm them with a base of information to guide progress into the changing future of this complex industry.

Overview of the Book

This book is intended to provide a current state of human factors in the aviation industry. Simply reviewing the current state of affairs in this field, however, does not adequately represent what industry personnel, researchers, and the aviation community at large are experiencing. In fact, if a one-word summary were required to describe human factors within the aviation industry at this moment in time, there is really only one relevant response: change.

The chapters in this book are organized into five sections. Following this introductory chapter, the next section focuses on the organizational perspective. At the organizational level, these chapters address human factors issues that occur at a macro level. As an industry composed of High Reliability Organizations (HROs), this section delves into the quantitative and qualitative conceptions of safety, and addresses the organization's responsibility for promoting a climate of safety.

In Chapter 2, Tom Sheridan discusses the origin and fundamental ideas of systems from a quantitative perspective. Using this perspective, he describes a number of relevant system models of control, decision making, information, and reliability. By providing history, definition, and normative modeling examples, he makes a case for how developing "systems thinking" encourages engineers to design with the human in mind.

Amy Pritchett (Chapter 3) follows this by emphasizing the consideration of safety as a function of risk mitigation distributed across technology, operational concepts and procedures, and effective human performance. Using the design of a collision avoidance system, she outlines how system "negatives" such as risk, error, and failures as well as positive human contributions such as shared representations, situation awareness, adaptation, and preemptive construction of the operation environment can help guide design using the system safety perspective.

In Chapter 4, Manoj Patankar and Edward Sabin address safety culture. By focusing on previously unaddressed issues like cultural evolution and the emergent role of leadership, they provide descriptions of safety culture measures and how these measures reflect safety culture. The resulting framework emphasizes the importance of change in values, leadership strategy, attitudes, and behaviors to the improvement of safety culture.

In Chapter 5, Sidney Dekker and David Woods provide a discussion on the importance of reliability in HRO. They address previous engineering conceptions of reliability, and how these may not fully encompass the concept of organizational reliability. The concept of Resilience Engineering is proposed as a way to address safety oversight such as overconfidence and performance demands that can interfere with recognition of organizational safety concerns.

Pilot and Crew Performance Issues

After providing a basis for a macro understanding of human factors in aviation, the third section narrows the focus to individual pilot and crew performance issues. A common thread in these chapters is in their agreement that changes that have and are occurring in the cockpit have altered the pilot and crew roles in the aviation system. The first four chapters of this section are heavily focused on specific concepts that influence performance.

In the first chapter in this section (Chapter 6), Kathleen Mosier suggests the idea that pilot tasking has evolved from a primarily kinesthetic task to a much more cognitive task as a result of technological advances. She discusses the hybrid ecology that results from the combination of naturalistic and electronic elements in the cockpit. Given the increased importance of NextGen in the aviation evolutionary process, she describes how facilitation of both correspondence and coherence are critical to foster cognitive sensibility in the modern cockpit.

Chapter 7, by Michael Vidulich, Christopher Wickens, Pamela Tsang, and John Flach, examines how the demands of information processing, like attention, are shifting with new aviation technology. They discuss how the balance between essential skill and human bias influences performance in aviation, as well as how information-processing limitations drive development of concepts like mental workload and situation awareness to help facilitate design and system assessment for current and future aviation systems.

In Chapter 8, Frank Durso and Amy Alexander further discuss the relationship between workload and situation awareness in relation to performance. They provide a framework made up of a situation understanding, a workload, and a critical management subsystem to describe the relationship of these concepts to each other and to performance. They address the dissociative nature of these concepts and make recommendations for how these might help predict performance.

In Chapter 9, Eduardo Salas, Marissa Shuffler, and Deborah DiazGranados provide an update to Foushee and Helmreich's chapter (1988) on team interaction and flight crew performance. The chapter addresses the advancement of crew performance research and outlines teamwork behaviors that impact crew performance. Through this, they describe ways currently utilized to help improve crew performance, most notably with CRM.

The next three chapters in this section delve into the measurement, training, and analysis of individual and crew performance.

John Bent and Kwok Chan (Chapter 10) address the importance of improving training to keep up with unprecedented growth in the aviation industry. They combine discussion on how growth is affecting industry safety standards with how it threatens training and safety. Based on this, they provide a detailed list of both hardware and software implications for the myriad of simulation-based training objectives in aviation.

In Chapter 11, Key Dismukes focuses his discussion on human error. More specifically, he discusses the measurement and analysis of human error in the cockpit as it applies to highly skilled pilots as opposed to novices. By addressing these various sources of error information, Dismukes contends that development of countermeasures are contingent on pursuing a more comprehensive understanding of error in this context.

Chapter 12, by Kevin Gluck, serves as an introduction to the concept of cognitive architectures and its utility in the aviation industry. Gluck provides a definition of cognitive architecture and a guide for deeper investigation into important works on cognitive architecture. He describes current computational simulation and mathematical modeling techniques for developing cognitive architectures and identifies a number of challenges to be addressed for improving cognitive architecture in aviation.

The final chapter in this section (Chapter 13) by Melissa Mallis, Siobhan Banks, and David Dinges, talks about the effects of fatigue in aviation. The biological influence of the circadian system and the homeostatic sleep drive are discussed in terms of how they influence fatigue in terms of work factors. Based on this, the authors discuss the challenges that fatigue offers and present several innovative approaches to address fatigue management.

Human Factors in Aircraft Design

In any human factors discussion on complex systems, the conversation would be incomplete without mention of the technology that flight crews interact with. The fourth section of this book focuses on the design of aircraft. In the first chapter in this section,

Chapter 14, Michael Curtis, Florian Jentsch, and John Wise concentrate on how technological advance has changed the nature in which information is displayed in the cockpit. They discuss how source, type, function, and environmental integration of displayed information influences pilot processing. Their discussion mirrors Mosier's discussion about the transition to higher cognitive demand with new displays and discusses how several near-future changes in aviation will influence display design.

Thomas Ferris, Nadine Sarter, and Christopher Wickens (Chapter 15) follow this with a concentration on the impact of cockpit automation on the aviation system. After discussing the current theoretical background on using automation, the authors mention how automation leads to pilot automation interaction breakdowns. Additionally, they outline a number of unintended consequences of automation implementation like workload imbalance, deskilling, and trust issues. Using what is currently known about automation, they include a discussion on how automation will influence future systems that result from NextGen and other near-future aviation industry objectives.

In Chapter 16, Alan Hobbs discusses the importance of the increasingly relevant topic of unmanned aircraft systems (UAS) into the present and future of the aviation industry. He posits that poor human-system integration design, not technological hindrance, are the main concern for UAS development. He organizes his discussion around UAS-relevant operation issues, including teleoperation, design of ground control stations, transfer of control, airspace issues, and maintenance challenges.

Integration and crew station design are the focus of the final chapter of this section, Chapter 17, by Alan Jacobsen, David Graeber, and John Wiedemann. After providing an evolutionary history of both military and commercial crew station design, they describe the competing flight crew station requirements that must be considered in the design process. In light of the increasing complexity of the system and information that needs to be integrated, the authors discuss a number of methods and tools to aid in crew station design.

Vehicles and Systems

To support the notion that there are no limits to the impact of human factors in all of aviation, the final section of this book addresses several specific perspectives that are underrepresented in the aviation human factors discussion. The first two chapters continue the discussion on the air operations perspective, but they are taken from slightly different viewpoints.

The opening chapter in this section, by William Hamman (Chapter 18), provides an airline pilot's perspective on the human factors challenges in the modern cockpit. The chapter raises concern over the effect that changes in the industry are having on the pilot population. Hamman discusses how reduced support for safety reporting programs like Aviation Safety Reporting System (ASRS), Flight Operational Quality Assurance (FOQA) and ASAP in combination with a growing demand for capable pilots is especially concerning. Increased cockpit complexity and limited training resources are leading to reliance on an increasingly inexperienced and overworked pilot fleet.

Chapter 19, by Stephen Casner, addresses the human factors issues associated with general aviation. More specifically, the chapter focuses on personal flying, flight training, and business flying. The chapter discusses several human factors issues similar to those faced in commercial and military flight as well as unique challenges that one would not likely encounter in the other two fields of research.

The next two chapters address human factors in ground operations. In Chapter 20, Ann-Elise Lindeis presents a discussion about the critical issues associated with air traffic management. After providing an effective overview of the air traffic controller–pilot relationship, she discusses the relevant human factors issues associated with effectively maintaining communication, navigation, and surveillance in this relationship. She describes how identifying operational issues through error reporting, air traffic control (ATS) investigations, and observation of normal operations can guide the improvement of ATS.

In Chapter 21, Barbara Kanki closes out the discussion on vehicles and systems with a commentary on aviation maintenance human factors. The chapter breaks the timeline of human factors in aviation maintenance into two eras, from 1988 to 1999 and from 2000 to the present. Using numerous accident report examples, Kanki effectively outlines the important progress made in this area of aviation human factors and points out the importance of sustaining maintenance error management and maintenance error in safety of flight to improve aviation maintenance output.

To complete this edition, Chapter 22, by Dino Piccione, Paul Croise, and Tom McCloy, provides a more detailed commentary on what NextGen and its worldwide counterparts hold for the future of aviation. The aviation industry is on the cusp of major changes in how flight crews and ground personnel interact to maintain separation in the skies.

A FINAL COMMENT

Ultimately, the aim of this book is to orient readers to the current state of human factors in aviation, and, with this, to prepare them for the challenges that are forthcoming in future generations of aviation. We hope that this new edition inspires scientists and those in practice to reach new heights in the application of human factors principles to aviation. We hope that this edition shows the progress, evolution, and maturation of the field. We hope that this edition educates, challenges, and guides those who believe that the science of human factors has much to say about our aviation system. Time will tell. We look forward to the third edition.

References

Aviation Safety Network. (n.d.) *Accident description: Avianca flight 410.* Retrieved July 1, 2009, from <http://aviation-safety.net/database/record. php?id=19880317-0/>

Bailey, R. E., Kramer, L. J., & Prinzel, L. J. (2007). Fusion of Synthetic and Enhanced Vision for All-Weather Commercial Aviation Operations. *NATO-OTAN*, 11.11–11.26.

Boehm-Davis, D. (Chair). (2009). Perspectives on human factors issues in NextGen. Panel discussion presented at the April 2009 International Symposium on Aviation Psychology, Dayton, OH. [panelists: Boehm-Davis, D., Carr, K., Lyall, B., & Piccione, D.]

Endsley, M. (1995). Measurement of situation awareness in dynamic systems. *Human Factors, 37*(1), 65–84.

European Commission. (2005). The SESAR program: Making air travel safer, cheaper and more efficient. Electronic memo retrieved May 10, 2009, from <http://ec.europa.eu/transport/air_portal/sesame/doc/2005_11_memo_sesar_en.pdf>

Federal Aviation Administration. (2009). What is NextGen? Retrieved March 16, 2009,from <http://www.faa.gov/about/initiatives/nextgen/defined/what/>

Foushee, H. C. (1984). Dyads and triads at 35,000 feet: Factors affecting group process and aircrew performance. *American Psychologist, 39*(8), 885–893.

Foushee, H. C., & Helmreich, R. L. (1988). Group interaction and flight crew performance. In E. L. Wiener & D. C. Nagel (Eds.), *Human factors in aviation* (pp. 189–228). London: Academic Press.

Guzzo, R. A., & Dickson, M. W. (1996). Teams in organizations: Recent research on performance and effectiveness. *Annual Review of Psychology, 47*, 307–338.

Klinect, J. R., Wilhelm, J. A., & Helmreich, R. L. (1999). Threat and error management: Data from line operations safety audits. In *Proceedings of the Tenth International Symposium on Aviation Psychology* (pp. 683–688). Columbus, OH: The Ohio State University.

Orlady, H. W., & Foushee, H. C. (1987). *Cockpit Resource Management Training, (NASA CP-2455)*. Moffett Field, CA: NASA Ames Research Center.

Parasuraman, R., & Riley, V. (1997). Humans and automation: Use, misuse, disuse, abuse. *Human Factors, 39*(2), 230–253.

Parasuraman, R., Sheridan, T. B., & Wickens, C. D. (2000). A model for types and levels of human interaction with automation. *IEEE Transactions on Systems, Man, and Cybernetics-Part A: Systems and Humans, 30*(3), 286–297.

Podczerwinski, E., Wickens, C. D., & Alexander, A. L. (2001). *Exploring the "Out of Sight, Out of Mind." Phenomenon in Dynamic Settings across Electronic Map Displays* (Tech. Rep. ARL-01-8/NASA-01-4). Savoy, IL: University of Illinois at Urbana-Champaign, Aviation Research Lab.

Prince, C., & Jentsch, F. (2001). In E. Salas, C.A. Bowers, & E. Edens (Eds.), *Aviation crew resource management in training with low fidelity devices.* (pp. 147–164). Mahwah, NJ: Lawrence Erlbaum.

Salas E., Bowers, C. A., & Edens, E. (Eds.), (2001). *Improving Teamwork in Organizations.* Mahwah, NJ: Lawrence Erlbaum (pp. 147–164).

Salas, E., Bowers, C. A., & Rhodenizer, L. (1998). It is not how much you have but how you use it: Toward a rational use of simulation to support aviation training. *International Journal of Aviation Psychology, 8*(3), 197–208.

Salas, E., Burke, C. S., Bowers, C. A., & Wilson, K. A. (2001). Team training in the skies: Does crew resource management (CRM) training work? *Human Factors, 43*(4), 641–674.

Sarter, N., & Woods, D. (1995). How in the world did we ever get into that mode? Mode error and awareness in supervisory control: Situation awareness. *Human Factors, 37*(1), 5–19.

Sarter, N., Woods, D., & Billings, C. (1997). Automation surprises. In G. Salvendy (Ed.), *Handbook of human factors/ergonomics* (2nd ed.) (pp. 1926–1943). New York: Wiley.

Taylor, H. L., Lintern, G., Hulin, C. L., Talleur, D. A., Emanuel, T. W., & Phillips, S. I. (1999). Transfer of training effectiveness of a personal computer aviation training device. *International Journal of Aviation Psychology, 9*(4), 319–335.

Van Erp, J. (2005). Presenting directions with a vibro-tactile torso display. *Ergonomics, 48,* 302–313.

Wiener, E. L. (1989). *Human factors of advanced technology ("glass cockpit") transport aircraft (NASA Contractor Rep. No. 177528).* Moffett Field, CA: NASA Ames Research Center.

Wiener, E. L., & Curry, R. E. (1980). Flight deck automation: Promises and problems. *Ergonomics, 23,* 995–1011.

Wickens, C., & Carswell, C. (1995). The proximity compatibility principle: Its psychological foundation and relevance to display design. *Human Factors, 37*(3), 473–494.

Woods, D., & Watts, J. (1997). How not to have to navigate through too many displays. In M. Helander, T. Landauer, & P. Prabhu (Eds.), *Handbook of human-computer cooperation* (2nd ed.) (pp. 617–650). New York: Elsevier.

PART II

ORGANIZATIONAL PERSPECTIVES

The System Perspective on Human Factors in Aviation

Thomas B. Sheridan

DoT Volpe National Transportation Systems Center
and Massachusetts Institute of Technology
Sheridan@mit.edu 1–617–244–4181

INTRODUCTION

This chapter provides reviews the system perspective in terms of its origins and fundamental quantitative ideas. An appreciation of these basic concepts adds rigor to analysis and synthesis of human-machine systems, and in particular to such systems in aviation. Other chapters in this volume consider aviation systems from a more qualitative perspective in consideration of safety, training, and so on.

The terms *systems, systems approach, systems thinking,* and so on are found to such a degree in current technical literature that they have almost lost their meaning. This chapter represents an effort to remind the reader of the meaning of "system," where it comes from, and what it implies for research, design, construction, operation, and evaluation in aviation, especially with regard to the human role in aviation.

Human factors professionals, pilots, and operational personnel in air traffic management and related practitioners who know about "systems" only as a general and often vague term for something complex can benefit from knowing a bit of the history, the people, and the quantitative substance that underlies the terminology.

The chapter begins by defining what is meant by a system, then discusses the history of the idea, the major contributors and what they contributed, and what made the systems idea different from previous ideas in technology. It goes on to give examples of systems thinking applied to design, development, and manufacturing of aviation systems in consideration of the people involved. Salient system models such as control, decision, information, and reliability are then explicated.

What Is a System?

A system has been defined variously as: "a set of interacting or interdependent entities, real or abstract, forming an integrated whole"; "a collection of parts which interact with each other to function as a whole"; and "a set of regularly interacting or interrelating groups of activities." Systems can be man-made or natural. The component parts can exist at any structural level: molecules, crystals, chemical substances, mechanical parts, biological cells, assemblies of mechanical parts, organs or organisms, vehicles, buildings, humans, cities, human organizations, nations, planets, galaxies, and so on. One can think of systems organized at various levels, where the "system" at a lower level becomes a component part at a next higher level (Figure 2.1). The level one chooses depends entirely on one's purpose.

The component entities, parts, or activities must be identifiable as must the whole set—to distinguish the component parts from each other and to distinguish the system from its environment. At the boundaries of each component are one or more input variables, where the input variables to that component have some significant causal relation to the output variable from that component. These are shown in Figure 2.2 by a block diagram, where the arrows are variables. Note that each block has only a single output, but that output can be a function of one or several input variables. An output from one component can be the input to one or many other components, as shown.

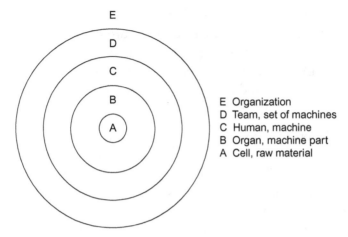

FIGURE 2.1   Levels at which systems are typically bounded and modeled.

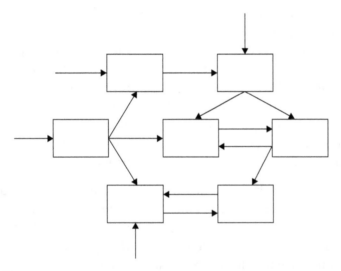

FIGURE 2.2   Component transformations (blocks) and variables (arrows) comprising a system.

The variables can be physical variables such as temperature, pressure, force, distance, voltage, current, and so on. They can also be more abstract entities such as supply, demand, stimulus, response, cooperation, competition, and so on—that are tractable insofar as they are operationally defined and measurable. The blocks represent transformations between variables, arrows representing causality.

Whether a particular variable is an input or an output to a given component depends on how other variables are defined. For example, consider a mechanical spring. If an applied force is considered an input (cause), then the displacement is the output (effect). By contrast, if the displacement is considered the input, then the force is the output.

For any given (defined, bounded) system the variables that have no explicit causes within the system are called *exogenous* or *independent* variables (inputs to the system from outside), whereas the others caused from within the system are *endogenous* or *dependent* variables. A system that is significantly dependent on its environment (characterized by exogenous inputs) is called an *open system;* otherwise, it is called a *closed system.*

The blocks in Figure 2.2, representing transformations between input and output can take many forms (including, e.g., words, pictures, computer programs) but are most useful when expressed in mathematical form. In a simplest case, the transformation is just an indication of whether any given input increases the output or decreases it. More generally, the transformation can be any deterministic relation (e.g., a time delay, a linear or nonlinear differential equation) or a probabilistic relation (e.g., a given value of the input determines the probability of occurrence of the output).

Don't real physical or human subsystems have multiple outputs? Yes, of course, but in modeling one is interested in functional transformations of variables, so requiring but a single output is simply a modeling convenience—and discipline. For example, a spatial output variable can be position, rate, and/or acceleration with respect to time. But the latter two variables are transformations of position that need to be made explicit in the model by additional time-derivative transformations.

It is important to keep in mind that any "system" is never more than a representation (model) of some characteristics of some segment of reality. It is not reality per se. That is a trap that too many engineers fall into when they become accustomed to working with a particular type of model. They forget about the many variables that have been left out of the model for convenience, and those that have not even been thought of. Where specified variables are sometimes called "unknowns," the latter may be called "unknown unknowns" (UNK-UNKS)!

What Is Systems Theory, and What Is Its History?

Systems theory is an interdisciplinary field of science concerned with the nature of complex systems, be they physical or natural or purely mathematical.

Although one can trace its roots to philosophers and economists of the early 19th century and even back to Plato, many systems theoretical developments were motivated by World War II (and the technological development of that period: servomechanisms for targeting guns and bombs, radar, sonar, electronic communication). Just after the war, psychologists and physiologists became aware of some of the early theories and saw use for them in characterizing human behavior, and especially humans interacting with machines. For example, control theory was applied to human sensory-motor behavior. Norbert Wiener gave us *Cybernetics: Communication and Control in the Animal and the Machine* (1948), an early systems perspective on humans as machines and human-machine systems. His *Human Use of Human Beings* (1955) introduced the ethical component. His last book, *God and Golem Incorporated* (1964), which won a Pulitzer Prize, centered on his strong foreboding about the pitfalls of putting too much trust in computers: They do what they are told, not necessarily what is intended or what is best.

Claude Shannon (1949) published his first paper on information theory in 1948, and together with Warren Weaver (1959) offered a more popular explanation of the theory of information, making the point that information as he interprets it is devoid of meaning and has only to do with how improbable some event is in terms of receiver (e.g., human) expectation, that is, how surprised that receiver is. After Shannon, there was a rush to measure experimentally how many bits could be transmitted by speaking, playing the piano, or other activities. Miller's (1956) *Magical Number Seven, Plus or Minus Two* modeled the information content of immediate memory of simple physical light and sound stimuli. (Miller's famous paper is perhaps the most misquoted one in human factors by people falsely asserting that humans can only remember seven items. The caveat is that those items must lie on a single sensory continuum, such as pitches of sound, intensities of light, and so on.) Colin Cherry (1957) gave us an important book *On Human Communication* using a systems point of view.

The theory of signal detection, developed out of necessity to characterize the capability of radar and sonar to detect and distinguish from noise enemy aircraft and submarines was seen by psychologists as an alternative approach to psychophysics (Green & Swets, 1966). They showed how absolute thresholds are not absolute: they can be modified by the trade-off between relative costs for false positives and missed signals. Signal detection theory has become a mainstay of modeling human sensing and perception.

Ludwig Bertalanffy (1950, 1968) was probably the first to call for a general theory of systems. G. J. Klir (1969) reinforced these ideas. Walter Cannon (1932) had already articulated ideas of homeostasis (feedback, self-regulation) in his important physiological systems book *The Wisdom of the Body*. C. West Churchman (1968) provided another popular book on how to think with a systems perspective. J. G. Miller (1978) wrote a well-known book on living systems, whereas Odum (1994) more recently characterized the ecological systems perspective. Sage and Rouse (1999) offer a management perspective on systems thinking.

Karl Deutsch (1963), a political scientist, wrote *The Nerves of Government*, from a system point of view. Jay Forester's *Industrial Dynamics* (1961) proposed modeling industrial enterprise using simple system models with variables connected by integrators and coefficients. The Club of Rome extended Forester's models to modeling nations and global systems in terms of population, energy, pollution, and so on (Meadows et al, 1972).

Stemming from efforts by the physicist (Philip Morse Morse & Kimball, 1951) to apply mathematical thinking to strategic and tactical decision making during World War II, the field of *operations research* (OR) was born and continues to this day as an active system modeling discipline within industry government and academe.

OR specializes in applications of linear and dynamic programming, queuing theory, game theory, graph theory, decision analysis, and simulation in which one is trying to model a real system and find a best solution in terms of variables that can be controlled, given assumed values of those that cannot be controlled.

The Human Factors Society started publishing its journal *Human Factors* in 1959, and aviation was has already a key topic. J. C. R. Licklider (1961) was given the premier spot as the first author in the first issue of *Transactions on Human Factors in Electronics* of the Institute of Electrical and Electronics Engineers (IEEE), the first engineering society to establish a special journal on the systems approach to human factors. His article "Man-Computer Symbiosis" has become a classic, predicting many aspects of human-computer interaction that have come to pass decades later. That journal has evolved into *Transactions on Systems, Man and Cybernetics*.

Today, systems models are used by engineers for analysis, design, and evaluation for all sorts of physical systems: from electrical and electronic circuits to highway and rail transportation systems: aeronautical and space systems; biological systems at the cell and organic level; hospital systems; business systems; and military command and control systems.

Systems models are almost always normative, meaning that they say how the system *should* behave if it were to follow the model. They suggest mathematical mechanisms that necessarily are simplifications of the real world and deal with only a restricted set of variables. It is always instructive to discover how the real world differs from any system model, which then provides a basis for improving the model.

AIRCRAFT SYSTEMS OF RELEVANCE
AND THE HUMAN COMPONENT

In this section, several aviation-related human-machine systems are considered qualitatively, together with some discussion of how systems thinking contributes to design and operation. In the next section, selected quantitative systems analysis tools are discussed.

Single Aircraft Control

The Basic Human-Machine System

Figure 2.3 (left) shows a single operator (e.g., pilot) observing information from a machine display (e.g., aircraft) and also from the machine's environment (e.g., out-the-window), and responding with control actions. One could also represent a single air traffic controller whose actions are instructions to an aircraft. In either case, a very simple systems breakdown might be what is shown in the block diagram (right).

The human functions are broken into sensation (all the neurophysiological functions of sensing visual, auditory, and haptic information from both the machine and the environment lumped together), cognition (including allocation of attention, perception, memory, and decision), and action (muscle responses to operate discrete controls such as buttons, switches, and pedals, to adjust continuous controls, and to speak).

One can make a finer analytical breakdown, but within the human body one is hard-pressed to identify and bound the component elements so as to know what are the interconnecting variables between sensation and cognition and between cognition and action. What a person sees in a display, for example, can be elicited by experiment, and what a person decides can be similarly determined.

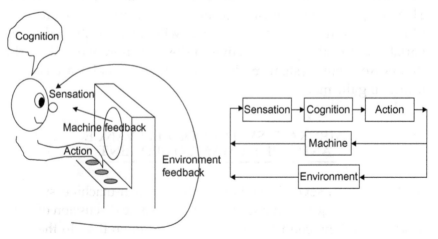

FIGURE 2.3   System consisting of single human, machine and environment.

Supervisory Control

Automation began to invade the aircraft many years ago with the first autopilots, and as so-called glass cockpits emerged with the B757/767 and the A317/320 human factors professionals as well as systems engineers recognized that automation was not all a blessing (Wiener & Curry, 1980). Automation (computer) control changes the nature of the human task, and there may be many control options. So it has become increasingly important to distinguish direct control from *supervisory control*, and the forms it may take. By direct control, we mean that a command is sent directly to the machine with no intervention by a computer; in supervisory control, the human command is sent to a computer, and the computer processes the human command using its own programmed intelligence to determine the final command to the aircraft of other machine. For example, to take an aircraft to a new altitude a pilot can continuously adjust pitch and throttle to raise the nose and increase thrust and eventually level off at the new altitude. Alternatively, the pilot can punch into the autopilot a new altitude and the computer will direct pitch and thrust to achieve that new altitude according to a preprogrammed strategy. The first approach is direct control, the second is supervisory control.

The pilot looking out the window would be direct display, while looking at the altimeter would be computer-mediated (where a broad definition of "computer" is used here: radar or pressure altitude instrumentation.

Figure 2.4 shows a human operator (pilot or traffic controller) having to allocate both his senses (usually vision) between direct and supervisory control, whereas in the supervisory control case for both displays and controls there is automation that mediates what the human sees or does.

Although the distinction between direct and supervisory control may seem clear-cut, it is not. In older and simpler aircraft where the control yoke or stick was directly coupled by cables to the aerodynamic surfaces and most of flight was by visual flight rules one can assert that flying was mostly direct control. Today, even in general aviation aircraft, control is *fly-by-(electrical)wire*, and a plethora

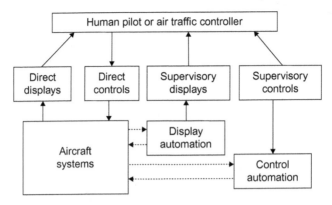

FIGURE 2.4 Human pilot or controller sharing between direct and supervisory displays and controls.

of computer-mediated displays are available for flying under both instrument flight rules (IFR) or visual flight rules (VFR) flying. The many and various ways that human interact with automation mean that a simple binary distinction between manual and supervisory control does not work. There is a continuum (see later discussion of systems methods), and indeed the "levels" can be regarded as multidimensional.

Multiple Nested Control Loops

In aviation, there are multiple levels of autopilot-mediated control (which is only one major function of the sophisticated computer) called the flight management system. One can identify (and model) an "inner" control loop, which represents, for different degrees of freedom, pitch, roll, and yaw. The control loop whereby the pilot commands a new altitude and/or heading, and the aircraft does what it needs to do to get there may be called "intermediate." At an even higher level of control the pilot can key in waypoints or the proper latitude and longitude of a distant airport and the aircraft will use its navigational inertial and/or GPS instrumentation (Figure 2.5).

Air Traffic Control and NEXTGEN

Air traffic control is a multiperson multiaircraft system of considerable complexity. In fact, without the tools of system analysis and

Loop	Command variable	Feedback variable
Inner	Pitch, roll, yaw	Aerodynamic surfaces
Intermediate	Altitude, heading	Altimeter, compass
Outer	Waypoint, airport	Latitude, longitude

FIGURE 2.5 Command (reference) and feedback (measured) variable for three nested loops in controlling an aircraft.

design, it would be almost impossible to understand, build, and manage a modern air traffic control system.

How the Current System Works

All readers will be familiar with the control tower, inhabited by several different kinds of controllers with different responsibilities. Here an FAA *local controller* is responsible for aircraft in final approach and landing and in line up and departure from runways. An FAA *ground controller* is responsible for taxi operations. At air carrier airports a *ramp controller* (usually provided by airline or airport authority at larger airports) is responsible for aircraft and vehicle movement in the ramp area. Each airline has at the airport an *airline operations office* that together with the FAA and pilot decides on pushback and gate assignment. And in larger (towered) airports there are *clearance delivery* and other specialists for traffic management.

Aircraft descending to airports and climbing to cruise altitude from airports in a radar service area are controlled from a terminal radar control facility (TRACON). These controllers must deal with the complex patterns of aircraft descending from different directions to get in line for one or more runways, sometimes multiple parallel runways and sometimes crossing runways, plus the aircraft departures from these runways and climb-out to their cruising altitudes and initial transitions to their respective destinations in terminal airspace. Normally such aircraft are to be separated by 3 miles horizontally and 1000 feet vertically (in terminal airspace) unless approved for visual separation. The TRACON controller's primary task is to keep all IFR and VFR aircraft sufficiently separated, to provide sequencing services for incoming aircraft and routings for departure.

Finally at the 20 *regional centers* throughout the United States (Air Route Traffic Control Centers [ARTCCs]), there are many *en route controllers*, each responsible for traffic in multiple *sectors* within that region. These aircraft fly at assigned (cruise) altitudes (typically between 18,000 and 40,000 feet) and on assigned air routes between waypoints. These controllers give instructions to the aircraft in their sectors to maintain 5 miles horizontal and either 1000 or 2000 feet vertical separation (depending on an aircraft's equipage) to provide traffic and weather advisories, and hand them off to the controller in the next sector.

Figure 2.6 diagrams the interperson interactions to control a single aircraft within an en route sector, a terminal sector, or under tower control. Not shown are the ties to multiple other aircraft. The heavy solid lines indicate the relatively "tight" control (meaning corrections are continuous or spaced at close time intervals). The corrections by the assigned air traffic controller are at somewhat longer intervals. Interventions by various other individuals (controller supervisor in the airport's TRACON or ARTCC, traffic flow manager in TRACON or ARTCC, each airline's own operations personnel are at still longer intervals. Finally there is a single command center in Virginia to plan and approve routings and restrictions, to coordinate operations when bad weather or traffic

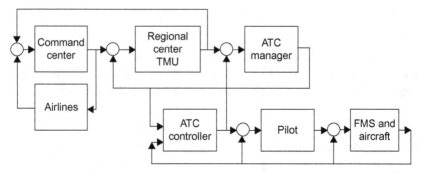

FIGURE 2.6   The various human agents who interact in air traffic control.

demand constrains operations in large regions, or some emergency occurs (as did happen on 9/11/2001when almost all domestic U.S. air travel was rerouted to Canada).

The communication and coordination between all these individuals, by both voice and computer messaging, poses major systems problems. They must be concerned about the capacity of airports to handle to traffic. They must be concerned about backing up or diverting traffic as it exceeds flow limits and bad weather intervenes. The systems challenge is to analyze the inputs and outputs and processing delays and errors of all these human/machine components and their respective display/decision aids to find best ways for the system to work.

There is a variety of computer-based decision aiding already installed or being tested in air traffic control systems. For example, the center TRACON automation system (CTAS) includes a traffic manager advisor for scheduling arriving traffic, a descent advisor to identify a trajectory that meets the scheduled arrival times, and a final approach spacing tool to make final (and minor) modifications in spacing between those arriving aircraft. These tools are used by controllers in both en route centers and TRACONS. Further information about human factors in air traffic control is found in Smolensky and Stein (1998).

Next Generation Air Transportation System (NEXTGEN)

NEXTGEN is a joint enterprise of the Federal Aviation Administration, National Aeronautics and Space Administration, Department of Defense, Department of Homeland Security, and several other government agencies as well as industry and academe. It is a planning, research and development effort to prepare the United States for civilian air transportation in the time frame of 2020–2025, when air travel demand may double, but runway and terminal facilities will be only marginally greater than exist now (NextGen Concept of Operations 2.0, 2007).

Various key technological capabilities are already being implemented: ADS-B, an aircraft-to-aircraft-to ground digital communication and surveillance system, substitutes the much greater surveillance accuracy of GPS as compared to radar. ASDE-X, a high resolution ground surveillance system based on multilateration radar, triangulates to determine aircraft position on the airport surface. Computer-planned and monitored four-dimensional trajectories of aircraft provide minute-by-minute reference trajectories. Digital communication between pilots and ground controllers (*data-link*) complements voice communication and is used for clearances and other detailed information displayed visually. Also there is much improved weather modeling and prediction, better systemwide availability of information of many sorts, and a variety of other new automation systems and decision/display aids to air traffic management, both over land (both airborne and airport surface) and over the oceans.

Plans include placing a greater responsibility on pilots for self-separation from other aircraft, greater use of smaller and currently underused airports on the periphery of metropolitan areas, and more flexibility in trajectories to conserve fuel.

Typical of system thinking is the greater use of the *flight object*. A flight object is a packet of information including the flight number, operating airline, departure and arrival airports, schedule, trajectory, and other information that are key to controlling the flight. This information is what is sent from the transponder or other data link transceiver, displayed to air traffic control, and increasingly processed by automated computer agents. The terminology, and the basic idea, is consistent with object-oriented programming.

Figures 2.7 and 2.8 provide an example (from developmental human-in-the-loop tests) of one system development meant to improve capability of local and ground controllers at airports. Currently these controllers depend on their ability to see the details of activity from control towers, but in bad weather and at night their vision is severely impaired. Using some combination of precision radar (ASDE-X), GPS, and digital transponder technology, it is possible to know within a few feet the location and identification of aircraft, independent of day/night and mostly independent of weather (precipitation has effects). This means that high-resolution displays can be provided to show at each second in detail all the aircraft on taxiways, runways, and for about 10 miles away both landing and departing. Through

FIGURE 2.7   Local and ground controller workstations.

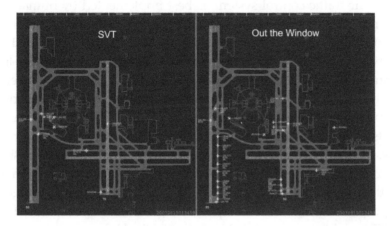

FIGURE 2.8   Comparison of traffic in the same poor weather scenario when protoype controller display is used (left) versus when controller depends on out-the-window view (right). Note line of aircraft lined up on taxiway waiting for take-off in the out-the-window case.

alphanumeric tags and color coding, their identification information, aircraft type, speed, altitude, clearance status, destination, and so on can be shown from a viewpoint as though the controller were ideally positioned directly above the airport.

Figure 2.7 shows (left-to-right) the supervisor, local and ground controller, each having in addition to the airport surface display a display of electronic flight strips (below) and a shared display of the queues of aircraft waiting at gates for pushback and aircraft arriving but not yet under local control.

Figure 2.8 illustrates what happens in the same poor weather scenario in which the only difference is that (at left) the controllers are able to utilize the new display tool to keep traffic moving smoothly (small bright symbols are aircraft) where (at right) with control from the tower in bad weather aircraft back up in the departure taxiway leading to the left runway.

Human-System Integration

The design and development of an aviation system, whether it be the aircraft itself, some component piece of avionics, an airport, or a major air traffic control system, is best implemented by human system integration (HSI). It might be said that HSI is the process of incorporating the efforts of human factors professional and the methods of experimentation, hardware/software design, human-in-the-loop simulation, and testing/evaluation into much larger engineering enterprise. It obliges the project manager to have a representation of the project goals and constraints that can be understood by all participants, that is updated periodically, and can serve as a record of progress. HSI may be characterized by a model in which progress is spiral with its axis as time whereby all the actors are consulted iteratively for their input and approval ("buy-in") at successive project stages— to keep everyone "in-the-loop" and "on-board." Pew and Mavor (2007) provide an extensive description of the HSI process.

Aircraft Manufacturing and Maintenance

Background

Systems thinking has been used in the aircraft manufacturing sector for many years and traces its roots back to the "scientific management" and the motion and time studies of Frederick Winslow Taylor (1911) and Frank and Lillian Gilbreth. Peter Drucker (1974) asserted that "Frederick W. Taylor was the first man in recorded history who deemed work deserving of systematic observation and study." His *Principles of Scientific Management* (1911) was an early example of applying systems thinking to human work, especially in manufacturing.

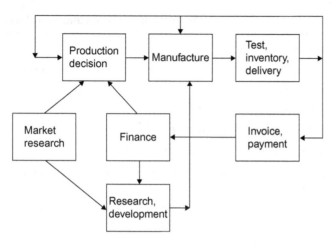

FIGURE 2.9   Interacting elements of an aircraft production system.

However, the labor movement was very critical of Taylor's efforts to systematize and, they claimed, accordingly dehumanize human work, and "Taylorism" has become a pejorative term in labor circles.

Modern manufacturing systems are now heavily infiltrated with computer-aided design, computer-aided planning, computer-aided accounting, and various forms of materials handling and automation including robots. The interrelation of these elements can be characterized by the diagram in Figure 2.9. Not shown are many exogenous inputs and outputs of hardware, software, people, money, and information.

The "Lean" Concept

The lean concept is often attributed to Toyota automobile manufacturing innovation (Ohno, 1988) and to a study of the automobile industry (Womack et al., 1990). More recently, the aircraft industry has made extensive use of it, particularly in manufacture of military aircraft. Ohno, the "father" of the Toyota production system, has stated simply that the goal is to increase efficiency by reducing waste. In 1992, the U.S. Department of Defense (DoD) initiated the Lean Aircraft Initiative, a consortium including the U.S. government, Massachusetts Institute of Technology (MIT), and approximately 20 industry sponsor companies. Hoppes (1995) summarizes the features of lean manufacturing in Table 2.1.

TABLE 2.1   Factory Operations Features of the Lean Concept (Hoppes, 1995)

—elimination of waste

—use empowered work teams and process improvement teams to identify and eliminate waste

—reduce floor space

—eliminate nonvalue added production steps

—reduce support staff (self-inspection, etc.)

—empower floor-level employees (operators) and award on group/company-level performance

—empower and give employees the right tools (capable processes, ability to stop line and make suggestions, etc.) to improve quality

—eliminate inventory through bottleneck management, kanban systems, and Just-in_Time (JIT)

—reduce cycle time

—tie design departments closely to manufacturing (concurrent design)

—manufacturing is closely tied to the rest of the value chain (suppliers, distributors, and customers)

—foster cooperation and forge relationships with suppliers

—help suppliers reduce costs and improve quality

—integrate sales and distribution with production so that the customer gets the product quickly and manufacturing responds to the customers' demand

—serving the customer

—respond quickly to changes in customer demand

—work toward continuous improvements affecting quality and cost

—continuous improvement

—trace defects quickly to cause and put irreversible corrective actions into place

—identify methods to remove waste from the production process

—empower and provide incentives for employees to make continuous improvements at all levels of the organization

—flexibility

—use quick changeover tooling (for setup time reduction)

—reduce number of job classifications, and cross-train employees (need employees with general job skills that can be used broadly)

—minimize investment in specific/nonflexible tooling

Systems thinking is included in practically every feature of Table 2.1, where product and process are analyzed into components and inter-related with respect to dynamic order and time constraints; factory layout, product configuration, and other spatial constraints; personnel; failure and reliability; and cost.

One could draw a similar diagram for the process of aircraft maintenance and repair, where elements include routine inspection, pilot reporting, warehousing of and transportation of parts, duty stations and availability of maintenance personnel, ground movement of aircraft to and from maintenance hangars, information dissemination between airline operations offices, maintenance hangars, air traffic management, finance and accounting, and so on.

The "systems" mentioned here have been characterized at a very gross level. Within any one element one can identify subelements and the interactions between them.

SYSTEMS ENGINEERING METHODS AND APPLICATION TO AVIATION HUMAN FACTORS

This section discusses at a very elementary level three very different models that exemplify systems engineering methodology as applied to human-machine systems, namely those for control, decision, information, and system reliability. Although not comprehensive, this set of models provides a grounding in the system perspective. There are many other system theories, too many to mention here. They vary not only according to their structure and type of mathematics but also as to the purposes to which they are put. The reader is referred to Sage and Rouse (1999), Sheridan (1974, 1992, 2002), and Parasuraman et al. (2000) for more extensive reviews of these methods.

Control System Theory

The Basic Idea of Feedback Control

Control usually means feedback control. Figure 2.10 shows a feedback control system consisting of a controlled *process* F_p (e.g., aircraft) being driven by the manipulated variable from a *control*

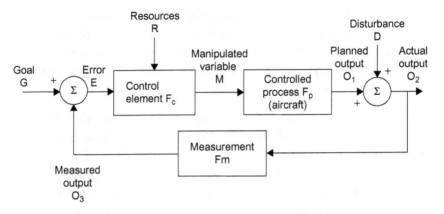

FIGURE 2.10 Standard single loop feedback control system.

element F_c (e.g., human or autopilot), which in turn is responding to the difference between a *goal* state G and a *measurement* O_3 of the actual system output state O_2 (e.g., aircraft pitch, roll, or yaw). F_c includes not only human or machine intelligence but also energy resources. Actual system output state O_2 is the planned output state of the aircraft distorted by any disturbance D due to turbulence, and so on. Assuming the actual output state is an accurate $(O_3 = O_2)$ measurement, where the functional relations F_c and F_p can be treated as algebraic coefficients (which is allowable for Laplace transformed functions but that rationale is not developed here), then

$$O_2 = O_1 + D = F_p F_c (G - O_2) + D$$

from which O_2 can be determined to be

$$O_2 = [F_p F_c / (F_p F_c + 1)]G + [1/(F_p F_c + 1)] D$$

Ideally, one would like O_2 to equal G and D to have no influence. It is evident that if $F_p F_c$ were much greater than one that would be true; the first term in square brackets would go to one and the second term in square brackets would go to zero.

However, feedback control is not quite so simple. The problem is that both F_p and F_c are generally complex dynamic operations approximated by differential equations, and arbitrarily increasing the gain of F_c (where the dynamic characteristics of the aircraft or other controlled process are givens) will make the system unstable. Control then becomes a balancing act to design a control element (an F_c dynamic equation) with enough gain and other properties to minimize the effect of D while making O_2 best conform to the goal state G. This includes both good transient response (quick but with little overshoot), good frequency response (tracking of periodic inputs with little phase lag), and insensitivity to unwanted disturbance inputs. Techniques for achieving this are available in any standard text on control systems theory.

The Human Transfer Function

During the 1950s the U.S. Air Force sponsored a major effort to determine the *transfer function* (differential equation relating the output, i.e., joystick position) of the human operator (the pilot) to the error input (discrepancy in pitch, roll, or yaw relative to a given reference setting on the attitude indicator). The reason for this was that the dynamic equations of the aircraft itself were known, but because the pilot was part of the control loop, to predict system stability the equation for pilot must be put into the overall system analysis. This led to a number of real-human-in-the-loop simulation experiments in target tracking with various unpredictable ("noise") disturbances and controlled element (aircraft) dynamics. McRuer and Jex (1967) were able to generalize this work with a precedent-breaking idea, namely, that whatever the controlled element dynamics (within a reasonable range) the human operator adjusts his own transfer function, so that the resulting forward loop dynamics, the combination (F_c F_p), approximates a single integrator plus a time delay (the human reaction time of 0.2 seconds, which is inherent in the human neuromuscular system). This has the effect of making the closed loop system (goal variable G to planned output O_1) behave as a "good servo"—stable and with minimal overshoot response to sudden transient inputs G (or disturbances D). Historically, this is one of the rare achievements in human factors where an equation predicts human behavior all the way from sensory input to motor output. (There are many mathematical models of visual and hearing sensitivity, muscle strength, memory, etc., but they are only components of behavior.)

Model-Based Control

Another technique that forms the basis of "modern" or "optimal" control but can also be used as a metaphor for the adaptive, learning, resilient worker, team, or organization is the Kalman observer (Kalman, 1960). The idea is to establish a functional dynamic model (boldface box in the center of Figure 2.11) of the physical system that is the controlled process (boldface cloud on the right), which is at the same time being operated, and use that as a basis for deciding on control actions (rather than simple feedback). The trick is to input the same control decision commands (1) as are being imposed on the actual system to a computer model the output of which (2) in turn serves as the basis for deciding on the next control decision commands. That same model output is compared with the output of the actual system and any discrepancy (3) is used to update the computer model. The dashed line divides the real world from the computer processing. The action distortion and the sensory distortion represent the delays, noise, and so on that may exist unavoidably in the interaction with the real world. By estimating the delay or bias in the measurement and including such a function on the left side, the actual measurement distortion can to some extent be cancelled. And by estimating and replicating on the left

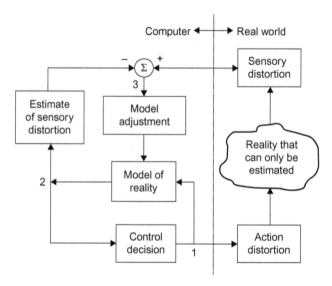

FIGURE 2.11   Model-based (optimal) control.

side the measurement noise the procedure provides a proper pace of model adjustment. The mathematics of using Kalman observers in control can be found in texts on modern control theory (1960).

Model-based control (Kalman observer) theory (Baron & Kleinman, 1969) has been used (to provide another model of the human controller, applied, e.g., to aircraft landings, and extended to include attention allocation among multiple instruments (Baron et al., 1980).

The Supervisory Control Framework

Supervisory control was defined earlier in this chapter. Given a hopefully better understanding by the reader of nested loops, mentioned earlier, as well as simple closed loop and model-based control just described, we take a closer look at supervisory control with respect to what functions are inherent for the human partner and what functions are inherent for the computer.

Figure 2.12 diagrams a sequence of five human functions:

1. *Plan* what the computer (automation) should be asked to do (goals to achieve and constraints to be accommodated). The

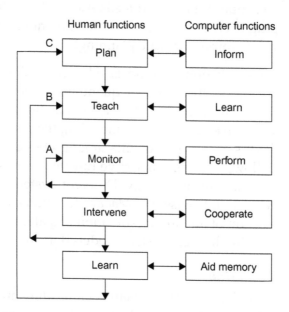

FIGURE 2.12   The supervisory control framework, showing the five supervisory functions and the associated computer functions.

computer's role at this step is to *inform* the human of relevant information and if need be to perform "what would happen if . . . " simulations.

2. *Teach* (program the computer) the goals and constraints. This may involve both symbolic (alphanumerics, special terminology) as well as analogic (positioning) adjustments. The computer's role here is to *learn*, and to give feedback to the human that it understands correctly.

3. *Monitor* the computer's execution of what has been taught, being sensitive to discrepancies and abnormalities. The computer's role is to actually *perform* the assigned task.

4. *Intervene* as necessary (based on observed discrepancy, abnormality, or failure) to modify the program or completely abort. The computer's role is to *cooperate* by gracefully phasing over from executing previous to new commands.

5. *Learn* from experience. The computer can help to *aid memory* by keeping track of what the human did and suggesting what might have been done differently and better

Note the nesting of control loops. Monitoring is a continual (continuous or intermittent) task, so it feeds back to itself. Intervening requires looping back to do more teaching. Learning from experience loops back to improve future planning.

Table 2.2 suggests theoretical models relevant to the five supervisory functions, and the types of quantitative models that are appropriate for each. The table also notes that the *Plan* function really consists of three subfunctions: modeling the physical system to be controlled, satisficing trade-offs among objectives (see discussion that follows regarding objective functions and satisficing), and formulating a control strategy. Different system models are appropriate to each. Likewise, the *Monitor* function can be said to consist of allocation of attention, estimation of state, and detection of failure, each with different system models that are appropriate.

Levels of Automation

It is fallacious to assume that a system is either fully automated or fully manual. In general one can define multiple degrees of automation, from the computer offering no assistance to the human, to

TABLE 2.2 Theoretical Models Relevant to the Five Supervisory Functions

Function	Appropriate Type of Quantitative Model
Plan	
Model physical system	System analysis/differential equations
Satisfice trade-offs among objectives	Multi-attribute utility theory
Formulate control strategy	Control theory, optimization theory
Teach	Information and coding theories
Monitor	
Allocate attention	Sampling theory, Markov network analysis
Estimate state	Estimation theory
Detect failure	Signal detection, Bayesian analysis, pattern recognition
Intervene	Decision theory
Learn	Human and machine learning theories

TABLE 2.3 A Scale of Levels of Automation

1. The computer offers no assistance; the human must do it all.
2. The computer suggests alternative ways to do the task.
3. The computer selects one way to do the task and
4. executes that suggestion if the human approves, or
5. allows the human a restricted time to veto before automatic execution, or
6. executes the suggestion automatically, then necessarily informs the human, or
7. executes the suggestion automatically, then informs the human only if asked.
8. The computer selects the method, executues the task, and ignores the human.

(after Sheridan, 2002)

"full" automation with no human involvement at all, with various shades in between as noted in Table 2.3. Since the notion of levels of automation was first introduced (Sheridan & Verplank, 1978), there have been various other scales proposed. Indeed, the number of levels, definitions, and even whether "degree of automation" is one-dimensional (surely it is not) are subjective judgments.

It is important to note that the number of "levels" of automation and the definition of categories is really quite arbitrary. Insofar as a task can be broken into the sequence of information *acquisition, analysis and display, decision, and execution of action,* one can assert that each of these different functions can be more or less automated independently of the others (Parasuraman et al., 2000). For example, one might say that in air traffic control TRACONS and ARTCCs, *information acquisition,* much of which comes in from radar and weather stations, is highly automated, although of course controllers and pilots talk to each other directly. Digital data-link systems will intervene more and more into that communication channel. *Analysis and display* of that information is becoming more automated with the spate of decision aids that have populated TRACONs and ARTCCs. Decisions are still made by humans at all levels, except perhaps for the traffic collision alert system (TCAS) advisory/resolution where pilots are trained to follow it unless there is a compelling reason not to do so. Execution of action by the controller per se is a matter of what is said to the pilot, or is keyed into a data-link communication system. Presumably in NEXTGEN all of these functions will see some further automation.

A final caveat with regard to levels of automation: Such scales and classification schemes are useful for system analysis and design (as are any models) in helping analysts and designers be more specific in their thinking. However, unless and until such representations become predictive models of performance (and to date they are not), it cannot be said that they determined design solutions. Sheridan (2002) and Sheridan and Parasuraman (2006) provide a recent review of research in human-automation interaction.

Decision Making

Another class of system model is about decision making. Humans make many decisions, most of them (depending on how any count is made) at low levels of the nervous system that do not even reach consciousness. Rasmussen's well-known skill-rule-knowledge classification is relevant here (the reader is assumed to be familiar with the general idea, so a diagram will not be needed). Skill-based decisions tend to be continuous, as with neuromuscular movements, and are mostly made by reflex and without conscious thought. They are analogous to continuous servomechanism control. Rule-based decisions are discrete, analogous to simple "do-loop" computer programs, and are mostly conscious, although well-learned and often used decision can also be done with little thought. Knowledge-based decisions require consideration of alternatives and novel circumstances. In discussing human decision making from a systems perspective, it is assumed that the behavior is at a knowledge-based level (we are not interested in neuromuscular control of movements), and there is no explicit lookup-table or algorithm for directly generating the decision.

Formal Decisions under Known Conditions

Formally, from a systems theory perspective, optimizing a decision is a matter of simultaneously solving two equations:

1. An equation interrelating all the salient variables of the problem, and
2. An equation that provides a scalar expression of goodness or badness (choose one) as a function of all the salient variables. Formally, this is called an *objective function.*

To consider a very simple example, let us suppose that we are concerned with planning a flight, where departure must be at or later than T_d and arrival at or earlier than T_a. T_{am} is minimum arrival time for shortest routing and maximum airspeed. The latter terms impose the dotted line constraint for any departure later than T_d. That then means that all acceptable solutions are within the triangular space in Figure 2.13. Ignore the gray dots for the moment. Now suppose the objective function is expressed as a penalty for later arrival plus a penalty for longer transit time (more total fuel burn). The minimum cost is then at the dark dot shown. However note that later departure could be better if a different criterion were used (see below under "satisficing").

As with most application of systems theory, life is usually more complex than simple examples like this, but the formalities do

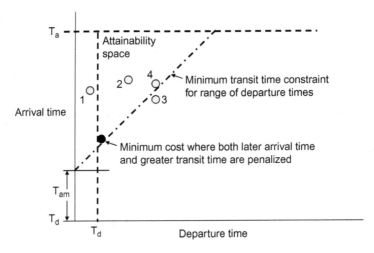

FIGURE 2.13   The flight planning problem space.

discipline one's thinking. First of all, the equation(s) characterizing all the salient variables are not easy to come by, and the more
variables that are included the more difficult that task—that is, the
task of modeling any system, and with a small number of salient
variables one can at least make some pretty good approximations.

Where does the objective function G (usually called a *multiattribute
utility function* by economists) come from? It is inherently subjective, of course, and must necessarily be the result of an elicitation
process, first recommended by Von Neumann and Morganstern
1944). It is inferred from a series of judgments to discover at what
value of some sure consequence is the judge indifferent to a given
probability of some other consequence. See Sheridan (2002) for a
somewhat more detailed explanation of the procedure. Because
utility trade-offs are personal matters, the additional question is:
whose objective (or multiattribute utility) function to use, and how
to combine the functions of multiple interested parties.

Satisficing
Short of finding the formal maximum goodness solution based on
solving equations that are almost impossible to find, a perhaps more
feasible approach is to do what is called *satisficing*. An explanation
of the procedure is most easily explained through another example,

see Figure 2.13 again. Assume the constraints in the flight planning example are entered into a computer-based *satisficing aid*. The points just inside the triangle along the diagonal are obviously the minimum time trip solutions (called the *Pareto optimal frontier*). But suppose there are weather, convenience, and other delaying factors that cannot be well articulated quantitatively. A first impulse might be to depart very early at point 1, but the computer would quickly point out that departure must be later than T_d. Then the planner might test whether point 2 would work, and the computer would say yes, but you could leave a bit later and schedule arrival bit earlier if you wanted. So the planner might then try point 3, and the computer would say no, but a very close point is attainable and would still allow the same departure time as 3 and get you there as early as possible, namely point 4. After several such interactions the planner could presumably hone in on something that was satisfactory. Finding a global unique optimum point is usually not practical, but this procedure allows for utilizing available quantitative data plus some consideration of the nonquantifiable factors to reach a "satisficing" solution (March & Simon, 1958; Charny & Sheridan, 1989). Although the simple problem in this example could have been solved without a computer, many others have more complex circumstances that are simply too much for the planner to weigh and consider, and a several-step satisficing approach may be much more efficient.

Decisions under Uncertainty

Deciding what action to take when the result of that action is uncertain poses a different kind of problem, and depends upon how much risk one is willing to take (e.g., think of buying stock). The situation can be illustrated again by a simple example.

Consider that there are two exogenous unpredictable circumstances that can occur, X and Y, and it is known that they occur with probabilities 0.7 and 0.3 respectively. There are three possible alternative decisions or actions, A, B, and C. The payoff for each action depends on which environmental circumstance is true, as shown in the table at the right of Figure 2.14. Perhaps surprising to some, there is no best strategy for maximizing payoff. It depends on one's proclivity to take risks.

	Circumstances		Payoffs		
	P(X)=0.7	P(Y)=0.3	Expected value	Worst possible	Best possible
A	6	4	(5.4)	4	6
B	5	5	5.0	(5)	5
C	7	0	4.9	0	(7)

(Actions A, B, C)

FIGURE 2.14   A simple decision matrix illustrating how what is "optimum" depends on the criterion selected.

The table at right indicates the expected value (average, given the probabilities) of each action, and also the worst possible and best possible payoffs for each action taken. Under an expected value criterion decision A (circled) is best. But now consider the worst possible column. A very conservative decision maker would consider the worst that can happen under each action alternative and choose that alternative with the least downside risk, that is, the best of the worst. This is called a *minimax* criterion. This is action B. An inveterate risk taker (fewer of them than minimaxers) might choose the option that has the greatest possible payoff, action C. Thus we see that the best action to take can be completely different depending on one's decision criterion.

Signal Detection

When any signal (or stimulus) is detected (or responded to) correctly, that is called a *hit* or *true positive*. When there is a response to the signal but it is incorrect, that is called a *false alarm* or *false positive*. When the signal is there but not detected (there is no response), that is called a *miss* or false negative. If there is no signal and no response, that is called a *true negative*. The four stimulus-response contingencies can be combined into a *truth table* as in Figure 2.15.

In the signal detection paradigm the strength of the signal is considered corrupted by noise, so that the probability of hit is less than one, and the probability of false alarm is greater than zero. If probabilities of hit and false alarm are cross-plotted as in Figure 2.16 any

		Truth: S present	Truth: S absent
Decision	Correct	True positive (hit)	True negative
	Incorrect	False negative (miss)	False positive (false alarm)

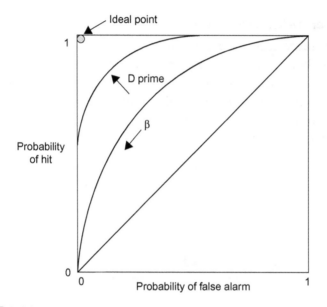

FIGURE 2.15 Truth table for signal detection.

FIGURE 2.16 Plot of ROC curves.

decision maker, whether human or machine, can be characterized by a curve, called a receiver operating characteristic (ROC). The straight diagonal is the ROC for a random decision maker, and the upper left corner is that for a perfect decision maker.

A parameter called D prime is the detectability. It is based on an underlying model consisting of two Gaussian probability density functions, one for deciding signal, the other for deciding no-signal, both plotted along a signal strength axis. The function for deciding signal is offset by d′ at greater signal strength. Given any D prime ROC curve, a second parameter called β is a point along that curve specifying the tendency to decide that the signal is present, given the chance of a false alarm. In the underlying model it is the value of hypothetical signal strength above which one should decide that the signal is present.

The signal detection model has found popular use in evaluating human response to warning and alarm systems.

Information

Information Transmission

For any imperfect communication of discrete messages between humans, between machines, or between humans and machines (in either direction) the messages an be categorized into what was sent and what was received, with a matrix of probabilities for joint x,y occurrences (within the cells) and marginal probabilities of any x or any y, as in Figure 2.17.

According to Shannon's definition of information, one can write Information sent: $H(x) = \Sigma_x P(x) \log_2 [1/P(x)]$

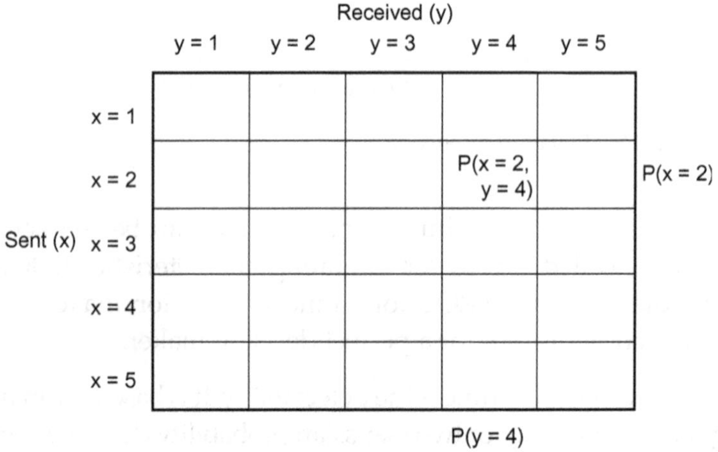

FIGURE 2.17   Matrix of probabilities for messages sent and received in noisy communication. Examples of a joint probability P(x,y) and marginal probabilities P(x) and P(y) are shown.

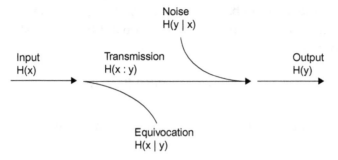

FIGURE 2.18   Components of information in a communication system.

Information received: $H(y) = \Sigma_y P(y) \log_2 [1/P(y)]$

Equivocation: $H(x|y) = \Sigma_{xy} P(x,y) \log_2 [1/ P(x|y)] = \Sigma_{xy} P(x,y) \log_2 [P(y)/P(x,y)]$

Noise: $H(y|x) = \Sigma_{xy} P(x,y) \log_2 [1/P(y|x)] = \Sigma_{xy} P(x,y) \log_2 [P(x)/P(x,y)]$

Transmission: $H(x{:}y) = H(x) \text{—} H(x|y) = H(y) \text{—} H(y|x)$

Figure 2.18 illustrates these relationships and how they combine. This form of systems model was very popular in the late 1950s following Shannon and Weaver (1959), but apparently has lost favor within the human factors in recent years for reasons unknown.

Information Value

An underappreciated model in the writer's opinion is that of information value, first discussed formally by Raiffa and Schlaiffer (1961). Information value is the marginal benefit of providing some state information minus the marginal cost of providing that information. That seems obvious enough, but depending on the situation S_i the benefit B of taking some action A_j can be very different, so one must average benefit over different states according to their probability.

The average benefit B_1 of knowing the each state S_i perfectly and taking the action that yields the greatest benefit for that state is

$$B_1 = \Sigma_i P(S_i) \, max_j \, [B(A_j|S_i)]$$

Whereas the average benefit B_2 of having to select that action that is best on average because only the probabilities of the state values are known is

$$B_2 = max_j [\Sigma_i P(S_i) B(A_j | S_i)]$$

The marginal value of providing the information about the state of S is then

$$Information\ value = B_1 - B_2 - cost$$

Systems Models of Cognition and Mental Load

In recent years, there has been increasing effort to model more complex aspects of cognition and intelligence. The whole movement of artificial intelligence (which is too extensive to be reviewed here) seeks not so much to model human brain function as to perform acts that are deemed intelligent. But computer science has given rise to what is often called *cognitive science* which to some researchers means computer-based modeling of human thinking, even though there is great debate about whether the animal brain is anything like that of a computer.

Most current cognitive science modeling is rule-based in a structure programming sense rather than taking the form of simple equations of other mathematical expressions. The work of Newell (1990) and Johnson-Laird (1993) are most often cited. The term *mental model* has crept into the literature (Moray, 1997). Most mental models have utilized "crisp" rules (or *production rules*), but fuzzy logic (Zadeh, 1965) suggests the appropriateness of fuzzy rules, where the variables overlap one another, much as the meanings of words do in human communication.

Closer to the aviation application, for example, a computer-based mental model called AIR MIDAS (Pisanich & Corker, 1995) represents information requirements, decision, communication and motor processes of flight crews in responding to controller instructions. Stochastic interruptions can be fed into the model to predict decision making, workload, and response to clearances.

System Reliability

Safety and Risk

The third class of system models is that used for reliability analysis. This is important in human-operated aviation systems because humans make errors and machines fail, and when they do there can be very serious consequences. Reliability is close to being synonymous with safety, and safety has always been a prime objective of human factors, especially of aviation.

There is no absolute criterion for what is safe. Safety is what is acceptable. The common plea for "100% safe" makes no sense, because errors, failures, and accidents, large or small, will sometimes happen. As more effort is put into safety the marginal cost increases. The cost to achieve 100% is unimaginable. The more sensible idea is to reduce the consequences of human errors, machine failures and accidents by making recovery easy and likely.

Often, human errors are due to misunderstandings and miscommunications between humans and automation. Degani (2001) explores how these confusions arise in the context not only of aircraft automation but also many other human-machine interactions, simple and complex.

What about *risk*? Risk is usually defined as a product of failure probability of an event and the consequences of that failure. More generally, risk R is the probability $P(E)$ of an undesirable event times the sum of the products of the probability of certain consequences (after whatever recovery may or may not have occurred) contingent upon that event $P(X_i|E)$ and the cost C_i of those consequences,

$$R = P(E)\sum_i [P(X_i|E) C_i]$$

Traditional Risk Analysis Methods

Traditional risk analysis has tried to associate given undesirable events with their probabilities of occurrence, with their consequences, and with precursor events that may be causal or at least warn of trouble so that amelioration and remediation can occur.

Risk analysis pertaining to humans, *human reliability analysis* (HRA), tries to predict the probability of failures of sequences of actions that humans must take in performing procedural tasks (Swain & Guttman, 1983). Such analyses require estimations of the probability of successfully completing each step and the assumption that those probabilities are independent, so that a simple product gives the desired result for the whole sequence. Such predictions have met with criticism because of the difficulty in making valid probability estimates of human actions, where error probabilities are very small and success probabilities are very close to 1.

Perhaps more useful are methods such as cause-consequence analysis, failure modes and effects analysis, Bayesian and Markov network models, event trees, and fault trees. The main benefit of using such methods is that that the analyst must think through the various possibilities for human and machine error and make qualitative judgments of what factors have the greatest effect.

For example, a fault tree (see Figure 2.19) is a model of the logical conditions that must obtain in order that each successive higher order goal be met. It requires the analyst to think through the necessary conditions (logical ANDs if more than one) and sufficient conditions (logical ORs if more than one) in a tree pattern as shown. For example, both C and D are required for A to happen, while any of F, G, or H is sufficient for B to happen. Probabilities can be associated with the links as well.

Resilience Engineering

Where traditional risk analysis stems from quantitative analysis based on retrospective consideration of human errors and machine failures, a newer approach (Hollnagel et al., 2006) called *resilience engineering* places the emphasis on preparation for future and unforeseen incidents, recovery before serious consequences occur, and return to a stable state. Its claim is that events at the "sharp end" where the specific human operator actions are typically where blame is placed, are wrong-headed. Resilience engineering tends to put the emphasis on analysis of procedures, management policies, the state of preparedness, and the degree of *safety culture* that exists—events at the "blunt end." While the contest for

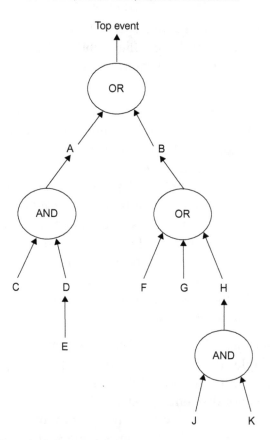

FIGURE 2.19   A fault tree.

resources between these two competing approaches will continue, they both have their compelling benefits. Both should be utilized in balance and to some affordable degree.

SUMMARY, THE FUTURE, AND SOME KEY ISSUES

Some definitions, history, and substantive modeling ideas have been presented, the purpose being to acquaint the reader with how these ideas together constitute "systems thinking" and how they relate to human factors in aviation. Only the basics are described above. All of these models can and have been further embellished to fit particular applications.

Today, the systems perspective is so much a part of aviation projects, whether in aircraft design, flight and air traffic control, maintenance or management, that the term has almost lost any specific meaning.

The mathematical models, analytical tools, and simulation methods consistent with systems thinking continue to be refined, using ever more powerful computational tools. The market place demands both increased performance and increased safety, so there is essentially no choice other than to incorporate systems thinking. The degree to which systems are *human-centered* (Billings, 1997) and what the term means remains an issue to be worked out and defended by human factors professionals. This author, in many years of teaching engineers, has tried to emphasize that technological design is ultimately a matter of satisfying human needs, although engineers are too quick to consider only that they are designing *things* rather than *relationships to people.*

All the models described here are *normative,* meaning that they are relatively simple idealizations of how people really behave and do not include the full and subtle complexities of behavior. One group of behavioral scientists rejects the use of normative models for this reason, but they have no quantitative alternative and therefore have little to offer the engineering system designer. Most engineering models are likewise simple approximations of physical reality, but nevertheless have proven very useful.

As the complexity of technology increases the human role must change to keep up, and human limitations seem to be blamed more and more relative to the technology for any mishaps. Human behavior is not easily amenable to mathematics and simple predictive models. Nevertheless, there are now pressures to combine models from various disciplines into interdisciplinary models, for example, models that might combine aviation technology (aerodynamics, electronic communication, control) with economics and with human factors. This is sometimes referred to as *systems of systems.* It is a big order, but as computers become more powerful and knowledge advances, we will see such models develop. They may not be predictive in a narrow sense, and perhaps not generalizeable beyond some specific applications, but they will serve the purpose of getting the engineers, economists, human factors professionals, and project managers to share assumptions and appreciate the points of view of the others.

References

Baron, S., & Kleinman, D. L. (1969). The human as an optimal controller and information processor. *IEEE Transactions on Man-Machine Systems, MMS-10*(1): 9–17.

Baron, S., Zacharias, G., Muralhidaran, R., & Lancraft, R. (1980). PROCRU: a model for analyzing flight crew procedures in approach to landing. In *Proceedings of the 8th IFAC congress*, Tokyo, 15:71–76. See also NASA Report CR152397.

Bertalanffy, L. (1950). An outline of general systems theory. *British Journal for the Philosophy of Science*, 2(2).

Bertalanffy, L. (1968). *General systems theory: Foundations, Development, Application.* New York: George Braziller.

Billings, C. E. (1997). *Aviation automation: The Search for a Human-centered Approach.* Mahwah, NJ: Erlbaum.

Cannon, W. (1932). *The Wisdom of the Body.* New York: W.W. Norton.

Charny, L., & Sheridan, T. B. (1989). Adaptive goal setting in tasks with multiple criteria. In *Proceedings of the 1989 International Conference on Cybernetics and Society.*

Cherry, C. (1957). *On Human Communication.* Cambridge, MA: MIT Press.

Churchman, C. W. (1968). *The Systems Approach.* New York: Laurel.

Degani, A. (2001). *Taming HAL: Designing Interfaces Beyond 2001.* New York: Palgrave.

Deutsch, K. (1963). *The Nerves of Government.* New York: Free Press.

Drucker, P. (1974). *Management: Tasks, Responsibilities, Practices.* New York: Harper & Row.

Forester, J. W. (1961). *Industrial Dynamics.* Cambridge, MA: Productivity Press.

Green, D. M., & Swets, J. A. (1966). *Signal Detection Theory and Psychophysics.* New York: Wiley.

Hollnagel, E., Woods, D. D., & Leveson, N. (Eds.), (2006). *Resilience Engineering: Concepts and Precepts.* Burlington, VT: Ashgate.

Hoppes, J. C. (1995). *Lean Manufacturing Practices in the Defense Aircraft Industry.* MS thesis, Department of Aeronautics and Astronautics. Cambridge, MA: MIT.

Johnson-Laird, P. N. (1993). *Human and Machine Thinking.* Mahwah, NJ: Erlbaum.

Kalman, R. E. (1960). A new approach to linear filtering and prediction problems. *Journal of Basic Engineering, Transactions of the ASME, 82D*, 33–45.

Klir, G. J. (1969). *An Approach to General Systems theory.* New York: Van Nostrand Reinhold.

Licklider, J. C. R. (1960). Man-computer symbiosis. *Institute of Radio Engineers Transactions on Human Factors in Electronics, HFE-1*, 4–11.

March, J. G., & Simon, H. A. (1958). *Organizations.* New York: Wiley.

McRuer, D. T., & Jex, H. R. (1967). A review of quasi-linear pilot models. *IEEE Trans. Human Factors in Electronics, HFE-4*(3), 231–249.

Meadows, D. H., Meadows, D. L., Randers, J., & Behrens, W. W. (1972). *The Limits to Growth.* New York, NY: The Club of Rome and Universe Books.

Miller, G. A. (1956). The magical number seven, plus or minus two: Some limits on our capacity for processing information. *Psychological Review, 63*, 81–97.

Miller, J. G. (1978). *Living Systems.* New York: McGraw Hill.

Moray, N. (1997). Models of models of mental models. In T. B. Sheridan & T. Van Lunteren (Eds.), *Perspectives on the Human Controller* (pp. 271–285). Mahway, NJ: Erlbaum.

Morse, P. M., & Kimball, D. (1951). *Methods of Operations Research.* Cambridge, MA: MIT Press.

Newell, A. (1990). *Unified Theories of Cognition.* Cambridge, MA: Harvard University Press.

NextGen Concept of Operations 2.0. (2007). Washington, DC: FAA Joint Planning and Development Office. Available at <http://www.jpdo.gov/library/NextGen_v2.0.pdf/>.

Odum, H. (1994). *Ecological and General Systems: An Introduction to Systems ecology.* Boulder: Colorado University Press.

Ohno, T. (1988). *Toyota production system.* Cambridge, MA: Productivity Press.

Parasuraman, R., Sheridan, T. B., & Wickens, C. D. (2000). A model for types and levels of human interaction with automation. *IEEE Transactions on Systems, Man and Cybernetics, 30*(3), 286–297.

Pew, R. W., & Mavor, A. S. (2007). *Human-System Integration in the Systems Development Process.* Washington, DC: National Academies Press.

Pisanich, G., & Corker, K. (1995, April). A predictive model of flight crew performance in aurtomated air traffic control and flight management operations. *Proceedings of the 8th international Symposium on Aviation Psychology.* Columbus: Ohio State University.

Raiffa, H., & Schlaiffer, R. (1961). *Applied Statistical Decision Theory.* Cambridge, MA: Harvard University Graduate School of Business Administration.

Sage, A. P., & Rouse, W. B. (Eds.), (1999). *Handbook of Systems Engineering and Management.* New York: John Wiley.

Shannon, C. E. (1949). Communication in the presence of noise. *Proceedings of the Institute of Radio Engineers, 37*, 10–22.

Shannon, C., & Weaver, W. (1959). *The Mathematical Theory of Communication.* Urbana: University of Illinois Press.

Sheridan, T. B. (1974). *Man-Machine systems: Information, Control and Decision Models of Human performance.* Cambridge, MA: MIT Press.

Sheridan, T. B. (1992). *Telerobotics, Automation and Human Supervisory Control.* Cambridge, MA: MIT Press.

Sheridan, T. B. (2002). *Humans and Automation.* New York: John Wiley.

Sheridan, T. B., & Parasuraman, R. (2006). Human-automation interaction. Chapter 2. In R. S. Nickerson (Ed.)*Reviews of Human Factors and Ergonomics: Vol. 1.* Santa Monica, CA: Human Factors and Ergonomics Society.

Smolensky, M. W., & Stein, E. S. (1998). *Human Factors in Air Traffic Control.* New York: Academic Press.

Swain, A. D., & Guttman, H. E. (1983). *Handbook of Human Reliability Analysis With Emphasis on Nuclear Power Plant Applications* Sandia National Labs/NUREG CR-1278. Washington, DC: Nuclear Regulatory Commission.

Taylor, F. W. (1911). *Principles of Scientific Management*. New York: Harper and Rowe.

Von Neumann, J., & Morganstern, O. (1944). *Theory of Games and Economic Behavior*. Princeton, NJ: Princeton University Press.

Wiener, E. L., & Curry, R. E. (1980). Flight deck automation: promises and problems. *Ergonomics, 23*, 995–1011.

Wiener, N. (1964). *God and Golem Incorporated*. Cambridge, MA: MIT Press.

Wiener, N. (1948). *Cybernetics*. New York: John Wiley.

Wiener, N. (1955). *The Human Use of Human Beings*. New York: Doubleday.

Womack, J., Jones, D., & Roos, D. (1990). *The machine that changed the world*. New York: Harper Perennial.

Zadeh, L. (1965). Fuzzy sets. *Information and Control, 8*, 338–353.

3

The System Safety Perspective

Amy R. Pritchett

Georgia Tech School of Aerospace Engineering

INTRODUCTION AND SOME DEFINITIONS

System (sĭs′tǝm) n. A group of interacting, interrelated, or interdependent elements forming a complex whole. [Late Latin systēma, systēmat-, from Greek sustēma, from sunistanai, to combine : sun-, syn- + histanai, set up, establish.]

Other chapters of this book focus on the myriad aspects of aviation human factors. This chapter intends to contribute a "systems perspective" but, given the often varied and vague use of the term, let us start by examining its definition to scope the discussion. First, note the word "group"—we shall not focus on an individual aspect of technology or human performance, but instead discuss the process of identifying the set of most relevant aspects. Second, note the emphasis on "interacting, interrelated, or interdependent elements"—beyond assembling a "grab-bag" of separate aspects the systems perspective represents how they relate to each other. Third, aviation systems definitely comprise "a complex whole," which can only be comprehended by its human participants if the "system perspective" serves as a lens illuminating the most meaningful representations. Finally, examining the etymology of "system," note the emphasis on action—an effective systems perspective is defined by its ability to improve design.

Likewise, this chapter seeks to include safety into this system perspective. Aviation accidents are often catastrophic—that is, their costs (especially including loss of life) can far exceed the value of the functioning system. As noted by Reason (1997), a negative view of safety is as a limit on "production"—in this case, limiting new aviation operations in the name of safety. Viewed positively, "safety" is the mitigation or elimination of risks; correspondingly, this chapter examines how risk mitigation can be distributed across technology, operational concepts and procedures, and effective human performance. Historically, effective human performance has been nearly synonymous with safety, especially given the significant role aviation confers on individuals' expertise. As aviation systems become more automated and proceduralized, effective human performance will remain critical, but the focus may shift from being principally on individual performance to the combined system performance of human-machine interaction and dispersed teams.

The "system safety perspective" examined in this chapter, then, will focus on identifying the set of aspects or components of a system relevant to safety. We start with considering the sometimes fuzzy definition of boundary between system and boundary, given their synergy and interaction. We will then discuss methods examining safety issues within systems, especially those of increasing complexity. We will additionally discuss how human performance can improve safety when properly supported within the system. Throughout, design of an airborne collision avoidance system will be used as a running example, building up to a summary of the most important considerations in applying a "system safety perspective."

DEFINING A SYSTEM BOUNDARY

Common system engineering practices start with drawing a crisp boundary delineating what is "in" and "out" of the system; such methods generally assume that important influences on system behavior can be crisply defined. As discussed by Sheridan in this volume, such a crisp delineation enables application of several powerful models of important system dynamics. However, in aviation a wide range of environmental factors can be important and a crisp delineation is not always sufficient. Therefore, this section will discuss the range of factors impacting system performance, starting from the most constant and profound and building to those which, depending on the purpose of the analysis, may be viewed loosely or partly.

As an example, consider the introduction of the Traffic Alert and Collision Avoidance System (TCAS). This system uses a specialized radar transponder to estimate range (accurately) and bearing (inaccurately) to other aircraft, and to broadcast altitude and coordinate any avoidance maneuvers between aircraft. Pilots are provided a horizontal traffic situation display (altitude shown in text), which may be large or small, and that may be part of a multifunction display also portraying navigation information, convective weather, terrain and/or required vertical speed. Upon a precautionary "traffic alert," an aural "Traffic, Traffic" is announced; with a full "resolution advisory" a vertical maneuver is commanded both aurally (e.g., "Climb, climb") and on a visual display of vertical speed (RTCA, 1983). Although its sensors and algorithms have been steadily improved, the system can still provide contextually inappropriate alarms (especially on closely spaced parallel approaches), and is limited in the types of maneuvers that they can command (notably, vertical maneuvers only). Proposed improvements (or new systems) may change the maneuver and alerting algorithms and even automate the avoidance maneuver entirely.

The Organism Itself

Let's start our discussion with the organism itself, placing the human smack in the system's center. As the other chapters in this book discuss, many aspects of human performance have been studied extensively, providing a solid foundation for many design decisions. For example, our understanding of perceptual and attentional mechanisms identifies fundamental properties in display design (cf. Wickens & Holland, 2000).

However, human performance spans many "levels" of behavior, some of which are less predictable. Although we may know how to design a flightdeck display for legibility, we can not always predict when (or whether) a pilot will choose to look at it, and how its information will be interpreted and acted upon given pilot training, immediate context and competing tasks. Thus, we also need to account for how the human controls his or her response to context. For some tasks, the human may be seen as an adaptive element, predictable only in terms of what performance they will drive the system towards. For example, McRuer and Jex (1967) recognized that the pilot would vary his or her behavior to cause a predictable "pilot-and-aircraft-combined" behavior. Thus, the field of manual control, rather than attempting to isolate the human component of system behavior, has focused on the "handling qualities" of the vehicle facilitating this adaptive performance (cf. Hess, 1997).

Such metaphors of "control" and "adaptation" have also been proposed for cognitive functions. For example, Hollnagel's Cognitive Control Model describes the organization of cognitive behaviors as a response to immediate resources (including subjectively available time, information availability, and knowledge) and demands (including task demands). Hollnagel (1993) describes a continuum from the unstressed strategic mode (with full information search, team coordination and optimal decision-making strategies), to the progressively more-stressed tactical mode (with rule-based information-search and team coordination, and with satisficing decision strategies) and then the opportunistic mode (dominated by naturalistic behaviors relying on intuition and effective cue-utilization), reaching the extreme of the scrambled mode (also called "panic'). Thus we can identify which models (e.g., rational versus naturalistic decision making) best describe expected behavior— and which one(s) the design should seek to support.

Simon used the analogy of an ant's track across a beach to describe how a simple organism (such as an ant) may exhibit a complex behavior when placed in a complex ecology. In aviation, neither the organism nor the ecology are simple. As illustrated in Table 3.1, some behaviors are internally predictable, but others are predictable more as a response to context, such as whether a pilot agrees with a TCAS resolution advisory. Thus, the following subsections expand our system boundary further to identify the aspects of the ecology (environment) which will impact behavior.

TABLE 3.1   Case Study: The Role of the Organism in Design of Aircraft
Collision Avoidance Systems

Many human factors studies have examined human performance issues with
aircraft collision systems. Some focused on the pilot's perception of the alert,
establishing guidelines for the saliency of alerts and the exact selection of
unambiguous aural commands.[i,ii,iii] Others examined the traffic situation display,
developing standards for symbology and display format. The precautionary
alert carefully directs pilots' attention to the traffic situation display and, if
possible, visual acquisition (if the pilot was not already aware of the potential
traffic conflict), mitigating the impact of different cognitive control modes. These
display features have been standardized so that the industry uses a common,
well-established standard (RTCA, 1983). This corpus has helped ensure that the
pilot understands which alert has been issued; assuming that the pilot agrees
with the commanded maneuver, he or she is then in a position to follow it.

[i] Stanton, N. A., & Edworthy, J. (Eds.). (1999). *Human Factors in Auditory Warnings.*
Aldershot, UK: Ashgate.
[ii] Veitengruber, J. E. (1977). Design criteria for aircraft warning, caution, and
advisory alerting systems. *Journal of Aircraft, 15*(9), 574–581.
[iii] Noyes, J. M., & Starr, A. F. (2000). Civil aircraft warning systems: Future
directions in information management and presentation. *International Journal
of Aviation Psychology, 10*(2), 169–188.

Physical Environment

Expertise may be defined as efficient adaptation to the demands of
the environment. Consider the task of aircraft control. The stick and
rudder directly control aircraft attitude (pitch, roll, and yaw), which
can change quickly; thus, display of aircraft attitude must be central
to the pilot's vision. Over a slightly longer interval pitch attitude
will integrate into vertical speed and deviation from an assigned
altitude, and roll attitude will integrate into heading; thus, display
of aircraft vertical speed, altitude and heading should be positioned
around the attitude display. Over an even longer interval, aircraft
heading and speed will combine to establish a track over the ground
relative to an assigned course; thus, navigation displays need to further surround the central cluster. Thus, modeling the physical environment (in this case, using flight mechanics) can often provide a
systematic means to inform display design (in this case, the "basic
T" layout of the primary flight instruments). Expertise may be
defined by how well the operator mirrors his or her behavior to the
environment, such as an "instrument scan" starting from the central

attitude display and air traffic controllers' visual search strategies mirroring airspace structures to filter out aircraft irrelevant to a search for potential conflicts (Rantenan & Nunes, 2005).

Many cognitive engineering techniques formally or systematically model fundamental constraints within the physical environment which expert behavior will need to control. Cognitive work analysis starts by itemizing the physical form of the environment as the lowest level of abstraction, and then analyzes the functions that they will perform relative to the high-level goals of the system, with intermediate levels of abstraction progressively describing the physical functions into "abstract functions" and "general functions" (Rasmussen, Pejtersen, & Goodstein 1994; Vicente, 1999). These models can be annotated with observed or potential sequences of actions. In a safety-critical environment such as aviation, these actions need to span both ideal situations and unexpected conditions. These models can identify information requirements and help generate new display concepts (e.g., Gonzalez Castro et al., 2007; Sperling, 2005; Van Dam, Mulder, & Van Paassen, 2008).

Experts may actively choose to construct their immediate environment to support their memory and representations. For example, pilots may place pointers to airspeed limits, target headings and altitudes on their displays, place an empty cup over the flap handle or fold over the page of an incomplete checklist—all as part of planning for upcoming events, prompting correct resumption of interrupted events, or tracking an assigned route of flight. Similarly, air traffic controllers may write on flight progress strips and then organize them spatially on their desk according to urgency and proximity, even holding the most urgent in their hands. Thus, as illustrated in Table 3.2, the physical environment is not just the stimulus for action but also a fluid and valuable part of cognition; during design we need to examine the human's ability to operate on the environment in support of their cognitive activities (Hutchins, 1995; Kirlik, 1998).

The Team

A team may be defined as a collection of individuals operating together towards a set of shared goals (Sperling, 2005). Team operations demand more of team members than just their individual

TABLE 3.2 Case Study: The Role of the Physical Environment in Design of Aircraft Collision Avoidance Systems

A pilot's judgment of the physical environment (i.e., assessment of a traffic situation) can conflict with the timing of the alert or with the commanded avoidance maneuver, one potential contributor to a higher rate of non-conformance to alerts than anticipated during design.[iv,v] For example, studies have noted situations where pilots (and other participants in simpler simulator studies) reported forming different judgments than the collision alerts; these included perceiving alerts as too early or conservative, and believing (based on the horizontal situation display) that a horizontal maneuver component should be added.[vi,vii,viii,ix,x] The displays are typically spatial, yet conflict alerts need to account for the predicted time of conflict. Thus, displays may be augmented to illustrate timing considerations or the "hazard space," or to explicitly show the spatial equivalent of a temporal alerting criteria relative to a potential conflict as it changes in response to factors such as closure rate (e.g., Pritchett & Vándor, 2001).

Pilots' interactions with the traffic situation display are often physical, with pilots pointing to or resting their finger on the traffic symbol of interest. Various highlighting mechanisms can help pilots with this physical interaction. In the established TCAS traffic situation display, the symbols change in response to proximity and hazard; in research prototypes, additional mechanisms allow the pilot to explicitly interact with the display, such as selecting aircraft of interest to highlight, or a mechanism for entering the identifier of an aircraft that they are to follow.

TCAS is only one of many alerting systems in the modern flightdeck. Thus, the industry is developing standards for prioritizing and integrating the many alerting systems such that an aircraft in crisis will only have one alert issued at a time, focusing on the highest priority system.[xi] This illustrates the extent that even a safety-enhancing system can not be "slapped into" the flightdeck without considering it as part of a broader ecology.

[iv] Mellone, V. J. (1993, June). Genie out of the bottle? *ASRS Directline*, 4.

[v] Williamson, T., & Spencer, N. A. (1989). Development and operation of the traffic alert and collision avoidance system (TCAS). *Proceedings of the IEEE, 77*(11), 1735–1744.

[vi] Bass, E. J., & Pritchett, A. R. (2008). Human-automated judgment learning: A methodology to investigate human interaction with automated judges. *IEEE Transactions on Systems, Man and Cybernetics.*

[vii] Bisantz, A., & Pritchett, A. R. (2003). Measuring judgment interaction with displays and automation: A lens model analysis of collision detection, *Human Factors, 45*, 266–280.

[viii] Pritchett, A. R. (1999). Pilot performance at collision avoidance during closely spaced parallel approaches, *Air Traffic Control Quarterly, 7*(1), 47–75.

[ix] Vallauri, E. (1995). *Operational Evaluation of TCAS II in France* (CENA/ R95-04). Toulouse: Centre D'études de la Navigation Aérienne.

[x] Vallauri, E. (1997). *Suivi de la Mise en Oeuvre du TCAS II en France en 1996* [1996 Survey of TCAS II implementation in France] (CENA/ R97-16). Toulouse: Centre D'études de la Navigation Aérienne.

[xi] Proctor, P. (1998, April 6). Integrated cockpit safety system certified. *Aviation Week and Space Technology*, 61.

TABLE 3.3   Case Study: The Role of the Team in Design of Aircraft
Collision Avoidance Systems

Historically, only air traffic controllers (via radar displays) had the knowledge
to reliably detect potential conflicts and command avoidance maneuvers. Thus,
the roles of pilot and controller were dictated and supported by their available
information. The introduction of the traffic situation display changed this
balance. In a simulator study, pilots and controllers were found to undertake
their roles and methods of cooperation differently (Farley et al., 2000); other
changes, some cooperative and some not, have also been reported to the Aviation
Safety Reporting System.[xii]

More significantly, TCAS is a system intended for situations (time-critical aircraft
conflicts) that should never arise when the pilot-controller team is functioning
properly. Thus, while pilots are normally instructed to follow a controller's clearances,
the issuance of a TCAS alert is intended to modify the team roles and responsibilities
such that the pilot then follows the TCAS commands despite any conflicting air traffic
commands, coordinating with air traffic controllers only once safety permits.

[xii]Mellone, V. J. (1993, June). Genie out of the bottle? *ASRS Directline*, 4.

actions, as illustrated in Table 3.3. One demand is team maintenance, by which team members affirm (and, in some instances,
modify) their roles and responsibilities. Many air traffic control
activities serve a team maintenance function, with controllers and
pilots explicitly confirming their relationship through handoffs as
an aircraft transitions from one sector to the next and through clearances which explicitly define their relationship. For example, in
accepting clearance for a visual approach the pilot assumes responsibility for separation from aircraft on parallel approaches, proper
spacing behind lead aircraft and their wake, and intercepting and
tracking the approach course down to land. Such team maintenance may, in the future, need to be provided with more explicit
mechanisms should, for example, all or some of the flight crew may
be located remotely, communication systems automatically change
frequencies without the current interaction of a handoff between
sectors, or controllers of small "tower-less airports" be placed at
remote stations where they will not have direct observation of cues
such as seeing whether a landing aircraft is exiting a runway.

Another demand is situation awareness about the team, including
keeping track of "who is doing what," "who knows what," and

"how everyone is doing." Depending on team structure all the members may participate fully; for example, a flight crew of two generally works to stay aware of each others' activities as a form of redundancy and error checking. In other cases, this maintenance is the explicit function of a supervisor or manager; for example, an air traffic supervisor's function is generally to stay aware of special demands on controllers, their workload, and concerns with fatigue, overload or time on duty. As with team maintenance, such situation awareness about the team is a natural, social process for co-located personnel that can be obstructed when they are placed distally and/or their interactions are mediated by technology.

The final demand is maintenance of team situation awareness, that is, a collective state of awareness about the physical environment. While commonly called "shared" situation awareness, it is crucial to note that "shared" can mean "held in common" or "divvied up." In aviation, team members generally have specialized roles and distinct responsibilities. In such teams, expecting situation awareness to be "held in common" can cause excessive information gathering and monitoring compared to the information needed for each team member's assigned tasks, lead to different interpretations, and even lead to role conflict and reduced team maintenance. Conversely, situation awareness "divvied-up" according to each team member's role can lead to improved information sharing (when team members know explicitly who needs what information) and improved overall team and system performance (e.g., Sperling, 2005; Sperling & Pritchett, 2006). Thus, effective team situation awareness should be measured relative to the roles of team members and their commensurate tasks (Farley et al., 2000).

The Procedural Environment

In aviation, the roles and responsibilities of human operators are generally defined by explicit and implicit procedures. Procedures are used broadly here as "prescribed patterns of activity"; this definition spans a wide range of sources from institutionalized (but informal) work practices to regulations, and a wide range of detail from loose descriptions of best practices to itemized checklists (cf. Ockerman & Pritchett, 2000). Thus, procedures can—and often should—be modeled explicitly as being within a system's boundary.

Procedures scaffold or guide behavior into safe, effective courses of action. In theory, an expert should be so well accommodated to their operating environment that he or she would not need procedures. However, in aviation several practicalities demand procedures. The first is the need for coordinated behavior between individuals. For example, in airline operations the first officer and captain may never have met each other before a flight, yet they can interact immediately by following procedures. Likewise, standard air traffic procedures allow pilots and controllers to function together globally. Of particular note to a system perspective is the converse, when one operator's procedures change or when someone does not understand another's procedures. For example, the introduction of sophisticated flight management systems enabled optimal descents not always supported by published air traffic procedures; at this time, unmanned aircraft systems (UAS) are limited in their ability to execute air traffic procedures, and thus their operation in controlled airspace is also limited.

A second practical demand for procedures is to scaffold behavior while expertise is developing. For example, training often starts with a description of the work—a procedure—which student pilots (or experienced pilots transitioning to a new aircraft or a new operation) then apply through the progressive stages of learning. Successful execution of the procedure is often considered sufficient for a pilot to fly—further learning occurs "on-the-job." Likewise, the inherent structures of procedures can provide a useful basis for structuring interfaces (e.g., Gonzalez Castro et al., 2007) and automated systems (e.g., Davis & Pritchett, 1999).

The third demand for procedures is to guide behavior in abnormal and emergency situations. An analyst, using multiple methods and exploring all foreseeable options, can rigorously design and test a safe course of action so that the operator in a stressed environment has a clear path to follow. A large set of such situations are considered explicitly during design, with the most common and severe situations trained extensively, others trained partly (e.g., only in ground school), and others perhaps provided in a handbook.

Procedures can never be perfect. When the environment mirrors the assumptions underlying a well-designed procedure, the safest course is to follow the procedure but, perversely, when the procedure is flat-out wrong, or the environment is wrong for it, the safest course of action is to *not* follow it! Thus, whether to rely on a procedure is itself a decision the operator must make in context, as illustrated in Table 3.4. Likewise, operators in aviation often need to decide how to manage interruptions or disruptions to procedures, such as assessing whether one step can be skipped or delayed (e.g., McFarlane & Latorella, 2002).

So far we have connoted procedures with action. Finally, let us note the relationship between procedures and information: procedures

TABLE 3.4 Case Study: The Role of the Procedural Environment in Design of Aircraft Collision Avoidance Systems

A general view of air traffic procedures is that pilots should follow their assigned route of flight; if any flight crew doesn't follow their assigned route and a potential conflict arises, a pair of aircraft will be issued TCAS-commanded maneuvers which procedure dictates both should follow. Of particular interest during design, however, are cases where everyone following procedures still leads to conflicts. For example, early in its introduction, TCAS was found to generate conflicts in specific airspace configurations where, if everyone conformed to their assigned routes no conflicts would occur but TCAS, based solely on linear projects of immediate routes, identified that a conflict was physically possible should one aircraft not level off or execute a turn to follow its route. In some cases, the airspace structure was modified to reduce such alerts. In other cases, the pilots are in the unfortunate situation of deciding when to follow the TCAS maneuvers; for example, in a busy terminal area pilots may be instructed to leave the TCAS on until they are on a landing approach with parallel traffic, at which point they are to turn the TCAS to a "traffic alerting only" mode. In such cases, where the guidance for pilots is to *always* follow the commanded maneuver *except* when they shouldn't, very clear and specific articulation of these exceptions is warranted, along with the easy ability to identify these exceptions.

Pilots' interpretation of the traffic situation display is also strongly impacted by their knowledge of associated air traffic procedures. For example, pilots tend to articulate traffic on a situation display not in terms of range and bearing, but in terms of which route they are on and whether they might be asked to act in some way relative to that aircraft; the "party line" available from current voice communication procedures also provides a mechanism by which pilots can interpret clearances issued to other aircraft in terms of likely subsequent clearances to themselves.

effectively distribute a shared expectation that guides use of real-time information displays. For example, in a flight simulator study of pilot self-spacing behind other aircraft, when the standard terminal arrival procedure published expected speeds along the arrival route, less importance was assigned to the real-time display of lead aircraft speed, and both own aircraft and lead aircraft speed were often described relative to the procedure (Pritchett & Yankosky, 2003). Thus, information requirements need to be assessed relative to the normative information provided by procedures, and changes in either an information display or a procedure may impact the use of the other.

The Shared Abstraction—Expectation, Training, and Representations

Another important aspect of expertise is the ability to form expectations about upcoming events, including the actions of others. A pilot familiar with an aircraft can expect a drop in airspeed resulting from extending flaps; a pilot familiar with an airport can expect air traffic control clearances. As illustrated in Table 3.5, these expectations can build not only on procedures as just noted but also on experience with the events, and on "lore" spread

TABLE 3.5   Case Study: The Role of Expectation and Representation in Design of Aircraft Collision Avoidance Systems

Pilots' representations of air traffic flows into busy airports are fed by several sources: a generic picture of arrival and departure flows (e.g., the standard landing pattern) formalized by constructs such as "downwind," "base leg," and the "outer marker"; published arrival and departure routes for that airport; and personal experience with and word-of-mouth about an airport's idiosyncrasies. A pilot monitoring a traffic situation display may reasonably expect to be cleared to turn base at the point providing an appropriate spacing behind another aircraft, and may expect the aircraft ahead to slow to landing speed at the outer marker. Thus, the traffic situation display is not read "in a vacuum" but instead framed within a mental construct. Changing the air traffic flows—especially profound changes in constructs such as free flight, novel route structures or other constructs proposed for NextGen operations—may temporarily interrupt this ability, thus changing pilots' use of traffic situation displays and responses to traffic alerts.

through formal training, informal mentoring, and even general media such as trade magazines and discussion forums.

When these expectations are correct, they support the operator's ability to predict, plan and prepare for events. For example, pilots may adjust the schedule by which they extend flaps according to their expectation about an airport's arrival operations, and may complete other tasks such as the landing checklist in anticipation of later being busy responding to air traffic clearances. This provides a method for balancing workload and pre-gathering important information. Likewise, it provides for error checking, such as pilots' inquiring about clearance to land once past the point where they would expect it.

However, when these expectations are incorrect, they can establish a resistant form of error often termed expectancy bias (although the utility of this concept needs also to examine the fault in the expectation as discussed by Klein [1999, p. 85]). For example, training material coaches pilots to make sense of rapid-fire air traffic control communications (via static-filled voice radio) by anticipating the clearances commensurate with each phase of flight (e.g., the training manual *Air Plane Talk*). This enables pilots to interpret routine air traffic control communications, but biases their interpretation of unusual, ambiguous or hard-to-hear communications. Thus, communicating events contrary to expectation can be particularly difficult, often requiring repeated, redundant and persistent methods of relaying unexpected circumstances.

A similar effect stems around a community that abstracts important aspects of the environment and tasks using a "shared representation," formally defined here as the mental organization of information used to support sense making about conditions in (and effect of actions on) the operating environment. Multiple representations may be possible of the same aspect of reality—for example, a pilot may describe pitch angle as controlling vertical speed and throttle as controlling airspeed, while also understanding a separate, equally valid energy management representation that links pitch and throttle. Because these representations provide a powerful basis for sense making, innovative technologies or operating procedures may need extra effort to promote true, deep understanding of novel underlying representations. One telling example is the transition from the "steam-gauge" cockpit to the glass cockpit. Several representations were challenged by this transition, ranging from making sense of new display formats and information, to the role of the pilot as a dynamic controller versus a system manager; such shifts in underlying representation can be stressful and take a long time to incorporate. Changes to air traffic operations may likewise call for such large shifts in representation; these will likely be justified by the operational improvements they will achieve, but may need to be carefully and consciously articulated to human operators.

The Socio-Technical System—Society, Culture, and Values

Stepping even further from the flightdeck and air traffic controller's station, aviation can be viewed as a socio-technical system in which social and cultural values can strongly influence behavior. For example, Hofstede's Cultural Dimensions describe several factors such as power distance (the extent to which the culture as a whole expects an unequal relationship between leaders and followers), individualism versus collectivism, and uncertainty avoidance (cf. Hofstede, 2001). Although these dimensions do not predict idiosyncrasies (and aviation operators are not always representative of the general population of their country or region), these dimensions (and other cultural descriptors) have been used to describe aggregate cultural factors that should be considered during design and evaluation. Of particular note, the first dimension, power distance, has been cited in studies of crew resource management as describing whether multiple crew organizations fully provide redundancy (e.g., the culture would encourage or discourage a first officer from questioning a possible error by the captain) and whether flightdeck designs support all cultures (e.g., Harris & Li, 2008). Another important aspect in socio-technical systems is a culture's attitude specifically toward automation and reliance on technology. As illustrated in Table 3.6, these attitudes can shape the extent to which operators will tend to accept and rely on new technological capabilities.

TABLE 3.6 Case Study: The Role of Culture in Design of Aircraft Collision Avoidance Systems

On 21:35 (UTC) July 1, 2002, two aircraft (a DHL Boeing 757 and a Bashkirian Tupolev 154) collided over Überlingen, Germany, near the Swiss border. Through a number of circumstances, the air traffic controller missed a conflict between two aircraft until shortly before TCAS issued collision avoidance maneuvers to each aircraft. The controller was unaware of which maneuver had been issued to which aircraft, and issued clearances opposite to the TCAS maneuvers. Cultural aspects with power distance and attitudes to technology are frequently cited as factors in the ensuing accident; the DHL pilots (British and Canadian) executed the TCAS maneuver, ignoring the air traffic controller's commands, while the Bashkirian pilots (Russian) ignored the TCAS and followed the air traffic controller's clearance.

EVALUATING FOR SAFETY

Defining the system boundary to include all system aspects import to design (as just covered in the previous section) provides the basis to examine the system for safety. This section describes several general approaches to such examination: (1) predicting what can go wrong, (2) examining for resilience and flexibility in the face of the unexpected, (3) viewing safety as an emergent effect, and (4) mitigating risk by distributing it among technology, procedures, and personnel.

Predicting What Can Go Wrong

One approach to system safety is to ask and reask two questions throughout the evolution of a design: "How should the system work?" "What could go wrong, and how should the system respond?" The result is a normative representation suitable for several purposes: evaluating for safety, designing procedures and technology for all conceivable situations, and training the operators for all conceivable situations.

Examining the first purpose in more detail, a number of methods exist for evaluating system safety. Reliability-based methods seek a mathematical representation of probability of failure or accident. For example, fault tree analysis decomposes all conceivable events which could block successful performance or, conversely, lead to

TABLE 3.7   Case Study: Fault-Tree Analysis of TCAS

During the initial design of TCAS an extensive fault-tree analysis was performed for TCAS, which continues to be examined through the current day. The analysis includes failure probabilities for important components (obtained through classic reliability analyses), as well as probabilities of a collision avoidance maneuver not being successful when all components are functional and the maneuver is properly executed due to factors such as difficult conflict geometry (estimated through extensive fast-time simulations of conflict scenarios with expected variability in component behavior).

Of particular note here, the TCAS standards expect the pilot's reactions to be within 5 seconds of the alert, to initiate a vertical maneuver with a 0.25g pitching maneuver (if the command requires a maneuver—some, such as "maintain climb" instead constrain any changes from current profile), and to maintain the maneuver for its entire duration. Early analyses estimated a probability of 5% that these specifications would not be met exactly;[xiii] following initial implementation, non-conformance was self-reported to the Aviation Safety Reporting System (ASRS) at a rate exceeding 20%.[xiv]

[xiii] Williamson, T., & Spencer, N.A. (1989). Development and operation of the traffic alert and collision avoidance system (TCAS). *Proceedings of the IEEE, 77*(11), 1735–1744.

[xiv] Mellone, V. J. (1993, June). Genie out of the bottle? *ASRS Directline, 4*.

---

an accident. (See Table 3.7 for an example.) Other methods include the use of dynamically colored Petri nets, Markov models of failure sequences, and other system-theoretic bases, often providing a finer resolution and allowing for dynamic sequencing of events. These methods have evolved from formal studies of reliability, providing a repeatable, mathematical basis for examining the contributions of each component in the system. For technical elements, a solid basis for describing failure probabilities can generally be found, including physics-based modeling and extensive testing-to-failure of each component. However, to include human performance explicitly these analyses, attempts have been made to isolate and quantify human performance in the same manner as technological performance. This often assumes human follows a set of mechanistic steps, each of which with an assigned probability of failure, that is, human error. To this end, a number of human error and reliability models have been developed (e.g., NRC, 2000).

For the high levels of safety demanded of aviation, such a full and comprehensive examination of all conceivable safety issues is a

valuable and necessary step (as for whether it is a "sufficient" process, please see the next section). This examination is hampered,
however, when it expects human performance to be neatly described
in the same manner as technological performance in terms of likely
faults and associated probabilities. Not only are probabilities of
human error difficult to predict with certainty (especially for new
operations), but actual human performance is generally much more
complex and situation-dependent than can be captured in a simple
action sequence. For example, pilots may predict (and act to resolve)
a potential aircraft conflict long before the alert sequence described
in a fault tree analysis; decision processes and actions may be iter-
ated upon in a cycle not captured in a linear sequence; and multiple,
interacting actions may be taken simultaneously.

Instead, such mechanistic modeling may be "turned on its head":
rather than predicting a singular risk estimate from a combination
of failure and error probabilities, one may explore the sensitivity of
final outcome to predictable variations. For example, rather than
saying "What is the probability of collision if the pilot does not
properly execute the commanded avoidance maneuver 5% of the
time?" the reverse (and potentially more useful) question could be
"With what frequency and within what tolerance does the system
need the pilot to execute the avoidance maneuver to meet required
safety levels?" thus providing a tangible expectation for behavior
which can then be monitored. Likewise, such thorough analysis
can be valued as a systematic process by which conceivable risks
can be identified and mitigations deliberately evaluated, even
when they can not be explicitly quantified.

Examining for Resilience and Flexibility in the Face of the Unexpected

Although a thorough analysis of conceivable events is *necessary*
for high system safety, it is not *sufficient*. Once conceivable events
are addressed, the remaining challenge is to enable the system
to accommodate, adapt to, and deal with the inconceivable and
unexpected.

A presumptuous approach assumes that an analyst can predict all
relevant events *a priori;* such an attitude leads naturally to a system

design that attempts to drive out all sources of variation, seeking technology that is failure-proof or fail-safe, and expecting human performance to strive to the same standard. Evaluations of air traffic concepts of operation, for example, may use reductions in variance as their measure of performance. However, an operating environment that is optimized for ideal conditions can be remarkably brittle in response to disturbances (see Table 3.8 for an example). Thus, variance is good when it provides the flexibility for expert responses to exogenous variation in the environment. Such flexibility also provides the opportunity for learning about and exploring the operating environment in support of expert understanding of, and accommodation to, its dynamics and constraints (e.g., Reason, 1990).

Aviation operators are explicitly trained on procedure following and the application of technologies. However, corresponding "meta-skills" are also needed to recognize when procedures and technologies are inappropriate. When procedures and technologies are known to be flakey, such meta-skills are constantly reinforced and their execution institutionally supported. However, in other situations the application of such meta-skills may run

TABLE 3.8   Case Study: Change in Dominant Collision Mechanisms as Navigation Variation is Reduced

Consider the collision risk of two aircraft on parallel approaches. Two sources of variation may be explicitly modeled in aircraft position side-to-side on approach: the variance in the aircraft navigation around the programmed route, reasonably modeled as a Gaussian distribution with variance σ^2; and the programming of the incorrect approach into the autoflight system, reasonably modeled as a discrete error with probability p.

When σ^2 is large the dominant mechanism in creating a near midair collision is navigation error—even when the incorrect approach is programmed by one pilot, two aircraft accidentally on the same approach are not guaranteed to conflict.

However, as the variance in navigation, σ^2, is reduced through the introduction of new technologies, the probability of conflict is driven by programming error. At an extreme, with perfect navigation the probability of conflict is that of either (but not both) pilots programming the wrong approach, that is, $2p-p^2$, regardless of spacing between the approaches. Thus, a reduction in one compelling form of variance may enable another form to become dominant and intransigent.

into several blocks. One is the underlying knowledge needed to assess whether an intricate technology or procedure is appropriate in context; studies suggest that the explicit presentation of such aspects of "procedure context" and "boundary conditions" can improve system safety (e.g., Ockerman & Pritchett, 2002, 2004). Other blocks stem more from organizational and cultural factors in which system failures are not expected or non-normative behavior is not condoned (e.g., Reason, 1997).

Likewise, as described by Reason's "Swiss Cheese" model (Reason, 1997), in very safe systems accidents occur when multiple barriers breakdown. An organization may condone one controller covering another's sector in addition to their own during a break, for example, and expect controllers to use backup communication and radar systems during routine maintenance; a distracted controller may not notice and then effectively coordinate about a looming conflict; in the resulting looming loss of separation a pilot from one organization may be trained to follow the resolution advisory generated by the technology (TCAS) and the pilot from another organizational culture may elect to follow the maneuver generated by the human air traffic controller (furthering the accident description given in Table 3.6). Any particular accident may be prevented by one of these "levels" being strengthened—overall safety is enabled when all of them are evaluated and coordinated.

On a more conceptual level, another aspect of system design is supporting *resilience* (cf. Hollnagel, Woods, & Leveson, 2006). Resilience focuses on maintaining control of a situation, even in the face of unexpected disturbances. Such resilience demands flexibility in the system, including several potential strategies for its operation, and recognizes that humans may adapt their cognitive strategies rather than apply a single pattern of behavior in all circumstances.

Multiple formal methods exist for examining safety, but they remain normative, prescriptive, and typically depend on forms most amenable to failures in technology. Thus, an additional, complementary approach to safety currently being proposed and tried in a number of aviation organizations (and other safety-critical domains) focus on examining assumptions and corrective mechanisms (e.g., Leveson, 2003; Rushby, 2008). These generally provide an explicit listing of the assumptions made at all stages of the design. Conflicts in assumptions frequently arise—conflicts between designers of different technology components, between designers of technologies and operating procedures (e.g., current difficulties in operating unmanned aircraft systems [UAS] within air traffic control procedures), within operations (e.g., the ambiguous use of different units to describe fuel resulting in 1983 in a fuel-starved Boeing 767, Air Canada flight 143, gliding to land at Gimli Manitoba) and between initial designs with subsequent modifications or maintenance protocols made, sometimes, years later (e.g., potential use of the TCAS traffic situation display for new "self-spacing" procedures beyond its original "situation awareness only" intent). Safety cases, then, can be applied during design to evaluate system assumptions not only for conflicts internally, but also with proposed operations and with established "safety philosophy" statements during design. In addition, they can support longitudinal study as to whether the assumptions hold true once applied in actual operations or subsequently modified or extended.

Safety as an Emergent Effect

At the start of this discussion of safety evaluation we discussed "predicting what can go wrong" through classic methods based largely on decomposition of the system, and the aggregation of conceivable component failures or errors into an estimate of safety. While useful, there are two notable limitations to this approach: first, as just discussed, we are limited in our ability to conceive all relevant problems, disturbances and failures. Second (the topic of this subsection), many aspects of safety may be *emergent,* which we will use formally here to describe phenomena observable at one level of abstraction that can not be predicted at a different level of abstraction. For example, safety issues may arise at the level of a traffic flow that may not be predicted (or even explained) by looking at the activities of the pilots and controllers in isolation, and may arise despite everyone acting according to their procedures.

Thus, emergence represents a challenge to predicting overall system safety. This challenge is generally the greatest for operations that are decentralized and yet tightly coupled. Historically, most airspace operations were centralized within each sector. Conversely, operations can be decentralized when, for example, aircraft are given authority to decide their own route or to fly relative to other aircraft. Likewise, operations can be tightly coupled when, for example, aircraft trajectories are constrained by those of neighboring aircraft.

This combination of tightly coupled (especially driven by high traffic density) and decentralized is representative of many future operations; thus, the challenge of predicting emergent behaviors is, well, emerging. Even without the ability to predict the exact manifestation of emergent safety issues, some mitigating factors maybe generally hypothesized. The first is redundancy between agents to purposefully create self-correcting behaviors within the organization (or at least early detection of deviant behaviors). The second is maintaining some "slack" in the specification of behavior so that, for example, one aircraft flying slightly slower than expected does not lead to immediate, progressively increasing speed decreases in all subsequent aircraft. The third is improved "look-ahead" information; just as "feedforward" information tends to stabilize feedback control loops, predictive information can allow agents to time and temper their response such that each progressively reduces rather than increases the "ripple-effect" from the original disturbance.

Risk Mitigation: Distributing Risk among Technology, Procedures, and Personnel

Mitigation of risks, whether specifically predicted, generally feared, or actually observed, can be distributed between some mix of technology, procedures and personnel. One type of technological mitigation of risks focuses on "forcing functions," that is, technological functions or features that encourage less error-prone behaviors. For example, an engine start/shut-down switch may have a cover and a flap mechanism may need to be pulled up before it can be moved forward or aft. These forcing functions, however, can also engender unintended work-arounds or side-effects.

Of immediate interest is the drive for more "automation," a term generally used to describe machines providing functions normally attributed to humans; the term is also often applied to novel machine functions that go beyond current human functions, such as automatic optimal global organization of traffic flows instead of air traffic controllers working individual sectors. In general, the term "more automated" is generally used to describe machines with increased capability. When such automation is demonstrated to provide safe, reliable behavior over all relevant operating conditions, it can serve as risk mitigation. However, in many cases the automation is only capable within a bounded set of operating conditions or has some identified limitations or failure modes; in such cases, the human operator is necessarily left partly "in-the-loop," with the responsibility to monitor the automation (and its operating environment) and intervene in the case of degraded operations. Repeated research has shown that humans are poor monitors of reliable automation due to complacency and workload so low that they are no longer engaged in the task (Endsley & Kiris, 1995; Parasuraman, Molloy, & Singh, 1993). This ability to monitor (and effectively intervene) is further aggravated when the automation becomes too complex for the human to comprehend (e.g., Pritchett, 2001); at an extreme, proposed automated functions beyond the capability of the human yet assuming human intervention in degraded situations are, on their face, matters of concern.

From the point of view of risk mitigation, *robustness* and *flexibility* may be better definers of improved automation. Flexibility is used here to describe the machine's ability to operate in (and, if necessary, adapt to) a wide range of operating conditions (e.g., Miller & Parasuraman, 2007). Robustness is used here to describe an attribute generally recognized as useful in a human team member but not often recognized in automation design—that is, the ability to not only perform a task (no matter how simple) but also to detect and report back when any factor prevents completing the task. Such robustness removes the need for constant monitoring, truly freeing up the human for other tasks.

Procedures, like technology, are designed *a priori* and thus are limited on their face to the designers' ability to predict future operations. When the design is correct, strict and detailed, procedures reduce their human executors to "meat servos" where automation could instead be employed. However, procedures need not be detailed but can instead, as noted in the previous section, provide risk mitigation by acting as scaffolds on behavior and coordinating multiple agents; one such example are air traffic regulations generally specifying right-of-way protocols without attempting to specify the details of all avoidance maneuvers for all situations. Likewise, procedures, properly applied, can promote flexible and robust behavior by their human executors: flexible when they explain their goals, intent, and the rationale behind their steps in a manner promoting their adaptation to an unusual situation; robust when the human can easily comprehend and compare their intended "boundary conditions" to the immediate situation (e.g., Ockerman & Pritchett, 2000, 2004).

Finally, humans in aviation are generally referred to as "the pointy end of the stick," that is, the ultimate source of mitigation against risks that were not predicted or for which no other risk mitigation has been implemented. Fully supporting human risk mitigation is more fully discussed in the next section, but some general principles can be noted here. First, a conscious decision to distribute a large amount of risk mitigation onto the human operators needs to ensure they have the knowledge (including training), immediate information and control authority to respond to problems. Second, a conscious decision to distribute risk mitigation away from the human operator needs to recognize that technology and procedures are designed according to predictions that may be violated or incomplete, likely leaving the human operator with a significant role. Third, the work assigned to the human operator needs to be carefully considered relative to their work environment, lest it be susceptible to predictable forms of human error.

In many cases, the risk mitigation strategy is implicitly left to the technology designers, with procedures and training coming afterwards in response. While technology cost is often cited as a driver in the decisions, recurring operational costs (e.g., the need for inefficient procedures or additional pilot training) can also be profound and, in some cases, can drive or limit technological innovation.

In addition, a formal discussion of risk mitigation strategy during technology design can not only better inform the procedure and training design but also can serve as a vital component of a "safety case," as discussed previously.

SUPPORTING THE HUMAN CONTRIBUTION TO SAFETY

So far, our discussion of the system safety perspective has focused on predicting and evaluating the negative side of safety—risks, failures, and errors. Such a view implies that "normal" is safe without any human effort, and the level of safety can only go down from normative operations. However, many things can "go right" in operations through human contributions to safety, even in the face of the unusual or unexpected. Thus, this section seeks to review these important human contributions, and methods by which they may be supported.

Shared Representations

A human operator in an aviation system may find his- or herself translating between multiple representations as part of their every-day work. Some of these are implicit in procedures and technologies. One notable example centers on the different representations of "flight" implicit in air traffic procedures versus those implicit in many autoflight systems. For example, some air traffic control clearances can require sifting through multiple "pages" in the system interface and corresponding air traffic phraseology to differing terms used in the interface. While these differences in representation are associated with problems, the positive converse is also true: designing systems to have functions and interfaces built on an established representation can reduce erroneous or cumbersome translations (e.g., mode confusion and improper programming in autoflight systems) and reduce training time (e.g., Riley et al., 1998).

Another source of multiple representations stem from the multiple agents involved in aviation operations. A pilot's representation of arrival and approach may focus on energy management and achieving a "stabilized approach" with the correct flap settings and approach speed, while a controller's representation may focus on a well-spaced stream of aircraft, each properly spaced for maximum airport capacity; given these differing representations, the actions and desires of each other may appear conflicting. By contrast, with effective communication of their representations (e.g., through cross-training), they may spontaneously identify opportunities where they can aid each other, and provide a source of redundancy and error-checking in support of safety (e.g., Volpe et al., 1996).

Situation Awareness

Using Endsley's classic definition (1995), full situation awareness is comprised of three levels: perception of relevant aspects of the environment (Level 1); comprehension of these aspects relative to goals (Level 2); and projection of future states of the environment (Level 3). Building on our earlier discussion, this situation awareness must comprise several factors—not only the physical environment but also the procedural environment and the status of the team. Commonly, development of systems for situation awareness focuses on Level 1 by presenting immediate information about the current state of the system; the processes of comprehension (Level 2) and projection (Level 3) are left to the operator. Increasingly, systems can also aid explicitly in the interpretation of information (for example, by supplementing a traffic situation display with an explicit presentation of hazard) and projection of its meaning (for example, by providing a continuous display of developing conflict hazard relative to an alerting threshold by which required maneuver time and generation of an alert can be projected) (e.g., Pritchett & Vándor, 2001).

Situation awareness (individual and team) is widely recognized as a leading contributor to the human contribution to safety, for it provides the basis for effective decision making, proper use of procedures and automation when they are appropriate and effective detection of circumstances where they are not appropriate, effective team interactions and team maintenance, and informed expectations. Achieving it requires careful design, as it is as susceptible to information overload as it is to a lack of information and because the information suitable to the operator's task, as noted in the preceding section, must be presented in a manner supporting the operator's representations as needed (and when needed) for their work.

Adaptation

As the ratio of task demands relative to resources increases, perceived workload does not increase linearly until the point of saturation. Instead, as noted earlier in our discussion of cognitive control, human operators can modify their strategies for organizing their cognitive activities, including prioritizing tasks and adopting different methods for decision making, information gathering, and team coordination. Such adaptation is also a vital component of resilient system behavior in the face of disturbances.

Studies suggest that total system performance is enhanced when the operator's cognitive control mode is supported, versus a common implicit design philosophy of using technology to enable or enforce "strategic" behavior in even the most stressed situation (e.g., Feigh & Pritchett, 2006). Thus, supporting effective human performance may require examining likely cognitive control modes and creating corresponding design modes to support each.

Preemptive Construction of the Operating Environment

Finally, an expert and informed operator can do more than just respond to events as they come up—as noted earlier, he or she can actively construct their environment to improve safety. Some of these activities can be simple and directed; for example, pilots may place physical reminders in the environment of upcoming events or required actions. Other activities may extend the process of sense-making and abstraction from the mind to the environment by spatially organizing physical artifacts. Thus, construction of the environment can be a tangible manifestation of a planning process in which, for example, charts and checklists are laid out in sequence during rehearsal and annotated with important upcoming activities.

Glossy pictures of modern flightdecks and air traffic controller stations suggest a clean, sterile environment filled with computer screens. Such an environment should be designed with care that either: the physical artifacts used as reminders or as tangible manifestations of plans can still be manipulated appropriately, or that their full contributions to cognition can be replicated via the computer systems.

DESIGN VIA A "SYSTEM SAFETY PERSPECTIVE"

This chapter has described a system safety perspective to examine the complex, safety-critical nature of aviation operations. This perspective is necessary to address our own human frailties; if we designers were omnipotent and omniscient, we could consider all factors fully and simultaneously. Thus, a system perspective is best applied with humility and a good sense of humor.

The first question is what should be considered "in the system" or "relevant to system performance," a decision that must be made relative to the analysis at hand, starting with the coarsest effects and refining from there as need demands and time allows. In doing so, several important considerations must be kept in mind. One is the range of factors that can impact behavior and performance: core psychological and physiological capabilities; the physical ecology and procedural environment that expert behavior will seek to accommodate; the social dynamics underlying team performance; and the broadest cultural influences guiding attitudes towards authority and the use of technology. Addressing them all requires knowledge and methods arising from different research and design communities. While not all may need to be extensively considered, none should be ignored out of hand.

Another consideration in setting the system boundary is ensuring a complete viewpoint of the tasks under consideration. In the aircraft collision avoidance case study referred to throughout here, for example, the pilot's task is commonly described as "responding to the alert and executing the commanded avoidance maneuver." However, in context, the tasks involved with implementing collision avoidance systems go much further. Some expand to more profound views of the alerts and commands—the pilot considering whether they agree with the system's output and whether, in their immediate context, they feel it is best to comply (or not). Others extend to tasks generated at less extreme times, such as monitoring a new traffic situation display, developing a hither-to impossible situation awareness of traffic (including interpreting and projecting the situation), developing and potentially acting upon expectations about upcoming events, and modifying their interaction with air traffic control based on the availability of new information. Finally, the collision avoidance system needs to be integrated into the complete flightdeck system in which other displays, alerts, and commands may compete for attention, requiring an active strategy for cognitive control and for managing tasks and interruptions.

Once a system boundary is defined, safety must be evaluated, risk mitigated, and the human contribution to safety supported. Given the high safety levels already found in aviation, no one method or perspective is sufficient by itself. The negative side of safety—risk—must be viewed both through systematically exploring conceivable failures and errors and through properly distributing risk among technology, procedures, and human performance. Safety must be viewed at the system level as well as the individual level for emergent effects and resilience. Finally, the human factors community also has a unique role for also articulating, emphasizing and designing to support the human contribution to safety.

References

Davis, S. D., & Pritchett, A. R. (1999). Alerting system assertiveness, knowledge, and over-reliance. *Journal of Information Technology Impact, 3*(1), 119–143.

Endsley, M. R. (1995). Toward a theory of situation awareness in dynamic situations. *Human Factors, 37*, 32–64.

Endsley, M. R., & Kiris, E. (1995). The out-of-the-loop performance problem and level of control in automation. *Human Factors, 37*, 381–394.

Farley, T. C., Hansman, R. J., Endsley, M. R., Amonlirdviman, K., & Vigeant-Langlois, L. (2000). Shared information between pilots and controllers in tactical air traffic control. *Journal of Guidance, Control, and Dynamics, 23*(5), 826–836.

Feigh, K. M., & Pritchett, A. R. (2006). Design of multi-mode support systems for airline operations. *HCI-Aero: International conference on human-computer interaction in aeronautics*. Seattle, WA.

Gonzalez Castro, L. N., Pritchett, A. R., Johnson, E. N., & Bruneau, D. P. J. (2007). Coherent design of uninhabited aerial vehicle operations and control stations. *Proceedings of the annual meeting of the Human Factors and Ergonomics Society*. Baltimore, MD.

Harris, D., & Li, W.-C. (2008). Cockpit design and cross-cultural issues underlying failures in crew resource management. *Aviation, Space, and Environmental Medicine, 79*(5), 537–538.

Hess, R. A. (1997). Unified theory for aircraft handling qualities and adverse aircraft–pilot coupling. *Journal of Guidance, Control and Dynamics, 20*(6), 1141–1148.

Hofstede, G. (2001). *Culture's consequences: Comparing values, behaviors, institutions, and organizations across nations* (2nd ed). Thousand Oaks, CA: Sage Publications.

Hollnagel, E. (1993). *Human reliability analysis: Context and control.* London, UK: Academic Press.

Hollnagel, E., Woods, D. D., & Leveson, N. (2006). *Resilience engineering: Concepts and precepts.* Aldershot: Ashgate.

Hutchins, E. L. (1995). *Cognition in the wild.* Cambridge, MA: MIT Press.

Kirlik, A. (1998). Everyday life environments. In W. Bechtel. & G. Graham (Eds.), *A companion to cognitive science* (pp. 702–712). Malden, MA: Blackwell.

Klein, G. (1999). *Sources of power: How people make decisions.* Cambridge, MA: MIT Press.

Leveson, N. G. (2003). A new approach to hazard analysis for complex systems. *International conference of the System Safety Society.* Ottawa.

McFarlane, D. C., & Latorella, K. A. (2002). The scope and importance of human interruption in human-computer interaction design. *Human-Computer Interaction, 17*(1), 1–61.

McRuer, D. T., & Jex, H. R. (1967). A review of quasi-linear pilot models. *IEEE Transactions on Human Factors in Electronics, 3,* 231–249.

Miller, C. A., & Parasuraman, R. (2007). Designing for flexible interaction between humans and automation: Delegation interfaces for supervisory control. *Human Factors, 49*(1), 57–75.

NRC (Nuclear Regulatory Commission). (2000). *Technical basis and implementation guidelines for a technique for human event analysis (ATHEANA).* Rockville, MD: U.S. Nuclear Regulatory Commission.

Ockerman, J. J., & Pritchett, A. R. (2000). A review and reappraisal of task guidance: Aiding workers in procedure following. *International Journal of Cognitive Ergonomics, 4*(3), 191–212.

Ockerman, J. J., & Pritchett, A. R. (2002). Impact of contextual information on automation brittleness. *Annual meeting of the Human Factors and Ergonomics Society.* Baltimore, MD.

Ockerman, J. J., & Pritchett, A. R. (2004). Improving performance on procedural tasks through presentation of locational procedure context: An empirical evaluation. *Behavior and Information Technology, 23*(1), 11–20.

Parasuraman, R., Molloy, R., & Singh, I. (1993). Performance consequences of automation-induced "complacency". *International Journal of Aviation Psychology, 3*(1), 1–23.

Pritchett, A. R. (2001). Reviewing the roles of cockpit alerting systems. *Human Factors in Aerospace Safety, 1*(1), 5–38.

Pritchett, A. R., & Vándor, B. (2001). Designing situation displays to promote conformance to automatic alerts. *Proceedings of the annual meeting of the Human Factors and Ergonomics Society.*

Pritchett, A. R., & Yankosky, L. J. (2003). Pilot-performed guidance in the national airspace system. *Journal of Guidance, Control and Dynamics, 26*(1), 143–150.

Radio Technical Commission for Aeronautics (RTCA). (1983). *Minimum operational performance standards for traffic alert and collision avoidance system (TCAS)
airborne equipment, (RTCA / DO—185)*. Washington, DC: RTCA.

Rantenan, E. M., & Nunes, A. (2005). Hierarchical conflict detection in air traffic
control. *International Journal of Aviation Psychology, 15*(4), 339–362.

Rasmussen, J., Pejtersen, A., & Goodstein, L. (1994). *Cognitive systems engineering*.
New York: Wiley.

Reason, J. (1990). *Human error*. New York: Cambridge University Press.

Reason, J. (1997). *Managing the risks of organizational accidents*. Aldershot: Ashgate.

Riley, V., DeMers, B., Misiak, C., & Schmalz, B. (1998). The cockpit control language: A pilot-centered avionics interface. *Proceedings of HCI-Aero: The international conference on human-computer interaction in aerospace.*

Rushby, J. (2008). How do we certify for the unexpected? *Proceedings of the AIAA
guidance, navigation and control conference.*

Sperling, B. K. (2005). *Information distribution in complex systems to improve team
performance* Doctoral thesis. Georgia Institute of Technology Available at
http://smartech.gatech.edu/handle/1853/6975.

Sperling, B. K., & Pritchett, A. R. (2006.) Information distribution to improve
team performance in military helicopter operations: An experimental study,
Interservice/industry training, simulation, and education conference (I/ITSEC).

Van Dam, S. B. J., Mulder, M., & Van Paassen, M. M. (2008). Ecological interface
design of a tactical airborne separation assistance tool. *IEEE Transactions on
Systems, Man and Cybernetics, Part A, 38*(6), 1221–1233.

Vicente, K. (1999). *Cognitive work analysis: Towards safe, productive and healthy
computer-based work*. Mahwah, NJ: Erlbaum.

Volpe, C. E., Cannon-Bowers, J. A., Salas, E., & Spector, P. E. (1996). The impact of
cross-training on team functioning: An empirical investigation. *Human Factors,
38*, 87–100.

Wickens, C. D., & Holland, J. G. (2000). *Engineering psychology and human performance* (3rd ed). Upper Saddle River, NJ: Prentice Hall.

4

The Safety Culture Perspective

Manoj S. Patankar, Ph.D.
Edward J. Sabin, Ph.D.
National Center for Aviation Safety Research
Saint Louis University

INTRODUCTION

The aviation industry is already noted for its very low accident rate of about one in one million operations. However, by the year 2025, the air transportation system is projected to become more complex: the volume is projected to grow exponentially; the systemic complexity is projected to increase with the addition of Very Light Jets, Unmanned Aerial Vehicles, and super–large aircraft such as the Airbus 380; and the air navigation infrastructure is projected to become more dependent on air/space-based systems rather than ground-based systems. Therefore, even the current accident rate is likely to produce an unacceptable frequency of accidents somewhere around the year 2020–2025. Thus, safety in the aviation industry must improve beyond the current level.

Judging by the past improvements in aviation safety, the decline in airliner accidents from the 1930s through the 1980s was largely attributable to the improvements in technology. As the safety benefits from technology began to saturate, the field of human factors emerged, giving rise to the study of the human-machine interface as well as human-human interactions and their impact on systemic safety. Since the late 1990s, increased attention has been focused on organizational aspects of the aviation industry. Specifically, studies have focused on the impact of organizational policies, processes, and practices on the creation of latent failures and behavioral norms that ultimately influence safety performance. Hence, there has been an increased emphasis on the study of safety culture.

The growing interest in understanding the current state of safety culture is closely associated with the need to transform the culture into a more desirable state. Managers want to know the current state of safety culture in their organization and how they can improve it. One reason that managers across a wide array of aviation organizations are interested in safety culture assessment and change is that the International Civil Aviation Organization (ICAO) has made it mandatory for all Member States (countries) to implement a Safety Management System (SMS), which includes improvement of safety culture (ICAO, 2006). National regulatory agencies such as the Federal Aviation Administration, the U.K. Civil Aviation Administration, and Transport Canada are required to implement SMS within their own organizations. Likewise, airlines, airports, and aviation maintenance organizations are also expected to implement their respective SMS programs. Embedded in each of these programs is the requirement to improve the prevailing safety culture.

Similarly, other industries like health care, nuclear power, and construction are also seeking assessment and improvement of safety culture (INPO, 2004; Choudhry, Fang, & Mohamed, 2007; Flin, 2007; Pronovost et al., 2009)

A review of the state of research in the area of safety culture, reveals the following key issues:

- The understanding of safety culture is still evolving; consequently, the tools used to assess or characterize safety culture are also developing.
- There are very few published longitudinal studies demonstrating a clear link between deep-rooted cultural factors and observable safety-oriented behaviors.

- There is an emerging body of literature emphasizing the role of leadership in influencing and institutionalizing a positive safety culture.

This chapter attempts to address these issues and offers a framework for the assessment and improvement of safety culture in the aviation industry.

DEFINITION OF SAFETY CULTURE

The term *safety culture* first appeared in the report on the 1986 Chernobyl disaster, where the errors and violations of the operating procedures that contributed to the accident were seen by some as being evidence of a poor safety culture at the plant (Fleming & Lardner, 1999). Since then, the safety culture concept has been used in reports pertaining to several well-known accidents such as the Piper Alpha oil platform explosion in the North Sea, the Clampham Junction rail disaster in London (Cox & Flin, 1998), the Canadian airline accident in Dryden (Maurino, Reason, Johnston, & Lee, 1999), and the NASA tragedies—*Challenger* (Vaughn, 1996) and *Columbia* (CAIB, 2003). Now, almost everyone involved in safety program management in high-consequence industries, where safety tends to be an explicit value (e.g., aviation, health care, chemical manufacturing, and construction), is interested in safety culture. The popularity of the term safety culture has led to the prolific use of this concept ahead of the science of describing and assessing it. This section explores the structural aspects of the term safety culture and concludes with a working definition and states of safety culture.

Safety Culture is a composite term wherein the two components, *safety* and *culture*, can be defined independently from a variety of perspectives. For example, regulatory bodies and insurance companies tend to refer to safety in terms of acceptable levels of risk (Wells & Chadbourne, 2000); engineers tend to refer to safety in terms of failure modes and effects (Bahr, 1997); psychologists tend to refer to safety in terms of individual, group, and organizational causal or contributory factors resulting in the errors or failures

(Flin et al., 2000; Zohar, 2000); and systems theorists tend to view safety as a product of multiple forces both macro- and micro-effects that contribute toward a safe or successful outcome of the entire system (Ericson, 2005).

The latest insight into the lessons learned from High Reliability Organizations (HROs) is leading toward an emphasis on resilience or systemic adaptability to cope with hazards and minimize failures (Hollnagel, Woods, & Leveson, 2006; Dekker & Woods in this book). This approach offers a more dynamic perspective: safety, particularly in HROs, is less a matter of individual component reliability and more a matter of overall systemic ability (resilience or adaptability) to cope with threats, errors, and failures. Therefore, resilience engineers express safety as a measure of systemic adaptability.

Culture has been studied by anthropologists, social scientists, and organizational scientists (see Smith, 2001, for a review of major contributions to cultural theory). Generally, anthropologists study culture to describe a given group of people in terms of their habitat, language, customs and traditions, legends and heroes, food and clothing, interactions within the community as well as interdependencies with other communities (Bernard & Spencer, 2002). Social and organizational scientists, by contrast, have studied culture from perspectives that emphasize quality of life, organizational effectiveness, and safety performance (Hofstede, 1984; Helmreich & Merritt, 1998; Taylor and Patankar, 1999).

There are two fundamental aspects of safety culture: *components* and *dynamics*. *Components* refer to four constituent parts that collectively describe safety culture: values, leadership strategies, attitudes, and performance (Cooper, 2002; Seo, 2005; Rundmo & Hale, 2003; Desai, Roberts, & Ciavarelli, 2006; Yu, Sun, & Egri, 2008; Wu, Chen, & Li, 2008; Zohar, 2000; Neal & Griffin, 2002). *Dynamics* refers to the interaction between the constituent parts that yields a dominant cultural state. Dynamics include interactions between the individual, group, and organization that can incubate latent problems leading to failure or can produce resilience that can compensate for errors and stop an error trajectory from culminating in failure (Hollnagel, Woods, & Leveson, 2006). Choudhary et al. (2007) describe these interactions as the "dynamic, perpetual, multifaceted, and holistic nature of safety culture." Cooper (2000) also describes culture dynamics as "the interactive relationship between psychological, behavioural, and organizational factors."

In response to the *dynamics* in organizational behavior, Karl Weick observed that "the environment that the organization worries about is put there by the organization" (1979). This simple yet powerful observation is supported by Bandura's (1986), model of reciprocal determinism wherein people and their environment influence each other perpetually. In safety culture literature, aviation organizations, naval aircraft carriers, and nuclear power plants have been the benchmark of safety performance and referred to as High Reliability Organizations (HROs) (Roberts, 1993). However, one of the most important characteristic of an HRO—the ability to adapt to new situations or hazards—was not fully explained. Dekker and Woods (in this book) use the concept of adaptive systems to explain HROs' distinctive feature as the ability to maintain systemic safety by dynamically adjusting to hazardous influences (internal and external). *Dynamics* also refers to the influence of values, strategies, and attitudes on safety behaviors or performance. Emerging studies show that there is an influential link between these aspects of safety culture (Zohar, 2000; Neal & Griffin, 2002; Cooper & Phillips, 2004; Desai, Roberts, & Ciavarelli, 2006; Wu, Chen, & Li, 2007).

The four parts of safety culture are presented in the form of a pyramid (see Figure 4.1). At the base of the pyramid are the foundational safety values, next are the organizational factors (safety leadership strategies), followed by the attitudes and opinions (safety climate), and at the top of the pyramid are safety behaviors (or safety performance).

- Safety Values—In high-consequence industries such as aviation, health care, nuclear power, and chemical manufacturing, safety of the individual worker as well as safety of the system needs to be an enduring value in the organization. This means that safety issues must be integrated in the organization's business

FIGURE 4.1   The safety culture pyramid.

plan and daily practice. Examples of value-affirming statements are as follows:

- "Safety is our passion. We are world leaders in aerospace safety" (FAA, 2009)
- "It is not okay for me to work in a place where it is okay for people to get hurt." (Knowles, 2002)
- "Patient safety must be a main priority for the CEO, the board, and all employees within the system." (Hendrich et al., 2007)

- Safety Leadership Strategies—Include the organizational mission, policies, procedures, employee evaluation tools, reward and penalty systems, and leadership practices. The personal values of corporate founders often tend to become enduring values of the organization, which are then translated into strategies. There is also an emerging body of literature on the influence that current leaders can exert in the shaping of safety culture (Rundmo & Hale, 2003; Choudhry et al., 2006; Wu et al., 2007). Some examples of safety strategies are as follows:
  - Aviation Safety Action Program (FAA, 2002)
  - Line Operations Safety Audit (FAA, 2006)
  - Safety Management System (ICAO, 2006)

- Safety Climate—Safety attitudes typically offer a temporal snapshot of employee perceptions, reactions, and opinions about safety policies, procedures, practices, and leadership. Attitudinal measures assess individual and group-level perceptions of the overall safety of the organization. Typically, attitudes are evaluated by safety climate survey questionnaires. There are numerous safety climate questionnaires in use. The following are key studies that present an overview of different safety climate questionnaires and their common features.
  - Flin, Mearns, O'Conner, and Bryden (2000): Management, Safety System, Risk, Work Pressure, and Competence
  - Wiegmann, Zhang, von Thaden, Sharma, and Mitchell (2002): Organizational Commitment, Management Involvement, Employee Empowerment, Reward Systems, and Reporting Systems
  - Guldenmund (2007): Organization-level, Group-level, and Individual-level analysis of Risks, Hardware Design and Layout, Maintenance, Procedures, Manpower Planning, Competence, Commitment, Communication, and Monitoring and Change.
- Safety Performance—Safety performance is the collective outcome of observable safety behaviors such as successful error recoveries, systemic safety improvements in response to error or hazard reports, and uncontained errors resulting in incidents or accidents. The typical methods used to measure safety performance are as follows:
  - Major accident investigation reports from appropriate governing or investigating agencies such as the National Transportation Safety Board (NTSB), Federal Aviation Administration (FAA), and the Occupational Safety and Health Administration (OSHA)
  - Periodic industry-level status updates by groups such as reports from the Aviation Safety Reporting System
  - Internal monitoring of safety performance through daily or weekly reports of equipment damage, personnel injury, rework of maintenance due to an error in the original work, and so on.

Based on the literature presented in this chapter, we believe the following:

- A state of safety culture always exists in a given organization or industry;
- The organization's dominant state of safety culture may be positive or negative;
- There is a dynamic, adaptable, bidirectional influence between the state of safety culture and its associated organizational, psychosocial, and technical components;
- Values, leadership strategies, attitudes, and performance are linked; and
- Safety culture can be improved through planned interventions.

Therefore, in high-consequence settings such as aviation, we define safety culture as:

> A dynamically balanced, adaptable state resulting from the configuration of values, leadership strategies, and attitudes that collectively impact safety performance at the individual, group, and enterprise level. Simply stated, safety culture is a dynamic configuration of factors at multiple levels that influences safety performance.

States of Safety Cultures

The dynamic configuration of values, leadership strategies, and attitudes result in four discernable dominant states of safety culture; secretive culture, blame culture, reporting culture, and just culture (see Figure 4.2).

Historically, the common understanding in aviation has been that with high professional responsibility, comes high probability of blame. When a mishap occurs, the general tendency in many national and organizational cultures has been to blame the individual

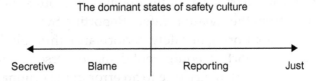

FIGURE 4.2   The dominant states of safety culture.

responsible for the last action prior to the mishap. In some countries, as described by Patankar and Taylor (2004), this practice is so prevalent that the licensed aircraft maintenance professionals have accepted the blame culture as a professional risk—they get significantly higher wages for holding the license and so they are expected to take the blame if one needs to be assigned. Clearly, such blame culture does not encourage communication of systemic problems or latent failures. Many commercial aviation organizations, as well as the FAA, have recognized this reality and are striving toward a "Just Culture" (Reason, 1997; Marx, 2001).

A Just Culture does not provide *carte blanche* forgiveness of the guilty party if an error occurs. Just Culture is based on the premise that as responsible professionals, pilots, aircraft mechanics/engineers, or air traffic controllers are expected to perform to certain basic professional and ethical standards. If they perform within those standards and commit an error, the error may not result in a disciplinary or punitive action. By contrast, if the performance is in violation of those standards, the error is classified as reckless or as intentional disregard for safety; hence, it may be subject to a disciplinary action.

Reason (1997) describes a Just Culture as an atmosphere of trust in which people are encouraged, even rewarded, for providing essential safety-related information, and where the line is clearly drawn between acceptable and unacceptable behavior. In moving from Blame Culture to Just Culture, it is essential to have a robust and functioning reporting system—hence the Reporting Culture. Such a reporting system allows for individual employees to report their own errors or systemic hazards without fear of reprisal. The level of employee-management trust at a given organization is the most critical factor in determining or predicting the success of such a system. (Examples of reporting systems include the Aviation Safety Action Program and the Aviation Safety Reporting System.)

Recently, Patankar and Sabin (2008) identified and described the presence of a fourth dominant state of safety culture, which they named "Secretive Culture." In such a culture, the employees as well as managers are known to either hide their own errors or to falsify performance data to mask the errors. In such cases, the employee-management trust tends to be low and the likelihood of punitive actions for all errors is high.

ASSESSMENT OF SAFETY CULTURE

The assessment of safety culture has been a longstanding challenge primarily because of an incomplete understanding of the safety culture construct and its influence on safety performance (Pidgeon, 1998). When the term *safety culture* was used in the 1987 Chernobyl report, it was neither clearly defined nor fully understood. In retrospect, it appears that the term was used to describe behavioral aspects of the employees at Chernobyl and possibly some observable aspects of corporate policies, procedures and practices. Since then, the term *safety culture* has gained tremendous popularity. However, formal studies have been largely limited to attitudinal aspects, which technically provide only a temporal snapshot of safety *climate* (Wiegmann et al., 2002).

Choudhry, Fang, and Mohamed (2007) maintain that the preoccupation of many researchers with safety climate neglects other critical issues such as safety behavior, safety management systems, and environmental factors. Likewise, Guldenmund (2007) argues that little scientific progress was made in the last decade to better understand safety culture. In particular, he maintains that "questionnaires have not been particularly successful in exposing the core of an organisational safety culture" (p. 723). Previous research has frequently focused on attempting to identify the structure of safety climate (e.g., attitudes) while often neglecting underlying developmental, functional, dynamic, and transformational aspects of safety culture (e.g., core values, assumptions, strategies). Choudhry et al. (2007) reviewed models to demonstrate the dynamic, perpetual, multifaceted, and holistic nature of safety culture. They also recognized the importance of identifying and studying safety subcultures within a given organization. Pidgeon's (1998) discussion of the "organizational construction of acceptable risk" and the roles played by power, politics and learning in organizational settings related to safety remind us of the need to better understand these dynamic forces.

Cooper (2000, 2002) discussed understanding cultural dynamics based on the reciprocal interplay among psychological, situational and behavioral factors and recommends a goal-oriented approach to better understand safety phenomena. Subsequently, Cooper and Phillips (2004) presented empirical evidence for the complexity of the relationships between safety climate and actual safety behavior. They found that changes in safety climate did not necessarily reflect changes in safety behavior and vice versa.

In a review of the various methodologies used to study safety culture, van den Berg and Wilderom (2004) discussed the need for more creative and varied methodologies; additional longitudinal research; further evaluation of safety culture's relationship with organizational performance; increased attention to the assessment of change programs; and specification of the relationships between leadership style and culture. Studies of the relationship between leadership and safety performance certainly deserve more attention. Leadership research by Barling, Loughlin, and Kelloway (2002) is encouraging: they found that occupational injuries could be predicted by safety-specific transformational leadership.

Assessment Methods

Given the multifaceted, multilevel complexity of safety culture, both qualitative as well as quantitative methods are necessary to produce a comprehensive understanding. Figure 4.3 presents four applicable analytic techniques and their corresponding empirical results. These main methodological approaches include: case, survey, qualitative, and quasi-experimental analyses. Each of these approaches yields information about a specific layer of the safety culture pyramid. Together, a multimethod approach (e.g., Di Pofi, 2002) enables triangulating of results from multiple sources (e.g., Paul, 1996) and ultimately yields a comprehensive understanding of the safety culture construct and its different states.

Safety culture assessment methods

- Assessment methods
  - Case analysis
    - Successful recoveries, undesirable events, incidents, and accidents
  - Survey analysis
    - Attitude and opinion questionnaires
  - Qualitative analysis
    - Field observations
    - Artifact analysis
    - Interviews and focus groups
    - Dialog

  - Quasi-experimental analysis

- Results
  - Retrospective findings
    - Systemic analysis of events, best practices, and lessons learned

  - Current safety climate
    - Employee perceptions of safety

  - Safety culture type
    - Description of existing safety policies, procedures, and practices
    - Stories of past safety successes and failures
    - Understanding of group dynamics
    - Surfacing underlying values and unquestioned assumptions

  - Establishing an intervention's causal effect

FIGURE 4.3   Safety culture assessment methods.

Case Analysis

Behaviors of individuals and/or groups that are related to the occurrence of unusual events, incidents or accidents attract attention from investigative and regulatory agencies. A case method is most commonly used in the retrospective analysis of undesirable events as well as successful recoveries from an impending disaster (e.g., pilot Sullenberger's successful landing of US Airways flight 1549 in the Hudson River in January 2009 after a disabling bird strike).

In the early years of accident investigation, the causal factors identified in retrospective case analysis were primarily technical and hence interventions were mostly focused on improving the reliability of the technology. As the aviation industry matured and its technology improved, safety attention was focused on individual human factors and team performance issues (e.g., Crew Resource Management and Maintenance Resource Management) in addition to technical factors. Consequently, accident investigation reports reflected issues concerned with crew coordination, communication, fatigue, and adherence to policies and procedures. More recently, attention has been focused on the additional role of organizational factors in accidents, yielding a corresponding concern with organizational safety culture.

Examples of this retrospective and case-based approach in aviation include investigations by the NTSB and NASA (e.g., NTSB 2002; NASA 2006). This retrospective case approach is also found in health care and includes Morbidity and Mortality conferences used to review medical error and undesirable patient outcomes (e.g., Gordon 1994; Pierluissi, Fischer, Campbell, & Landefeld 2003; Rosenfeld 2005).

Survey Analysis

In recent years, due in large part to the popularity of the term *safety culture*, there has been a strong interest in measuring safety attitudes and opinions. The dominant assessment technique has been a survey questionnaire. While many instruments are labeled safety *culture* surveys, most are measuring safety *climate* (i.e., attitudes). Among the numerous safety climate/culture questionnaires several instruments of particular relevance are those designed by the following researchers: Hofstede (1984); Helmreich, Fouchee, Benson, and Russini (1986); Westrum (1993); Taylor, (1995); Helmreich and Merritt (1998); Ciavarelli (1998); Gaba, Singer, Bowen, and Ciavarelli (2003); IOMA (2003); Patankar and Taylor (2004); Patankar (2003); Wiegmann et al. (2003); and Gibbons, von Thaden, and Wiegmann (2004). These scales have been through a number of iterations and customizations as the researchers have tested the validity and relevance of their instruments in specific domains such as flight, maintenance, medicine, and general organizational safety climate.

These researchers developed their respective instruments to suit specific measurement needs as well as to match the industry or professional community that was subject of their study. These instruments often include measures of employee-management trust, effectiveness of error/hazard reporting systems, communication/coordination, and relational supervision. Flin, Mearns, O'Connor, and Bryden (2000) reviewed 18 safety climate surveys. They identified several themes that were most commonly assessed in surveys of safety climate, including: management, the safety system, and risk. In a recent review of safety climate surveys in health care, Flin (2007) reported that many of the existing instruments have not been developed using standard psychometric methods to establish scale reliability and validity.

Qualitative Analysis

A more comprehensive analysis of safety culture should include qualitative methods such as field observations, artifact analysis, interviews, focus group discussions, and dialog with key individuals in the organization (e.g., Maxwell 2004).

Field Observations Field observation may include naturalistic or participant observation. In naturalistic observation, the goal of the research is to observe participants in their natural setting and avoid any intervention or interference with the normal course of events. Participant observation is the primary research approach of cultural anthropology. Here the researcher develops an intensive relationship with the participants by typically interacting over an extended period of time. Ethnography represents a particular approach to field observation by describing human culture from a holistic perspective. This approach is described in the writings of Hammersley and Atkinson (1983), Erickson and Stull (1997), and Spradley (1979).

Artifact Analysis

Cultural anthropologists have developed sophisticated and detailed techniques of describing cultures based on their artifacts. While most of the artifacts are in the form of physical articles used by the community or symbols, drawings, and paintings created by the community, some artifacts are in the form of stories that convey core values and key experiences so that the future generations learn from the past (Bernard & Spencer, 1996). In organizational safety culture, there are similar artifacts.

Artifacts are an important element of the overall cultural assessment because they are created by the people who are part of the organization, and the design of such artifacts is therefore reflective of the organization's culture. According to anthropologists, artifacts include physical evidence as well as key stories, legends, and myths that are passed down through generations. Examples of such artifacts, in safety culture, include the following:

- Company publications—mission statement, safety policies, safety performance reports, safety training manuals, accident/incident reporting forms, error-reporting forms, policy or procedure change protocols, accident/incident reports, posters, warning/caution signs, and so on
- Personal protective safety equipment actually used by the employees—goggles, masks, hearing protection, fluorescent vests, hard hats, steel-toe shoes, and so on
- Safety markings in the facility—symbols, words, and demarcation lines clearly identifying hazards or hazardous areas as well as required safety equipment (hard hats, goggles, masks, etc.)
- Employee training and certification programs—safety awareness training, specific behavioral training to reduce errors and injuries, first-responder type training, incorporation of safe practices in technical training, human and team performance issues, system safety programs, and so on (e.g., Crew Resource Management and Maintenance Resource Management)
- Symbols of safety initiatives—logos, stickers, t-shirts, lanyards, awards/medals, certificates, and so on
- Hazard or error reporting programs—specific nonpunitive programs to encourage all employees to self report their errors and report specific hazards (e.g., Aviation Safety Action Program and Aviation Safety Reporting System)
- Peer observation programs—designed to raise the awareness of safety threats during normal operation and proactive, nonpunitive analysis of ability to mitigate such threats (e.g., Line Operations Safety Audit)
- Stories of past experiences—what makes heroes and legends in the organization? What happened when someone pointed out a serious safety issue—was the individual punished, was the safety issue resolved, is there a general sense of positive achievement or is there a bitterness that signifies distrust? Are safety champions valued or are they viewed as barriers to productivity? How have critical safety challenges been

addressed when they threatened to compromise productivity? Are there any well-recognized safety champions—on the employee side (including labor union representatives) as well as the management side? What are the senior employees telling the junior employees?

- Methods and tools used to chronicle, analyze, and proactively solve safety problems.

Interviews and Focus Group Discussion   Typically, interviews and focus group discussions have been used by safety researchers to develop a basic understanding of the vocabulary, norms, key safety challenges, organizational policies, and procedures. Systematic procedures and best practices can be used to design, facilitate, and analyze information from focus groups and interviews (Krueger & Casey 2008; Schwarz 2002; Stewart, Shamdasani, & Rook 2006).

Such qualitative data are also useful in developing customized survey instruments that can be distributed to a larger audience to collect data on attitudes and opinions. Additionally, stories collected from interviews and group discussions can reveal heroes and legends that symbolize key cultural experiences. These stories tend to reveal deeply rooted meanings and help expose underlying values and assumptions.

Dialog   Dialog is a special technique of engaging individuals in deep, serious discussions aimed at revealing underlying values and unquestioned assumptions. These discussions are not likely to occur unless there is sufficient trust and openness between the interviewer and the participants. Once established, dialog can be an extremely powerful mechanism to better understand the values and assumptions that are at the foundation of the existing safety culture.

Schein (1999) presents a classic description of using dialog to reveal hidden forces and processes in organizational settings. Isaacs (1999) also provides a detailed description of this powerful form of communication in his text *Dialogue and the Art of Thinking Together*. Senge (2006) describes the importance of surfacing and examining fundamental assumptions (mental models); his *Fifth*

Discipline Fieldbook (Senge, Kleiner, Roberts, Ross, & Smith 1994) offers a practical approach to building these skills. Argyris (1992) describes the true complexities of dealing with individual and organizational defensive routines that occur when deeply held values and assumptions are threatened.

Quasi Experimental Analysis

Applied field research does not have the same ability to manipulate, randomize and control variables as experimental laboratory research. However, effective quasi-experimental designs (Campbell and Stanley 1963) are available for use in field setting to study the linkage between variables. Safety culture researchers have called for better empirical evidence to test hypothesized relationships between safety climate, safety culture, safety behavior, and organizational performance (e.g., Pidgeon 1998; van den Berg & Wilderom 2004). Wiegmann et al. (2002) suggests that organizational psychology is particularly attuned to the importance of testing causal relationships due to its interest in implementing change interventions to improve safety culture. It is especially recommended that more research be conducted to confirm the theoretical relationship between safety culture variables and also to establish the effectiveness of change and development programs aimed at improving safety performance.

Connecting Safety Climate, Culture, and Behavior

It has been established by Johnson (2007) that safety climate surveys can provide reliable assessment of the prevailing psychosocial conditions and organizational conditions; further these conditions have been known to impact the safety-oriented behaviors of employees (Neal & Griffin, 2002, Cooper & Phillips, 2004). In addition to the specific psychosocial aspects of the extant safety culture, Arezes and Miguel (2003) note that there are a wide variety of performance factors such as lost time injuries, number of accidents/incidents, and damage events that could be used as indicators of safety. While a decline in such numbers may indicate that the safety of the operation is increasing, there is no guarantee that the trend will continue. Also, there may be an implicit incentive to underreport incidents in order to maintain the positive image of the organization. Nonetheless, performance measures can be correlated with attitudinal changes that are in turn related to specific safety culture change efforts (Patankar & Taylor, 2004) and subsequently used to demonstrate the financial benefits of the change program. Patankar and Taylor (2004, Ch. 8) demonstrate that both positive as well as negative financial return on investment are possible, depending on how the change program is managed.

SAFETY CULTURE TRANSFORMATION

In a study of acculturational and anthropological theory, Voget (1963) noted the following characteristics of a culture: it is a state of dynamic equilibrium achieved by a range of individual variables, their relationships with one another, their individual and collective values and motivations, their relationships with their environment. As such, "change begins with alterations in the kind, rate, and intensity of interactions that link individuals to the significant institutionalized patterns of the system." Therefore, in order to change the state of a safety culture from, say blame culture to reporting culture, one has to change the kind, rate, and intensity of the interactions among individuals such that they are consistent with those espoused in a reporting culture. Also, Voget (1963) describes *cultural change* as an evolutionary, and therefore irreversible, process.

Why Does One Change Culture?

Mitroussi (2003) argues that when the survival of the organization is at stake, its people will be more willing to give up old values and practices and take up new ones. So, a safety culture might be amenable to change if there is an appropriately intense and urgent internal or external threat. In the case of aviation, the ICAO requirement to implement a Safety Management System and improve safety culture might serve as a sufficient external threat. However, it is also well recognized in the aviation industry that safety is a business and operational necessity. Although the specific business benefits of discrete safety programs have not been fully established, the impact of catastrophic accidents on the overall financial health of an airline is clear (e.g. Eastern Airlines, Pan Am, TWA, ValuJet, etc.).

How Does One Change Culture?

As presented in this chapter, most of the safety culture studies have focused on the *concept* of safety culture. "Exactly how to create a safety culture is not clear, although many agree that it will include continuous organizational learning from 'near miss' incidents as well as accidents" (Ringstad & Szameitat, 2000).

Theoretically, cultural change efforts may start from the bottom up or from the top down. In safety culture transformation, in particular, there is a need for both bottom-up and top-down alignment. While leaders need to establish policies and provide resources to sustain the policies, all the employees need to participate in the actual transformation effort. Knowles (2002) demonstrates the significance of the top leader in the organization to step out of his/her comfort zone, face the safety facts, and sincerely pledge to make the necessary changes.

The cultural change direction pursued by the top management needs to be translated for middle managers because they play a critical role in this cultural transformation processes. Guth and MacMillan (1985) claim that the degree to which middle managers support the new initiatives will determine the rate at which the initiatives will be successful or if they would be successful at all. Applying Guth and MacMillan's perspective to a specific example, if the top leaders are committed to move their organization from a blame culture to a just culture by introducing a nonpunitive error reporting system, the middle managers need to agree that (a) changing the organizational culture is the right corporate objective, (b) implementation of a non-punitive error reporting system is the right strategy, and (c) the corporate objective is consistent with their individual self-interest.

Hendrich et al. (2007) report the journey of Ascension Health toward a corporate goal of "zero preventable deaths and injuries." They present the following key steps in accomplishing their cultural change: "establishing a sense of urgency, creating a guiding coalition, and developing the Destination Statement II." The Destination Statement II vividly describes the desired safety culture. While all of the steps involved in achieving the transformation—and characterizing the transformation as an ongoing journey rather than a destination—are important, building a coalition to rally in support of the desired culture is consistent with findings from other domains (Guth & MacMillan, 2007; Fernández-Muñiz, Montes-Peón, & Vázquez-Ordás, 2007).

(Patankar and Sabin (2008) describe a safety culture transformation effort in the Technical Operations domain of the Air Traffic Organization of the FAA. Based on the pre-post interventions surveys, they establish that a nonpunitive error reporting system could be used to shift away from the predominantly blame-oriented safety culture. This study also demonstrated the importance of longitudinal studies that use a quasi-experimental approach to illustrate the effects of an intervention.

How Long will It Take to Change the Safety Culture?

The question of how long it would take to change a given culture is common among managers, but the response depends on the scale (how many people need to change their behavior) at which such a change is desired. Based on Moore's (1991) classification, it will depend on the relative proportion of early adopters, skeptics and critics of the change program. Also, there is a notion of "organizational readiness" for change—described in terms of preconditions such as the levels of compliance with standard operating procedures, collective commitment to safety, individual sense of responsibility toward safety, and employee-management relationship (Patankar, 2003). Literature on fostering cultural change suggests that focused and deliberate efforts are essential to craft such a change (Sabin, 2005).

Additionally, there are leverage factors such as level of awareness in the greater community, regulatory requirements/pressures, business survival factors, industry standards, and so on. When local change efforts transition toward organization-wide changes and mature into a policy/regulatory change, the change program tends to achieve a much higher degree of stability. For example, if we trace the evolution of the Crew Resource Management (CRM) program (Wiener, Kanki, & Helmreich, 1993) and the Maintenance Resource Management (MRM) program (Taylor & Patankar, 2001), one glaring difference as to why the CRM program has been institutionalized and not the MRM program, is that CRM training is a regulatory requirement. Also, the global awareness and acceptance of the CRM program was greatly enhanced by practitioner heroes such as Captain Al Haynes, multiple airline pilot unions, and the operational leaders of many airlines who are pilots. The pilots had their own share of challenges in overcoming the deep-rooted culture of command and control, but large-scale efforts leveraged by these factors have led to a time when most novice pilots are familiar with the key concepts of CRM and expect their future airline jobs to demand refined practice of CRM techniques. (Patankar, Block, & Kelly, 2009).

In contrast, the maintenance community suffered significantly from not having the MRM training as a regulatory requirement. As the financial strength of most U.S. airlines declined in the "post 9/11" era, MRM programs were disbanded at many of the airlines and champions of such programs lost their jobs. Well-recognized spokespersons like past National Transportation Safety Board Member John Goglia continued to promote MRM programs, and with the new group of maintenance personnel in charge of safety, many of the companies have turned their attention to the maintenance version of the Aviation Safety Action Program (ASAP). This is an FAA-endorsed program and has demonstrated its effectiveness in effecting specific changes at both organization as well as industry level. Therefore, it seems the maintenance industry is set to "leapfrog", build the ASAP programs, and incorporate the MRM concepts in the comprehensive solutions that are generated in the ASAP recommendations. (Patankar, Block, & Kelly, 2009).

CONCLUSION: SYNTHESIS OF SAFETY CULTURE METHODS AND OUTCOMES

Looking back at the safety culture pyramid and the definition of safety culture, three key aspects come to light:

- An organization, as a dynamically balanced system, may exist in any of the four states of safety culture—secretive, blame, reporting, or just.
- If an organization seeks to change its cultural state toward a just culture, there needs to be a purposeful alignment across all four components of the safety culture—values, leadership strategies, attitudes, and behaviors.
- The time taken to change the culture depends on (a) effective articulation of the safety values and communication of the need (intensity and urgency) to change the culture and (b) the collective engagement at all levels of the organization to make the necessary changes in strategies, processes, and policies in order to align with the safety values.

Figure 4.4 presents a synthesis of safety culture components, methods and outcomes. Safety culture is presented here as a pyramid with four components: safety values, safety leadership strategies, safety attitudes, and safety performance. In order to truly and fully characterize safety culture, the state of each component, the interconnection across the four components, the dynamic balance between the four components, and the cumulative influence of values, safety leadership strategies, and safety attitudes on safety performance must be examined.

Future Directions

The following directions are recommended:

- Improve the scientific methods and measures that provide the fundamental data to understand safety culture
- Improve safety culture assessment in organizations by using multiple methods that go beyond the safety climate surveys
- Commit to longitudinal studies that employ a multimethod approach

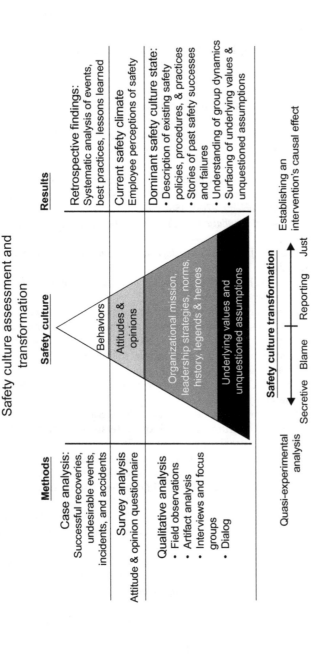

FIGURE 4.4   Safety culture assessment and transformation.

- Seek to discover the influence of various components of safety culture on the ultimate performance of the system
- Systematically study the impact and interactions to improve safety culture
- Share best practices, lessons learned, and research findings.

References

Arezes, P., & Miguel, A. (2003). The role of safety culture in safety performance measurement. *Measuring Business Excellence, 7*(4), 20–29.

Argyris, C. (1999). *On Organizational Learning*. Oxford, UK: Blackwell Publishers.

Bahr, N. (1997). *System Safety Engineering and Risk Assessment: A Practical Approach*. New York: Taylor & Francis.

Barling, J., Loughlin, C., & Kelloway, E. (2002). Development and Test of a Model Linking Safety-Specific Transformational Leadership and Occupational Safety. *Journal of Applied Psychology, 87*(3), 488–496.

Bandura, A. (1986). *Social foundations of thoughts and action*. Englewood Cliffs, NJ: Prentice Hall.

Bernard, A., & Spencer, J. (1996). *Encyclopedia of social and cultural anthropology*. London: Routledge.

Campbell, D., & Stanley, J. (1963). *Experimental and Quasi-Experimental Designs for Research*. Chicago: Rand McNally.

CAIB. (2003). *Report of the Columbia Accident Investigation Board*, Vol. 1. Retrieved from <http://caib.nasa.gov/news/report/volume1/default.html> 30.08.05.

Choudhry, R., Fang, D., & Mohamed, S. (2007). The nature of safety culture: A survey of the state-of-the-art. *Safety Science, 45*(10), 993–1012.

Ciavarelli, A. (1998). Assessing organizational accident risk. *Presented at the Prevention First '98: Onshore and Offshore Spill Prevention Symposium*, Long Beach, CA, September 9–10.

Cooper, M. D. (2000). Towards a model of safety culture. *Safety Science, 36*(2), 111–136.

Cooper, D. (2002). Safety Culture: A model for understanding and quantifying a difficult concept. *Professional Safety, 47*(6), 30–36.

Cooper, M. D., & Phillips, R. A. (2004). Exploratory analysis of the safety climate and safety behavior relationship. *Journal of Safety Research, 35*(5), 497–512.

Cox, S., & Flin, R. (1998). Safety culture: Philosopher's stone or man of straw? *Work and Stress, 12*(3), 189–201.

Desai, V., Roberts, K., & Ciavarelli, A. (2006). The relationship between safety climate and recent accidents: behavioral learning and cognitive attributions. *Human Factors, 48*(4), 639–650.

Di Pofi, J. (2002). Organizational diagnostics: integrating qualitative and quantitative methodology. *Journal of Organizational Change Management, 15*(2), 156–168.

Ericson, C. (2005). *Hazard Analysis Techniques for System Safety*. Hoboken, NJ: John Wiley & Sons, Inc.

Erickson, K., & Stull, D. (1997). *Doing Team Ethnography: Warnings and Advice*. Thousand Oaks, Ca: Sage.

FAA (2002). Aviation Safety Action Program Advisory Circular and Guidance AC 12–66B. Washington, DC: Federal Aviation Administration. Retrieved from

<http://www.faa.gov/safety/programs_initiatives/aircraft_aviation/asap/policy/> 14.11.05.

FAA (2006). Line Operations Safety Audit AC 120–90. Washington, DC: Federal Aviation Administration. Retrieved from <http://www.airweb.faa.gov/> 4.04.09.

FAA (2009). FAA Flight Plan 2009–2013. Washington, DC: Federal Aviation Administration. Retrieved from <http://www.faa.gov/about/plans_reports/media/flight_plan_2009–2013.pdf> on April 4, 2009.

Fernández-Muñiz, B., Montes-Peón, J., & Vázquez-Ordás, C. (2007). Safety culture: Analysis of the causal relationships between its key dimensions. *Journal of Safety Research, 38,* 627–641.

Fleming, M., & Lardner, R. (1999). Safety culture—the way forward. *The Chemical Engineer,* 16–18.

Flin, R. (2007). Measuring safety culture in healthcare: a case for accurate diagnosis. *Safety Science, 45,* 653–667.

Flin, R., Mearns, K., O'Connor, P., & Bryden, R. (2000). Measuring safety climate: Identifying the common features. *Safety Science, 34,* 177–192.

Gaba, D., Singer, S., Bowen, J., & Ciavarelli, A. (2003). Differences in safety climate between hospital personnel and naval aviators. *Human Factors, 45*(2), 173–185.

Gibbons, A., von Thaden, T., & Wiegmann, D. (2004). Exploration of the correlation structure of a survey for evaluating airline safety culture. Technical Report AHFD-04–06/FAA-04–3. University of Illinois, Urbana-Champaign.

Gordon, L. (1994). *Gordon's guide to the surgical Morbidity and Mortality conference.* Philadelphia, PA: Hanley and Belfus.

Guldenmund, F. (2007). The use of questionnaires in safety culture research—an evaluation. *Safety Science, 45*(6), 723–743.

Guth, W., & MacMillan, I. (1986). Strategy implementation versus middle management self-interest. *Strategic Management Journal, July/August,* 313–327.

Hammersley, M., & Atkinson, P. (1983). *Ethnography: Principles and Practice.* London: Tavistock Books.

Harper, M., & Helmreich, R. (2003). Creating and maintaining a reporting culture. In R. Jensen (Ed.) Proceedings of the Twelfth International Symposium on Aviation Psychology. 496–501, April 14–17, Dayton, OH.

Helmreich, R., Fouchee, C., Benson, R., & Russini, W. (1986). Cockpit Resource Management: Exploring the attitude-performance linkage. *Aviation Space & Environmental Medicine, 57*(12), 1198–1200.

Helmreich, R., & Merritt, A. (1998). *Culture at work in aviation and medicine: National, organizational and professional influences.* Aldershot, UK: Ashgate.

Hendrich, A., Tersigni, A., Jeffcoat, S., Barnett, C., Brideau, L., & Pryor, D. (2007). The Ascension Health journey to zero: lessons learned and leadership perspective. *The Joint Commission Journal on Quality and Patient Safety, 33*(12), 739–749.

Hofstede, G. (1984). *Culture's consequences: International differences in work-related values* (abridged ed.). Beverly Hills, CA: Sage.

Hollnagel, E., Woods, D., & Leveson, N. (2006). *Resilience engineering: Concepts and precepts.* Aldershot, UK: Ashgate Publishing Ltd.

ICAO. (2006). Safety Management Manual [Document Number 9859/AN460]. Montreal: International Civil Aviation Organization. Retrieved from <http://www.icao.int/icaonet/dcs/9859/9859_1ed_en.pdf> 01.07.08.

INPO. (2004). *Principles for a Strong Nuclear Safety Culture*. Atlanta, GA: Institute of Nuclear Power Operations.

IOMA. (2003). Are you focusing enough on your safety culture? Safety Director's Report p. 1. Retrieved from <http://www.ioma.com/pub/SADR/2003_08/553094–1.html> 22.02.05.

Isaacs, W. (1999). *Dialogue and the Art of Thinking Together*. New York: Currency Doubleday.

Johnson, S. (2007). The predictive validity of safety climate. *Journal of Safety Research, 38,* 511–521.

Knowles, R. N. (2002). *The Leadership Dance: Pathways to Extraordinary organizational effectiveness*. Niagara Falls, NY: The Center for Self-Organizing Leadership.

Krueger, R. A., & Casey, M. A. (2008). *Focus Groups: A practical guide for applied research*. Thousand Oaks: Sage.

Marx, D. (2001). Patient safety and the "just culture": A primer for health care executives. Retrieved from <http://www.mers-tm.net/support/Marx_Primer.pdf> 11.10.04.

Maurino, D., Reason, J., Johnston, N., & Lee, R. (1997). *Beyond Aviation Human Factors*. Aldershot, UK: Ashgate.

Mitroussi, K. (2003). The evolution of safety culture of IMO: A case of organizational culture change. *Disaster Prevention and Management, 12*(1), 16–23.

Moore, G. (1991). *Crossing the Chasm: Marketing and Selling High-Tech Products to Mainstream Customers*. New York: HarperCollins.

NASA. (2006). NASA Procedural Requirements for Mishap and Close Call Reporting, Investigating, and Recordkeeping. Retrieved from <http://nodis3.gsfc.nasa.gov/npg_img/N_PR_8621_001B_/N_PR_8621_001B_.pdf> 04.04.09.

Neal, A., & Griffin, M. (2002). Safety climate and safety behaviour. *Australian Journal of Management, 27,* 67–75.

NTSB. (2002). National Transportation Safety Board Aviation Investigation Manual Major Team Investigations. Retrieved from <http://www.ntsb.gov/Aviation/Manuals/MajorInvestigationsManual.pdf> 03.04.09.

Patankar, M. (2003). A study of safety culture at an aviation organization Okalahoma City, OK: FAA Academy. *International Journal of Applied Aviation Studies, 3*(2), 243–258.

Patankar, M., Block, E., & Kelly, T. (March 5, 2009). *Integrated systems approach to improving safety culture*. Dallas, TX: Presented at the Maintenance Aviation Safety Action Program Information Sharing Meeting.

Patankar, M., Bigda-Peyton, T., Sabin, E., Brown, J., & Kelly, T. (2005). A Comparative Review of Safety Cultures. Report prepared for the Federal Aviation Administration. Available at <http://hf.faa.gov>.

Patankar, M., & Sabin, E. (2008). Safety Culture Transformation Project Report. Report prepared for the Federal Aviation Administration. Available at <http://hf.faa.gov>.

Patankar, M. S., & Taylor, J. C. (2004). *Risk Management and Error Reduction in Aviation Maintenance*. Aldershot, UK: Ashgate Publishing Limited.

Paul, J. (1996). Between-method triangulation in organizational diagnosis. *The International Journal of Organizational Analysis, 4*(2), 135–153.

Pidgeon, N. (1998). Safety culture: Key theoretical issues. *Work & Stress, 12*(3), 202–216.

Pierluissi, E., Fischer, M., Campbell, A., & Landefeld, S. (2003). Discussion of medical error in Morbidity and Mortality conferences. *Journal of the American Medical Association, 290*(21), 2838–2842.

Pronovost, P., Goeschel, C., Marsteller, J., Sexton, B., Pham, J., & Berenholtz, S. (2009). Framework for patient safety research and improvement The American Heart Association. *Circulation, 119*, 330–337.

Reason, J. (1997). *Managing the risk of organizational accidents.* Aldershot, UK: Ashgate.

Roberts, K. H. (Ed.). (1993). *New Challenges to Organizations: High Reliability Understanding Organizations.* New York: Macmillan.

Rosenfeld, J. (2005). Using the Morbidity and Mortality Conference to Teach and Assess the ACGME General Competencies. *Current Surgery, 62*(6), 664–669.

Ringstad A. J., & Szameitat, S. A. (2000). Comparative Study of Accident and Near Miss Reporting Systems in the German Nuclear Industry and the Norwegian Offshore Industry. *Proceedings of the Human Factors and Ergonomics Society* 44th, pp. 380–384.

Rundmo, T., & Hale, A. (2003). Managers' attitudes towards safety and accident prevention. *Safety Science, 41*, 557–574.

Sabin, E. (2005). Promoting the transfer of innovations. In M. Patankar, & J. Ma, (Eds.)*Proceedings of the Second Safety across High-onsequence Industries onference: Vol. 1* . St. Louis, MO: Parks College of Engineering, Aviation and Technology.

Seo, D. (2005). An explicative model of unsafe work behavior. *Safety Science, 43*, 187–211.

Schein, E. H. (1999). *Process Consultation Revisited: Building the helping relationship.* Reading, MA: Addison-Wesley.

Schwarz, R. (2002). *The Skilled Facilitator* (4th ed.). San Francisco, CA: Jossey-Bass.

Senge, P. M. (2006). *The Fifth Discipline: The Art and Practice of the Learning Organization.* New York: Currency Doubleday.

Senge, P. M., Kleiner, A., Roberts, C., Ross, R. B., & Smith, B. J. (1994). *The Fifth Discipline Fieldbook: Strategies and Tools for Building a Learning Organization.* New York: Currency Doubleday.

Smith, P. (2001). *Cultural Theory: An introduction.* Malden, MA: Blackwell Publishers, Inc.

Spradley, J. (1979). *The Ethnographic Interview.* New York: Wadsworth.

Stewart, D., Shamdasani, P., & Rook, D. (2006). *Focus Groups: Theory and Practice.* Thousand Oaks, CA: Sage.

Taylor, J. (1995). Effects of communication & participation in aviation maintenance. In R. Jensen (Ed.), *Proceedings of the Eighth International Symposium on Aviation Psychology* (pp. 472–477). Columbus: The Ohio State University.

Taylor, J., & Patankar, M. (1999). Cultural factors contributing to the successful implementation of macro human factors principles in aviation maintenance. In R. Jensen (Ed.), *Proceedings of the Tenth International Symposium on Aviation Psychology.* Columbus: Ohio State University.

Taylor, J., & Patankar, M. (2001). Four generations of MRM: evolution of human error management programs in the United States. *Journal of Air Transportation World Wide, 6*(2), 3–32.

van den Berg, P., & Wilderom, C. (2004). Defining, measuring, and comparing organizational cultures. *Applied Psychology: An International Review, 53*(4), 570–582.

Vaughn, D. (1996). *The Challenger Launch Decision: Risky technology, culture, and deviance at NASA*. Chicago: University of Chicago Press.

Voget, F. (1963). Cultural change. *Biennial Review of Anthropology, 3*, 228–275.

Weick, K. (1979). *The Social Psychology of Organization*. New York: Random House.

Wells, A., & Chadbourne, B. (2000). *Introduction to Aviation Insurance and Risk Management* (2nd ed.). Malabar, FL: Krieger Publishing Company.

Westrum, R. (1993). Cultures with requisite imagination. In J. Wise & D. Hopkin (Eds.), *Verification and Validation: Human Factors Aspects* (pp. 401–416). New York: Springer.

Wiegmann, D., Zhang, H., von Thaden, T., Sharma, G., & Mitchell, A. (2002). *Safety Culture: A Review* (Technical Report ARL-02-3/FAA-02-2). Atlantic City, NJ: Federal Aviation Administration.

Wiegmann, D., Zhang, Z., von Thaden, T., Sharma, G., & Gibbons, A. (2003). Development and initial validation of a safety culture survey for commercial aviation. Federal Aviation Administration Technical Report AHFD-03-3/FAA-03-1. University of Illinois at Urbana-Champaign.

Wiener, E. L., Kanki, B. G., & Helmreich, R. L. (1993). *Cockpit resource management*. San Diego: Academic Press.

Wu, T., Chin, C., & Li, C. (2008). A correlation among safety leadership, safety climate and safety performance. *Journal of Loss Prevention in the Process Industries, 21*, 307–318.

Yu, M., Sun, L., & Egri, C. (2008). Workplace safety climate assessment based on behaviors and measurable indicators. *Process Safety Progress, 27*(3), 239–247.

Zohar, D. (2000). A group-level model of safety climate: testing the effect of group climate on micro accidents in manufacturing jobs. *Journal of Applied Psychology, 85*, 587–596.

5

The High Reliability Organization Perspective

Sidney W.A. Dekker
Lund University School of Aviation, Ljungbyhed, Sweden

David D. Woods
The Ohio State University, Columbus, Ohio, USA

INTRODUCTION

High reliability theory describes the extent and nature of the effort that people, at all levels in an organization, have to engage in to ensure consistently safe operations despite its inherent complexity and risks. It is founded on an empirical research base that shows how safety originates in large part in the managerial and operational activities of people at all levels of an organization. The high reliability organizations (HROs) perspective is relevant here, since aviation has done an effective job in institutionalizing and systematizing its learning from incidents and accidents. HRO, however, tries to go further—pulling learning forward in time, studying how managerial and operational activities can encourage the exploration and exchange of safety-related information. The aim is to pick up early signs that trouble may be on the horizon and then be able to make modifications without having to wait for the more obvious signs of failure in the form of incidents or accidents. HROs are able to stay curious about their own operations and keep wondering why they are successful. They stay open-minded about the sources of risk, try to remain complexly sensitized to multiple sources of safety information, keep inviting doubt and minority opinion, and stay ambivalent toward the past, so that confidence gleaned from previous results are not taken as a guarantee of future safety (Weick, 1993).

This chapter first considers some of the origins of HRO, then addresses the "reliability" part of its label as applied to aviation safety in part on the basis of an example of a systems accident, and concludes with how Resilience Engineering represents the action agenda of HRO (Hollnagel et al., 2006). With an emerging set of techniques and models to track how organizations learn, adapt and change without waiting for major failures, Resilience Engineering introduces ways to indicate where overconfidence in past results may be occuring, where minority viewpoints may risk getting downplayed, and where acute performance or production demands may trump chronic safety concerns. The reason such instrumentality is important for aviation is both its high safety and its complexity: accidents have long ceased to be the result from single component failures. Rather, they emerge from the system's organized complexity (Amalberti, 2001). It takes more than tracking individual component behavior to anticipate whether aviation systems can keep coping with change and complexity.

HRO: SOME ORIGINS

Through a series of empirical studies, HRO researchers have found that through leadership safety objectives, the maintenance of relatively closed systems, functional decentralization, the creation of a safety culture, redundancy of equipment and personnel, and systematic learning, organizations can achieve the consistency and stability required to effect nearly failure-free operations (LaPorte and Consolini, 1991). Some of these categories were very much inspired by the worlds studied—naval aircraft carrier air operations, for example (Rochlin, LaPorte, and Roberts, 1987). There, in a relatively self-contained and disconnected closed system, systematic learning was an automatic by-product of the swift rotations of naval personnel, turning everybody into instructor and trainee, often at the same time. Functional decentralization meant that complex activities (like landing an aircraft and arresting it with the wire at the correct tension) were decomposed into simpler and relatively homogenous tasks, delegated down into small workgroups with substantial autonomy to intervene and stop the entire process independent of rank. HRO researchers found many forms of redundancy—in technical systems, supplies, even decision-making and management hierarchies, the latter through shadow units and multiskilling.

When HRO researchers first set out to examine how safety is created and maintained in such complex systems, they took an approach that can be found in parts of aviation human factors today. They focused on errors and other negative indicators, such as incidents, assuming that these were the basic units that people in these organizations used to map the physical and dynamic safety properties of their production technologies, ultimately to control risk (Rochlin, 1999). The assumption turned out wrong: they were not. Operational people, those who work at the sharp end of an organization, hardly defined safety in terms of risk management or error avoidance. Ensuing empirical work by HRO, stretching across decades and a multitude of high-hazard, complex domains (aviation, nuclear power, utility grid management, navy) would paint a more complex, and in many ways a more constructive picture with safety not being the *absence* of negatives, but rather the *presence* of certain activities to manage risk. HRO began to describe how operational safety—how it is created, maintained, discussed, mythologized—should be captured as much more than the control of negatives. As Rochlin (1999, p. 1549) put it,

> the culture of safety that was observed is a dynamic, intersubjectively constructed belief in the possibility of continued operational safety, instantiated by experience with anticipation of events that could have led to serious errors, and complemented by the continuing expectation of future surprise.

The creation of safety, in other words, involves a belief about the possibility to continue operating safely (Woods and Cook, 2003). This belief is built up and shared among those who do the work every day. It is moderated or even held up in part by the constant preparation for future surprise—preparation for situations that may challenge people's current assumptions about what makes their operation risky or safe. And yes, it is also a belief punctuated by encounters with risk. But errors, or any other negatives, function at most as the narrative spice that keeps the belief flavorful and worth sharing. They turned out not to be its main substance.

Intriguingly, the label "high reliability" grew increasingly at odds with the findings this school produced. What was a research effort to examine how high-risk systems can produce high-reliability outcomes despite their inherent danger (i.e., measured in terms of reducing negatives, or failure events), transmogrified into a discovery of safety as a reflexive social construct that challenged virtually all available methodological, ontological and theoretical guidance available at the time. Safety, HRO concluded, does not exist "out there," independent from the minds or actions of the people who create it through their practice, simply to be discovered, laid bare, by those with the right measuring instrument. Knowing about safety cannot be synonymous with a tabulation of "objective" measures from real-world performance. And, indeed, the predictive value of such measures is generally quite disappointing. While ensuring consistent and reliable component performance (both human and machine) has been a hugely important contributor to the successful safety record of aviation to date, there are limits to this approach, particularly when it comes to avoiding complex system accidents that emerge from the normal functioning of already almost totally safe transportation systems (Amalberti, 2001).

Reliability and its Effects on Safety

To be sure, safety is not the same as reliability. A part can be reliable, but in and of itself it cannot be safe. It can perform its stated function to the expected level or amount, but it is context, the context of other parts, of the dynamics and the interactions and cross-adaptations between parts, that make things safe or unsafe. Reliability as an engineering property can be expressed as a component's failure rate or probablilities over a period of time. In other words, it addresses the question of whether a component lives up to its prespecified performance criteria. Organizationally, reliability is often associated with a reduction in variability, and concomitantly, with an increase in replicability: the same process, narrowly guarded, produces the same predictable outcomes. Becoming highly reliable may be a desirable goal for unsafe or moderately safe operations (Amalberti, 2001). The guaranteed production of standard outcomes through consistent component performance is a way to reduce failure probability in those operations, and it is often expressed as a drive to eliminate errors and technical breakdowns.

In moderately safe systems, such as chemical industries, driving or chartered flights, approaches based on reliability can still generate significant safety returns (Amalberti, 2001). Regulations and safety procedures have a way of converging practice onto a common basis of proven performance. Collecting stories about negative near-miss events (errors, incidents) has the benefit in that the same encounters with risk show up in real accidents that happen to that system. There is, in other words, an overlap between the ingredients of incidents and the ingredients of accidents: recombining incident narratives has predictive (and potentially preventive) value. Finally, developing error-resistant and error-tolerant designs helps prevent errors from becoming incidents or accidents.

The monitoring of performance through operational safety audits, error counting, flight data collection, and incident tabulations has become institutionalized and in many cases required by legislation or regulation. The latest incarnation, an integrative effort to make both safety management and its inspection more streamlined with other organizational processes, is known as the Safety Management System (SMS), which is now demanded in most Western countries by regulators. Safety management systems typically encompass a process for identifying hazards to aviation safety and for evaluating and managing the associated risks, a process for ensuring that personnel are trained and competent to perform their duties and a process for the internal reporting and analyzing of hazards, incidents and accidents and for taking corrective actions to prevent their recurrence. The SMS is also about itself; about the bureaucratic accountability it both represents and spawns. Regulators typically demand that an SMS contains considerable documentation containing all safety management system processes and a process for making personnel aware of their responsibilities with respect to them. Quality assurance and safety management within the airline industry are often mentioned in the same sentence or used under one department heading. The relationship is taken as non-problematic or even coincident. Quality assurance is seen as a fundamental activity in risk management. Good quality management will help ensure safety. This idea, together with the growing implementation of SMS, may indeed have helped aviation attain even stronger safety records than before, as SMSs help focus decision makers' attention on risk management and safety aspects of both organizational and technological change, forcing an active consideration and documentation of how that risk should be managed.

One possible downside is that pure quality assurance programs (or reliability in the original engineering sense) contain decomposition assumptions that may not really be applicable to systems that are overall as complex as aviation (see Leveson, 2006). For example, it suggests that each component or subsystem (layer of defense) operates reasonably independently, so that the results of a safety analysis (e.g., inspection or certification of people or components or subsystems) are not distorted when we start putting the pieces back together again. It also assumes that the principles that govern the assembly of the entire system from its constituent subsystems or components is straightforward. And that the interactions, if any, between the subsystems will be linear: not subject to unanticipated feedback loops or nonlinear interactions.

The assumptions of such a reliability (or quality assurance) approach imply that aviation must continue to strive for systems with high theoretical performance and a high safety potential. A less useful portion of this notion, of course, is the elimination of component breakdowns (e.g., human errors), but it is still a widely pursued goal, sometimes suggesting that the aviation industry today is the custodian of an already safe system that needs protection from unpredictable, erratic components that are its remaining sources of unreliability. This common sense approach, says Amalberti (2001), which indeed may have helped aviation progress to the safety levels of today, is perhaps less applicable to a system that has the levels of complexity and safety already enjoyed today. This is echoed by Vaughan (1996, p. 416):

> ...we should be extremely sensitive to the limitations of known remedies. While good management and organizational design may reduce accidents in certain systems, they can never prevent them ... technical system failures may be more difficult to avoid than even the most pessimistic among us would have believed. The effect of unacknowledged and invisible social forces on information, interpretation, knowledge, and—ultimately—action, are very difficult to identify and to control.

As progress on safety in aviation has become asymptotic, further optimization of this approach is not likely to generate significant safety returns. In fact, there could be indications that continued linear extensions of a traditional-componential reliability approach could paradoxically help produce a new kind of system accident at the border of almost totally safe practice (Amalberti, 2001, p. 110):

> The safety of these systems becomes asymptotic around a mythical frontier, placed somewhere around 5×10^{-7} risks of disastrous accident per safety unit in the system. As of today, no man-machine system has ever crossed this frontier, in fact, solutions now designed tend to have devious effects when systems border total safety.

The aviation accident described in the following section may illustrate some of the challenges ahead in terms of thinking about what reliability (or HRO) really should mean in aviation. Through a concurrence of functions and events, of which a language barrier was a product as well as constitutive, the flight of a Boeing 737 out of Cyprus in 2005 may have been pushed past the edge of chaos, into that area in nonlinear dynamic behavior where new system behaviors emerged that could be difficult to anticipated using a logic of decomposition. The accident encourages us to consider HRO for its ability to monitor higher-order system properties: the system's ability to recognize, adapt to, and absorb disruptions that fall outside the disturbances it was designed to handle.

An Accident Perhaps Beyond the Reach of Traditional Reliability

On August 13, 2005, on the flight before the accident, a Helios Airways Boeing 737−300 flew from London to Larnaca, Cyprus. The cabin crew noted a problem with one of the doors, and convinced the flight crew to write that the "Aft service door requires full inspection" in the aircraft logbook. Once in Larnaca, a ground engineer performed an inspection of the door and carried out a cabin pressurization leak check during the night. He found no defects. The aircraft was released from maintenance at 03:15 and scheduled for flight 522 at 06:00 via Athens, Greece to Prague, Czech Republic (AAISASB, 2006).

A few minutes after taking off from Larnaca, the captain called the company in Cyprus on the radio to report a problem with his equipment cooling and the takeoff configuration horn (which warns pilots that the aircraft is not configured properly for takeoff, even though it evidently had taken off successfully already). A ground engineer was called to talk with the captain, the same ground engineer who had worked on the aircraft in the night hours before. The ground engineer may have suspected that the pressurization switches could be in play (given that he had just worked on the aircraft's pressurization system), but his suggestion to that effect to the captain was not acted on. Instead, the captain wanted to know where the circuit breakers for his equipment cooling were so that he could pull and reset them.

During this conversation, the oxygen masks deployed in the passenger cabin as they are designed to do when cabin altitude exceeds 14,000 feet. The conversation with the ground engineer ended, and would be the last that would have been heard from Flight 522. Hours later, the aircraft finally ran out of fuel and crashed in hilly terrain north of Athens. Everybody on board had been dead for hours, except for one cabin attendant who held a commercial pilots license. Probably using medical oxygen bottles to survive, he finally had made it into the cockpit, but his efforts to save the aircraft were too late. The pressurization system had been set to manual so that the engineer could carry out the leak check. It had never been set back to automatic (which is done in the cockpit), which meant the aircraft did not pressurize during its ascent, unless a pilot had manually controlled the pressurization outflow valve during the entire climb. Passenger oxygen had been available for no more than 15 minutes, the captain had left his seat, and the co-pilot had not put on an oxygen mask.

Helios 522 is illustrative, because nothing was "wrong" with the components. They all met their applicable criteria. "The captain and First Officer were licensed and qualified in accordance with applicable regulations and Operator requirements. Their duty time, flight time, rest time, and duty activity patterns were according to regulations. The cabin attendants were trained and qualified to perform their duties in accordance with existing requirements" (AAISASB, 2006, p. 112). Moreover, both pilots had been declared medically fit, even though postmortems revealed significant arterial clogging that may have exacerbated the effects of hypoxia. And while there are variations in what JAR-compliant means across Europe, the Cypriot regulator (Cyprus DCA, or Department of Civil Aviation) complied with the standards in JAR OPS 1 and Part 145. This was seen to with help from the U.K. CAA, who provided inspectors for flight operations and airworthiness audits by means of contracts with the DCA. Helios and the maintenance organization were both certified by the DCA.

The German captain and the Cypriot co-pilot met the criteria set for their jobs. Even when it came to English, they passed. They were within the bandwidth of quality control within which we think system safety is guaranteed, or at least highly likely. That layer of defense—if you choose speak that language—had no holes as far as our system for checking and regulation could determine in advance. And we thought we could line these subsystems up linearly, without complicated interactions. A German captain, backed up by a Cypriot co-pilot. In a long-since certified airframe, maintained by an approved organization. The assembly of the total system could not be simpler. And it must have, should have, been safe.

Yet there was a brittleness of having individual components meet prespecified criteria which became apparent when compounding problems pushed demands for crew coordination beyond the routine. As the AAISASB observed, "Sufficient ease of use of English for the performance of duties in the course of a normal, routine flight does not necessarily imply that communication in the stress and time pressure of an abnormal situation is equally effective. The abnormal situation can potentially require words that are not part of the "normal" vocabulary (words and technical terms one used in a foreign tongue under normal circumstances), thus potentially leaving two pilots unable to express themselves clearly. Also, human performance, and particularly memory, is known to suffer from the effects of stress, thus implying that in a stressful situation the search and choice of words to express one's concern in a non-native language can be severely compromised. ...In particular, there were difficulties due to the fact that the captain spoke with a German accent and could not be understood by the British engineer. The British engineer did not confirm this, but did claim that he was also unable to understand the nature of the problem that the captain was encountering" (pp. 122–123).

The irony is that the regulatory system designed to standardize aviation safety across Europe, has, through its harmonization of crew licensing, also legalized the blending of a large number of crew cultures and languages inside of a single airliner, from Greek to Norwegian, from Slovenian to Dutch. On August 14, 2005, this certified system may not have been able to recognize, adapt to, and absorb a disruption that fell outside the set of disturbances it was designed to handle. The "stochastic fit" (see Snook, 2000) that put together this crew, this engineer, from this airline, in this airframe, with these system anomalies, on this day, outsmarted how we all have learned to adapt, create and maintain safety in an already very safe industry. Helios 522 testifies that the quality of individual components or subsystems cannot always effectively predict how they can recombine to create novel pathways to failure (see Dekker, 2005).

Emergence and Resilience

Helios 522 in a sense represents the temporary inability to cope effectively with complexity. This is true, of course, for the cockpit crew after climbing out from Larnaca, but this is even more interesting at a larger system level. It was the system of pilot and airline certification, regulation, in an environment of scarcity and competition, with new operators in a market role which they not only fulfill but also help constitute beyond traditional Old Europe boundaries—that could not recognize, adapt to, and absorb a disruption that fell outside the set of disturbances the system was designed to handle (see Rochlin, 1999; Weick et al., 1999; Woods, 2003; 2005; Hollnagel et al., 1996). The "stochastic fit" (see Snook, 2000) or functional resonance (Hollnagel, Woods, and Leveson, 2006) that put together this crew, from this airline, in this airframe, with these system anomalies, on this day, in a way challenged how an industry learned to adapt, create and maintain safety when it was already very safe.

It could be interesting to shift from a mechanistic interpretation of complex systems to a systemic one. A machine can be controlled, and it will "fail" or perform less well or run into trouble when one or more of its components break. In contrast, a living system can be disturbed to any number of degrees. Consequently, its functioning is is much less binary, and potentially much more resilient. Such resilience means that failure is not really, or can't even really be, the result of individual or compound component breakage. Instead, it is related to the ability of the system to adapt to, and absorb variations, changes, disturbances, disruptions and surprises. If it adapts well, absorbs effectively, then even compound component breakages may not hamper chances of survival. United 232 in July 1989 is a case in point. After losing control of the aircraft's control surfaces as a result of a center engine failure that ripped fragments through all three hydraulic lines nearby, the crew figured out how to maneuver the aircraft with differential thrust on two remaining engines. They managed to put the crippled DC-10 down at Sioux City, saving 185 lives out of 293.

Simple things can generate very complex outcomes that could not be anticipated by just looking at the parts themselves. Small changes in the initial state of a complex system (e.g., a Cypriot and German pilot, rather than, say, two Cypriot ones) can drastically alter the final outcome. The underlying reason for this is that complex systems are dynamically stable, not statically so (like machines): instability emerges not from components, but from concurrence of functions and events in time. The essence of resilience is the intrinsic ability of a system to maintain or regain a dynamically stable state (Hollnagel, Woods, and Leveson, 2006).

Practitioners and organizations, as adaptive systems, continually assess and revise their approaches to work in an attempt to remain sensitive to the possibility of failure. Efforts to create safety, in other words, are ongoing. Not being successful is related to limits of the current model of competence, and, in a learning organization, reflects a discovery of those boundaries. Strategies that practitioners and organizations (including regulators and inspectors) maintain for coping with potential pathways to failure can either be strong or resilient (i.e., well-calibrated) or weak and mistaken (i.e., ill-calibrated). Organizations and people can also become overconfident in how well-calibrated their strategies are. High-reliability organizations remain alert for signs that circumstances exist, or are developing, in which that confidence is erroneous or misplaced (Rochlin, 1993; Gras, Moricot, Poirot-Delpech, and Scardigli, 1994). This, after all, can avoid narrow interpretations of risk and stale strategies (e.g., checking quality of components).

Resilience is the system's ability to effectively adjust to hazardous influences, rather than resist or deflect them (Hollnagel, Woods, and Leveson, 2006). The reason for this is that these influences are also ecologically adaptive and help guarantee the system's survival. Engaging crews from different (lower-wage) countries makes it possible to keep flying even with oil prices at record highs. But effective adjustment to these potentially hazardous influences did not occur at any level in the system in this case. The systems perspective, of living organizations whose stability is dynamically emergent rather than structurally inherent, means that safety is something a system does, not something a system has (Hollnagel, Woods, and Leveson, 2006; Hollnagel, 2009). Failures represent breakdowns in adaptations directed at coping with complexity (Woods, 2003). Learning and adaptation as advocated by HRO are ongoing—without it, safety cannot be maintained in a dynamic and changing organizational setting and environment. As HRO research found, this involves multiple rationalities, reflexivity and self-consciousnesses, since the ability to identify situations that had the potential to evolve into real trouble (and separate them from the ones that did not) is in itself part of the safe operation as social construct. Differently positioned actor-groups are learning, and are learning different things at different times—never excluding their own structure or social relations from the discourse in which that learning is embedded (Rochlin, 1999).

Ensuring Resilience in High-Reliability Organizations

The HRO perspective has given credence to the notion of safety as something that an organization does, not something that an organization has. How can we collapse some of these research results into useful guidance for organizations in aviation and elsewhere? How can we keep an organization's belief in its own continued safe operation curious, open-minded, complexly sensitized, inviting of doubt, and ambivalent toward the past? Resilience is in some sense the latest action agenda of HRO, with some of the following items:

Not taking past success as guarantee of future safety. Does the system see continued operational success as a guarantee of future safety, as an indication that hazards are not present or that countermeasures in place suffice? In their work, HRO researchers found how safe operation in commercial aviation depends in part on front-line operators treating their operational environment not only as inherently risky, but also as actively hostile to those who misestimate that risk (Rochlin, 1993). Confidence in equipment and training does not take away the need operators see for constant vigilance for signs that a situation is developing in which that confidence is erroneous or misplaced (Rochlin, 1999). Weick (1993) cites the example of Naskapi Indians who use caribou shoulder bones to locate game. They hold the bones over a fire until they crack and then hunt in the directions where the cracks point. This means future decisions about where to hunt are not influenced by past success, so the animal stock is not depleted and game does not get a chance to habituate to the Indians' hunting patterns. Not only are past results not taken as reason for confidence in future ones—*not* doing so actually increases future chances of success.

Distancing through differencing. In this process, organizational members look at other incidents or failures in other organizations or subunits as not relevant to them and their situation (Cook and Woods, 2006). They discard other events because they appear to be dissimilar or distant. But just because the organization or section has different technical problems, different operational settings, different managers, different histories, or can claim to already have addressed a particular safety concern revealed by the event, does not mean that they are immune to the problem. Seemingly divergent events can represent similar underlying patterns in the drift toward hazard.

Fragmented problem solving. It could be interesting to probe to what extent problem-solving activities are disjointed across organizational departments, sections or subcontractors, as discontinuities and internal handovers of tasks increase risk (Patterson, Roth, Woods, Chow, and Gomez, 2004). With information incomplete, disjointed and patchy, nobody may be able to recognize the gradual erosion of safety constraints on the design and operation of the original system (Woods, 2005). HRO researchers have found that the importance of free-flowing information cannot be overestimated. A spontaneous and continuous exchange of information relevant to normal funtioning of the system offers a background from which signs of trouble can be spotted by those with the experience to do so (Weick, 1993; Rochlin, 1999). Research done on handovers, which is one coordinative device to avert the fragmentation of problem-solving (Patterson et al., 2004) has identified some of the potential costs of failing to be told, forgetting, or misunderstanding information communicated. These costs, for the incoming crew, include:

- Having an incomplete model of the system's state;
- Being unaware of significant data or events;
- Being unprepared to deal with impacts from previous events;
- Failing to anticipate future events;
- Lacking knowledge that is necessary to perform tasks safely;
- Dropping or reworking activities that are in progress or that the team has agreed to do;
- Creating an unwarranted shift in goals, decisions, priorities, or plans.

The courage to say no. Having a person or function within the system with the authority, credibility and resources to go against common interpretations and decisions about safety and risk (Woods,

2006). A shift in organizational goal trade-offs often proceed gradually as pressure leads to a narrowing focus on some goals while obscuring the trade-off with other goals. This process usually happens when acute goals like production/efficiency take precedence over chronic goals like safety. If uncertain "warning" signs always led organizations to make sacrifices on schedule and efficiency, it would be difficult to meet competitive and stakeholder demands. By contrast, if uncertain "warning" signs are always rationalized away the organization is acting much riskier than it realizes or wishes. Sometimes people need the courage to put chronic goals ahead of acute short term goals. Thus it is necessary for organizations to support people when they have the courage to say "no" (e.g., in procedures, training, feedback on performance) as these moments serve as reminders of chronic concerns even when the organization is under acute pressures that easily can trump the warnings (see Dekker, 2007, about how to create a Just Culture). Resilient systems build in this function at meaningful organizational levels, which relates to the next point.

The ability to bring in fresh perspectives. Systems that apply fresh perspectives (e.g., people from another backgrounds, diverse viewpoints) on problem-solving activities seem to be more effective: they generate more hypotheses, cover more contingencies, openly debate rationales for decision making, reveal hidden assumptions (Watts-Perotti & Woods, 2009). In HRO studies of some organizations constant rotation of personnel turned out to be valuable in part because it helped introduce fresh viewpoints in an organizationally and hierarchically legitimate fashion (Rochlin, 1999). Crucially important here is also the role of minority viewpoints, those that can be dismissed easily because they represent dissent from a smaller group. Minority viewpoints can be blocked because they deviate from the mainstream interpretation which will be able to generate many reasons the minority view misunderstands current conditions and retards the organizations formal plans (Woods, 2006b). The alternative readings that minority viewpoints represent, however, can offer a fresh angle that reveals aspects of practice that were obscured from the mainstream perspective (Starbuck and Farjoun, 2005). Historically, "whistleblowers" may hail from lower ranks where the amount of knowledge about the extent of the prob-lem is not matched by the authority or resources to do something about it or have the system change course (Vaughan, 1996). Yet in risky judgments we have to defer to those with technical exper-tise (and have to set up a problem-solving process that engages those practiced at recognizing anomalies in the event).

All of this can serve to *keep a discussion about risk alive* even (or especially) when everything looks safe. One way is to see whether activities associated with recalibrating models of safety and risk are going on at all. Encouraging this behavior typically creates forums where stakeholders can discuss risks even when there is no evidence of risk present in terms of current safety statistics. As Weick (1993) illustrates, extreme confidence and extreme caution can both paralyze people and organizations because they sponsor a closed-mindedness that either shuns curiosity or deepens uncer-tainties (see also DeKeyser and Woods, 1990). But if discussions about risk are going on even in the absence of obvious threats to safety, one could get some confidence that an organization is investing in an analysis, and possibly in a critique and subsequent update, of its models of how it creates safety.

Knowing the gap between work-as-imagined and work-as-practiced. One marker of resilience is the distance between operations as management imagines they go on and how they actually go on. A large distance indicates that organizational leadership may be mis-calibrated to the challenges and risks encountered in real oper-ations. Also, they may also miss how safety is actually created as people conduct work, construct discourse and rationality around it, and gather meaning from it (Weick et al., 1999; Dekker, 2006).

Monitoring of safety monitoring (or meta-monitoring). In develop-ing their safety strategies and risk countermeasures, organizations should invest in an awareness of the models of risk they believe in and apply. This is important if organizations want to avoid stale coping mechanisms, misplaced confidence in how they regulate or check safety, and if do not want to miss new possible path-ways to failure. Such meta-monitoring would obviously represent an interesting new task for regulators in aviation worldwide, but it applies reflexively to themselves, too. The most important

ingredient of engineering a resilient system is constantly testing whether ideas about risk still match with reality; whether
the model of operations (and what makes them safe or unsafe) is
still up to date—at every level in the operational, managerial and
regulatory hierarchy.

High Resilience Organizations

Over the past two decades, high reliability research has begun
to show how organizations can manage acute pressures of performance and production in a constantly dynamic balance with
chronic concern for safety. Safety is not something that these organizations have, it is something that organizations do. Practitioners
and organizations, as adaptive systems, continually assess and
revise their work so as to remain sensitive to the possibility of failure. Efforts to create safety are ongoing, but not always successfully so. An organization usually is unable to change its model of
itself unless and until overwhelming evidence accumulates that
demands revising the model. This is a guarantee that the organization will tend to learn late, that is, revise its model of risk only
after serious events occur. The crux is to notice the information
that changes past models of risk and calls into question the effectiveness of previous risk reduction actions, without having to wait
for complete clear cut evidence. If revision only occurs when evidence is overwhelming, there is a grave risk of an organization
acting too risky and finding out only from near misses, serious
incidents, or even actual harm. The practice of revising assessments of risk needs to be continuous.

High reliability organization research is, and will always be, a
work in progress, as its language for accommodating the results,
and the methodological persuasions for finding and arguing
for them, evolves all the time. It is already obvious, though, that
traditional engineering notions of reliability (that safety can be
maintained by keeping system component performance inside
acceptable and prespecified bandwidths) have very little to do
with what makes organizations highly reliable (or, rather, resilient). As progress on safety in aviation has become asymptotic,
further optimization of this reliability approach is not likely to

generate significant safety returns. In fact, adhering to it may partly become constitutive of new kinds of system accidents, as illustrated by the Helios 522 case in this chapter. Failure in aviation today is not really, or not in any interesting or predictively powerful way, the result of individual or compound component breakage. Instead, it is related to the ability of the industry to effectively adapt to, and absorb variations, changes, disturbances, disruptions, and surprises.

Resilience Engineering is built on insights derived, in part, from the HRO work described here (Weick et al., 1999; Sutcliffe & Vogus, 2003). It is concerned with assessing organizational risk, that is the risk that holes in organizational decision making will produce unrecognized drift toward failure boundaries. While assessing technical hazards is one kind of input into Resilience Engineering, the goal is to monitor organizational decision making. For example, Resilience Engineering would monitor evidence that effective cross checks are well-integrated when risky decisions are made or that the organization is providing sufficient practice at handling simulated anomalies (and what kind of anomalies are practiced).

Other dimensions of organizational risk include the commitment of the management to balance the acute pressures of production with the chronic pressures of protection. Their willingness to invest in safety and to allocate resources to safety improvement in a timely, proactive manner, despite pressures on production and efficiency, are key factors in ensuring a resilient organization. The degree to which the reporting of safety concerns and problems is truly open and encouraged provides another significant source of resilience within the organization. Assessing the organization's response to incidents indicates if there is a learning culture or a culture of denial. Other dimensions of organizations which could be monitored include:

Preparedness/Anticipation: is the organization proactive in picking up on evidence of developing problems versus only reacting after problems become significant?

Opacity/Observability—does the organization monitor safety boundaries and recognize how close it is to "the edge" in terms of degraded defenses and barriers? To what extent is information about safety concerns widely distributed throughout the organization at all levels versus closely held by a few individuals?

Flexibility/Stiffness—how does the organization adapt to change, disruptions, and opportunities?

Successful, highly reliable aviation organizations in the future will have become skilled at the three basics of Resilience Engineering:

1. detecting signs of increasing organizational risk, especially when production pressures are intense or increasing;
2. having the resources and authority to make extra investments in safety at precisely the times when it appears least affordable;
3. having a means to recognize when and where to make targeted investments to control rising signs of organizational risk and rebalance the safety and production trade-off.

These mechanisms will produce an organization that creates foresight about changing risks before failures and harm occur.

References

Air Accident Investigation and Aviation Safety Board (AAIASB). (2006). Aircraft accident report (11/2006): Helios Airways flight HCY522, Boeing 737–31S at Grammatiko, Hellas on 14 August 2005. Athens, Greece: Helenic Republic Ministry of Transport and Communications.

Amalberti, R. (2001). The paradoxes of almost totally safe transportation systems. *Safety science, 37*, 109–126.

Cook, R. I., Woods, D. D. (2006). Distancing through Differencing: An Obstacle to Learning Following Accidents. In E. Hollnagel, D. D. Woods, and N. Leveson (Eds.), *Resilience engineering: concepts and precepts* (pp. 329–338). Aldershot, UK: Ashgate.

De Keyser, V., & Woods, D. D. (1990). Fixation errors: Failures to revise situation assessment in dynamic and risky systems. In A. G. Colombo and A. Saiz de Bustamante (Eds.), *System reliability assessment* (pp. 231–251). The Netherlands: Kluwer Academic.

Dekker, S. W. A. (2005). *Ten questions about human error: A new view of human factors and system safety*. Mahwah, NJ: Lawrence Erlbaum Associates.

Dekker, S. W. A. (2006). Resilience Engineering: Chronicling the emergence of a confused consensus. In E. Hollnagel, D. D. Woods, & N. Leveson (Eds.), Resilience engineering: concepts and precepts (pp. 77–92). Aldershot, UK: Ashgate.

Dekker, S. W. A. (2007). Just Culture: Balancing safety and accountability. Aldershot, UK: Ashgate.

Gras, A., Moricot, C., Poirot-Delpech, S. L., & Scardigli, V. (1994). *Faced with automation: The pilot, the controller, and the engineer (trans. J. Lundsten)*. Paris: Publications de la Sorbonne.

Hollnagel, E. (2009). *The ETTO principle, efficiency-thoroughness tradeoff: Why things that go right sometimes go wrong*. Aldershot, UK: Ashgate.

Hollnagel, E., Leveson, N., & Woods, D. D. (Eds.), (2006). *Resilience engineering: Concepts and precepts*. Aldershot, UK: Ashgate.

LaPorte, T. R., & Consolini, P. M. (1991). Working in Practice but not in Theory: Theoretical Challenges of High-Reliability Organizations. *Journal of public administration research and theory, 1,* 19–47.

Leveson, N. (2006). *A new approach to system safety engineering*. Cambridge, MA: Aeronautics and Astronautics, Massachusetts Institute of Technology.

Patterson, E. S., Roth, E. M., Woods, D. D., Chow, R., & Gomez, J. O. (2004). Handoff strategies in settings with high consequences for failure: Lessons for health care operations. *International journal for quality in health care, 16*(2), 125–132.

Reason, J. T. (1990). *Human error*. Cambridge, UK: Cambridge University Press.

Rochlin, G. I., LaPorte, T. R., & Roberts, K. H. (1987). The self-designing high-reliability organization: aircraft carrier flight operations at sea. Naval War College Review Autumn 1987.

Rochlin, G. I. (1993). Defining high-reliability organizations in practice: A taxo-nomic prolegomenon. In K. H. Roberts (Ed.), *New challenges to understanding organizations* (pp. 11–32). New York: Macmillan.

Rochlin, G. I. (1999). Safe operation as a social construct. *Ergonomics, 42,* 1549–1560.

Snook, S. A. (2000). *Friendly fire: the accidental shootdown of us black hawks over northern Iraq*. Princeton, NJ: Princeton University Press.

Starbuck, W. H., & Farjoun, M. (Eds.), (2005). *Organization at the limit: lessons from the columbia disaster*. London: Blackwell Publishing.

Sutcliffe, K., & Vogus, T. (2003). Organizing for resilience. In K. S. Cameron, I. E. Dutton, & R. E. Quinn (Eds.), *Positive organizational scholarship* (pp. 94–110). San Francisco: Berrett-Koehler.

Vaughan, D. (1996). *The challenger launch decision: risky technology, culture and devi-ance at NASA*. Chicago: University of Chicago Press.

Watts-Perotti, J., & Woods, D. D. (2009). Cooperative Advocacy: A Strategy for Integrating Diverse Perspectives in Anomaly Response. *Computer supported cooperative work: the journal of collaborative computing, 18*(2), 175–198.

Weick, K. E. (1988). Enacted sensemaking in crisis situations. *Journal of manage-ment studies, 25*(4), 305–317.

Weick, K. E. (1993). The collapse of sensemaking in organizations: The Mann Gulch disaster. *Administrative science quarterly, 38*(4), 628–652.

Weick, K. E., Sutcliffe, K. M., & Obstfeld, D. (1999). Organizing for high reliability: Processes of collective mindfulness. *Research in organizational behavior, 21,* 13–81.

Woods, D.D (2003). Creating foresight: How resilience engineering can transform NASA's approach to risky decision making. *US Senate Testimony for the Committee on Commerce, Science and Transportation,* John McCain, chair. Washington, DC, October 29 2003. http://csel.eng.ohio-state.edu/podcasts/woods/.

Woods, D. D. (2005). Creating foresight: Lessons for resilience from *Columbia*. In W. H. Starbuck & M. Farjoun (Eds.), *Organization at the limit: NASA and the columbia disaster* (pp. 289–308). Malden, MA: Blackwell.

Woods, D. D. (2006a). Essential characteristics of resilience for organizations. In E. Hollnagel, D. D. Woods, & N. Leveson (Eds.), *Resilience engineering: Concepts and precepts*. Aldershot, UK: Ashgate.

Woods, D. D. (2006b). How to design a safety organization: Test case for resilience engineering. In E. Hollnagel, D. D. Woods, & N. Leveson (Eds.), *Resilience engineering: concepts and precepts* (pp. 315–324). Aldershot, UK: Ashgate.

Woods, D. D. (2009). Escaping failures of foresight. *Safety science, 47*(4), 498–501.

Woods, D. D., & Cook, R. I. (2003). Mistaking error. In M. J. Hatlie & B. J. Youngberg (Eds.), *Patient safety handbook* (pp. 95–108). Sudbury, MA: Jones and Bartlett.

Woods, D. D., Dekker, S. W. A., Cook, R. I., Johannesen, L., & Sarter, N. (in press). *Behind Human Error* (2nd ed). Aldershot, UK: Ashgate.

PART III

PILOT AND CREW
PERFORMANCE ISSUES

6

The Human in Flight: From Kinesthetic Sense to Cognitive Sensibility

Kathleen L. Mosier
San Francisco State University

INTRODUCTION

Technological advances since the early days of flight have significantly transformed the aircraft cockpit and have altered the relationships among the human pilot, the aircraft, and the environment. Consistent with technological advances in aviation—many of which occurred after publication of the Wiener and Nagel (1988) volume—the role of the pilot has evolved from one characterized by sensory, perceptual, memory, and motor skills (Liebowitz, 1988) to one characterized primarily by cognitive skills. The flightdeck has evolved into a *hybrid* ecology comprised of both naturalistic and electronic elements. The environment is deterministic in that much of the uncertainty has been engineered out through technical reliability, but it is naturalistic in that conditions of the physical and social world—including ill-structured problems, ambiguous cues, time pressure, and rapid changes—interact with and complement conditions in the electronic world. Cues and information may originate in either the naturalistic (external, physical) environment or the deterministic systems (internal, electronic).

Different cognitive strategies and goals are required for dealing with each side of the hybrid ecology. Correspondence, or empirical, objective accuracy, is the primary goal in the naturalistic world. A correspondence strategy in the flying task involves evaluating probabilistic cues in the natural world (multiple fallible indicators; see, e.g., Brunswik, 1956; Hammond, 1996; Wickens & Flach, 1988) to formulate judgments with reference to it. In contrast, coherence, or rationality and consistency in judgment and decision making, is the primary goal in the electronic world. Using a coherence strategy, a pilot might evaluate the information displayed inside the cockpit to ensure that system parameters, flight modes, and navigational displays are consistent with each other and with what should be present in a given situation. In the hybrid ecology of the modern cockpit, input from both sides must be integrated to evaluate situations and make decisions. In this environment, both visual and kinesthetic sensing and, to a greater extent, cognitive sensibility are critical to the safety of humans in flight.

The goals of this chapter are: (1) to trace the technological evolution of the aircraft cockpit and of the flying task; (2) to describe issues inherent in the *naturalistic* side of the hybrid ecology, the need for correspondence, or accuracy in perception and judgment, in dealing with external environmental factors such as ambiguity and probabilism of cues, and the dangers of errors in correspondence; (3) to describe issues inherent in the *electronic* side of the hybrid ecology, including the need for analytical and consistent use of information, the need for coherence, or rational use of data and information in dealing with the internal, electronic environment, and the dangers of coherence errors; and (4) to discuss the integration of the two sides of the hybrid ecology and challenges for the design of Next Generation (NextGen) aircraft.

THE EVOLUTION OF THE AIRCRAFT COCKPIT AND OF THE PILOT'S TASK

Piloting an aircraft used to be a very physical- and sensory-oriented task. First generation aircraft were highly unstable, and demanded constant physical control inputs (Billings, 1996). The flight control task was a "stick and rudder" process involving knowing the characteristics of the aircraft and sensing the information necessary for control (Baron, 1988). Early aviation research therefore focused heavily on troubleshooting manual control and operational problems.

The senses—especially sight—were critical for problem diagnosis and navigation as well as for spotting other aircraft or obstacles. Judgments in early days of aviation were made via sensory—visual and kinesthetic—perception of the natural, physical world, in what has been referred to as "contact" flying (Hopkins, 1982). The emphasis was on accurate judgment of objects in the environment—height of obstacles in terrain, distance from ground, severity of storm activity in and around clouds, location of landmarks—and accurate response to them (e.g., using the controls to maneuver around obstacles or storm clouds, or to make precise landings). Features of the environment and of the available cues impacted the accuracy of judgments. Clear weather and concrete, easily discernable cues facilitated judgment. Pilots could easily discern a 5-mile reporting point when it was marked by a tall building. Murky weather, darkness, or ambiguous cues hindered judgment. The reporting point would be harder to find when the building that marked it was covered in fog or overshadowed by a long string of similar buildings. As pilots gained experience, more accurate perception resulted in more accurate response.

Pilots avoided situations that would put the accuracy of their senses in jeopardy, such as clouds or unbroken darkness. Early airmail pilots often relied on railroad tracks to guide their way, and mail planes had one of the landing lights slanted downward to make it easier to follow the railroad at night. Later, a system of beacons and gas lights created a 902-mile illuminated airway for nighttime flight. In 1929, Lieutenant James H. Doolittle of the U.S. Air Corps completed a historic 15-minute flight guided only by instruments (an altimeter, a gyrocompass, and an artificial horizon) and special radio receivers, and demonstrated that flight could be conducted entirely without visual reference to the outside world (Orlady & Orlady, 1999). This was a milestone on the road to instrument-guided flight.

Most facets of the flying task have become less sensory-oriented than in the past. As aircraft evolved, flying became physically easier as parts of control task were automated (e.g., through use of an autopilot), and automated systems began to perform many of the flight tasks previously accomplished by the pilot. The demand for all-weather flight capabilities resulted in the development of instruments that would supposedly compensate for any conditions that threatened to erode pilots' perceptual accuracy. Conflicts between visual and vestibular cues when the ground was not visible could lead to spatial disorientation and erroneous control inputs—but an artificial horizon, or attitude indicator, within the cockpit could help resolve these conflicts. Limitations of night vision could be overcome with navigational displays. These were first steps in the transformation of the cockpit ecology. Soon, more and more information placed inside the aircraft supplemented or replaced cues outside the aircraft (e.g., altitude indicator, airspeed indicator, alert and warning systems) and vastly decreased reliance on perceptual (i.e., probabilistic) cues. The readings inside the cockpit provided more accurate data than could be gleaned from the senses, and pilots became increasingly reliant on them.

As the aviation domain matured, much of the related research was geared toward defining and overcoming human limitations in terms of detection and recognition, night vision, and visual-vestibular interaction (Liebowitz, 1988). The perception and human information processing focus that dominated aviation research for many years was consistent with this era of aviation history. Limitations of the human as *perceiver* such as the degradation of vision at night and difficulties in detecting and recognizing large airplanes at a distance (e.g., Leibowitz, 1988), or as *information processor* such as attention and memory limitations (e.g., Wickens and Flach, 1988) were key topics of research. In the modern cockpit, many of these human limitations were addressed by instruments and technological aids.

Figure 6.1 illustrates the advances in automation in the aircraft cockpit since the 1930s. Note that in each successive generation, more data from the outside environment are brought into the cockpit and displayed as highly reliable and accurate information, reducing pilot dependence on ambiguous and probabilistic cues. With each

Automation and Technology

Winnie Mae	1st Generation	2nd Generation	3rd Generation	4th Generation	Next generation planned
CONTROL AUTOMATION 3-Axis autopilot	CONTROL AUTOMATION 3-Axis autopilot yaw damper	MANAGEMENT AUTOMATION Area navigation systems (RNAV) INFORMATION AUTOMATION Flight director VHF navigation Configuration warning systems Malfunction alerts CONTROL AUTOMATION Automatic spoilers Autoland systems Integrated flight systems	MANAGEMENT AUTOMATION Flight management system (FMS – program flight from take-off to landing) INFORMATION AUTOMATION First "Glass cockpit" Primary flight display Navigation displays (moving map) Multifunction displays Weather radar System/sub-system status displays Collision avoidance system (TCAS) Integrated alerting systems CONTROL AUTOMATION Mode control panel	MANAGEMENT AUTOMATION Flight management system (FMS – program flight take-off to landing) Electronic checklist INFORMATION AUTOMATION "All-glass" cockpit Primary flight display Navigation displays (moving map) Multifunction displays Weather radar System/sub-system status displays Collision avoidance system (TCAS) Integrated alerting systems Windshear displays Datalink displays CONTROL AUTOMATION "Fly-by-wire" (no tactile feedback) Integrated systems operation	MANAGEMENT AUTOMATION Easier FMS interfaces Direct FMS-ATC computer communication Error monitoring and trapping Improved electronic checklist INFORMATION AUTOMATION Electronic library–'paperless cockpit' Satellite navigation Digital data link communication 'Big picture' integrated displays Enhanced head-up displays Enhanced or synthetic vision systems CONTROL AUTOMATION Low-visibility taxi guidance High-precision in-trail guidance in terminal areas Automated collision avoidance maneuvers Automated wind shear avoidance maneuvers
Winnie Mae (Wiley Post's around-the-world flight, 1933) Lockheed 14	**1st Generation** DeHavilland Comet Boeing 707 Douglas DC-8 Douglas DC-9	**2nd Generation** Boeing 727, Boeing 737-100,200 Boeing 747-100-300 DC-10, L-1011 Airbus A-300	**3rd Generation** Boeing 767/757, 747-400 McDonnell-Douglas MD-80 Airbus A-310, 300-600 Fokker F-28-100 MD-11 (transition to 4th Gen)	**4th Generation** Airbus A-319/320/321 Airbus A-330, 340 Boeing 777,787	**Next generation planned enhancements**

FIGURE 6-1 Evolution of technology in civil air transport. Adapted from Fadden (1990) and Billings (1996).

technological advance, the pilot has less reason to look outside the cockpit and more reason to focus on systems and displays within it. In fourth-generation aircraft, all the information required to fly the aircraft and navigate from A to B can be found inside the cockpit, and the need for kinesthetic sensing or searching for cues outside the cockpit is greatly diminished. In fact, with reliable instruments aircraft can operate in low- or no-visibility conditions. Exact location in space can be read from cockpit displays, whether or not visual cues are visible outside of the cockpit. In some aircraft even the need for tactile sensing has been reduced and tactile feedback eliminated as "fly-by-wire" controls, which provide little or no tactile feedback on thrust setting, have replaced conventional hydraulically actuated control columns. Fourth-generation aircraft such as the Airbus A-319/320/321/330/340 and the Boeing 777/787 are qualitatively different entities than early generation aircraft such as the Boeing 707, and issues and requirements for flying them derive not only from sensory and external (naturalistic) aspects of flying, but also from cognitive and internal (electronic/deterministic) factors. Table 6.1 outlines the characteristics of each side of the hybrid ecology, each of which is discussed in the sections that follow.

CORRESPONDENCE AND THE NATURALISTIC WORLD

Aviation has traditionally been described as a correspondence-driven domain in that it exists within and is subject to the constraints of the natural environment (Vicente, 1990), including dynamic, changing condition, ambiguous cues, ill-structured problems, and time pressure (Zsambok & Klein, 1997). Correspondence, or accuracy, in this natural environment involves integration of multiple probabilistic cues. Much of the applied work that has been done on expert processes in aviation has focused on intuitive and sensory-driven correspondence strategies. Klein's model of expert Recognition-Primed Decision Making (e.g., Klein, 1993, 2000; Zsambok & Klein, 1997), for example, describes expertise as the ability to identify critical cues in the environment, to recognize patterns of cues, and to understand the structural relationships among cues. According to this model, expert pilots look for familiar

TABLE 6-1   The Hybrid Ecology of the Modern Flightdeck

Naturalistic World	Electronic World
Sensory and kinesthetic	Cognitive
Correspondence strategies and goals	Coherence strategies and goals
Ambiguity	Reliability
Intuitive processes	Analytical processes
Probabilistic cues	Deterministic data and information
Expertise affords intuitive "short cuts"	Analysis required at all levels of expertise

patterns of relevant cues, signaling situations that they have dealt with in the past, and base their responses on what they know "works" (e.g., Klein, 1993; Klein, Calderwood, & Clinton-Cirocco, 1986). The expert pilot may check to see if the view out the window is as expected with respect to patterns of surrounding and runway lights, and whether the cues match what he or she has previously encountered at this location or point in the flight. Or, the expert pilot may scan the sky ahead, intuitively gauging a safe distance from clouds, estimating their density and the horizontal visibility, and picking out a safe route through them.

Experienced pilots are typically highly competent in correspondence strategies. They are adept at assessing cue validity within specific situational contexts, and they are better than inexperienced pilots at predicting the outcome of a given decision or action. Their expectations have been shaped by a wide array of experiences, and they utilize these experiences to assess patterns of cues. Expertise offers a great advantage in correspondence judgments, as expert pilots are able to quickly recognize a situation from patterns of probabilistic cues, and may be able to use intuitive judgment processes under conditions that would demand analysis from a novice. For example, novice pilots may need to use a combination of computations and cues outside of the aircraft to figure out when to start a descent for landing. Experienced pilots in contrast may look outside the cockpit window and intuitively recognize when the situation "looks right" to start down.

Correspondence Errors

Correspondence errors typically entail misreadings of probabilistic cues or failure to integrate the cues appropriately. One important probabilistic element of flight is weather information. Weather has been identified as a factor in 15% of fatal GA (General Aviation) accidents (Coyne, Baldwin, & Latorella, 2001) and is a significant contributor to airline accidents and incidents. It is also a salient factor in plan continuation errors, in which crews continue with a plan of action despite the presence of cues suggesting that the original plan is not longer optimal and should be modified (Orasanu, Martin, & Davison, 2001). Cockpit weather displays are "…not sufficiently specific or accurate to allow pilots to distinguish hazardous-looking but benign weather from truly hazardous weather" (Dismukes, Berman & Loukopoulos, 2007). Pilots must rely on supplemental information from outside sources and from their own senses. Dismukes et al. (2007) analyzed major U.S. air carrier accidents between 1991 and 2000 in which the National Transportation Safety Board (NTSB) cited crew errors as having a central role. Several of these as well as many other aviation incidents and accidents can be traced to crew misjudgments of weather conditions, or inappropriate responses to weather situations.

For example, based on poor information and incomplete and misleading cues from ATC, the crew of US Air 1016 undertook an approach into severe storm activity and encountered a microburst at about 200 feet above the ground, with an accompanying increase in airspeed. They responded by initiating a right turn and a missed approach, a reaction that would be appropriate for avoiding a windshear, and in doing so

> "…entered the downdraft and tailwind (performance-decreasing) portion of the microburst wind field, and the airplane began to sink toward the ground despite its nose-up attitude and climb thrust setting. …In order for the crew to recognize that they needed to switch from a normal missed approach to the windshear recovery procedure, they would have had to note and interpret additional cues resulting as the aircraft's performance began to deteriorate—decreasing airspeed and poor climb performance—but this would have required the pilots to integrate and interpret information from multiple sources while under workload, time pressure, and stress" (Dismukes et al., 2007, p. 17).

Instead, they misinterpreted the cues, the captain ordered "Down, push it down" (probably to increase airspeed), and they were unable to recover from the resultant steep sink rate in time to avoid impact.

This accident demonstrates the difficulty of competent correspondence judgments when cues and information are highly ambiguous, missing, or misleading. Poor information from ATC led the crew to expect better weather at the airport than was actually the case, and they were not aware of the severity of the weather threat. Because their perceptions of the location of the precipitation cell were outdated, they executed a missed approach with a turn to the right, carrying the aircraft into the downdraft portion of the microburst. Notably, the aircraft's onboard windshear aural warning did not activate, and the absence of this cue—which would have been expected in a windshear event—may have contributed to the crew's slowness in diagnosing the situation. The lack of turbulence, which would have been consistent with all of the company's windshear training scenarios, was also a confusing factor. In this event, information in the electronic cockpit was not helpful for diagnosing the situation. The crew was forced to rely on probabilistic cues and their ability to use them effectively was compromised by cue ambiguity and rapidly changing external conditions.

Spatial Disorientation

One of the most important functions of instruments in early flight was to help pilots combat the correspondence error of spatial disorientation (SD) and associated illusions. Many accidents occurred in aviation before the source and compelling nature of false signals from the vestibular system were properly appreciated, and the phenomenon of SD remains the primary and most dangerous sensory pitfall in modern flight.

Spatial orientation refers to the perception of one's body position in relation to a reference frame (Young, 2003), and spatial *dis*orientation is a perceptual problem in which a pilot is unable to correctly interpret aircraft position, motion, attitude, altitude or airspeed in relation to points of reference or to the earth (Newman, 2007). Basically, it is the inability to tell "which way is up" (FAA, 1983).

Comprehensive descriptions of SD illusions resulting from sensory misperceptions can be found in Newman (2007) or Young (2003).

The most common SD illusions result from visual phenomena, and pilots are most susceptible to them when visual cues are inadequate or are absent, as in bad weather or at night. The vestibular and proprioceptive (seat-of-the-pants) systems may also be involved, as all three of these systems contribute information to help us determine our horizontal and vertical position relative to the earth. The expression "one cannot fly by the seat of the pants" reflects the possibility of erroneous sensations from the vestibular system or from body pressure when the ground is not visible. Our sensory systems:

> "… are not designed to operate in the three-dimensional environment of flight. In that environment, it is possible to operate independently of the normal visual cues (as in bad weather or night flying) and both the magnitude and applied direction of gravity can be altered. The complex motion environment of flight thus dramatically increases the risks of SD by exposing the physiological limitations of the normal human orientation system" (Newman, 2007, p. 6).

Three types of SD have been identified and categorized (Gillingham and Previc, 1996; Newman, 2007; Young, 2003):

- Type I (unrecognized) is the most dangerous type of SD. In Type I SD, pilots have no notion that they are disoriented. Because pilots are not aware that their perception is faulty, they will not take corrective action and may fly the aircraft into the ground.
- Type II (recognized) SD is less dangerous because pilots are aware that their perception is not reliable. Often there is conflict between the aircraft's true orientation (as indicated by the instruments) and the pilot's sense of motion or orientation. This type of SD is associated with the old adage "believe your instruments," as that is the primary route to recovery from SD. Type II SD is quite common among aviators, and many if not most pilots experience it at some point in their careers (Newman, 2007).
- Type III (incapacitating) SD mentally and physically overwhelms pilots, creating a sense of confusion about orientation, severe motion sickness, helplessness, and/or inability to control the aircraft. The pilot may be aware of disorientation, but is unable to combat it because normal cognitive processes have broken down. The pilot who is experiencing this type of SD

may freeze and become incapable of making control inputs, or may fight the aircraft with inappropriate control inputs.

Spatial disorientation as a cause of mishaps has been most frequently tracked in military operations. According to recent reviews of this phenomenon, SD is still a common problem in military aviation, accounting for 6–32% of major accidents and 15–69% of fatal accidents in military forces (Newman, 2007). The U.S. Air Force loses on average five aircraft with aircrews each year due to SD (Ercoline, DeVilbiss, Yauch, & Brown, 2000). The risk of SD is particularly high in helicopters and fighter/attack aircraft, and is increased at night. Experience does not insulate pilots from SD, although it may help them recognize its onset, and the data suggest that a second crewmember does not protect against it (Lyons, Ercoline, O'Toole, & Grayson, 2006). Data for the incidence of SD in civil aviation are lacking. Pilots who recover from it do not report it, and it is difficult to positively conclude that a particular fatal accident was due to SD occurring.

The solution to the problem of visual-vestibular interaction is that pilots must *never* rely on sensations from their vestibular systems or from the pressure exerted on their bodies. "Pilots must learn to rely on their instruments and to disregard their body sensations no matter how compelling they might be" (Liebowitz, 1988, p. 101). Researchers are still looking for the best displays to curtail the incidence of SD (e.g., Wickens, Self, Andre, Reynolds, & Small, 2007). Head-up displays (HUDs), which present essential flight parameters on a clear screen directly in back of the windscreen (so that the pilot does not have to look down for instrument readings), have significantly reduced SD situations associated with shifting frames of reference and head movements (Young, 2003). Note that the solutions to this naturalistic, sensory problem are found in the electronic world, reinforcing the importance of cognitive sensibility over kinesthetic sensing.

COHERENCE AND THE ELECTRONIC WORLD

Although sensory and naturalistic issues still exist in modern, high-tech aircraft, the flying task today is more cognitively than physically demanding, and is much more a matter of cognitive

sensibility than of kinesthetic sensing. In contrast to earlier aviators, glass cockpit pilots can spend relatively little of their time
looking out the window or manipulating flight controls, and most
to all of it focused on integrating, updating, and utilizing tech-
nology-generated information inside the cockpit. This represents
a profound change in the pilots' operational environment and
an alteration and intensification of the cognitive requirements to
function successfully within it:

> The development and introduction of modern automation technology has
> led to new cognitive demands. ... The result is new knowledge require-
> ments (e.g., understanding the functional structure of the system), new
> communication tasks (e.g., knowing how to instruct the automation to
> carry out a particular task), new data management tasks (e.g., knowing
> when to look for, and where to find, relevant information in the system's
> data architecture), and new attentional demands (tracking the status and
> behavior of the automation as well as the controlled process). (Amalberti &
> Sarter, 2000, p. 4)

The data that pilots utilize to fly can, in most cases, be found on
cockpit display panels and CRTs, and are qualitatively different from the cues used in correspondence judgments. They are
data rather than *cues* in that they are precise, reliable indicators of
whatever they are designed to represent. In the electronic, deterministic environment of the cockpit, the primary task of pilot is to
supervise and monitor systems and information displays to ensure
consistency, or coherence, of the "world" and to restore it when
disruptions occur. The task is primarily a cognitive task, and the
control required is primarily cognitive control. To a great extent,
interactions with the environment have been supplanted by interactions with electronic systems, and the importance of handling
skills has been replaced by an expectation of management skills
requiring rule-based and knowledge-based cognitive control.

Coherence strategies involve the electronic side of the hybrid cockpit ecology and typically require analytical processing across all
levels of expertise. Managing the hybrid ecology of the modern
high-tech aircraft is a primarily coherence-based, complex, cognitively demanding mental task, involves data, rationality, logic, and
an analytical mode of judgment. The more advanced the aircraft,

the higher the demand for coherence competence, that is, for an understanding of aircraft systems and the ability to achieve and maintain consistency among indicators, and for a more analytical type of cognitive processing than in correspondence judgments. Pilots of high-tech aircraft must discern and set correct flight modes, compare display data with expected data, investigate sources of discrepancies, program and operate flight computer systems, and evaluate what a given piece of data means when shown in a particular color, in a particular position on the screen, in a particular configuration, in a particular system mode. Competence in coherence also entails knowledge of potential pitfalls that are the artifacts of technology such as mode errors (i.e., executing a function or command that is not appropriate for the current system mode; Sarter, Woods, & Billings, 1997), hidden data, or noncoupled systems and indicators, recognition of inconsistencies in data that signal a lack of coherence, and an understanding of the limits of electronic systems as well as their strengths and weaknesses.

Importantly, expertise does not offer all of the same advantages to the pilot in the electronic world as it does in the naturalistic world. Information in deterministic environments cannot be managed intuitively, and relying on expert intuition may not only be insufficient but may also be counter-productive. In comparison with naturalistic cues, technological data and information are not amenable to intuitive shortcuts or pattern recognition processes, but rather have to be assessed analytically to ensure that they form a cohesive, coherent representation of what is going on. Once interpreted, data must be compared with expected data to detect inconsistencies, and, if they exist, analysis is required to resolve them before they translate into unexpected or undesired behaviors.

Some researchers have begun to explore the issue of information search and information use in the electronic cockpit. In a recent study we found that time pressure, a common factor in airline operations, had a strong negative effect on the coherence of diagnosis and decision making, and that the presence of contradictory information (noncoherence in indicators) heightened these negative effects. Overall, pilots responded to information conflicts by taking more time to come to a diagnosis, checking more information, and performing more double-checks of information. However, they were significantly less thorough in their information search when pressed to come to a diagnosis quickly than when under no time pressure. This meant that they tended to miss relevant information under time pressure, resulting in lower diagnosis accuracy. These results confirm both the need for coherence in judgment and diagnosis and the difficulty of maintaining it under time pressure (Mosier, Sethi, McCauley, Khoo, & Orasanu, 2007).

Coherence Errors

Coherence errors typically entail failures to note and process relevant data or detect something in the electronic "story" that is not consistent with the rest of the picture. Coherence errors have real and potentially catastrophic consequences in the physical world because they are directly linked to correspondence goals. Parasuraman and Riley (1997), for example, cite controlled flight into terrain accidents in which crews failed to notice that their guidance mode was inconsistent with other descent settings, and flew the aircraft into the ground. "Automation surprises," or situations in which crews are surprised by control actions taken by automated systems (Sarter et al., 1997), occur when pilots have an inaccurate judgment of coherence—they misinterpret or misassess data on system states and functioning (Woods & Sarter, 2000). Mode error, or confusion about the active system mode (e.g., Sarter & Woods, 1994; Woods and Sarter, 2000), is a type of coherence error that has resulted in several incidents and accidents. Perhaps the most well-known of these occurred in Strasbourg, France (Ministre de l'Equipement, des Transports et du Tourisme, 1993), when an Airbus 320 confused approach modes:

> It is believed that the pilots intended to make an automatic approach using a flight path angle of $-3.3°$ from the final approach fix. ... The pilots, however, appear to have executed the approach in heading/vertical speed mode instead of track/flight path angle mode. The Flight Control Unit setting of " -33" yields a vertical descent rate of $-3300\,\text{ft/min}$ in this mode, and this is almost precisely the rate of descent the airplane realized until it crashed into mountainous terrain several miles short of the airport. (Billings, 1996, p. 178).

In another illustrative example (Dismukes et al., 2007), the crew of AA 903 did not notice that the autothrottle had disconnected during descent so that when they leveled off, it did not produce the extra thrust needed to maintain straight and level flight. When the aircraft entered a turn at its holding fix the autopilot attempted to maintain altitude, and as the airspeed decreased to 177 knots the aircraft went into a stall. This caused a sharp roll to the right, autopilot disconnect, uncontrolled pitch and roll oscillations, and a descent of several thousand feet before the crew could recover.

The aircraft, an A300, had two cockpit indicators for the status of the autothrottle system, but neither of these was very salient and the disconnect went undetected. Airspeed indicators provided critical information as the speed decreased, but the pilots missed this information because they instead focused on the ADI (attitude director indicators) to control their attitude. The pilots didn't realize they were in a stall because it occurred at a lower angle of attack than the aircraft's design criterion, so the stick shaker (indicating imminent stall) did not activate immediately. The autothrottle disconnect was not designed to be accompanied by an aural warning or a master caution light, which may have misled the crew further as pilots expect most systems to be tied into the master caution/warning signal. Sensory cues provided little insight into the cause of their state, as the behavior of the aircraft was similar to an "autothrottle undershoot that would self-correct" (Dismukes et al., 2007, p. 202), and the oscillations and steep descent made the crew think they were in turbulence or windshear conditions.

To further complicate diagnosis, the pilots' primary flight display (PFD) blanked out for 2–3 seconds during their recovery. Because the aircraft systems must constantly refresh and reinterpret data, the PFDs of the A300 are designed to blank out when the aircraft undergoes extreme or rapidly changing attitudes to avoid giving erroneous data. This meant that critical information was removed precisely at the time when pilots needed it the most (Dismukes et al., 2007). In this event, cues in the external environment were not helpful for diagnosing the situation. The crew did not identify the system parameter that was out of sync with the desired flight mode and navigational status and their ability to diagnose the situation and restore level flight was compromised by the resultant inability to track the source of their predicament.

In the AA903 event, all the information needed to diagnose the situation was available in the cockpit, but critical diagnostic data were not detected by the crew and therefore were not factored into their judgment processes. Dismukes and his colleagues concluded that a contributor to the AA903 loss of control was the "… automation interface, which in some cockpit systems does not adequately support the complex monitoring processes required of pilots supervising automated systems" (p. 202), and noted that maintaining mode awareness and detecting unexpected changes in automated mode are difficult tasks for crews.

As an example of system complexity, Sherry and his colleagues (Sherry, Feary, Polson, Mumaw, & Palmer, 2001) decomposed the functions and displays of the vertical navigation system (VNAV) and of the flight mode annunciator (FMA—indicates system mode) of the electronic cockpit. They found that the selection of the VNAV in the descent and approach phases of flight results in the engagement of one of *six* possible trajectories—and that these trajectories will change autonomously as the situation evolves. Moreover, the same FMA display is used to represent several different trajectories commanded by the VNAV function and the interface does not provide the necessary information to establish or reinforce correct mental models of system functioning (Sherry et al., 2001). It is not surprising, given these sources of confusion, that coherence errors occur.

MANAGING THE HYBRID ECOLOGY

Integrating the Naturalistic and Electronic Worlds

The modern flight environment demands strategies and judgments that are both coherent (rational and consistent) in the electronic cockpit and correspondent (accurate) in the physical world. Managing the electronic cockpit entails attention to and integration of all appropriate and relevant sensory cues and electronic information. Weather displays, for example, offer imperfect data on the severity of weather cells and should be supplemented when possible with probabilistic cues such as reports from preceding flights and ground-based weather radar (Dismukes et al., 2007). Out-the-window views provide corroboration for cockpit indications and verify the correctness of the electronic settings (mode, flight path, etc.). Sensory cues such as smoke or sounds can often provide critical input to the diagnosis of high-tech system anomalies. By contrast, sensory information may also lead pilots astray if it is inconsistent with system indicators, as is the case for SD. Pilots must also recognize when contextual factors impose physical constraints on the accuracy of electronic prescriptions. It does no good, for example, to make a perfectly "coherent" landing with all systems in sync if a truck is stalled in the middle of the runway.

Choosing a Strategy

The choice of a correspondence cue-based strategy or a coherence system-based strategy may depend on contextual conditions. A pilot, for example, may rely on analytical processing of cockpit information to ensure coherence if visibility is obscured, or may choose to disengage the automatic aircraft functions to "hand fly" an approach via correspondence if visibility is good. Jacobson and Mosier (2004) noted in pilots' incident reports that different judgment strategies were mentioned as a function of the context within which a decision event occurred. During traffic problems in good visibility conditions, for example, pilots tended to rely on intuitive correspondence strategies, reacting to one or two probabilistic cues; for incidents involving equipment problems, however, pilots were more likely to use more analytical coherence decision strategies, checking and cross-checking indicators to formulate a diagnosis of the situation.

Importantly, because the hybrid ecology of the high-tech flight environment is characterized by highly reliable deterministic systems, coherence can often be a means to and a surrogate for correspondence, and accuracy can be ensured when a coherent state is present. In fact, in the modern cockpit *the primary route to correspondence in the physical world is through coherence in the electronic world*. Aircraft systems and sensors in the automated cockpit accomplish most of the "correspondence" tasks if they are in a coherent state—that is *as long as* system parameters, flight modes, and navigational displays are consistent with each other and with what should be present in a given situation. For example, if tasks are programmed properly, flight management automation will ensure they are performed without fail by automated systems. Exact distance from the ground can be read from the altimeter as long as the barometric pressure is set correctly. Aircraft orientation can be discerned from attitude indicators even in the event of faulty proprioceptive and vestibular cues. Today's aircraft can fly from A to B with zero outside visibility—once initial coordinates are accurately programmed into the flight computer, navigation can be accomplished without any external reference cues. When flight crews have ensured that *all* information inside the cockpit paints a consistent picture of the aircraft on the glide path, they can be confident that the aircraft *is* on the glide path. The pilots do not need to look out the window for airport cues to confirm it, and, in fact, visibility conditions often do not allow them to do so. They do, however, need to examine data with an analytical eye, decipher their meaning in the context of flight modes or phases, and interpret their implications in terms of outcomes.

CHALLENGES FOR NEXTGEN AIRCRAFT

Other chapters in this volume (e.g., Chapters 15 and 16) will address more fully the range of issues associated with automation design and system displays. In terms of cognitive sensibility, however, designers of NextGen aircraft will need to address the issue of matching system design with cognitive requirements. Important design factors are display salience, placement, mode, and format.

Display Salience

If a critical issue in managing in the hybrid cockpit ecology is detection of and attention to relevant information, one possible design solution would be to make this information very salient. Dismukes et al. (2007) suggest that "the best solution currently available [for promoting mode awareness] is to provide highly salient warnings whenever a system changes modes …" (p. 203).

However, increasing the salience of particular cues or data may not be the best way to foster coherence, as salient cues draw attention to themselves and away from other indicators, and foster a reliance on the salient cue to the exclusion of others.

A byproduct of the overreliance on salient cues is *automation bias*, a flawed decision process characterized by the use of (typically salient) automated information as a heuristic replacement for vigilant information seeking and processing. This flawed judgment process has been identified as a factor in professional pilot judgment errors (e.g. Mosier, Skitka, Dunbar, & McDonnell, 2001; Mosier, Skitka, Heers, & Burdick, 1998). Two classes of technology-related errors commonly emerge in hybrid decision environments: (1) omission errors, defined as failures to respond to system irregularities or events when automated devices fail to detect or indicate them; and (2) commission errors, which occur when decision makers incorrectly follow a automation-based directive or recommendation without verifying it against other available information, or in spite of contradictions from other sources of information. The automation bias phenomenon illustrates the danger of noncoherent judgment in the electronic environment, and suggests that making some information salient is only part of the solution to managing the cockpit. In fact, the design of many high technology systems may encourage decision makers—including experts—to focus on the most salient, available cues and to make intuitive, heuristic judgments before they have taken into account a broader array of relevant cues and information. System designers must recognize this tendency and provide safeguards against it.

Display Placement and Mode

One of the challenges of the high technology cockpit is avoiding potential visual overload, as most data are acquired visually. This is particularly important with respect to status changes and warnings. The failure of pilots to notice mode transitions and mode changes, for example, may be due to their visual display, which is not preemptive enough to command the crew's attention (Wickens, 2003a). Interestingly, some solutions to this issue involve reengaging the senses—warnings or alarms presented in the auditory modality have been found to be more effective in interrupting ongoing visual monitoring in the electronic cockpit (Stanton, 1994; Wickens & Liu, 1988), and do not add to the array of data and information that must be absorbed visually.

Auditory alerts have also been implemented for traffic detection in modern aircraft. The standard TCAS (Traffic Alert and Collision Avoidance System) uses aural alerts as the level of potential danger rises, and a combination of visual and auditory commands if it detects that the likelihood of collision is high (Wickens, 2003b). Three-dimensional auditory cues have also been found to result in faster traffic detection than conventional head-down displays (Begault & Pittman, 1996). Feedback may also be distributed across modes of presentation or provided in the tactile mode. The stick shaker, for example, is a traditional tactile warning of impending aircraft stall. Sklar and Sarter (1999) found that tactile cues were also better than visual cues for detecting and responding to uncommanded mode transitions.

Other solutions have involved changing the format and accessibility of visual cues and information. For example, the HUD superimposes symbology representing aircraft trajectory parameters (e.g., altitude, airspeed, flight path) on the pilot's external view (e.g., Pope, 2006; Young, 2003), enabling easy monitoring of both electronic and naturalistic cues and information. (It should be noted that there are other potential performance trade-offs associated with the HUD—see, e.g., Fadden, Ververs, & Wickens, 1998).

Display Format

Technology in the electronic cockpit may be either a facilitator of or an obstacle to cognitive processing. Often, pictorial representations exploit human intuitive pattern-matching abilities, and allow quick detection of some out-of-parameter system states. This design philosophy is consistent with the goals of workload reduction and information consolidation—and, indeed, many features of cockpit displays do foster the detection of disruptions to a coherent state. However, these displays also foster intuitive rather than analytical cognition, and in doing so set up the expectation that the cockpit can be managed in an intuitive fashion. This is an erroneous assumption.

In their efforts to provide an easy-to-use intuitive display format, designers have often buried the data needed to retrace or follow system actions. "Intuitive" cockpit displays represent complex data, highly complex combinations of features, options, functions, and system couplings that may produce unanticipated, quickly propagating effects if not analyzed and taken into account (Woods, 1996). Technological decision-aiding systems, in particular, often present only what has been deemed "necessary." Data are preprocessed, and presented in a format that allows, for the most part, only a surface view of system functioning, and precludes analysis of the consistency or coherence of data. Within the seemingly intuitive displays reside numerical data that signify different commands or values in different modes. For example, mode confusion, as described in the A-320 accident above, often results from what looks, without sufficient analysis, like a coherent picture. The cockpit setup for a flight path angle of −3.3 deg in one flight mode looks very much like the setup for a −3300 ft/min approach in another flight mode.

Moreover, calculations and resultant actions often occur without the awareness of the human operator. System opaqueness interferes with the capability to track processes analytically (e.g., Woods, 1996; Sarter, Woods, & Billings, 1997). Paries and Amalberti (2000) have discussed the dangers of using intuitive, generalized responses when dealing with the "exceptions" that "can be found in the commands or the displays of all existing aircraft" (p. 277). The exceptions they cite involve the logic of the fight mode indications and behavior in relation to the context of the flight—the interpretation of display indications differs depending on the phase of flight and the aircraft. Highly coupled autopilot modes make things even more complex. Moreover, as the aircraft cockpit has evolved, much of the systems information has either been altered in format, not presented at all, or buried below surface displays. Woods and Sarter (2000), for example, in describing a typical sequence leading to an "automation surprise," note that "it seems that the crew generally does not notice their misassessment from the displays of data about the state or activities of the automated systems. The misassessment is detected, and thus the point of surprise is reached, in most cases based on observation of unexpected and sometimes undesirable aircraft behavior" (p. 331).

SUMMARY AND IMPLICATIONS FOR DESIGN

Providing transparency and traceability in displays is essential for eliciting appropriate cognitive strategies in the hybrid ecology of NextGen aircraft. Designers must recognize that the goals of cognition, as well as the cognitive tactics and strategies utilized to achieve them, are impacted by features and properties of the ecology. Woods and Sarter (2000), in describing the need for activity-centered system design, underlined the need to discover how "computer-based and other artifacts shape the cognitive and coordinative activities of people in the pursuit off their goals and task context" (p. 339). Hammond, Hamm, Grassia, and Pearson (1997) provided evidence that performance depends on the degree to which task properties elicit the most effective cognitive response.

Currently, a mismatch exists between cognitive requirements of the electronic cockpit and the cognitive strategies afforded by current systems and displays. On the one hand, system displays and opacity of system functioning foster intuition and discourage analysis; on the other hand, the complexity of automated systems makes them impossible to manage intuitively, and requires analytical cognitive processing. The opaque electronic interface and sequential, layered presentation of data inherent in cockpit systems may short circuit rather than support cognitive processing because they do not elicit appropriate cognitive responses. Recognizing this mismatch is the first step toward rectifying it.

Because the electronic cockpit contains a coherence-based, deterministic environment that represents the aircraft and its systems as well as features of the correspondence-based, naturalistic external world, a second step in design for NextGen operations is ensuring that systems are not only reliable in terms of correspondence (empirical accuracy), but are also interpretable in terms of coherence. Principles of human-centered automation prescribe that the pilot must be actively involved, adequately informed, and able to monitor and predict the functioning of automated systems (Billings, 1996). To this list should be added the requirement that the design of automation and automated displays elicits the cognition appropriate for accomplishing the pilot's role in the flying task. Questions that can guide this design requirement include:

- Does the system provide critical information to the pilots in a transparent and timely fashion?
- Does the design enable the pilot to understand what cognitive processes are required to utilize a system and how to accomplish goals? Does it facilitate the acquisition of an accurate mental model of the system?
- Does the design enable the recognition of coherence among systems and indicators? Does it facilitate attention to anomalies and inconsistencies in system setup?
- Does the design facilitate alternation between coherence and correspondence strategies when necessary (e.g., during a visual approach).
- Do surface features of the system induce the mode of cognition that will enable most effective task performance?
- Does the design ensure sufficient analytical activity without cognitive overload?

The last of these questions is perhaps the most difficult. In an environment that is heavily dependent on cognitive sensibility, a key issue for designers and manufacturers will be to determine how much information is needed to aid judgment and how this information should be displayed. The key will be to facilitate the processing of all relevant cues and information. Electronic systems could for example provide a roadmap for tracking their recommendations, help the decision maker interpret them, and point out sources of confirming or disconfirming information. Through appropriate design in the presentation of information and cues, technological aids could assist individuals by highlighting relevant data and information, helping them determine whether all information is consistent with a particular diagnosis or judgment, and providing assistance in accounting for missing or contradictory information.

In addition to presenting data and information appropriately, NextGen technology should be designed to enhance individual metacognition—that is, it should help people to be aware of how they are thinking and making judgments, and whether or not their process is appropriate, coherent, and accurate. A metacognitive intervention, then, would *prompt process* (e.g., *check engine parameters*). Metacognitive interventions may also take the form of training for process vigilance. Results of research on automation bias, for example, suggested that the perception that one is directly *accountable* for one's decision-making processes fosters more vigilant information seeking and processing, as well as more accurate judgments. Pilots who reported a higher internalized sense of accountability for their interactions with automation, regardless of assigned experimental condition, verified correct automation functioning more often and committed fewer errors than other pilots (Mosier et al., 1998). These results suggest that aiding the metacognitive monitoring of judgment processes may facilitate more thorough information search, more analytical cognition, and more coherent judgment strategies.

CONCLUSIONS

The modern flightdeck is a hybrid ecology characterized by naturalistic and electronic elements. Technology has created a shift in cognitive demands, and these demands cannot be met by strategies that have worked in previous generation aircraft. Flying in this ecology is primarily a cognitive task, and the human in this environment must exercise much more cognitive sensibility than kinesthetic sensing. Designers in the NextGen era need to recognize this and to meet the challenge of facilitating both correspondence and coherence—by not only providing the data required for correspondent, accurate judgment but also of presenting these data in a way that elicits and facilitates coherence in judgment. Most important, technology must foster cognitive sensibility, and must help pilots monitor their cognitive strategies by prompting *process* as well as providing information, by facilitating navigation within and between the naturalistic and electronic sides of the hybrid ecology, and by enabling pilots to adapt their cognition effectively to the demands of the environment.

References

Amalberti, R., & Sarter, N. B. (2000). Cognitive engineering in the aviation domain—Opportunities and challenges. In N. B. Sarter & R. Amalberti (Eds.), *Cognitive engineering in the aviation domain* (pp. 1–9). Mahwah, NJ: Lawrence Erlbaum Associates.

Baron, S. (1988). Pilot control. In E. L. Wiener & D. C. Nagel (Eds.), *Human factors in aviation* (pp. 347–386). San Diego, CA: Academic Press.

Begault, D. R., & Pittman, M. T. (1996). Three-dimensional audio versus head-down alert and collision avoidance system displays. *The International Journal of Aviation Psychology, 6*(1), 79–93.

Billings, C. E. (1996). *Human-centered aviation automation: Principles and guidelines* (NASA Technical Memorandum #110381). Moffett Field, CA: NASA Ames Research Center.

Brunswik, E. (1956). Perception and the representative design of psychological experiments. Berkeley: University of California Press.

Coyne, J. T., Baldwin, C. L., & Latorella, K. A. (2001). Pilot weather assessment: Implications for visual flight rules flight into instrument meteorological conditions. *The International Journal of Aviation Psychology, 18*(2), 153–166.

Dismukes, R. K., Berman, B. A., & Loukopoulos, L. D. (2007). The limits of expertise: Rethinking pilot error and the causes of airline accidents. Aldershot, UK: Ashgate.

FAA. (1983). *Pilot's spatial disorientation (AC 60–4A).* Washington, DC: Federal Aviation Administration.

Ercoline, W. R., DeVilbiss, C. A., Yauch, D. W., & Brown, D. L. (2000). Post0roll effects on attitude perception: The Gillingham Illusion. *Aviation, Space, and Environmental Medicine, 71*(5), 489–495.

Fadden, D. (1990). Aircraft automation changes. In: *Abstracts of AIAA-NASA-FAA-HFS symposium, challenges in aviation human factors: The national plan.* Washington, DC: American Institute of Aeronautics and Astronautics.

Fadden, S., Ververs, P. M., & Wickens, C. D. (1998). Costs and benefits of head-up display use: A meta-analytic approach. In *Proceedings of the 42nd annual meeting of the human factors and ergonomics society.* Santa Monica, CA: Human Factors and Ergonomics Society (pp. 16–20).

Gillingham, K. K., & Previc, F. H. (1996). Spatial orientation in flight. In R. DeHart (Ed.), *Fundamentals of aerospace medicine* (2nd ed.) (pp. 309–397). Baltimore, MD: Williams & Wilkins.

Hammond, K. R. (1996). Human judgment and social policy. New York: Oxford University Press.

Hammond, K. R., Hamm, R. M., Grassia, J., & Pearson, T. (1997). Direct comparison of the efficacy of intuitive and analytical cognition in expert judgment. In W. M. Goldstein & R. M. Hogarth (Eds.), *Research on judgment and decision making: Currents, connections, and controversies* (pp. 144–180). Cambridge: Cambridge University Press.

Hopkins, G. E. (1982). *Flying the line: The first half century of the air line pilots association*. Washington, DC: The Air Line Pilots Association.

Jacobson, C., & Mosier, K. L. (2004). Coherence and correspondence decision making in aviation: A study of pilot incident reports. *International Journal of Applied Aviation Studies, 4*(2), 123–134.

Klein, G. A. (1993). A recognition-primed decision (RPD) model of rapid decision making. In G. A. Klein, J. Orasanu, R. Calderwood, & C. E. Zsambok (Eds.), *Decision making in action: Models and methods* (pp. 138–147). Norwood, NJ: Ablex.

Klein, G. A. (2000). Sources of power: How people make decisions. Cambridge, MA: MIT Press.

Klein, G. A., Calderwood, R., & Clinton Cirocco, A. (1986). *Rapid decision making on the fire ground (KA-TR-84–41–7)* (Prepared under contract MDA903–85-G-0099 for the U.S. Army Research Institute, Alexandria, VA). Yellow Springs, OH: Klein Associates Inc..

Leibowitz, H. W. (1988). Human senses in flight. In E. L. Wiener & D. C. Nagel (Eds.), *Human factors in aviation* (pp. 83–110). San Diego, CA: Academic Press Inc..

Lyons, T., Ercoline, W., O' Toole, K., & Grayson, K. (2006). Aircraft and related factors in crashes involving spatial disorientation: 15 years of U.S. Air Force data. *Aviation, Space, and Environmental Medicine, 77*(7), 720–723.

Ministre de l'Equipement, des Transports et du Tourisme. (1993). *Rapport de la Commission d'Enquete sur l'Accident survenu le 20 Janvier 1992 pres du Mont Saite Odile a l/Airbus A320 Immatricule F-GGED Exploite par lay Compagnie Air Inter.* Paris: Author.

Mosier, K., Sethi, N., McCauley, S., Khoo, L., & Orasanu, J. (2007). What you don't know CAN hurt you: Factors impacting diagnosis in the automated cockpit. *Human Factors, 49*, 300–310.

Mosier, K. L., Skitka, L. J., Dunbar, M., & McDonnell, L. (2001). Air crews and automation bias: The advantages of teamwork?. *International Journal of Aviation Psychology, 11*, 1–14.

Mosier, K. L., Skitka, L. J., Heers, S., & Burdick, M. D. (1998). Automation bias: Decision making and performance in high-tech cockpits. *International Journal of Aviation Psychology, 8*, 47–63.

Newman, D. G. (2007). *An overview of spatial disorientation as a factor in aviation accidents and incidents* ATSB Transport Safety Investigation Report—B2007/0063. Canberra City, Australia: Australian Transport Safety Board.

Orasanu, J., Martin, L., & Davison, J. (2001). Cognitive and contextual factors in aviation accidents: Decision errors. In E. Salas & G. Klein (Eds.), *Linking expertise and naturalistic decision making* (pp. 209–225). Mahwah, NJ: Erlbaum.

Orlady, H. W., & Orlady, L. M. (1999). *Human factors in multi-crew flight operations*. Brookfield, Vermont: Ashgate Publishing Company.

Parasuraman, R., & Riley, V. (1997). Humans and automation: Use, misuse, disuse, abuse. *Human Factors, 39*, 230–253.

Paries, J., & Amalberti, R. (2000). Aviation safety paradigms and training implications. In N. B. Sarter & R. Amalberti (Eds.), *Cognitive engineering in the aviation domain* (pp. 253–286). Mahwah, NJ: Lawrence Erlbaum Associates.

Pope, S. (2006, January). The future of head-up display technology. *Aviation International News online*. Retrieved from <http://www.ainonline.com/news/

single-news-page/article/the-future-of-head-up-display-technology/> in December 2008.

Sarter, NB., Woods, D. D., & Billings, C. (1997). Automation surprises. In G. Savendy (Ed.), *Handbook of human factors/ergonomics* (2nd ed.) (pp. 1926–1943). New York: Wiley.

Sherry, L., Feary, M., Polson, P., & Palmer, E. (2001). What's it doing now? Taking the covers off autopilot behavior. *Proceedings of the 11th International Symposium on Aviation Psychology,* Columbus, OH.

Sklar, A. E., & Sarter, N. B. (1999). Good vibrations: Tactile feedback in support of attention allocation and human-automation coordination in event-driven domains. *Human Factors, 41,* 543–552.

Stanton, N. (1994). *Human factors of alarm design.* London: Taylor & Francis.

Vicente, K. J. (1990). Coherence- and correspondence-driven work domains: Implications for systems design. *Behavior and Information Technology, 9*(6), 493–502.

Wickens, C. D. (2003a). Pilot actions and tasks: Selections, execution, and control. In P. S. Tsang & M. A. Vidulich (Eds.), *Principles and practice of aviation psychology* (pp. 239–263). Mahweh, NJ: Lawrence Erlbaum Associates.

Wickens, C. D. (2003b). Aviation displays. In P. S. Tsang & M. A. Vidulich (Eds.), *Principles and practice of aviation psychology* (pp. 147–200). Mahweh, NJ: Lawrence Erlbaum Associates.

Wickens, C. D., & Flach, J. M. (1988). Information processing. In E. L. Wiener & D. C. Nagel (Eds.), *Human factors in aviation* (pp. 111–156). San Diego, CA: Academic Press.

Wickens, C. D., & Liu, Y. (1988). Codes and modalities in multiple resources: A success and a qualification. *Human Factors, 20,* 599–616.

Wickens, C. D., Self, B. P., Andre, T. S., Reynolds, T. J., & Small, R. L. (2007). Unusual attitude recoveries with a spatial disorientation icon. *The International Journal of Aviation Psychology, 17*(2), 153–166.

Wiener, E. L., & Nagel, D. C. (1988). Human factors in aviation. San Diego, CA: Academic Press.

Woods, D. D. (1996). Decomposing automation: Apparent simplicity, real complexity. In R. Parasuraman & M. Mouloua (Eds.), *Automation and human performance: Theory and applications* (pp. 3–18). Mahwah, NJ: Lawrence Erlbaum Associates.

Woods, D. D., & Sarter, N. B. (2000). Learning from automation surprises and "going sour" accidents. In N. B. Sarter & R. Amalberti (Eds.), *Cognitive engineering in the aviation domain* (pp. 327–353). Mahwah, NJ: Lawrence Erlbaum Associates.

Young, L. R. (2003). Spatial orientation. In P. S. Tsang & M. A. Vidulich (Eds.), *Principles and practice of aviation psychology.* Mahwah, NJ: Lawrence Erlbaum Associates.

Zsambok, C. E., & Klein, G. A. (1997). *Naturalistic decision making.* Mahweh,. NJ: Lawrence Erlbaum Associates.

7

Information Processing in Aviation

Michael A. Vidulich
[a]711th Human Performance Wing, Wright-Patterson
Air Force Base

Christopher D. Wickens,
[b]Alion Science Corporation, Boulder, Colorado

Pamela S. Tsang and John M. Flach
[c]Department of Psychology, Wright State University

INTRODUCTION

Understanding human information processing is vital to maximizing the effectiveness and safety of the aviation system. All too often, errors in human information processing have contributed to tragic aviation incidents. For example, on March 27, 1977, at the Tenerife airport, a KLM 747 pilot misinterpreted the communications from the air traffic control tower and started to take-off, resulting in a collision with a Pan Am 747 that was still on the runway (Cushing, 1994). The collision caused 583 fatalities and destroyed both aircraft. In contrast, on January 15, 2009, Captain C. B. Sullenberger experienced a double bird strike within a few minutes of taking off from New York's LaGuardia Airport (Batty, 2009). With the engines failing, Captain Sullenberger considered returning to LaGuardia or diverting to a nearer airport. Assessing that the aircraft could not make it to even the closest airport, he then glided the Airbus aircraft to the Hudson River and perfectly executed a difficult water ditching. All 155 people on board the aircraft survived. These two examples illustrate the extremes of the influence that human information processing can have upon aviation. Although human error can cause terrible tragedies sometimes, at other times, only the human can confront an unexpected challenge and adapt to meet it.

Impact of the Technological Evolution in Aviation on Human Information Processing

When Orville Wright piloted the first heavier-than-air flight in 1903, the mission of the pilot was simple. To be a success, the flight merely had to occur. No cargo was delivered, no passengers traveled anywhere, and no military objective was achieved. In the slightly more than a century that has passed since that first flight, the roles of aviation within human society have grown dramatically. Air travel in the United States has increased five times faster than the increase in the population (Adams and Reed, 2008). In 2006, FedEx's moved an average of 6 million packages a day with its Memphis hub handling about 400 flights per day by wide-body jet aircraft (Baird, 2006). Following the 9/11 attacks, the U.S. military had a 25-fold increase in its number of unmanned aerial vehicles (UAVs) by 2008 (Harrington, 2008).

During this period of explosive growth of aviation, the roles of the human within the aviation system have grown in number and evolved in character. In *The Pilot's Burden* (1994), Captain Robert Buck has written eloquently about the changing roles and responsibilities of the pilot in the years from the 1930s to the 1990s. Much of Buck's early piloting, including his setting of the junior transcontinental airspeed record in 1930, involved flying an open-cockpit, single-engine biplane, Pitcairn PA-6 Mailwing, with no radio and a minimal set of instruments. Buck described flying the Mailwing: "Imagine flying across New York City through the northeast corridor with no traffic, no ATC, no two-way radio, not a thing to think about except those key basics: fly the airplane, navigate, and avoid the terrain. It was a beautiful, simple life" (p. 8).

Buck became a professional airline pilot for TWA in 1937. He recounted that as aircraft engines increased in power, complexity, and number, the pilot's task of monitoring and control increased. New aircraft features such as wing flaps or thrust reversers added capability to the aircraft, but at the cost of more for the pilot to think about. New noise abatement regulations required that the pilot exert careful control of flight paths and throttle settings. Put simply, the increasing "burden" that Buck described was an increase in the amount and complexity of information processing that the pilot was required to perform.

The developments in aviation affected not only pilots, but the entire aviation system. For example, Gilbert (1973) documented the dramatic increase in the number of flights and consequently the number of opportunities for conflicts between flights from the 1930s to the 1970s that led to the establishment of the air traffic control system. The need for technological support such as radio, radar, and computerized automation increased as the authority of air traffic controllers expanded to manage minimum separation of more and more flights of faster moving aircraft ranging over increasingly long routes.

Today's aviation system is so tightly interconnected that when the flow of information is interrupted, the cascading effects can be extensive. For example, on August 26, 2008, one of the two computers that handle all flight plans for the U.S. airspace was given corrupted data and shut down. As the load from that center was transferred to the remaining center, the second center was overwhelmed (Tarm, 2008). During this interruption controllers had to input flight data into other computers manually and send them out to pilots via radio, causing 646 flight delays. However, it is noteworthy that as the computerized system failed, the humans were able to step in and keep the system running safely, albeit at reduced efficiency.

Despite the continual development of sophisticated sensing, data processing, and communicating technology applied to the real time control in both commercial and military aviation, human information processing capabilities remains a cornerstone of aviation safety and effectiveness. The beautiful, simple life described by

Buck consisted of a lone pilot experiencing flight through unmediated sensory information and exerting direct control of the aircraft via mechanical linkages to the control surfaces. By contrast, within the modern aviation system the pilot is one person among many and interacting with increasingly automated systems to achieve precisely defined goals within a multitude of constraints.

This chapter reviews the extant understanding of the fundamental capabilities that humans can contribute to aviation operations. The chapter examines how the pilots adapt to the changing aviation environment as information processing demands shifted from direct perceptual-motor control of the aircraft flight path and attitude to more complex and strategic, higher-level processing enabling the aircraft to safely operate in otherwise untenable regimes (e.g., night flight) and more crowded airspace. How limited human attention is mobilized, allocated, and optimized to meet the myriad challenges of the aviation environment is discussed. The interplay between the essential skills that enable performance and the human biases that can conspire to produce errors is examined. Last, the pragmatic need to characterize the demands of information processing has led to the development of the concepts of mental workload and situation awareness to help guide design and to assess the effectiveness of aviation systems. Considering the extant knowledge of human information processing and emerging technological trends, the chapter looks forward to a likely future aviation system.

INFORMATION PROCESSING IN EARLY AVIATION

Direct Perception

With the exception of a streamer on the wing to help early pilots to see the direction of the relative wind, there was little technology to help with the task of spatial orientation and control in the earliest days of flight. How was the pilot able to judge the motion of his craft relative to the aerodynamic requirements for stable flight and relative to his own goals (e.g., to land safely at a specific destination)? Many of these questions were raised and tentatively answered by Langewiesche's (1944) classic book *Stick and Rudder*. Langewieshe's intuitive analysis helped to lead Gibson (1966, 1979) to important insights about the nature of optical information that have radically changed our understanding of both the nature of visual information and the nature of human performance. The key insight was that the dynamic changes in visual perspective (what today we call "optical flow") have structure (e.g., optical invariants) that help to specify both the layout of the environment and the observer's motion relative to that layout. Gibson, Olum, and Rosenblatt (1955) were the first to describe this structure analytically and Gibson (1958/1982) offered general hypotheses about how the optical structure could provide feedback for guiding the control of locomotion.

It is interesting that empirical evaluation of many of Gibson's hypotheses about the visual control of motion had to wait for an important advance in aviation technology—the development of flight simulators with interactive, real-time visual displays. The flight simulator has become an important scientific instrument for studying human control of locomotion. This instrument allows the manipulation and control of the optical flow geometry and the measurement of human perceptual judgments and action. Beginning with the work of Owen and Warren (1987; Warren and Wertheim, 1990), there is now a rich empirical literature testing Gibson's hypotheses (e.g., Flach and Warren 1995; Hecht and Savelsburgh, 2004). This literature offers clear evidence linking structure in optical flow (e.g., global optical flow rate, edge rate, change in splay angle and depression angles, optical expansion, and the rate of optical expansion) to human control of locomotion (e.g., control of speed, altitude, and collision). It is important to note that this literature addresses both human abilities (e.g. to anticipate collisions) and the limitations of these abilities (e.g., misperceptions of speed resulting from changes of altitude above ground). Also, intuitions from the study of natural optical flow fields have inspired the design of modern graphical displays (e.g., Mulder. 2003a, 2003b, Amelink, Mulder, van Paassen, and Flach, 2005).

Not only did aviation challenge and expand our understanding of visual perception, but experiences with flying through clouds and other contexts drew attention to the vestibular system as an important source of information (and misinformation) relative to spatial orientation (Gillingham and Previc, 1996). Ocker and Crane's (1932) work on "blind flight" was instrumental in convincing aviation of the need for avionic displays to specify the state of the aircraft and to compensate for limitations of the vestibular system that could lead to dangerous illusions and misjudgments when optical flow information was not available (Previc and Ercoline, 2004).

Transformation and Computation

With the ever-expanding capability of the flying machine, especially in conditions with impoverished perceptual cues as in high altitude, at night, and during adverse weather, much supplemental information in symbolic, digital, or pictorial forms has to be furnished to enable the pilot to manage the flight. Relevant information now would have to be extracted from multiple sources and integrated into one coherent picture. Broadbent (1958), Miller, Galanter, and Pribram (1960), and Neisser (1967) were especially instrumental in bringing together developments in communication, computer science, and cybernetics to help describe and model human information processing. For example, Broadbent likened the human operator to a capacity-limited communication channel and put forth a human information processing model based on the information theory framework.

The information theory framework has proven to be useful. Hick (1952) and Hyman (1953) describe a linear relationship between reaction time to a signal and the amount of information to be resolved. As uncertainty, quantified in bits of information, increases with the number of response choices, reaction time will slow. In practice, this relationship suggests that greater response choices can better accommodate flexibility but will also cost response time.

But limitations of the information theory framework were soon recognized. One, the Hick-Hyman law is practical only when the amount of information to be processed can be quantified. Two, performance was found to be related to a host of additional factors (Wickens and Hollands, 2000). For example, Miller (1956) showed that short-term memory has a real but not a fixed limit defined by the number of bits of information. Rather, the capacity of short-term memory appeared to be limited by one's ability to chunk meaningful information strategically. But Broadbent (1959, p. 113) explained that the importance of the information theory framework was "the cybernetic approach . . . which has provided a clear language for discussing those various internal complexities which make the nervous system differ from a simple channel."

To illustrate, early signal detection studies with radar control showed that the same physical signal did not always elicit the same detection response from the operator (e.g., Mackworth, 1948). Green and Swets (1966) proposed that in addition to the sensory evidence of the signal, the response is influenced by a decision regarding the sufficiency of the evidence to indicate the presence of a signal. The decision itself is influenced by the larger information context, which includes the signal probability and consequences of the correct and incorrect responses.

Similarly, seemingly effortless direct perception often requires resolving ambiguity by making plausible inferences based on acquired knowledge. For instance, the increasing size of the retinal image of an object could signify an approaching airplane or an enlarging airplane. Because most objects do not expand on the spot, the increasing image size typically can be interpreted correctly as an approaching object. Perceptual ambiguities are particularly likely to occur when the three-dimensional (3D) world is represented on a two-dimensional (2D) display (Wickens, Liang, Prevett, and Olmos, 1996; Wickens, 2003). Knowing both the advantages and inherent ambiguities of 3D displays is invaluable for making judicious display decisions. For example, Ellis, McGreevy, and Hitchcock (1987) devised an ingenious solution for the 3D cockpit display of traffic information (CDTI) by attaching a post to each displayed aircraft. The post protruded from the ground at the aircraft's current geographical location and the aircraft's altitude was unambiguously specified by markers on the post.

As will become more evident in the ensuing higher-order processing section, input information is not merely transmitted from station to station until an output can be produced. On the way, much information transformation involving a number of internal processes like perceiving, allocating resources, retrieving and storing information, selecting and processing a response, and ultimately using the information to plan, to solve problems, and to make decisions occur. These transformations are neither static nor passive, but dynamic and adaptively strategic (Neisser, 1967; Moray, 1978). Whether mentally less demanding heuristics or effortful, deliberate, formal algorithms are used for making a decision would depend on the consequence of the decision, the skills of the decision maker, and whether time permits a deliberate process (Payne, Bateman, and Johnson, 1993).

HIGHER-LEVEL PROCESSING IN AVIATION SYSTEMS

Decision Making

Pilots make decisions with high frequency; most of these are routine, and indeed are hardly thought of as "decisions" at all, but rather more like procedures following (e.g., lower the landing gear, initiate a check list; Degani and Wiener, 1993). Typically the correct option to choose is so obvious, and its consequences so well anticipated that little cognitive effort is involved, once the appropriate time to make the decision has been identified. Rasmussen (1986) has referred to these as examples of *rule based behavior*, and such behavior typically fails only when the pilot forgets to perform the procedure often in times of high workload (Dismukes and Nowinski, 2007, Loukopoulos Dismukes and Barshi, 2009). While these errors in procedures-following are important (Degani and Wiener, 1993), they are not the focus of the current chapter.

In contrast, the current focus is on the class of decisions where two (or more) choice alternatives are plausible in the context, where evidence (cues) in the environment must be considered to drive the correct choice; when the outcome of one or the other choice cannot be predicted with certainty, and when harmful consequences could result from some of the possible outcomes. These define the properties of *risky decision making* which falls within the class of knowledge-based behavior (Rasmussen, 1986). For example, a pilot deciding to reject, rather than continue a takeoff upon diagnosing an engine failure, risks the possibility that the aircraft may overrun the runway. However, if thrust is applied to continue the take off, the airplane may fail to obtain necessary lift, and then stall (Inagake, 1999). One can identify numerous other examples of risky decision making in aviation: should the pilot shut down or leave running a suspected overheating engine? Should the VFR (visual flight rules) pilot continue or turn back in the face of possibly deteriorating weather (i.e., Instrument Meteorological Conditions [IMC])? Indeed, this last case has been found to be the number one cause of fatal accidents in general aviation (Wiegmann and Shappell, 2003).

Information processing operations. When analyzing such choices, it is evident that they must be proceeded by two prior information processing operations: a *situation assessment* (or diagnosis) is necessary to make the best choice; in the VRF→IMC decision above, this involves an accurate diagnosis of the visibility forward, and an understanding of the distance (and fuel remaining) required to fly back to the origin of the flight. But in order to provide an accurate situation assessment, *cue processing* is required, whereby evidence in the world is sought, attended and perceived. In the example above, such cues may come from forward visibility, consultation with on-board weather displays, a prior forecast, pilot reports, geographical location, and fuel status. An important linkage between cue seeking and situation assessment is the *diagnostic value* of a cue (Schriver, Morrow, Wickens, and Talleur 2008; Wickens and Hollands, 2000). Some cues are highly diagnostic; a view of flame coming out of an engine is diagnostic of overheating. But others are less so: momentary poor visibility ahead of the aircraft may or may not be diagnostic of poor weather ahead. Cue seeking and situation assessment often take place in an iterative closed-loop fashion, with a tentative assessment followed by further cue seeking to confirm or refute that assessment.

Psychologists and economists have studied these three processes of decision: cue seeking, situation assessment, and risky choice—from several different perspectives; both in generic form (Payne Bettman and Johnson, 1993; Einhorn and Hogarth, 1982; Hogarth,

1987; Lehto and Nah, 2006; Hoffman, Crandall and Shadbolt, 1998) and specifically in the case of aviation (Jensen; 1981; Orasanu, 1993; O'Hare, 2003; Schriver et al., 2008). We consider here, two overlapping views on pilot decision making.

Naturalistic decision making. (Zsambok and Klein, 1997; Orasanu and Fischer, 1997; McKinney, 1999) considers the expert pilot (or expert in other domains) making decisions in familiar settings. One hallmark of this analysis is the identification of *recognition primed decision* (RPD; Klein, 1989). Here the expert need not consult separate cues individually and integrate them through a workload intensive process to form a situation assessment; but instead recognizes the set of environmental features through direct perception based on extensive experience, so as to automatically classify it (e.g., engine overheat, clogged pitot tube). As a consequence, the choice of what to do in that situation is also relatively automatic: do what has proven successful in previous encounters with the same situation. A second hallmark of naturalistic decision making, characterizing the choice process of experts is *mental simulation* (Klein and Crandall, 1995). Here, given an assessment (reached through RPD), the pilot will run through a mental simulation of what will happen if the favored choice is made, given the assessed situation. If such simulation, based again on past experience, produces an acceptable outcome, it is immediately chosen. Hence (and this is particularly important in time-critical domains like aviation), all possible choice options need not be considered. An acceptable one can be rapidly picked. Mental simulations can be particularly effective in pre-mission planning when time constraints are far less severe. For example, Amalberti and Deblon (1992) found that pilots would simulate flights and structure their plans based on an awareness of their own capabilities. This careful planning can help to ensure that some problems are anticipated and solved in the planning stage, other problems might be avoided, and still others might be aided by ensuring that critical information is available in advance. As the old saw goes, "skilled pilots are able to avoid situations that require skilled piloting."

An alternative approach to decision analysis, but one that is not mutually exclusive, focuses on the decision capabilities of those who may not be experts; or on situations in which an expert encounters a new problem. It is often in these situations that decision making goes bad. While instances of bad decision making are infrequent compared to the multitude of good decisions, in safety critical domains such as aviation it is important to focus on the breakdowns in information processing that do occur on such infrequent occasions. Two examples are the above mentioned VFR flights into IMC and the tragic decision of the flight crew aboard an Air Florida transport in 1987 to take off from Washington National Airport with ice remaining on the wings, and insufficient power to lift off.

Heuristics and biases. We address these circumstances within the information processing framework of the three processes of decision making presented above (cue processing, situation assessment and choice), coupled with two different categorizations that researchers have applied to less-than-optimal decision making: heuristics and biases. Briefly, *decision heuristics* are "mental shortcuts" that are often used when time or cognitive resources are in short supply (constraints typical of the flight deck), and such heuristics usually provide a good outcome (correct decision; Tversky and Kahneman, 1974; Kahneman Slovic and Tversky, 1982; Gigerenzer, Todd, and the ABC Research Group, 1999). However, because their employment does not fully process all the information required for the best (optimal) assessment or decision, there may be instances in which their engagement leads to the wrong outcome. Development of the heuristics concept by Kahnneman and Tversky eventually led to a Nobel Prize in economics (Kahneman, 2003). *Decision biases* in contrast are systematic ways in which decision strategies may ignore features leading to the best decision (Wickens and Hollands, 2002). We describe as following, both biases and heuristics related to all three processes.

In cue seeking, the *"as if"* heuristic is one in which diagnosis is not based on weighing more diagnostic cues more than less diagnostic ones, but rather, on integrating all cues *as if* they were equally diagnostic (Johnson, Cavenaugh, Spooner, and Samet, 1973; Wickens and Hollands, 2000) by simply summing the cues in favor of one assessment versus the other, and choosing the assessed state that is the "winner" in this unweighted cue count.

Such a heuristic, normally simple, fast and of low mental work-load, can be inappropriate if the *salience bias* dominates cue selection. Here, information that is highly salient (loud, bright, in the forward view) tends to dominate, be more attended, and hence have a greater impact or weight on the situation assessment, than less salient information. For the VFR pilot, the view forward out of the cockpit is a highly salient cue for predicting weather (even though this may not be fully reliable, and hence may be incorrect).

Characteristics that make cues salient—attract attention—can be well defined and validated from attention research, as discussed later in the chapter. But in this regard, it is important to note here that the *absence of an event* can define a cue that is *not* very salient and people do not use very well the absence of cues to make positive diagnoses. This is often the case in troubleshooting or diagnosis, when, for example a light that does not go on, may be quite diagnostic of which system has failed (Hunt and Rouse, 1981).

In situation assessment, an *anchoring heuristic* (Tversky and Kahneman, 1974; Einhorn and Hogarth, 1982 and a *confirmation bias* (Einhorn and Hogarth, 1978) often go hand in hand, to degrade accurate assessment. The anchoring heuristic (one of three classic heuristics identified by Tversky and Kahneman, along with availability and representativeness) applies when cues become sequentially available over time. Such is typically the case in the decision to continue a flight into "iffy" weather or turn back. First, a weather report will be consulted on the ground; then, after takeoff, visibility will be used, possibly a pilot report will be heard from another aircraft, and perhaps an updated forecast will be consulted. When applying the *anchoring* heuristic to these sequentially updated cues (Perrin, Barnett, Walrath, and Grossman, 2001), the decision maker tends to anchor on the situation assessment supported by the first arriving cue (here the initial weather forecast), and be more reluctant than s/he should be to give subsequent arriving cues adequate weight in situation assessment. Such a bias is acceptable (and perhaps optimal) if those subsequent cues are consistent with the initial cue, but may be counterproductive if the later cues support the alternative assessment (e.g., that the weather may be deteriorating). Indeed, in the dynamic airspace a strong case can be made that later arriving cues *are* typically more reliable (and hence should be given greater, not less weight), simply because the passage of time degrades the reliability of all cues including the earliest one that was used to set the mental anchor.

Anchoring can feed back into the *confirmation bias* (Nickerson, 1998; Wickens and Hollands, 2000) by which people, having formed an initial state assessment (e.g., based on anchoring), have a tendency to seek and therefore find cues consistent with this assessment (confirming it to be true) rather than seeking cues to suggest that the assessment could be wrong. For example, when pilots have anchored their belief on a good forecast, they might look at and give great weight to the clear sky ahead, and not call for an updated forecast.

Finally, at choice, we note that different choice options can produce different outcomes depending not only on which option is chosen, but also on which diagnosed state or situation assessment actually turns out to be the case. In "risky" decisions, this uncertainty in state is always present; such as, not knowing with perfect certainty whether the weather *will* be good on the future flight path. Hence, for example, with two different choice alternatives, made in the face of two possible forecast states (each associated with a different likelihood), there could be costs and benefits associated with each of the four possible outcomes (continue or turn back, in the face of good or bad weather). The effective decision maker will explicitly consider each of these. The *elimination by aspects* heuristic (Tversky, 1972) sometimes employed in choice, when time pressure is great, is to avoid considering all possible states or alternatives; but only consider those alternatives that are most likely to be best. For example, in considering an alternative airport to divert to in bad weather, employing such a heuristic will lead the pilot to avoid considering all possible airports, and retain for consideration, only those that are closest to the current location; given that the aspect of geographical proximity (dictated by fuel remaining) is more important than other airport attributes, such as runway length and ATC support.

At choice, the *framing bias* is one that is often found to apply when the pilot is faced with two negative (unpleasant) possible choice alternatives; one a "sure loss" and the other a "risky loss." As a typical example, each alternative in the pilot's choice to turn back versus "drill through" potentially bad weather ahead contains a negative outcome. There will be a sure loss associated with turning back (the flight mission will definitely be delayed); but there will be a risky loss associated with continuing: a high probability of successful mission completion (and a good outcome), and a low probability of a very bad outcome and possible disaster. It turns out that when people consider two such options *framed* in this way (as a choice between negatives), they have a tendency to seek the risky alternative, even when the sure loss associated with the first option is less than the expected loss of the risky option (Kahneman and Tversky, 1984). Such a bias is called framing, because when the same two alternatives are *framed* in the positive (e.g., a sure guarantee of safety versus a high possibility of mission success), people tend to reverse their preference, and now go for the sure thing.

An extension of the framing bias is the *sunk cost* bias (Arkes and Blumer, 1985). Here people, having invested a great deal in a losing endeavor (the sunk cost) are more likely to continue with the endeavor, hoping that it will turn around and end okay (the risky option), rather than quit with a sure loss. One can easily see how this applies to the pilot desiring to continue a mission into bad weather rather than turn back, particularly if the mission is well underway.

Implications of Biases and Heuristics. Thus far, we have only presented a small sampling of the much larger number of biases and heuristics that have been both identified in decision research, and found to be applicable in applied settings (See Kahneman Slovic and Tversky, 1982; Wickens and Hollands, 2000). By focusing more text on these shortcomings than on naturalistic decision making, we do *not* mean to imply that pilot judgment and decision making is generally problematic or flawed. Instead, we wish to emphasize that to improve the already high levels of aviation safety still further, given that faulty judgment has been associated as the greatest human performance cause of fatal crashes, it is essential to understand the causes of these decision shortcomings. In decision making, such circumstances have paved the way for advances in decision training (e.g., Banbury, Dudfield, Hoermann, and Soll, 2007), for better displays to support cue processing and situation assessment, and for possible automated decision support systems (Layton, Smith, and McCoy, 1994); it is this role of automation that is treated in the chapter by Sarter, Ferris, and Wickens.

Complex Processing in Context

In 1978 Herbert Simon received the Nobel Prize in Economics for his work on organizational decision making. A central intuition of Simon's work was the recognition that human rationality is bounded. That is, humans do not seem to follow the prescriptions of normative models of economic decision making or rationality. Rather, they consider only a subset of the potentially relevant information and they tend to seek satisfactory, rather than optimal solutions (i.e., humans satisfice). In other words, as discussed in the previous section, people use heuristics (take short cuts), rather than carrying out the computationally difficult processes associated with most normative models of rationality. Rasmussen's (1986) Decision Ladder, shown in Figure 7.1, provides a useful format for visualizing these short cuts and the implications for human information processing. To make sense of research on experts in their natural domains, including his own observations of expert electronic troubleshooting, Rasmussen (1986) took the conventional single channel image of information processing stages and folded it in a way that suggests how experience with a task environment can create associations that interact with the information processes to shape performance.

A key insight that suggested this Decision Ladder is that Rasmussen discovered that the thought processes of the experts were characterized by associative leaps from one state of knowledge to another. Thus, the Decision Ladder includes both information processes and the states of knowledge that are the product of these processes. Rasmussen wrote:

> immediate associations may lead directly from one state of knowledge to the next. Such direct associations between states of knowledge is the typical process in familiar situations and leads to very efficient bypassing of

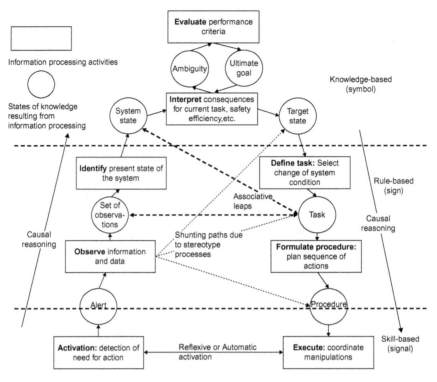

FIGURE 7-1 Rasmussen's (1986) Decision Ladder: this diagram illustrates how qualitatively different semiotic systems (signal, sign, and symbol) enable qualitatively different types of information processing (skill-, rule-, and knowledge-based).

> low-capacity, higher-level processes. Such associations do not follow logical rules; they are based on prior experience and can connect all categories of "states of knowledge." (1986, p. 69)

By folding the model, it was easier for Rasmussen to show these associative leaps as short cuts from one knowledge state to another. Thus, for example, an observation (e.g., an instrument that seems stuck) might not lead to a detailed causal analysis, but rather it might stimulate a memory from this morning when you forgot to plug in the coffee pot. Thus, triggering a simple task (e.g., check the power source). Note that the task was not triggered by an analytic process of interpreting the situation, evaluating multiple possible causes, and systematically eliminating those possibilities. Rather, as in recognition primed decision making, it was driven by an "automatic" association with a rather serendipitous event. If the morning had been different, the first hypothesis that would come to mind might have been different. Have you ever called the technician to fix a broken piece of equipment, only to have him come in and discover that it is unplugged? The key point is that there are many paths through the cognitive system and the specific path taken depends on both properties of the information processor (i.e., awareness) and properties of the situation. In essence, human information processing tends to be opportunistic and highly influenced by experience and context. This conclusion has been echoed by others (e.g., Suchman, 1987; Hutchins, 1995) who have emphasized that human cognition is "situated."

The shunts or short cuts shown in Figure 7.1 illustrate just some of the many possible associations between knowledge states. In principle, Rasmussen realized that any knowledge state can be associated with any other, and further these associations can go in any direction (e.g., in the process of formulating a procedure an observation may become more salient, or a hypothesis about the system state might be triggered). Thus, the Decision Ladder models the cognitive process as a richly connected associative network, rather than as a multistage, fixed-capacity communication channel.

Another important aspect of the Decision Ladder is that the "higher" processes (e.g., interpreting and evaluating) are the most demanding from a computational point of view. In other words, the higher processes represent the potential processing bottleneck. Thus, the shortcuts across the Decision Ladder represent specific hypotheses about how experts can bypass the limitations typically associated with the fixed-capacity model. The general hypothesis is that due to their extensive experience, experts have developed many associations that provide them with opportunities to bypass the intensive resource-limited processes associated with the higher levels, as in naturalistic decision making. These associations allow experts to "recognize" solutions, rather than analytically interpreting the situation and evaluating options as would be required of novices, who haven't learned the shortcuts yet.

This idea that human cognition involves the use of "short-cuts" is not unique to the expertise literature or to Rasmussen. In fact, this intuition is quite pervasive. However, it may not be easy to appreciate, since the terms used to frame this intuition can be different depending on the research context. For example, terms such as "automatic processing" (Shiffrin and Schneider, 1977), "direct perception" (Gibson, 1979), and "heuristic decision-making" (Tversky and Kahneman, 1974) each, in their own way highlight how humans are able to take advantage of situated task constraints (e.g., consistent mappings, optical invariants, and availability) to bypass the need for resource demanding higher-level processing.

It is an irony of language that knowledge-based behaviors are typically demanded in situations where people have the least experience or skill. As people get more experience with a domain, they will naturally become more aware and be more able to utilize the time-space patterns and the consistencies that are available. The first time you try to fly, the aircraft and the skyways are full of ambiguities that you have to resolve through knowledge-based interactions. With practice (trial and error) you discover the patterns and consistencies and come to rely more and more on skill- and rule-based modes of control. To the point where now you can sometimes find yourself operating in routine situations with little conscious awareness—you managed the routines automatically—while higher knowledge-based levels of information processing were engaged by more challenging puzzles—like route planning. The irony of the language is that the more experience or knowledge you have about a particular situation—the less dependent you will be on what Rasmussen labeled as knowledge-based modes of interaction.

For the same reasons that humans are able to take advantage of consistencies to "automate" tasks in the lower (skill- and rule-based) levels of the Decision Ladder, engineers have been able to design automated control systems that are increasingly able to manage these components of the aviation work domain. Thus, increasingly the role of the human operators (e.g., the pilot or air traffic control operator) has shifted from that of a manual controller to that of a supervisory controller whose primary function is to make high-level or risky decisions (e.g., set goals) and to solve problems (e.g., fault diagnosis). An example is when a pilot must decide whether or not to abort or continue takeoff upon diagnosing an engine failure. As discussed in the previous section, this type of risky decision-making or problem-solving demands Rasmussen's knowledge-based level of processing.

Norman's (1981) concept of action slips reflects the dark side of automatic processing at the skill- and rule-based levels in the same way that the earlier discussions of decision biases reflects the limitations of heuristics for decision making. Reason's (1990) Generic Error Modeling System (GEMS) explicitly links the research on heuristics with Rasmussen's Decision Ladder. As Reason (1990) notes "errors are ... the inevitable and usually acceptable price human beings have to pay for their remarkable ability to cope with very difficult informational tasks quickly and, more often than not effectively" (p. 148). In Reason's "fallible machine" model, processes such as *similarity matching* and *frequency gambling* that typically facilitate information processing can lead to errors in unkind situations. In essence, these processes use contextual similarity and frequency respectively as means to bypass more comprehensive searches of stored knowledge. While these short cuts reduce search time and often lead to satisfactory results, they are fallible and in some situations (e.g., exceptional cases) they will lead to less than satisfactory results.

Reason's GEMS model associates errors with the qualitative levels of processing (skill-, rule-, and knowledge-based) illustrated in the Decision Ladder. At the skill-based levels, errors are generally associated with either inattention or over-attention. At the rule-based level, Reason distinguishes between misapplications of generally good rules and application of bad rules. An example of the misapplication of a good rule is "first exception" errors. That is, when a person encounters the first situation where a general rule does not apply, the "strong-but-wrong rule will be applied" (1990, p. 75). It is expected that with experience rules will be revised to accommodate the exceptions, as experts learn from experience. An example of a bad rule is an encoding deficiency where potentially significant properties of the problem space are not recognized or perceived. Errors associated with development such as egocentrism and conservation may reflect failures of children to encode significant dimensions of a situation. With experience, people typically broaden their awareness to include more of the relevant dimensions.

The concepts of "bounded rationality" and "fallible machine" can be seen as either a glass that is half-empty or a glass that is half-full. As has been discussed earlier in the chapter, the study of heuristics and biases have contributed much to our understanding of the causes of sometimes dangerous failures of human judgment. However, from the perspective of naturalistic decision making and expertise, the emphasis tends to be on the ability of experts to leverage task constraints in ways that increase processing efficiency and that often, but not always, lead to satisfactory results.

RESOURCES FOR MEETING THE INFORMATION PROCESSING CHALLENGES IN AVIATION

Attention

Attention, in its most general form, characterizes the limits of processing information about multiple tasks and their perceptual and cognitive elements. Such elements can be objects or events in the environment (e.g., two plane symbols on a traffic display), ideas, thoughts, or plans in cognition and working memory (instructions from ATC and implications of a weather forecast), or tasks and goals (e.g., communicating while loading a flight plan in an FMS). Across these different entities (environmental events, cognitive elements, and tasks), it is possible to focus either on constraints imposed by the limits of attention, or techniques for overcoming these limits. We consider these two foci, as applied, in turn, to the environment, and to tasks (Wickens and McCarley, 2008).

Limits of attention to the environment. Examples where attentional failures have prevented pilots from noticing critical events are plentiful; psychologists speak of *selective attention,* as the vehicle whereby the eyes (for vision) or the focus of auditory attention select parts of the environment for processing, while filtering others. Selection is driven both by internal goals and by external events, as the latter is defined by *attention capture.* For example analysis of pilot scanning reveals the heavy influence of internal task goals and information relevance in determining the allocation of attention (Wickens, Goh, Helleberg, Horrey, and Talleur, 2003; Wickens; McCarley, Alexander, Thomas, Ambinder, and Zheng, 2008). Such allocation is also driven heavily by expectancy, such that pilots look to areas where they expect to see dynamic changes (Senders, 1965, 1983). At the same time, as noted in our discussion of decision making above, salient events can capture attention, even if they may not be expected, or relevant for the task at hand. And in contrast to attention capture, a recent history of the study of *change blindness* indicated that nonsalient events that should be noticed because of their importance, are not always noticed (Stelzer and Wickens, 2006; Sarter, Mumaw, and Wickens, 2007). One of the most critical roles of aviation alerts and alarms is to capture attention, through bottom-up processing, and direct it to important sources of information, even when these may not be part of the pilot's ongoing goal structure (Pritchett, 2002).

Divided attention to elements in the environment. Selective attention, as noted above, is serial. But there are four ways in which these serial limits can be overcome to foster desirable parallel processing in the high tempo domain of the cockpit or ATC workstation.

- Display integration. The pilot does not need two successive fixations to extract pitch and roll information from the moving horizon. Instead, our attention system processes in parallel two attributes of a single visual object (here slant and vertical position). Several other techniques of display integration have been used to foster parallel processing (Haskell and Wickens, 1993; Wickens and Andre, 1990; Wickens and Carswell, 1995).
- Automaticity. As noted earlier, extensive and consistent practice and familiarity with a visual image can trigger nearly attention-free, or "automatic" processing. For example, the skilled pilot knows from a rapid glance, whether she is on the appropriate glide path for landing from the desired convergence angle or trapezoidal shape of the runway. The pilot also knows that she will be automatically alerted when hearing her call sign spoken over the radio, not needing to continuously devote attention to "listen for" it. Such automaticity is associated with the concept of very low workload, a construct discussed below.

- Peripheral and ambient vision. The human visual system is constructed so that peripheral vision can process ambient cues such as motion flow and roll attitude using ambient vision of the outside world, very much in parallel with the focal vision typically used for object detection and reading print (Previc and Ercoline, 2000; See Leibowitz chapter).
- Multiple resources. To some extent, pilots are capable of processing sounds (and tactile input) in parallel with visual input, as if the different sensory systems posses separate or "multiple" attentional resources, an issue we address later in this chapter; but one that speaks favorably for the value of voice displays, in the otherwise heavily visual cockpit environment (Wickens, Sandry and Vidulich, 1983).

Selective attention to tasks and goals. Often pilots' tasks must be completed (or initiated) in sequence, just as selective attention to inputs is necessarily sequential for widely separated displays. For example verifying engine health cannot be done in parallel with solving a navigational problem. Thus not only does the eye constrain sequential processing of relevant inputs, but the brain also cannot "engage" in both cognitive tasks at once. The understanding of the many constraints in sequential task behavior is embodied in the field of *cockpit task management* (CTM, Chou, Madhavan, and Funk, 1996; Funk, 1991; Raby and Wickens, 1994; Dismukes and Nowinski, 2007; Loukopoulos, Dismukes, and Barshi, 2009). What are the rules of when one task dominates, or interrupts another? Or when particular tasks may be "shed" at high workload and then, perhaps forgotten to be resumed? Unfortunately, such breakdowns of appropriate task management in the cockpit happen with some regularity, sometimes leading to major accidents or incidents when tasks that should be of higher priority get "dropped" by interrupting tasks, and perhaps never resumed. A frequent example of such a dropped task is that of altitude monitoring whose abandonment can lead to controlled flight into terrain accidents. Well designed cockpit checklists (Degani and Wiener, 1993) can often assure that important procedural tasks are not dropped. But they are no guarantee, particularly in nonroutine situations, and need be augmented with CTM training.

Successful divided attention between tasks and mental processes. In contrast to failures of CTM, there are many instances in which parallel processing is possible; most noticeably when flying (tracking) while communicating. Two features seem to make this possible. (1) some tasks, such as routine closed loop control are sufficiently easy (automated, skill-based) as to consume little attention—thereby availing plenty of attention resources for concurrent tasks; (2) sometimes two tasks use different resources, literally related to different processing areas in the brain (Wickens et al., 1983; Wickens, 2008a). We have already referred to the different resources of auditory and visual processing; so too the spatial processes involved in so much of flying depend upon different resources from the verbal linguistic ones in communications, and action uses different resources from perception and cognition.

Collectively then, limits of attention constrain pilots' sampling of the environment, and multi-tasking ability. But understanding of when and how these constraints can be reduced; by display design and task structure, can go a reasonably long way to improving the pilots multitask information processing capabilities.

Expertise

As discussed earlier, experience plays a critical role in the approach to decision making and problem solving. The extant view is that expertise is largely acquired through extensive *deliberate practice* (e.g., Adams and Ericsson, 1992; Ericsson, 1996). Distinct from casual practice or routine performance, deliberate practice is intensive and involves self-monitoring and error correction with an intent to improve. Further, expert-level performance has to be actively maintained through continual deliberate practice. As Yacovone, Borosky, Bason and Alcov (1992) put it, "Flying an airplane is not like riding a bike. If you haven't done it recently, you might not remember how" (p. 60).

What happens as skill develops? Anderson (1983) proposed three stages of skill acquisition that roughly correspond to the three phases outlined by Fitts and Posner (1967). First is the acquisition of declarative knowledge—knowledge of facts and things. Fitts and Posner recounted Alexander Williams's technique to reduce the training time for novice pilots to be able to fly their first solo flight: "Williams conducted detailed discussions of each maneuver to be practiced, of the exact sequence of responses to be made, and of the exact perceptual cues to be observed at each step" (p. 11). The second stage is the acquisition of procedural knowledge—knowledge of how to perform various tasks. Production rules that specify specific actions for specific conditions are acquired. For example, during landing, if the airspeed is too high, then raise the flaps. Gross errors are gradually eliminated during this stage. In the third stage, the production rules become more refined and concatenated into larger rules that can produce action sequences rapidly. Upon recognizing the need to make an airspeed correction during landing, an experienced pilot now would have more elaborate production rules that involve a number of corrective actions such as raising the flaps and reducing thrust by an amount appropriate for the atmospheric conditions and landing phase.

Brain imaging data are now revealing both quantitative and qualitative changes in cortical activity as skill develops. Hill and Schneider (2006) noted that the most commonly observed brain change pattern for tasks that have an unvarying relationship between stimulus and response is reduced control network activation, while the level of perceptual motor activity is maintained. The control network comprises a set of cortical areas controlling goal processing, attention, and decision making. Gopher (1996) summarizes the skill acquisition process to be "marked by a gradual weaning from dependency on more general, slower, and less efficient controlled processes (i.e., rule-based processes), in favor of task specific "automatically accessed" representations and speeded operation procedures" (p. 24; i.e., skill-based processes). But few complex performances can be developed to the point of complete automaticity because a necessary condition for automaticity to develop is a consistent stimulus-response mapping (Schneider and Shiffrin, 1977) that is often precluded in the dynamic aviation environment.

That expert performance is supported by a vast amount of domain-specific knowledge is consistent with common brain change patterns after extensive training and an increase in the cortical tissue devoted to the task (Hill and Schneider, 2006). For example, Maguire et al. (2000) found that taxi drivers had larger posterior hippocampi than control subjects and the hippocampal volume correlated with the amount of time spent as a taxi driver, suggesting that the posterior hippocampus could expand to accommodate elaborate representation of navigational skills.

Importantly, expert's organization of their knowledge is fundamentally different from that of novice's. Expert problem representations are organized around principles that may not be apparent on the surface whereas novice representations are organized around literal objects and explicit events (Glaser, 1987). For example, Schvaneveldt et al. (1985) found expert Air Force instructors and National Guard pilots to have a better organized knowledge structure of air-to-air and air-to-ground fighter concepts than the trainees. In another study, Stankovic, Raufaste, and Averty (2008) proposed a model that used three horizontal distances as basis for aircraft conflict judgments. The model accounted for about 45% of the judgment variance of certified ATCs but only 20% of that of the trainees, suggesting that the two groups did not use the information in the same way.

Experts also have highly developed strategies that could circumvent certain normal processing limits. Ericsson and Kintsch (1995) proposed that a defining feature of advanced skill is the development of a long-term working memory (LTWM) that enables experts to relate a large amount of information to their highly-structured knowledge in long-term memory and to devise retrieval cues for eventual access to the information. In a simulated air traffic control task, Di Nocera, Fabrizi, Terenzi, and Ferlazzo (2006) found that senior military controllers recalled more flight strips than junior controllers when automation support was off.

The many aforementioned "intuitive" mental shortcuts used by experts can very well be accounted for by the large body of highly organized knowledge that experts posses, which enables them to readily see meaningful patterns, make inferences from partial information, continuously update their perception of the current situation, and to anticipate future conditions (Cellier, Eyrolle, and

Mariné, 1997; Durso and Dattel, 2006; Vidulich, 2003). For example, when making weather-related decisions, experienced pilots were found to be more efficient in acquiring weather-related data, better at integrating conflicting information, faster at recognizing deteriorating weather conditions, and more guided by developed strategies than less experienced pilots (Wiegmann, Goh, and O'Hare, 2002; Wiggins and O'Hare, 1995).

In addition, experts have metacognitive capabilities of knowing what one knows and does not know that allow them to plan and to apportion their resources more effectively (Glaser, 1987). In an ATC study, Seamster, Redding, Cannon, Ryder, and Purcell (1993) observed that developmental controllers tended to first solve problems of the highest priority that involved violation of the minimum separation standards and then problems of the next highest priority that involved deviation from standard operating procedures. In contrast, experienced controllers alternated between solving the violation and deviation problems. In another study, Amalberti, Valot, and Grau (1989) examined fighter pilots' ground mission preparation that entailed defining the flight path and choosing the procedures that they would apply in response to potential incidents. They found the expert and novice pilots adopted greatly different flight paths and plans for the same mission. (See also Bellenkes, Wickens, and Kramer, 1997; Tsang, 2007.)

Recent brain imaging data reveal that extensive practice can lead to a reorganization of active brain areas (Hill and Schneider, 2006). In a simulator study, Pérès et al. (2000) had expert and novice French Air Force pilots fly at two speeds. Dominant neuronal activation was observed in the right hemisphere as would be expected for a visual spatial task. Novices exhibited more intense and more extensive activation than experts. More interesting, for the high-speed (high workload) condition, the experts exhibited increased activation in the frontal and prefrontal cortical areas and reduced activity in the visual and motor regions, suggesting their focusing resources on the higher-level functions such as planning and attentional management. In contrast, the novices appeared to be engaged in nonspecific perceptual processing. Hill and Schneider (2006) proposed that different strategies are used as skills develop and new areas become active to perform the underlying processing. This advanced level of skill is distinct from the highly automated skills with consistent tasks in that no functional reorganization is evident with the latter.

In fact, Gopher (1996) pointed out that there are important aspects of high-level performance that do not develop from repetitious consistent stimulus-response mapping, but benefit from variations introduced during training. Gopher (2007) documented a number of training successes with the introduction of multiple changes in task emphasis. For example, Gopher, Weil, and Bareket (1994) found a 10-hour training of emphasis change on a video game, Space Fortress, increased the operational flight performance of the Israel Air Force flight cadets by 30%. Seagull and Gopher (1997) found that, following training with an additional secondary task, pilot's HMD flight performance equaled their flight performance under normal viewing condition and was superior to the HMD flight performance of pilots without the secondary task training. (Gopher, 2007) proposed that these training procedures force the trainees to explore response strategies and to develop the executive control skill to best utilize available resources.

The nature of expertise has several important implications. The crucial role of deliberate practice in the development of expertise implies that duration of job tenure (such as life-time accumulation of total flight hours) by itself will not be an accurate index of expertise. Indeed, it is found to have only a weak relationship with pilot performance and a nonsignificant relationship with accident rate (see Tsang, 2003). Rather, the continual development of expertise is subject to the availability of training opportunities and to one's effort to improve. Second, although expert performance is supported by automatic processing of the more elemental task performance, it would be a gross oversimplification to equate expert-level performance to automaticity. The expertise advantage goes beyond the recognition of familiar patterns and robotically quick retrieval of stored response productions. In fact, expertise is especially needed in novel, unanticipated, dynamic conditions where experts could apply their highly structured knowledge to strategically arrest the current situation and to reduce future workload.

For this level of performance to emerge, Gopher (2007) showed that a training strategy very different from that for attaining automaticity is needed.

EMERGENT PROCESSES—MENTAL WORKLOAD AND SITUATION AWARENESS

Given the impossibility of enumerating, much less assessing, all of the myriad information processing components involved in performing any complex task, it is not surprising that human engineering practitioners have adopted concepts and procedures that attempt to integrate the overall quality of the information processing undertaken by the human operator (Tsang and Vidulich, 2006; Vidulich, 2003). The two most common integrated measures are mental workload and situation awareness (Wickens, 2002). Mental workload is the mental "cost" of performing the information processing required by task performance. Inasmuch as discussions of mental workload are typically associated with notions of information processing that challenges, perhaps even exceeds, the available information processing capabilities of the operator, the most common cognitive construct to explain or understand mental workload is attention (Gopher and Donchin, 1986). Parasuraman, Sheridan, and Wickens describe mental workload as "the relation between the function relating the mental resources demanded by a task and those resources supplied by the human operator" (2008, pp. 146–147). In other words, attention provides the "power" to support information processing, in a way analogous to an engine providing horsepower to move an automobile or to electricity providing the power to a sound amplifier of a home entertainment system. In each case, there are levels of demand that are comfortable, higher levels of demand that are achievable but perhaps cause strain, and even higher levels of demand that cause breakdown.

Situation awareness, on the other hand, has been more strongly associated with the cognitive construct of memory (Endsley, 1995a, 1995b; Tsang and Vidulich, 2006; Vidulich, 2003). Situation awareness involves the interplay of information from perception interacting in working memory with knowledge and expertise from long-term memory to create a "picture" of the current situation in the world and to support the necessary decisions to achieve whatever goal is currently being pursued. More formally, situation awareness is defined as, "the perception of the elements in the environment within a volume of time and space, the comprehension of their meaning, and the projection of their status in the near future" (Endsley, 1995b, p. 36).

Both mental workload and situation awareness are attempts to produce expansive metrics that capture the overall quality and accuracy of the cognitive processing performed by the human operator dealing with the complex and dynamic aviation environment. In 2008 the *Human Factors* journal celebrated its 50th anniversary with a special issue with articles surveying the state of the art solicited from leading human factors researchers. Within this select set of articles both mental workload (Parasuraman and Wilson, 2008; Warm, Parasuraman, and Matthews, 2008; Wickens, 2008a) and situation awareness (Durso and Sethumadhaven, 2008; Wickens, 2008b) were prominent topics of multiple papers. Further, Parasuraman et al.'s (2008) review of the state of research in mental workload and situation awareness research concluded that although these constructs are theoretically distinct from performance, they have been strongly operationalized and contributed to the assessment and understanding of human performance and system effectiveness.

Consistent with Parasuraman et al.'s (2008) viewpoint, researchers investigating the impact of modern technology on the future of commercial and military aviation have found mental workload and situation awareness metrics to provide valuable insights. For example, Ruigrok and Hoekstra (2007) investigated the potential effects of transitioning from the relatively stringent traditional flight control rules to more liberalized rules. Under the more liberalized rules, pilots would have considerably greater freedom to adjust their own aircraft's flight path and air speed in accordance to their own interpretation of the air traffic situation. The goal of implementing the more liberalized rules is to exploit the greater flexibility to enable increased airspace capacity and thereby accommodate an anticipated increased demand for air travel. Ruigrok and Hoekstra (2007) reviewed the results of six human-in-the-loop experiments that examined the impact of various rule implementations on achievable traffic density and pilot workload. The results supported the implementation of more liberalized rules as a means for safely increasing air traffic density without inflicting excessive pilot workload. At the same time that this reallocation of trajectory choice to pilots does not appear to increase their workload, it has been found to decrease the situation awareness of a second critical class of airspace workers, the air traffic controllers.

Galster, Duley, Masalonis, and Parasuraman (2001) examined the mental workload and performance of air traffic controllers in a simulation that required the controllers to monitor self-separating aircraft. They found that the monitoring demands of higher levels of air traffic inflicted excessive mental workload and reduced situation awareness, the latter reflected by poorer aircraft conflict detection and recognition of self-separation events performed by the aircraft. Galster et al. (2001) concluded that additional automated decision support would be needed to ensure safe operation under more liberalized air traffic control rules. Similarly, Fothergill and Neal (2008) demonstrated that the level of workload strongly influenced the quality of conflict resolution strategies used by the air traffic controllers. Under conditions of low to moderate mental workload, the air traffic controllers provided pilots with instructions that resolved the conflict and were efficient for the aircraft. Under high levels of workload, the controllers provided more expedient strategies that resolved the conflict, but were less efficient.

These results underscore the fact that a complete system analysis will often require an examination of both mental workload and situation awareness. This is because system modifications that might have been intended to reduce mental workload could have unintended consequences on situation awareness or modifications intended to improve situation awareness might inadvertently inflict unacceptable levels of workload. Endsley (1993) highlighted the interrelated nature of mental workload and situation awareness and illustrated that any combination of mental work-load level and situation awareness level was possible. For instance, Wickens (2008b) detailed the importance of automation-induced interactions between mental workload and situation awareness as one of the fundamental challenges for new air traffic control sys-tems. As automation is employed to mitigate mental workload overloads, the controller's attention will be less actively engaged with the task and any memory for actions initiated by the auto-mation may be degraded. Thus, mitigating mental workload can provide a serious risk of failing to adequately support the control-ler's situation awareness. As the application of automation in such settings is expected to dramatically increase, research to support a fuller understanding of the variables that drive mental workload trade-offs with situation awareness and a fuller understanding of the mechanisms by which the variables influence these trade-offs is sorely needed.

In military settings, considerable interest appears to be develop-ing in using situation awareness as a guiding concept in the devel-opment of command and control systems. Vidulich, Bolia, and Nelson (2004) reviewed the history of air battle management sys-tems since World War I and demonstrated that improvements in the capability and effectiveness of such systems were generally associated with increasing the overall situation awareness present within the system.

Arguing that existing situation awareness measures might not be ideal for the assessment of command and control environments, Salmon, Stanton, Walker, and Green (2006) advocate the creation of new approaches with demonstrated validity and reliability for assessing team or shared situation awareness. To address some of the challenges in conceptualizing and assessing situation aware-ness in complex, networked command and control systems, Stewart et al. (2008) developed the concept of distributed situa-tion awareness as a system-oriented, rather than individual-level approach. By linking the assessment of situation awareness to the task structure created by a task analysis and representing the meaning of task elements based on information collected from interviews of subject-matter experts, Stewart et al. (2008) plan to produce a situation awareness conceptualization that will drive the design and assessment of command and control network structure and communication connectivity. Meanwhile, Salmon et al. (2006) advised combining several situation awareness measurement approaches in the evaluation of a command and control systems.

Mental workload and situation awareness have played central roles in the development and assessment of cockpit displays and controls in the past, and it appears that these concepts will continue to evolve and contribute to future developments in aviation systems.

FUTURE DIRECTIONS

Currently aviation has entered a revolutionary period. This revolution has been supported by two developments. First, future traffic and congestion problems are forecast to grow exponentially, particularly in certain regions of the world such as the Pacific Rim. In other regions, such as the east coast of the United States and over much of Europe, congestion appears to be so high that further growth, while demanded, is terribly constrained. Second, satellite navigation has enabled aircraft separation from other aircraft and from weather to be accomplished without dependence on air traffic control—the imprecision of its ground based radar, and often the serial nature of vectoring aircraft. Thus the first of these has created a demand; the second has suggested the possible solution of decentralized control of flight trajectories.

Together, these two developments have spawned the program of NEXTGEN (Joint Planning and Development Office, 2007) or, in Europe, PHARE (i.e., Program for Harmonized ATC Research in Europe). These programs encompass a large set of near-term, mid-term and far-term technological and procedural innovations, all of which have profound implications for pilots and controllers' information processing. On the one hand, more decentralized control of flight trajectories around weather and other traffic, imposes a considerably greater demand on the pilots attention and information processing: added displays may need to be consulted, and new decisions made (how do I maneuver around this traffic to avoid a potential conflict; Thomas and Wickens, 2006; 2008). Hence, pilot workload can be severely increased. On the other hand, it is assumed that pilots cannot accept these added duties without the support of considerable levels of automation; but here, too, as discussed above and in the chapter by Sarter et al.; added layers of automation can sometimes also increase workload, and may certainly decrease situation awareness.

While at first glance, the increase in pilot responsibility may seem to diminish workload and attention concerns for the controller, more careful consideration reveals that this is not the case. First, if the goal of such a responsibility shift is to put more planes in the air, and have them fly more locally desirable routes; the controller will now be faced with a more chaotic picture containing more aircraft. Such a picture may well contain a growing number of unmanned aerial vehicles (UAVs) as well. Furthermore, every scenario for the future airspace envisions the controller continuing to maintain *ultimate responsibility* for traffic separation and flow management, even as controllers may have reduced *authority* for trajectory management. As a consequence, the need for increased controller situation awareness is amplified; just as such awareness may be degraded, as other agents (pilots and automation) are making more decisions regarding flight trajectories (Galster et al. 2001). Thus the new displays, procedures, and automation tools invoked by NEXTGEN and PHARE, place continued demands on human factors personnel to understand pilot and controller information processing, and consider the changes in such processing engendered by these revolutions in the airspace.

References

Adams, M., and Reed, D. (2008). Rising costs reshaping air travel across the USA Retrieved August 23, 2008, from http://www.usatoday.com/travel/flights/2008–04–30-jet-fuel-high-fares_N.htm. *USA Today.com: Today in the Sky*.
Adams, R. J., and Ericsson, K. A. (1992). *Introduction to cognitive processes of expert pilots* (DOT/FAA/RD-92/12). Washington, DC: U.S. Department of Transportation, Federal Aviation Administration.

Amalberti, R., and Deblon, F. (1992). Cognitive modeling of fighter aircraft process control: A step towards an intelligent on board assistance system. *International Journal of Man-Machine Studies, 36*, 639–671.

Amalberti, R., Valot, C., and Grau, J-Y. (1989). Metaknowledge in process control 4–6 December. In L. Bainbridge and S. Reinantz (Eds.), *Cognitive process in complex tasks: Proceedings of the Workshop at Wilgersdorf.* Germany: TUV Rheinland.

Amelink, M. H. J., Mulder, M., van Passen, M. M., and Flach, J. M. (2005). Theoretical foundations for a total energy-based perspective flight path display. *The International Journal of Aviation Psychology, 15*, 205–231.

Anderson, J. R. (1983). *The architecture of cognition.* Cambridge, MA: Harvard University Press.

Arkes, H. R., and Blumer, C. (1985). The psychology of sunk cost. *Organizational Behavior and Human Performance, 35*, 129–140.

Baird, W. (2006 18). FedEx hub is controlled holiday chaos. Retrieved August 23, 2008, from http://www.ibtimes.com/articles/20061218/fedex-hub-is-controlled-holiday-chaos.htm. *International Business Times.*

Banbury, S., Dudfield, H., Hoermann, H. J., and Soll, H. (2007). FASA: Development and validation of a novel measure to assess the effectiveness of commercial airline pilot situation awareness training. *The International Journal of Aviation Psychology, 17*, 131–152.

Batty, D. (2009, January 18). Pilot tells of crash-landing as plane pulled from river. *Guardian.co.uk.* Retrieved March 5, 2009, from http://www.guardian.co.uk/world/2009/jan/18/new-york-plane-crash-salvage.

Bellenkes, A. H., Wickens, C. D., and Kramer, A. F. (1997). Visual scanning and pilot expertise: The role of attentional flexibility and mental model development. *Aviation, Space and Environmental Medicine, 68*, 569–579.

Broadbent, D. (1958). *Perception and communication.* New York: Oxford University Press.

Broadbent, D. E. (1959). Auditory perception of temporal order. *Journal of the Acoustical Society of America, 31*, 1539.

Buck, R. N. (1994). *The pilot's burden: Flight safety and the roots of pilot error.* Ames: Iowa State University Press.

Cellier, J.–M., Eyrolle, H., and Mariné, C. (1997). Expertise in dynamic environments. *Ergonomics, 40*, 28–50.

Chou, C. D., Madhavan, D., and Funk, K. (1996). Studies of cockpit task management errors. *The International Journal of Aviation Psychology, 6*, 307–320.

Cushing, S. (1994). *Fatal words: Communication clashes and aircraft crashes.* Chicago: University of Chicago Press.

Degani, A., and Wiener, E. L. (1993). Cockpit checklists: Concepts, design and use. *Human Factors, 35*, 345–360.

Di Nocera, F., Fabrizi, R., Terenzi, M., and Ferlazzo, F. (2006). Procedural errors in air traffic control: Effects of traffic density, expertise, and automation. *Aviation, Space, and Environmental Medicine, 77*, 639–643.

Dismukes, K., and Nowinski, J. (2007). Prospective memory, concurrent task management, and pilot error. In A. F. Kramer, D. A. Wiegmann, and A. Kirlik (Eds.), *Attention: From theory to practice* (pp. 225–236). New York: Oxford University Press.

Durso, F. T., and Dattel, A. R. (2006). Expertise and transportation. In K. A. Ericsson, N. Charness, P. J. Feltovich, and R. R. Hoffman (Eds.), *The Cambridge handbook of expertise and expert performance* (pp. 355–371). New York: Cambridge University Press.

Durso, F. T., and Sethumadhavan, A. (2008). Situation awareness: Understanding dynamic environments. *Human Factors, 50,* 442–448.

Einhorn, H. J., and Hogarth, R. M. (1978). Confidence in judgment: Persistence of the illusion of validity. *Psychological Review, 85,* 395–416.

Einhorn, H. J., and Hogarth, R. M. (1982). *Theory of diagnostic interference 1: Imagination and the psychophysics of evidence* (Technical Report no. 2). Chicago: University of Chicago School of Business.

Ellis, S. R., McGreevy, M. W., and Hitchcock, R. J. (1987). Perspective traffic display format and air pilot traffic avoidance. *Human Factors, 29,* 371–382.

Endsley, M. R. (1993). Situation awareness and mental workload: Flip sides of the same coin. In: *Proceedings of the seventh international symposium on aviation psychology.* Columbus: Ohio State University (pp. 906–911).

Endsley, M. R. (1995a). Measurement of situation awareness in dynamic systems. *Human Factors, 37,* 65–84.

Endsley, M. R. (1995b). Toward a theory of situation awareness in dynamic systems. *Human Factors, 37,* 32–64.

Ericsson, K. A. (1996). The acquisition of expert performance: An introduction to some of the issues. In K. A. Ericsson (Ed.), *The road to excellence* (pp. 1–50). Mahwah, NJ: Erlbaum.

Ericsson, K. A., and Kintsch, W. (1995). Long-term working memory. *Psychological Review, 105,* 211–245.

Fitts, P. M., and Posner, M. I. (1967). *Human performance.* Belmont, CA: Brooks/Cole.

Flach, J., and Warren, R. (1995). Low altitude flight. In P. Hancock, J. Flach, J. Caird, and K. Vicente (Eds.), *Local applications of the ecological approach to human-machine systems* (pp. 65–103). Hillsdale, NJ: Erlbaum.

Fothergill, S., and Neal, A. (2008). The effect of workload on conflict decision making strategies in air traffic control. In: *Proceedings of the human factors and ergonomics society 52nd annual meeting.* Santa Monica, CA: Human Factors and Ergonomics Society (pp. 39–43)

Funk, K. (1991). Cockpit task management: Preliminary definitions, normative theory, error taxonomy, and design recommendations. *The International Journal of Aviation Psychology, 1,* 271–285.

Galster, S. M., Duley, J. A., Masalonis, A. J., and Parasuraman, A. J. (2001). Air traffic controller performance and workload under mature Free Flight: Conflict detection and resolution of aircraft self-separation. *International Journal of Aviation Psychology, 11,* 71–93.

Gibson, J. J. (1958/1982). Visually controlled locomotion and visual orientation in animalsOriginally published in *British Journal of Psychology,* 49, 182–194. In E. Reed and R. Jones (Eds.), *Reasons for realism* (pp. 148–163). Hillsdale, NJ: Erlbaum.

Gibson, J. J. (1966). *The senses considered as perceptual systems.* Boston: Houghton Mifflin.

Gibson, J. J. (1979). *The ecological approach to human perception.* Boston: Houghton Mifflin.

Gibson, J. J., Olum, P., and Rosenblatt, F. (1955). Parallax and perspective during aircraft landings. *American Journal of Psychology, 68,* 372–385.

Gigerenzer, G., and Todd, P. M.and the ABC Research Group. (1999). *Simple heuristics that make us smart.* New York: Oxford University Press.

Gilbert, G. A. (1973). Historical development of the air traffic control system. *IEEE Transactions on Communications, COM-21*(5), 364–375.

Gillingham, K., and Previc, F. (1996). Spatial orientation in flight. In R. DeHart (Ed.), *Fundamentals of aerospace medicine* (2nd ed.) (pp. 309–397). Baltimore, MD: Williams and Wilkins.

Glaser, R. (1987). Thoughts on expertise. In C. Schooler and K. W. Schaie (Eds.), *Cognitive functioning and social structure over the life course* (pp. 81–94). Norwood, NJ: Ablex.

Gopher, D. (1996). Attention control: Explorations of the work of an executive controller. *Cognitive Brain Research, 5,* 23–38.

Gopher, D. (2007). Emphasis change as a training protocol for high-demand tasks. In A. F. Kramer, A. Kirlik, and D. Weigmann (Eds.), *Applied attention: From theory to practice* (pp. 209–224). New York: Oxford University Press.

Gopher, D., and Donchin, E. (1986). Workload: An examination of the concept(pp. 41–49). In K. R. Boff, L. Kaufman, and J. P. Thomas (Eds.), *Handbook of perception and human performance—Volume II: Cognitive processes and performance.* New York: Wiley.

Gopher, D., Weil, M., and Baraket, T. (1994). Transfer of skill from a computer game trainer to flight. *Human Factors, 36,* 1–9.

Green, D. M., and Swets, J. A. (1966). *Signal detection theory and psychophysics.* New York: Wiley.

Harrington, C. (2008 24). USAF faces further push to boost UAVs in Afghanistan, Iraq Retrieved August 23, 2008, from http://www.janes.com/news/defence/air/jdw/jdw080424_1_n.shtml. *Jane's Intelligence and Insight You Can Trust.*

Haskell, I. D., and Wickens, C. D. (1993). Two- and three-dimensional displays for aviation: A theoretical and empirical comparison. *The International Journal of Aviation Psychology, 3,* 87–109.

Hecht, H., and Savelsburgh, G. J. P. (2004). *Time to contact.* Amsterdam: Elsevier.

Hick, W. E. (1952). On the rate of gain of information. *Quarterly Journal of Experimental Psychology, 4,* 11–26.

Hill, N. M., and Schneider, W. (2006). Brain changes in the development of expertise: Neuroanatomical and neurophysiological evidence about skill-based adaptations. In K. A. Enicsson, N. Charness, P. J. Feltovich, and R. R. Hoffman (Eds.), *The Cambridge handbook of expertise and expert performance* (pp. 653–682). New York: Cambridge University Press.

Hoffman, R. R., Crandall, B. W., and Shadbolt, N. R. (1998). Use of the Critical Decision Method to elicit expert knowledge: A case study in cognitive task analysis methodology. *Human Factors, 40,* 254–276.

Hogarth, A. (1987). *Judgment and choice* (2nd ed.). Chichester: Wiley.

Hunt, R., and Rouse, W. (1981). Problem-solving skills of maintenance trainees in diagnosing faults in simulated power plants. *Human Factors, 23,* 317–328.

Hutchins, E. (1995). *Cognition in the wild.* Cambridge, MA: MIT Press.

Hyman, R. (1953). Stimulus information as a determinant of reaction time. *Journal of Experimental Psychology, 45,* 423–432.

Inagake, T. (1999). Situation adaptive autonomy: trading control of authority in human-machine systems. In M. Scerbo and M. Mouloua (Eds.), *Automation Technology and Human Performance.* Mahwah, NJ: Laurence Erlbaum.

Jensen, R. S. (1981). Prediction and quickening in perspective flight displays for curved landing approaches. *Human Factors, 23,* 355–363.

Johnson, E. M., Cavenaugh, R. C., Spooner, R. L., and Samet, M. G. (1973). Utilization of reliability estimates in Bayesian inference. *IEEE Transactions on Reliability, 22,* 176–183.

Joint Planning and Development Office. (2007, June 13). Concept of operations for the next generation air transportation office: Version 2. Retrieved August 23, 2008, from http://www.faa.gov/about/office_org/headquarters_offices/ato/publications/nextgenplan/resources/view/NextGen_v2.0.pdf.

Kahneman, D. (2003). A perspective on judgment and choice. *American Psychologist, 56,* 697–720.

Kahneman, D., Slovic, P., and Tversky, A. (Eds.), (1982). *Judgment under uncertainty: Heuristics and biases.* New York: Cambridge University Press.

Kahneman, D., and Tversky, A. (1984). Choices, values, and frames. *American Psychologist, 39,* 341–350.

Klein, G. (1989). Recognition primed decision making. *Advances in Man-Machine Systems Research, 5,* 47–92.

Klein, G., and Crandall, B. W. (1995). The role of mental simulation in problem solving and decision making. In P. Hancock, J. Flach, J. Caird, and K. Vicente, (Eds.)*Local applications of the ecological approach to human-machine systems: Vol. 2* (pp. 324–358). Hillsdale, NJ: Erlbaum.

Langewiesche, W. (1944). *Stick and Rudder.* New York: McGraw-Hill.

Layton, C., Smith, P. J., and McCoy, C. E. (1994). Design of a cooperative problem-solving system for en-route flight planning: An empirical evaluation. *Human Factors, 36,* 94–119.

Lehto, M. R., and Nah, F. (2006). Decision making models and decision support. In G. Salvendy (Ed.), *Handbook of human factors and ergonomics* (3rd ed.) (pp. 191–242). New York: Wiley.

Loukopoulos, L., Dismukes, K., and Barshi, E. (2009). *The multitasking myth: Handling complexity in real-world operations.* Averbury, VT: Ashgate.

Mackworth, N. H. (1948). The breakdown of vigilance during prolonged visual search. *Quarterly Journal of Experimental Psychology, 1,* 5–61.

McKinney, E. (1999). Flight leads and crisis decision making. *Aviation Space and Environmental Medicine, 64,* 359–362.

Maguire, E. A., Gadian, D. G., Johnsrude, I. S., Good, C. D., Ashburner, J., Frackowiak, R. S. J., and Firth, C. D. (2000). Navigation-related structural change in the hippocampi of taxi drivers. *Proceedings of the National Academy of Science United States of America, 97,* 4398–4403.

Miller, G. A. (1956). The magical number seven plus or minus two: Some limits on our capacity for processing information. *Psychological Review, 63,* 81–97.

Miller, G. A., Galanter, E., and Pribram, K. H. (1960). *Plans and the structure of behavior.* New York: Henry Holt and Company.

Moray, N. (1978). The strategic control of information processing. In G. Underwood (Ed.), *Strategies or information processing* (pp. 301–327). New York: Academic.

Mulder, M. (2003a). An information-centered analysis of the tunnel in the sky display, part one: Straight tunnel trajectories. *International Journal of Aviation Psychology, 13,* 49–72.

Mulder, M. (2003b). An information-centered analysis of the tunnel in the sky display, part two: Curved tunnel trajectories. *International Journal of Aviation Psychology, 13,* 131–151.

Müller, T., and Giesa, H.-G. (2002). Effects of airborne data link communication on demands, workload and situation awareness. *Cognition, Technology, and Work, 4,* 211–228.

Neisser, U. (1967). *Cognitive psychology.* New York: Appleton-Century-Crofts.

Nickerson, R. S. (1998). Confirmation bias: A ubiquitous phenomenon in many guises. *Review of General Psychology, 2,* 175–220.

Norman, D. A. (1981). Categorization of action slips. *Psychological Review, 88,* 1–15.

Ocker, W. C., and Crane, C. J. (1932). *Blind flight in theory and in practice.* San Antonio, TX: Naylor.

O'Hare, D. (2003). Aeronautical decision making: Metaphors, models, and methods. In P. S. Tsang and M. A. Vidulich (Eds.), *Principles and practice of aviation psychology* (pp. 201–237). Mahwah, NJ: Erlbaum.

Orasanu, J. (1993). Decision making in the cockpit. In E. L. Wiener, B. G. Kanki, and R. L. Helmreich (Eds.), *Cockpit resource management* (pp. 137–172). San Diego, CA: Academic Press.

Orasanu, J., and Fisher, U. (1997). Finding decisions in natural environments: The view from the cockpit. In C. E. Zsambok and G. Klein (Eds.), *Naturalistic decision making* (pp. 343–358). Mahwah, NJ: Erlbaum.

Owen, D. H., and Warren, R. (1987). Perception and control of self-motion: Implications for visual simulation of vehicular locomotion. In L. S. Mark, J. S. Warm, and R. L. Huston (Eds.), *Ergonomics and human factors; Recent research.* New York: Springer-Verlag.

Parasuraman, R., Sheridan, T. B., and Wickens, C. D. (2008). Situation awareness, mental workload, and trust in automation: Viable, empirically supported cognitive engineering constructs. *Journal of Cognitive Engineering and Decision Making, 2,* 140–160.

Parasuraman, R., and Wilson, G. F. (2008). Putting the brain to work: Neuroergonomics past, present, and future. *Human Factors, 50,* 468–474.

Payne, J. W., Bettman, J. R., and Johnson, E. J. (1993). *The adaptive decision maker.* Cambridge, UK: Cambridge University Press.

Pérès, M., Van De Moortele, P. F., Pierard, C., Lehericy, S., Satabin, P., Le Bihan, D., and Guezennec, C. Y. (2000). Functional magnetic resonance imaging of mental strategy in a simulated aviation performance task. *Aviation, Space, and Environmental Medicine, 71,* 1218–1231.

Perrin, B., Barnett, B. J., Walrath, L., and Grossman, L. (2001). Information order and outcome framing: An assessment of judgment bias in a naturalistic decision making context. *Human Factors, 43,* 227–238.

Previc, F. H., and Ercoline, W. R. (Eds.), (2004). *Spatial disorientation in flight.* Reston, VA: American Institute of Aeronautics and Astronautics.

Pritchett, A. R. (2002). Human-computer interaction in aerospace. In J. A. Jacko and A. Sears (Eds.), *The human-computer interaction handbook: Fundamentals, evolving technologies, and emerging applications* (pp. 861–882). Hillsdale, NJ: Erlbaum.

Raby, M., and Wickens, C. D. (1994). Strategic workload management and decision biases in aviation. *The International Journal of Aviation Psychology, 4,* 211–240.

Rasmussen, J. (1986). *Information processing and human machine interaction.* New York: North Holland.

Reason, J. (1990). *Human error.* New York: Cambridge University Press.

Ruigrok, R. C. J., and Hoekstra, J. M. (2007). Human factors evaluations of Free Flight issues solved and issues remaining. *Applied Ergonomics, 38,* 437–455.

Salmon, P., Stanton, N., Walker, G., and Green, D. (2006). Situation awareness measurement: A review of applicability for C4i environments. *Applied Ergonomics, 37,* 225–238.

Sarter, N., Mumaw, R., and Wickens, C. D. (2007). Pilots' monitoring strategies and performance on highly automated flight decks: an empirical study combining behavioral and eye tracking data. *Human Factors, 49,* 347–357.

Schneider, W., and Shiffrin, R. M. (1977). Contolled and automatic human information processing: I. Detection, search, and attention. *Psychological Review, 84,* 1–66.

Schriver, A. T., Morrow, D. G., Wickens, C. D., and Talleur, D. A. (2008). Expertise differences in attentional strategies related to pilot decision making. *Human Factors, 50.*

Schvaneveldt, R. W., Durso, F. T., Goldsmith, T. E., Breen, T. J., Cooke, N. M., Tucker, R. G., and De Maio, J. C. (1985). Measuring the structure of expertise. *The International Journal of Man-Machine Studies, 23,* 699–728.

Seagull, F. J., and Gopher, D. (1997). Training had movement in visual scanning: An embedded approach to the development of piloting skills with helmet mounted displays. *Journal of Experimental Psychology: Applied, 3,* 463–480.

Seamster, T. L., Redding, R. E., Cannon, J. R., Ryder, J. M., and Purcell, J. A. (1993). Cognitive task analysis of expertise in air traffic control. *International Journal of Aviation Psychology, 3,* 237–283.

Senders, J. W. (1965). Sampling behavior of human monitors. *IEEE Spectrum, 2*(3), 58.

Senders, J. W. (1983). Models of visual scanning processes. *Bulletin of the Psychonomic Society, 21,* 340.

Shiffrin, R. M., and Schneider, W. (1977). Controlled and automatic human information processing: II. Perceptual learning, automatic attending and a general theory. *Psychological Review, 84,* 127–190.

Stankovic, S., Raufaste, E., and Averty, P. (2008). Determinants of conflict detection: A model of risk judgments in air traffic control. *Human Factors, 50,* 121–134.

Stelzer, E. M., and Wickens, C. D. (2006). Pilots strategically compensate for display enlargements in surveillance and flight control tasks. *Human Factors, 48,* 166–181.

Stewart, R., Stanton, N., Harris, D., Baber, C., Salmon, P., Mock, M., Tatlock, K., Wells, L., and Kay, A. (2008). Distributed situation awareness in an airborne warning and control system: Application of novel ergonomics methodology. *Cognition, Technology, and Work, 10,* 221–229.

Suchman, L. A. (1987). *Plans and situated actions.* Cambridge: Cambridge University Press.

Tarm, M. (2008 August 27). One small hitch for FAA, one giant mess for fliers Retrieved August 30, 2008, from http://news.yahoo.com/s/ap/20080827/ap_on_bi_ge/faa_communication_breakdown. *Yahoo! News.*

Thomas, L. C., and Wickens, C. D. (2006). Display dimensionality, conflict geometry, and time pressure effects on conflict detection and resolution performance using cockpit displays of traffic information. *The International Journal of Aviation Psychology, 16,* 315–336.

Thomas, L. C., and Wickens, C. D. (2008). Display dimensionality and conflict geometry effects on maneuver preferences for resolving in-flight conflicts. *Human Factors, 50,* 576–588.

Tsang, P. S. (2003). Assessing cognitive aging in piloting. In P. S. Tsang and M. A. Vidulich (Eds.), *Principles and practice of aviation psychology* (pp. 507–546). Mahwah, NJ: Erlbaum.

Tsang, P. S. (2007). The dynamics of attention and aging in aviation. In A. F. Kramer, A. Kirlik, and D. Weigmann (Eds.), *Applied attention: From theory to practice* (pp. 170–184). New York: Oxford University Press.

Tsang, P. S., and Vidulich, M. A. (2006). Mental workload and situation awareness. In G. Salvendy (Ed.), *Handbook of human factors and ergonomics* (3rd ed.) (pp. 243–268). New York: Wiley.

Tversky, A. (1972). Elimination by aspects: A theory of choice. *Psychological Review, 79,* 281–289.

Tversky, A., and Kahneman, D. (1974). Judgment under uncertainty: Heuristics and biases. *Science, 185,* 1124–1131.

Vidulich, M. A. (2003). Mental Workload and Situation Awareness: Essential Concepts for Aviation Psychology Practice. In P. S. Tsang and M. A. Vidulich (Eds.), *Principles and practice of aviation psychology* (pp. 115–146). Mahwah, NJ: Erlbaum.

Vidulich, M. A., Bolia, R. S., and Nelson, W. T. (2004). Technology, organization, and collaborative situation awareness in air battle management: Historical and theoretical perspectives. In S. Banbury and S. Tremblay (Eds.), *A cognitive approach to situation awareness: Theory and application* (pp. 233–253). Aldershot, UK: Ashgate.

Warm, J. S., Parasuraman, R., and Matthews, G. (2008). Vigilance requires hard mental work and is stressful. *Human Factors, 50,* 433–441.

Warren, R., and Wertheim, A. H. (1990). *Perception and control of self-motion.* Mahwah, NJ: Erlbaum.

Wickens, C. D. (2002). Situation awareness and workload in aviation. *Current Directions in Psychological Science, 11,* 128–133.

Wickens, C. D. (2003). Aviation displays. In Tsang P S and Vidulich M. A (Eds.), *Principles and practice of aviation psychology* (pp. 147–200). Mahwah, NJ: Erlbaum.

Wickens, C. D. (2008a). Multiple resources and mental workload. *Human Factors, 50,* 449–455.

Wickens, C. D. (2008b). Situation awareness: Review of Mica Endsley's 1995 articles on situation awareness theory and measurement. *Human Factors, 50,* 397–403.

Wickens, C. D., and Andre, A. D. (1990). Proximity compatibility and information display: Effects of color, space, and objectness on information integration. *Human Factors, 32,* 61–77.

Wickens, C. D., and Carswell, C. M. (1995). The proximity compatibility principle: Its psychological foundations and its relevance to display design. *Human Factors, 37*, 473–494.

Wickens, C. D., Goh, J., Helleberg, J., Horrey, W. J., and Talleur, D. A. (2003). Attentional models of multitask pilot performance using advanced display technology. *Human Factors, 45*, 360–380.

Wickens, C. D., and Hollands, J. G. (2000). *Engineering psychology and human performance* (3rd ed.). Upper Saddle River, NJ: Prentice Hall.

Wickens, C. D., Liang, C. C., Prevett, T. T., and Olmos, O. (1996). Egocentric and exocentric displays for terminal area navigation. *International Journal of Aviation Psychology, 6*, 241–271.

Wickens, C. D., and McCarley, J. (2008). *Applied attention theory*. Boca Raton, FL: Taylor and Francis.

Wickens, C. D., McCarley, J. S., Alexander, A. L., Thomas, L. C., Ambinder, M., and Zheng, S. (2008). Attention-situation awareness (A-SA) model of pilot error. In D. Foyle and B. Hooey (Eds.), *Human performance modeling in aviation* (pp. 213–239). Boca Raton, FL: Taylor and Francis.

Wickens, C. D., Sandry, D., and Vidulich, M. (1983). Compatibility and resource competition between modalities of input, output and central processing. *Human Factors, 25*, 227–248.

Wiegmann, D. A., Goh, J., and O'Hare, D. (2002). The role of situation assessment and flight experience in pilots' decisions to continue visual flight rules flight into adverse weather. *Human Factors, 44*, 187–197.

Wiegmann, D. A., and Shappell, S. A. (2003). *A human error approach to aviation accident analysis: The human factor analysis and classification system*. Aldershot, UK: Ashgate.

Wiggins, M., and O'Hare, D. (1995). Expertise in aeronautical weather-related decision making: A cross-sectional analysis of general aviation pilots. *Journal of Experimental Psychology: Applied, 1*, 305–320.

Yacovone, D. W., Borosky, M. S., Bason, R., and Alcov, R. A. (1992). Flight experience and the likelihood of U.S. Navy mishaps. *Aviation, Space, and Environmental Medicine, 63*, 72–74.

Zsambok, C. E., and Klein, G. (1997). *Naturalistic decision making*. Mahwah, NJ: Erlbaum.

8

Managing Workload, Performance, and Situation Awareness in Aviation Systems

Francis T. Durso and Amy L. Alexander

Georgia Institute of Technology and Aptima, Inc.

On October 20, 2009, Delta 3401 was in cruise under Captain Kate Lewis with First Officer Kent Bleckley on her right. They were most of their way through a redeye flight from Salt Lake to Atlanta with 231 souls on board. Lewis' workload was low in this phase of flight, even lower with the autopilot. The First Officer had stepped out of the cockpit, letting the captain lose herself in the beauty of a new dawn. The flight was approaching its next waypoint. Lewis expected a 20° turn to the left at this waypoint, and just as expected the big six-seven bent toward the rising sun. But, by the time the First Officer returned Lewis had noticed the groundspeed had dropped by 20 knots, a surprise since her preview of the flight had not suggested there would be any headwinds. That surprise led to a search for answers. Just as air traffic control (ATC) was calling to enquire, there on the flight management system she saw that the wrong waypoint had brought the plane into a 50°, not 20°, turn and into the headwinds. Even though she had not been busy, her situation awareness had slipped long enough to allow a course deviation. Fortunately, she had caught it and so had ATC.

On the ground, Atlanta approach controller Rick Henry was waiting for Delta 3401 as well as a host of others due for the morning push. So far, his workload was low and he was able to stay ahead of things, even granting more than his usual number of pilot requests. Although he did not have much experience working this particular airspace, he knew that soon he would have to shed some of these optional tasks and focus on the push if he wanted to keep his workload manageable and his performance high.

The push came hard and fast that morning, and Henry knew his workload was near its peak; he thought about asking for some help. At that same time, our Delta crew was busy as well. Sterile cockpit procedures were in effect, but even with the assurance they would not be interrupted and would focus on the task at hand, an approach and landing at the nation's busiest airport is never trivial. Henry was focusing on an American flight that seemed to be having some serious trouble when 3401 made contact. He cleared Delta for approach into Atlanta.

"Roger approach, cleared for 8 Left," Bleckley read back the clearance and set the localizer.

Captain Lewis was busy, but noticed the runway was not what she usually used when she landed at Hartsfield. "Check that, Kent. They usually land us on 8 Right."

After a quick call to Henry who had successfully helped the young pilot on the American flight along, Bleckley reported to his captain, "You were right Captain, he meant 8 Right."

Our fictional flight captures many of the intriguing issues surrounding mental workload (WL), situation awareness (SA), and performance. A loss of understanding can occur when WL is low or WL is high. WL and SA are not the same construct, but they are clearly related. How should we view their relationship? Why do WL, SA, and performance disassociate, often making it difficult to predict one from the others?

The purpose of the current chapter is to review recent work in aviation systems by providing a framework from which one can view the interplay among WL, SA, and performance in complex systems. We describe a situation understanding subsystem, a workload subsystem, and a critical management subsystem. We then discuss what might be expected when considering WL, SA, and performance in the selection of displays, ending with a discussion of the role of strategies.

A MANAGEMENT FRAMEWORK

The research on WL (e.g., Kantowitz, 1986) and SA (e.g., Endsley, 1995b) has increased to the point that the interested reader now has a number of excellent treatises to consider (e.g., Durso and Gronlund, 1999; Durso, Rawson, and Girotto, 2007; Stanton, Salmon, Walker, Baber, and Jenkins, 2005; Tenney and Pew, 2006), including reviews that consider both (e.g., Tsang and Vidulich, 2006; Wickens, 2001). A myriad of measures for SA and WL, respectively, have been developed. Although the constructs are still challenged occasionally from some perspectives (e.g., Dekker and Hollnagel, 2004; Dekker and Woods, 2002), they are generally accepted and can be supported as valuable scientific constructs (Parasuraman, Sheridan, and Wickens, 2008).

Mental WL has been defined as "the relation between the function relating the mental resources demanded by a task and those resources supplied by the human operator" (Parasuraman, Sheridan, and Wickens, 2008). WL is an important construct in aviation. The potential for error caused by nonoptimal levels of WL and the subsequent consequences of these errors make the value of understanding optimal and non-optimal WL clear. It is important to recognize that "nonoptimal" can be either excessively high or excessively low levels of WL, as both are expected to contribute to pilot or controller error (e.g., Durso and Manning, 2008; Kantowitz and Casper, 1988). Given the inherent safety implications of workers who are overworked or understimulated, it is no wonder that many studies in previous years have focused on either measuring or manipulating WL (or both). In fact, barring performance itself, in the ATC literature, WL has been the construct most often to receive empirical scrutiny (Rantanen, 2004).

As for SA, the term seems to have originated in aviation (e.g., Endsley, 1989; Harwood, Barnett, and Wickens, 1988; Spick, 1988) and erupted on the research landscape after seminal papers by Endsley (1995a; 1995b). As is the case with WL, the potential for error caused by a poor level of SA and the consequences of those errors make the value of understanding SA clear. In aviation it is often called "having the picture." Losing the picture can have dramatic and devastating consequences from both the cockpit (e.g., controlled flight into terrain, CFIT, accidents, Woodhouse and Woodhouse, 1995) and the ATC station (more frequent and more severe operational errors, Durso et al., 1998). For these reasons, and because SA is likely to become more important as the cognitive nature of the tasks we ask operators to perform increases, SA has earned a central role in our consideration of controlling a modern, complex, dynamic system (e.g., Durso and Sethumadhavan, 2008; Wickens, 2008b)

We integrate WL and SA here by relying on a notion that can be traced to several others in the aviation psychology literature (e.g., Funk, 1991; Hart and Wickens, 1990; Helmreich, Wilhelm, Klinect, and Merritt, 2001; Lee et al., 2007; Loft et al., 2007; Sperandio, 1971; Tsang and Vidulich, 2006), simply that the pilot and controller are managing the situation. Management of the situation entails both a management of resources and a management of performance. We argue that at some level, the same cognitive mechanisms involved in management are in play whether they are managing performance or resources. The difference is the target of the management, the feedback loop to the management processes, and the strategies available to maintain the appropriate level of resources and the appropriate level of performance. This is different than a view that treats WL management and situation awareness as unrelated concepts. In our view, the same cognitive processes are involved, simply operating on different aspects of the situation.

There is reason to believe that understanding the situation is not merely understanding elements from the task environment. Niessen, Eyferth, and Bierwagen (1999) had controllers sort 30 radar screenshots and corresponding flight strips each depicting a traffic situations. From the resulting multidimensional scaling (MDS) solution, the authors made an intriguing suggestion that the amount of supervisory control and the amount of anticipatory control captured the primary dimensions controllers used to classify traffic situations. This suggests that classification is based on control strategies, not on physical characteristics of the traffic. Based on interviews with controllers, Seamster, Redding, Cannon, Ryder, and Purcell (1993), suggested that WL management strategies are a key feature in ATC. Durso and Dattel (2006) relate the argument to expertise. If WL management is a part of what is involved in understanding a situation, then the constructs of SA and WL are related in intricate, but understandable, ways. To the operator, understanding and managing the current WL and resources defines a situation just as do the buffeting crosswinds or the number of blips on the radar screen.

Figure 8.1 presents a schematic of the relationship among WL, SA, and performance. Operators are assumed to be managing a variety of factors in order to achieve a variety of goals. They attempt to keep WL within manageable levels while keeping performance high, despite threats to those goals from environmental changes such as increases in complexity and from increases in errors the operators themselves might make. In our view, understanding of both performance and WL is the front end of a management system that responds to what is understood by adjusting strategies to accomplish the variety of goals.

Situation Understanding Subsystem

Situation understanding entails situation assessment and situation awareness. Situation assessment refers to the processes in which operators engage to make sense of the situation, to understand it. Through situation assessment some limited part of the situation may be represented in consciousness. This conscious product of situation assessment is studied under the rubric situation awareness or SA. Both the assessment process, even though often operating beneath awareness, and the awareness of the product of understanding are

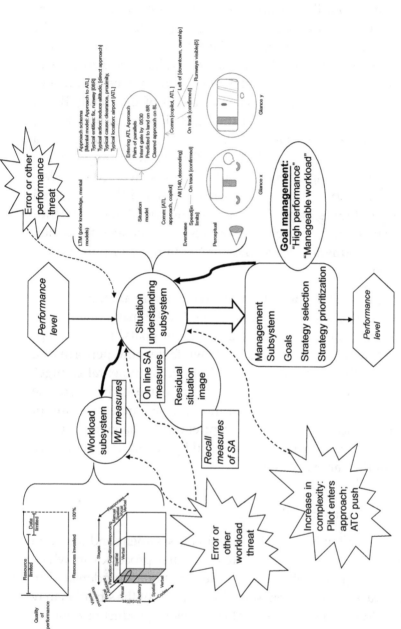

FIGURE 8-1 A schematic of the relationship between workload and situation understanding. The workload system comes from Wickens and colleagues. The situation understanding subsystem from Durso et al. (2007). Situation understanding includes both situation assessment and a memorial product of that assessment studied as situation awareness. Workload as well as performance and task events are processed by the situation understanding subsystem. Ultimately, depending on goals and the evaluation of current workload and performance levels, the pilot or controller selects and prioritizes strategies that keep workload manageable and performance high.

important if the construct is to help us understand how operators manage their WL and performance. At a macroscopic level, situation understanding seems clearly to involve an appreciation of what is currently happening—that plane is descending, that bogey is closing—and what is likely to happen in the immediate future—those planes are going to be in conflict, that maneuver will put me on his six. We contend that it also includes what is happening to the operator—I am getting confused, I am working too hard.

Researchers have proposed a variety of measures of SA (see Table 8.1). Subjective measures, such as the Situation Awareness Rating Technique (SART; Taylor, 1990) ask operators to judge their own level of SA. Such measures are best thought of as measures of meta-SA, an operator's awareness of their own awareness. Query methods are the most popular, and typically involve asking the operator to answer a series of questions much as one would ask a student to answer comprehension questions about a story they had just read. The most popular measure, the Situation Awareness Global Assessment Technique (SAGAT; Endsley, 1995), asks the

TABLE 8-1   SA Measures

Types of SA Measures	Description	Examples
Subjective measures	Meta-SA. Ratings of SA along a number of scales provided by operators or observers of the operators	SART, SALIANT, SA-BARS, SASHA-Q
Explicit queries	Operators are asked directly about some aspect of the situation. Recall methods remove the situation before querying the operator. Online methods keep the situation available. The various methods differ in how they query the participant.	SAGAT, SPAM, SASHA-L, SAVANT, SALSA, SAPS
Implicit performance	An event whose detection would be indicative of good SA is embedded naturally or by the experimenter into the scenario.	Detection of course deviation, location of bogey, awareness of low fuel level, conflict detection

questions after the situation has been removed (or blanked); recall accuracy is the measure of SA.

The Situation Present Assessment Method (SPAM; Durso, Hackworth, Truitt, Crutchfield, Nikolic, and Manning, 1998); (Durso and Dattel, 2004) is also a query method, but keeps the situation present while questions are asked. The time it takes to answer the question, as well as accuracy, are used to measure SA. Of interest to the current chapter, SPAM also allows the experimenter to measure workload. For example, Durso et al. (1998) used the SPAM procedure with en route controllers. Controllers routinely answer landlines as part of their job, but typically would only answer them when the primary task of separating and directing traffic allowed. One of the landlines in Durso et al. was the line on which controllers received the SPAM query. The time taken to answer the landline was used as a measure of workload and the time and accuracy to answer the question was used as a measure of SA.

Implicit performance measures involve placing an event into a scenario that is clearly indicative of SA. For example, most would agree that detecting a flight that deviates from a filed flight plan is a measure of SA. It is important to realize that implicit performance measures are *not* simply any measure of performance. The evident connection to understanding the situation must be present. Researchers often disagree on the extent to which a performance measure reflects only SA.

There are also efforts to use physiological measures of SA, such as eye movements. Although it is difficult to conclude from current measures that the operator understands the situation, rather than merely attending to it, advances in biotechnology hold promise that such measures may be on the horizon. For example, the burgeoning area of augmented cognition (e.g., Schmorrow and Kruse, 2004) attempts to determine the cognitive state of operators by measuring physiological indices.

Finally, a number of researchers have begun to refer to SA while measuring behaviors of the operators or aspects of the task rather than using a direct measure that reflects knowledge or understanding of the situation.

To us, however, "understanding" the situation in which pilots or controllers find themselves (Durso, Rawson, and Girotto, 2007) evokes much that is cognitive, and this is intentional. Perhaps other factors (such as personality) could affect the extent to which an operator is aware of what is happening, SA does indeed seem to have a large cognitive component. For example, Caretta, Perry, and Ree (1996) showed that the ability to predict SA of fighter pilots depended entirely on cognitive variables.

Although it has cognitive underpinnings, SA is more than a sum of numerous cognitive and personality tests. Durso, Bleckley, and Dattel (2006) tried to predict performance on the Air Traffic Scenarios Test (ATST), the low-fidelity simulation of ATC used in the FAA's AT-SAT selection battery. The question of interest regarding SA was the extent to which measures of SA could add to the ability to predict performance above and beyond traditional cognitive and personality variables. Durso et al. found that considering accuracy and latency of answers to SPAM queries could increase the ability to predict performance quite noticeably. For example, ATC errors could be predicted from personality (i.e., conscientiousness), spatial working memory, and the ability to find simple patterns in complex ones. When SPAM predictors were added, variance-accounted-for increased by 15%.

This result is interesting for a couple of reasons. On the one hand, this shows that SA when added to other variables can improve overall performance prediction, above and beyond traditional cognitive tests. On the other hand, it points to the fact that we do not know exactly what constitutes SA. If we did, then presumably a composite of simpler tests would have rendered the addition of SA redundant. Thus, this second sense is an invitation to researchers to develop a simple test that might reduce the value added by SA measures or, failing that, to acknowledge (situation) understanding as an atomic construct not devisable into constituent primitives. In fact, based on studies using WOMBAT SA scores (Roscoe, Corl, and Larouche, 2001), O'Hare and colleagues (O'Brien and O'Hare, 2007; O'Hare, 1997) have suggested that the "ability to coordinate and integrate . . . cognitive activities under temporal constraints and changing priorities" (O'Brien and O'Hare, 2007,

p. 1066) is that which distinguishes SA scores from the presumed underlying cognitive abilities.

The situation understanding subsystem as depicted in the Figure 8.1 insert has both a bottom-up and a top-down component (see Durso, et al., 2007). The operator is presumed to create an "event-base," which is a network of events integrated across glances. This event-base is not influenced by domain-specific knowledge and thus is a relatively veridical representation of the passing events. The construction of the event-base is limited by working memory capacity and so not all components of the event-base network survive from moment to moment. The top-down influences on this event base come from the operator's knowledge base, which is thought to comprise a variety of structures, including the mental model—an abstract "theory" of what the situation is and how it will unfold. The abstract mental model together with the actual environment produces a situation model. Often, a particular default value—the waypoint in our earlier flight—comes from the mental model, and not the environment, and thus the operator comes to expect the wrong future—the 20° turn, for example, from our fictional flight. More often, default values of the relatively abstract mental model are accurate. If the operator attends to the situation, slots in the mental model will have their default values replaced by values from from the actual situation, thus creating an accurate situation model based on both top-down (expectations) and bottom-up (reality) inputs.

There is evidence that developing a situation model depends on underlying cognitive mechanisms, some dependent on knowledge of a domain and some not. For example, Sohn and Doane (2004) measured domain-independent spatial working memory and knowledge-dependent long-term working memory in apprentice and expert pilots. Scores on an aviation-neutral rotation span task yielded the measure of working memory. Comparing delayed recall of pairs of cockpit displays that were either meaningfully or nonmeaningfully related yielded the aviation-relevant long term working memory scores. Finally, SA was measured by having participants indicate if a goal would be reached given a cockpit configuration. The findings suggested that as reliance on domain-specific long term working memory increases, reliance on general working memory decreases. Thus, expertise and the long-term working memory that is a concomitant part of it, frees SA from the bounds of a restricted working memory. Expertise seems to compensate for natural human limits, thus allowing SA to exceed what would otherwise be possible.

In our schematic, we acknowledge that as time passes the operator stores or remembers the various states of the situation and situation model in an episodic memory we refer to as the residual situation image. This residual situation image not only supplies the memorial record, but it serves as the context in which the next cycle of situation model and event-base takes place. The residual situation image is likely the underlying mechanism that accounts for findings that the relevance of aircraft parameters change over time (Hauss and Eyferth, 2003). This record of recent history thus supplies a richness to the understanding of the situation that would otherwise be difficult to capture. In our theory, the controller's or pilot's "picture" is this residual situation image.

Finally, another important part of SA is the ability to project the current situation into the future. This allows the operator to run mental simulations and to anticipate, expect, and plan. Such projections are not perfect: Boudes and Cellier (2000) found that controllers will tend to place aircraft ahead of the actual future positions. In addition, how much of one's SA is about the present and how much is about the projected future may depend on the individual (Durso et al., 1997). Durso et al. (1997) found that controllers who did well when answering SAGAT queries about the present performed more poorly in the sense that they had more control actions remaining to be performed at the end of the scenario, whereas those who answered more questions about the future had fewer remaining actions. Thus, future-oriented SA and not just any composite of SA predicted those controllers who had the least left to do.

SA is often used to select among different displays, devices, or procedures and attempts have been made to offer guidance to designers interested in SA (Endsley, Bolte, and Jones, 2003). Consider the following examples.

Weather was the focus of interest when comparing the Next Generation Radar (NEXRAD) system and an onboard system utilizing real-time sensors (Bustamante, Fallon, Bliss, Bailey, and Anderson, 2005). Whereas the NEXRAD system provides a broad view of the weather environment, the information is only updated at specific time intervals (e.g., every 5 minutes). Onboard sensors, by contrast, provide real-time information but are limited in scope due to sensor-range constraints. Pilot SA ratings decreased as they approached the weather event if the NEXRAD system information was not available. In other words, receiving real-time, detailed information from on-board sensors when approaching a weather event did not support understanding "the big picture" required for good SA as did NEXRAD.

A series of experiments conducted by Jenkins and Gallimore (2008) assessed the impact of three different configural helmet-mounted displays on pilot SA of aircraft attitude. The experiments ranged in fidelity from utilizing static screenshots to a dynamic simulation without outside world screens to a dynamic simulation with the outside world present. In the first experiment, which utilized static screenshots of configural displays, pilots were required to respond to queries regarding aircraft attitude. This explicit SA measurement technique proved sensitive to the salience of emergent features in the configural displays, the range of system information available relevant to task goals, and the ability to distinguish system states from one another.

Lorenz and colleagues (2006) compared the effects of an advanced airport surface display on pilot performance and SA to traditional navigation support under varied visibility and traffic conditions. A custom-designed subjective questionnaire was utilized to measure SA, revealing that the advanced display supported better SA compared to conventional navigation means. This increase in the operator's awareness of their own SA, their meta-SA, was accompanied by improvements in performance with the advanced display, particularly under low visibility conditions. While degrading visibility led to lower SA ratings, this effect was reduced when the advanced display was available.

Workload Subsystem

Interest in WL stems from the belief that WL is pivotal to the human operators' ability to manage complex systems effectively. Indeed, on the flightdeck, technology *together with* human factors design reduced WL to the point that commercial aircraft that routinely required three persons, now require two; fighter planes that routinely required two persons, now require one. As with flightdeck research, measures of WL in ATC have had dramatic impacts in the field, to the point of affecting real world staffing decisions (Vogt, Hagemann, and Kastner, 2006). When proponents of change in the National Airspace System (NAS) argue the system is too human-centric, at least part of what they are trying to express is that traffic throughput is thought to be limited by controller WL (Hilburn, 2004).

Understanding how resources are devoted to tasks is critical in the aviation domain given the implications of degraded pilot or controller performance. Pilots and controllers operate in extremely complex sociotechnical systems that by their very nature require a substantial amount of multitasking. The more activities pilots and controllers are required to conduct, the larger the drain on mental resources. Performance will break down to the extent that the drain on mental resources exceeds operator capacity.

Given the importance of WL and the number of studies devoted to understanding it, it is no surprise that there are a number of measures of it, from the physiological to the subjective (e.g., Stanton, Salmon, Walker, Baber, and Jenkins, 2005; Young and Stanton, 2001). Table 8.2 provides a summary of commonly used primary task performance, secondary task performance, physiological, and subjective measures of WL. Furthermore, WL measures vary in terms of whether they are designed to assess either residual capacity or overload within the context of the performance-resource function described later.

Although the subjective self-report of the controller or pilot remains a primary object of investigation, there is an increasing

TABLE 8.2 WL Measures

Types of WL Measures	Description	Examples
Primary task performance	Measures of performance on the main task of interest	Speed, accuracy, control activity
Secondary task performance	Measures of performance on additional tasks introduced for the purpose of measuring residual attention or capacity	Response time, monitoring count accuracy, time estimation, mental arithmetic, memory search, tracking, embedded
Physiological		
Circulatory system	Measures of covert response	Heart rate, heart rate variability, incremental heart rate, blood pressure
Respiratory system	Measures of cardiovascular activity Measures of breathing-related activity	Respiration rate
Visual system	Measures of ocular or eye-related activity	Pupil diameter, eye blink rate, eye blink duration, direction of gaze
Central nervous system	Measures of brain-related activity	Electroencephlograph (EEG), event related potentials (ERPs)
Endocrine system	Measures of hormone-related activity	Salivary cortisol
Subjective ratings		
Uni-dimensional/global	Ratings provided by operators or observers Raters provide a single workload rating value	Pilot Objective/Subjective Workload Assessment Technique (POSWAT)
Hierarchical	Raters make a series of decisions discriminating between alternatives	Cooper-Harper Handling Qualities Rating (HQR) Scale, Modified Cooper-Harper (MCH), Bedford Workload Scale
Multidimensional	Raters provide workload rating values on multiple subscales that are combined to provide a composite workload rating that accounts for rater biases	Subjective Workload Assessment Technique (SWAT), NASA-Task Load Index (TLX)

trend toward the use of physiological measures of WL. There are a number of physiological measures that are viable candidates to measure WL from cardiovascular ones, such as heart rate and heart rate variability, to ocular ones, such as eye blinks and pupil diameter. Heart rate variability and event-related brain potentials (ERPs) results in particular have been found to be most sensitive to the demands placed on pilots, supporting the use of physiological measures as indicators of WL (Sirevaag, Kramer, Wickens, Reisweber, Strayer, and Grenell, 1993; cf. Kaber, Perry, Segall, and Sheik-Nainar, 2007). Undoubtedly, improvements in technology are a catalyst for the trend toward physiological measures.

Physiological and subjective measures of WL often, but not always, correlate. Consider research on phases of flight. Several flightdeck studies in recent years have focused on characterizing WL via physiological and subjective measures across phases of flight. For example, Lee and Liu (2003) found high correlations between incremental heart rate and NASA-TLX ratings associated with varying WL demands across phases of flight. Other studies, however, have shown that the phases of flight (e.g., takeoff and landing) associated with the highest levels of WL according to physiological data are not always coupled with the highest subjective WL ratings (Wilson, 2002).

Efforts to combine various WL measures have led some researchers to untie the intellectual knots with modeling efforts. Kaber et al. (2007) asked undergraduates to control traffic with a low fidelity ATC simulator under different levels of automation. Physiological and subjective measures of WL as well as performance measures were taken. A genetic algorithm network model including secondary task performance and a simple heart rate measure was the best predictor of WL. Relatedly, Wilson and Russell (2003) used a backpropagation neural-network model to classify WL as "acceptable" or "overload" from standard physiological inputs such as EEG, heart rate, eye blink, and respiration.

A considerable body of WL research in aviation involves attempting to predict WL from some taskload variables. Generally speaking, WL is driven by such variables as task difficulty, number of concurrent tasks, temporal demand associated with tasks, and operator training (Wickens, 2008a). The taskload variables can be either preexisting, as when ATC research compares a "busy sector" with a less busy one, or it can be manipulated, as when flightdeck researchers increase turbulence or increase number of tasks. In flightdeck research, common manipulations of what might be called complexity or taskload include increasing the number of tasks (Morris and Leung, 2006), time pressure (Thomas and Wickens, 2006), level of turbulence and number of traffic entities (Alexander and Wickens, 2005), and number of conflict aircraft (Alexander, Wickens, and Merwin, 2005) associated with flight-related scenarios. In ATC research, airspace structure, traffic, and cognitive-behavioral surrogates, like number of keystrokes made by a controller, are used to predict WL (see Durso and Manning, 2008, for a review).

Is it possible to predict a measure of workload, say an operator's judgment of workload, from variables that can be gathered from the environment? Since the 1960s (Davis, Danaher, and Fischl, 1963), researchers have been trying to answer this question by developing a constellation of predictor variables ("complexity" variables) capable of indicating WL. Managing complexity is thought to be crucial to maintaining safety in ATC (Kirwan, Scaife, and Kennedy, 2001). Historically in ATC research, investigators have distinguished a number of types of complexity. Airspace complexity refers to factors such as sector shape, standard flows, critical points, small angle convergence routes, crossing points, and so on (e.g., Christien, Benkouar, Chaboud, and Loubieres, 2002; Histon, Hansman, Gottlieb, Kleinwaks, Yenson, Delahaye, and Puechmorel, 2002; Remington, Johnston, Ruthruff, Gold, and Romera, 2000). Traffic complexity refers to factors such as aircraft count, traffic density, nature of conflicts, speed, heading, and so on. The best single predictor of WL has proven to be a simple count of the number of aircraft for which the controller is responsible. More intricate investigations into aspects of traffic have been revealing. Kopardekar and Magyarits (2003) created a "dynamic density" metric of variables that was able to account for almost 40% of the variance in WL ratings, which surpassed the benchmark of aircraft count by 15%.Complexity that depends on the operator is often referred to as cognitive complexity but may include behavioral and communication measures as well. Using both measures of traffic complexity and cognitive complexity, as much as 77% of the variance in ratings of WL has been accounted (Manning, Mills, Fox, Pfleiderer, and Mogilka, 2002).

Such efforts to predict judgments of WL from a constellation of variables have have shown some success. How successful must they be? From a scientific perspective, the efforts have accounted for significant amounts of variance and given insights into the underlying causal structure. However, as Loft et al. (2007) note, the amounts of variance, though significant and enviable by most standards (e.g., 77%), are not near the amount that would be necessary to develop an effective aid.

The WL subsystem depicted in Figure 8.1 is modeled after work by Wickens on multiple resource theory (Wickens, 1980, 1984, 2008). The operator is presumed to have separate pools of resources defined by stages of processing, codes of processing, and modalities. As the operator goes about completing tasks, different resources are utilized to support the perceptual and cognitive components of tasks as compared to the resources required to support responding either manually or vocally—this is represented by the *stages of processing* dimension. Different resources are also utilized to support the *codes of processing* dimension such that spatial task-related activities use different resources than verbal activities across all stages of processing (i.e., perception, cognition, and responding). Within the perception stage of processing, auditory and visual *modalities* also use different resources. Finally, a fourth dimension indicates that focal vision and ambient vision utilize different *visual processing* resources.

Operator WL will decrease to the extent that task demand decreases on any single dimension, requiring less and less of that single pool of resources, or to the extent that task demands spread across multiple dimensions, utilizing nominal amounts of those separate pools of resources. The manner in which these task demands relate to performance can be represented by a performance-resource function, as shown in the inset of Figure 8.1. This hypothetical performance-resource function relates the quality of performance to the amount of resources invested, noting two distinct regions representing resource-limited and data-limited components. The resource-limited region indicates that task demand is less than the capacity of resources available, providing residual capacity for additional tasks. The data-limited region indicates that maximum-level performance has been reached and allocating additional resources to the task cannot improve performance further.

The performance-resource function is of great interest in our model because it can be used to highlight circumstances in which resources demanded exceed resources available, leading to breakdowns in performance. These are some of the times that operators must manage the situation in order to maintain appropriate levels of performance.

Management Subsystem

The notion that operators manage error, threats, resources, complexity, WL and so on, assumes that operators have access to a quiver of strategies that they can launch at appropriate times. Pilots, for example, adjust their task management strategies under high WL conditions with the most general, efficient strategy being that of adhering to an aviate-navigate-communicate-systems management (ANCS) hierarchy (Schutte and Trujillo, 1996). Controllers report that they behave differently when busy (D'Arcy and Della Rocco, 2001; Rantanen, Yeakel, and Steelman, 2006), become more conservative, take actions faster and earlier, plan and communicate more, prioritize differently, and so on. They also shed the less relevant tasks (e.g., Lee et al., 2007) and reduce the extent to which they treat flights as individual, instead relying on a more uniform strategy for all flights (Sperandio, 1971).

Sperandio (1971) noted the importance of strategies almost four decades ago. Modern versions of Sperandio's insights about strategies are apparent in work by Loft et al. (2007), Lee et al. (2007), and

Averty et al. (2004). The controller or pilot monitors and evaluates the work to be done and changes strategies, behavioral or cognitive. Strategy selection is influenced by both transient and long-term priorities, which in turn are influenced by metacognitive (subjective, often conscious) assessments of needed-time, own-capacity, anticipated demands, and so forth. If strategy choices are deemed successful they may even become codified, such as the sterile cockpit procedures our pilots entered during the approach to Atlanta.

Cockpit task management (Chou, Madhaven, and Funk, 1996; Dismukes, 2001; Funk, 1991) or strategic WL management (Hart and Wickens, 1990) are often referred to when discussing the specific strategies pilots use within the flightdeck. In this context, pilots must choose which tasks to perform and which to defer when multiple tasks compete for attention and cannot be conducted in parallel, a situation that often occurs under high WL or when off-nominal (i.e., unexpected) events occur. Strategy selection is influenced by optimal scheduling of tasks as determined by task importance, and has been codified as the ANCS hierarchy (Schutte and Trujillo, 1996). Aviating corresponds to flying the aircraft at the correct speed, bank angle, heading and altitude; navigating relates to keeping the aircraft on route along the appropriate waypoints to the correct destination; communicating captures verbal interactions with ATC; and systems management pertains to the operation of aircraft systems. It is generally agreed that prioritizing tasks in this manner is appropriate, and pilots often shed the lower-priority tasks (e.g., communicating and managing systems) when under extreme duress. Poor cockpit task management occurs when more important tasks (e.g., aviate) are allowed to be preempted or superseded by those of lesser importance (e.g., systems management; Wickens, 2002).

Typically, whether a pilot or controller changes strategy depends on the discrepancies between the current situation and the goal situation. Situation here refers both to the ongoing task and events that the operator controls and to the ongoing resources available and being allocated. Thus, the states here can refer to either state of the task, in which case adjustments will be made to improve performance, or to states of WL in which case adjustments will be made to improve resources expended.

In either case, the operator's understanding of the situation—either performance or WL—is critical to gaining an appreciation of how an operator manages either the task or the mental resources needed to perform the task. Understanding resources and their allocation allows WL to be regulated. Understanding the task allows the planes and the conflicts they may produce to be managed. Together, WL regulation and performance management—both driven by situation understanding—lead to efficient and successful threat and error management.

Strategy changes can be performance strategies, such as switching to speed changes rather than altitude changes, to resolve conflicts; or they can be WL strategies, such as shedding pilot requests for the next 10 minutes. The operator monitors levels of WL and performance, changing strategies to keep both within acceptable parameters, that is, to protect WL or to protect performance. These changes and their subsequent impact are recorded in the residual situation image which the operator can also use later to assess the effectiveness of recent strategy choices.

The operator can reflect on these cognitive events and can use them in different ways. The operator might assess the level of his or her own understanding, giving a meta-SA judgment (Durso et al., 2007). Such judgments can be captured by subjective rating techniques and are valuable for several reasons, although we want to be clear that such subjective judgments are measures of meta-SA, not SA; they are measures of how well you think you understand and not how well you do understand. Of importance in our discussion is that it is the meta-SA on which the operator bases assessments of performance and resource demands. It is meta-SA on which the operator decides to make major changes in strategy, such as asking for help or turning off a new automated aid.

An operator can interrogate the situation model or the residual situation image directly, pulling important pieces of information to communicate to a colleague or to the interested human factors

researcher using query methods of SA. If common cognitive mechanisms are being used to manage both WL and performance a rich
interplay of performance management and WL management can
occur. For example, our controller who had been giving particular
attention and customized service to each of the few flights knew that
a push was about to begin and that traffic load would increase dramatically. The decision to resort to standard operating procedures
reduced WL, and based on his assessment of recent performance
would not unduly influence performance. Further, the demands of
the monitoring tasks themselves can be considered in selecting performance and WL strategies.

PRACTICAL ISSUES: COMBINING WL, SA, AND PERFORMANCE

A difficulty in evaluating a piece of technology or comparing two
pieces of technology is that one can find disassociations among performance, WL, and SA. For example, the relationship between
performance and WL is not the commonsense, straightforward
relationship for which one might hope. Tsang and Vidulich (2006)
and Yeh and Wickens (1988) outline variables and conditions that
tend to lead to the disassociation of performance and WL (based
on subjective measures). Disassociations seem more likely when
WL is low, tasks are data-limited, effort is increased, the number
of tasks and difficulty disagree (e.g., comparing easy dual tasks
with a difficult single task), and the severity of resource competition between dual tasks increases (e.g., the man task demands
increase.)

There have been a number of examples of disassociations
between WL and performance reported in the aviation literature. Differences in performance can be observed while subjective
workload does not change. As examples, in Alexander, Wickens,
and Merwin (2005), a 2D coplanar cockpit display of traffic information (CDTI) with either a rear-view or side-view vertical profile orientation was compared to a 3D CDTI across a series of three
experiments. The first experiment involved minimal traffic, the
second involved an increase in traffic, and the third experiment

involved an increase in traffic as well as conflicts. The 2D copla-nar displays were consistently rated lower on the NASA-TLX than the 3D display, but performance with the 2D coplanar display was more affected by increases in the level of taskload than was the 3D display. Oman and colleagues (2001) compared pilot per-formance and WL across three vertical navigation displays and a more traditional lateral navigation display. Whereas flight tech-nical error data revealed that the graphic vertical navigation dis-plays supported superior performance, subjective ratings of WL were equivalent across display formats. On the other hand, per-formance can sometimes be insensitive, while workload varies. Saleem and Kleiner (2005) found differences in subjective WL rat-ings among visual flight rules (VFR)– and instrument flight rules (IFR)–rated pilots during approaches under different environmen-tal conditions (i.e., daytime/nighttime; clear weather/deteriorat-ing weather), yet objective flight performance measures did not differ between the two pilot groups. One might expect higher WL ratings to be associated with poor performance, but it appears that pilots compensated for demanding conditions by working harder to maintain acceptable performance levels. As another example, Researchers interested in data-link communications (Stedman et al., 2007) used a novel ATC task and found that speech and text had comparable effects on performance, but WL seemed to vary more with speech communications. The complexity of the rela-tionship between performance and workload is nicely illustrated in a study by Cummings and Guerlain (2007) on the control of Tomahawk missiles. Instead of showing an increase in WL and a corresponding decrease in performance, an inverted-U relation-ship was found between performance and the percentage of time that the operator was busy.

Thus, a common trend across several aviation studies is that experi-mental manipulations often impact performance without influ-encing subjective ratings of WL. In these cases, it may be that operators are able to tap automatic processing to improve perfor-mance in certain conditions without investing additional resources. Alternatively, there are also many cases where performance remains relatively constant across conditions while subjective ratings of WL indicate that one manipulation is more difficult than the other. In these cases, it appears that operators feel that they are investing more resources under certain conditions, but that extra effort is not translating into improved performance. In other words, operators may be compensating for more difficult conditions by investing more resources to attain similar performance levels as achieved in easier conditions.

SA has also been shown to be insensitive while performance and workload seem to vary. As examples, Lancaster and Casali (2008) measured performance, mental WL (subjective and objective), and SA of lone general aviation pilots using different data-link modalities: text, synthesized speech, digitized speech, and synthesized speech/text combination. Although the pilots had good SAGAT scores in the text-only data-link condition, the text-only condition was poor in workload (perhaps dangerously so) and in performance. Although a single display did not emerge as preferred based on each of the dependent measures, the authors supported the addition of speech to textual data-link displays. Schnell, Kwon, Merchant, and Etherington (2004) also took measures of performance, workload (NASA-TLX), and SA (SAGAT), this time to compare synthetic vision information systems (SVIS) with a conventional glass cockpit. Again, all of the measures, with the exception of SA, pointed toward one type of display—the SVIS. SAGAT performance on the SVIS was statistically comparable (although numerically worse) to the conventional displays.

Researchers should attempt not only to measure both WL and SA directly, but they should use more than one measure of each construct. Leveraging multiple types of WL and SA measures enables researchers and practitioners to capitalize on the relative strengths of the different measurement types, while making up for their inherent weaknesses. In addition, researchers should be sensitive to the fact that subjective judgments are not necessarily capturing the same dimensions of the construct as are objective measures, such as physiological measures of WL and explicit measures of SA.

Research has also established some clear relationships between performance and the cognitive variables WL and SA. Some relatively coarse changes in the environment can be expected to have greater consequences for WL and SA than others. For example, in ATC, all else being equal, the more planes under control the higher the workload; in the flightdeck, phase of flight is often a clear predictor. However, making finer-grained, more subtle predictions about whether a pilot will have better understanding in this situation or that, or whether a controller will be working harder in this sector or that sector, has eluded researchers. In order to make progress beyond today, researchers must consider the strategies that pilots and controllers use to protect workload and improve performance.

By using the accumulated strategies that come with experience, the pilot or controller can modify how they approach the task, perhaps increasing WL or perhaps protecting resources by using less resource-demanding strategies (Averty, Collet, Dittmar, Athenes, and Vernet-Maury, 2004; Lee, Mercer, and Callantine, 2007; Loft et al., 2007; Schutte and Trujillo, 1996; Schvaneveldt, Beringer, and Lamonica, 2001; Sperandio, 1971; 1978; Tsang and Vidulich, 2006). In other words, taskload impacts WL and performance, which in turn affects strategy selection, which in turn impacts taskload. There are many types of strategies. For example, there are task-shedding strategies in which an operator will no longer perform a task in an effort to reduce WL. When our fictional controller started denying pilot requests, he was task shedding. As mentioned previously, pilots tend (and are trained) to adopt a task-shedding strategy defined by the ANCS hierarchy when under high WL conditions, such as when unexpected events occur (Schutte and Trujillo, 1996).

Schutte and Trujillo (1996) also noted a number of strategies other than ANCS that may be utilized under off-nominal conditions. The first is similar to ANCS in that pilots prioritize activities based on a goal-driven paradigm, in this case associated with perceived severity. Through a perceived-severity strategy, pilots place the highest priority on what they perceive to be the most threatening problem. Two additional strategies were identified based on a stimulus-driven paradigm, where stimuli are defined as procedures or events/interruptions. A procedure-based strategy consists of prioritizing tasks for which there are well-defined procedures, while an event/interrupt-driven strategy involves migrating toward a particular task based on a specific event or interruption until another event or interruption disrupts that task.

It is important to note that strategies change over time, particularly as pilots engage in different cockpit task management activities. Schutte and Trujillo (1996) noted the use of different strategies when monitoring and assessing a situation compared to managing personal WL such that one pilot used the ANCS strategy to monitor the situation and a perceived-severity strategy to manage personal WL, while another did exactly the opposite. When confronted with an off-nominal event, performance was best when pilots employed a goal-driven combination of the ANCS strategy to monitor the situation and a perceived-severity strategy to manage personal WL.

Similarly, in ATC, Sperandio (1971; 1978) recognized that controllers manage the complexity through WL-reducing strategies. For example, one such simple strategy is to rely on standard operation procedures (Koros, Della Rocco, Panjwani, Ingurgio, and D'Arcy, 2003) when the situation becomes so complex that WL would be driven higher. Lee et al. (2007) showed that controllers shed the less essential tasks, like displaying routes when WL increased. We believe this explains why it is difficult to predict WL from taskload beyond straight forward, relatively coarse predictions—like number of aircraft to be controlled or phase of flight. This leads us to expect that the relationship between WL and performance should be difficult to discern without taking into account operators' strategies.

Loft et al. (2007) listed the control strategies studied in ATC during the past few decades. An inspection of that list suggests they fall into two broad types: control strategies that classify and control strategies that distinguish. We will refer to the strategies that lead to the treatment of individual objects as belonging to a group as amalgamation strategies, and those that lead to distinguishing among objects as differentiation strategies. A common amalgamation strategy is to group aircraft into streams or flows (e.g., Seamster et al., 1993), although there are other examples such as an "important" group of aircraft (Gronlund, Ohrt, Dougherty, Perry, & Manning, 1998). A common differentiation strategy is to attend selectively to altitude in conflict resolution (e.g., Bisseret, 1971), although again other examples have been studied, such as attending to critical points of previous conflicts (e.g., Boag, Neal, Loft, and Halford, 2006).

Although human factors researchers have supplied considerable guidance at a broad level for what impacts workload, situation awareness, and performance, in our view, considerably more work should be conducted on the strategies that pilots and controllers choose and the "triggers" that give rise to those strategy choices. We call for more work on the process of how such strategy choices are made. Managing a situation requires an understanding of the situation, and understanding a situation requires the operator to evaluate performance and the resources devoted to achieving that performance. A complete understanding of the subtleties of WL, SA, and performance will continue to elude aviation researchers until steps are taken to consider these variables from a management perspective in which strategies play a pivotal role. To accomplish this will require both rich empirical studies and sophisticated computational modeling efforts.

Authors' Note

The authors would like to thank Chris Wickens for comments on an earlier version of this chapter. Thanks also to Kaitlin Geldbach for help with preparing the manuscript.

References

Alexander, A. L., & Wickens, C. D. (2005). *3D navigation and integrated hazard display in advanced avionics: Performance, situation awareness, and workload* (Technical Report AHFD-05–10/NASA-05-2). Savoy, IL: Aviation Human Factors Division.

Alexander, A. L., Wickens, C. D., & Merwin, D. H. (2005). Perspective and coplanar cockpit displays of traffic information: Implications for maneuver choice, flight safety, and mental WL. *International Journal of Aviation Psychology, 15,* 1–21.

Averty, P., Collet, C., Dittmar, A., Athenes, S., & Vernet-Maury, E. (2004). Mental WL in air traffic control: An index constructed from field tests. *Aviation, Space, and Environmental Medicine, 75,* 333–341.

Bisseret, A. (1971). Analysis of mental processes involved in air traffic control. *Ergonomics, 14,* 565–570.

Boag, C., Neal, A., Loft, S., & Halford, G. S. (2006). An analysis of relational complexity in an air traffic control conflict detection task. *Ergonomics, 49,* 1508–1526.

Boudes, N., & Cellier, J. M. (2000). Accuracy of estimations made by air traffic controllers. *International Journal of Aviation Psychology, 10,* 207–225.

Bustamante, E. A., Fallon, C. K., Bliss, J. P., Bailey, W. R., & Anderson, B. L. (2005). Pilots' WL, situation awareness, and trust during weather events as a function of time pressure, role assignment, pilots' rank, weather display, and weather system. *International Journal of Applied Aviation Studies, 5,* 347–367.

Caretta, T. R., Perry, D. C., & Ree, M. J. (1996). Prediction of situational awareness in F-15 pilots. *International Journal of Aviation Psychology, 6,* 21–41.

Chou, C., Madhaven, D., & Funk, K. (1996). Studies of cockpit task management errors. *International Journal of Aviation Psychology, 6,* 307–320.

Christien, R., Benkouar, A., Chaboud, T., & Loubieres, P. (2003, June). *Air traffic complexity indicators & ATC sectors classified.* Paper presented at ATM 2003, Budapest, Hungary.

Cummings, M. L., & Guerlain, S. (2007). Developing operator capacity estimates for supervisory control of autonomous vehicles. *Human Factors, 49,* 1–15.

Davis, C. G., Danaher, J. W., & Fischl, M. A. (1963). *The influence of selected sector characteristics upon ARTCC controller activities* (Contract No. FAA/BRD-301). Arlington, VA: The Matrix Corporation.

Dekker, S. W. A., & Hollnagel, E. (2004). Human factors and folk models. *Cognition, Technology, and Work, 6,* 79–86.

Dekker, S. W. A., & Woods, D. D. (2002). MABA-MABA or abracadabra? Progress on human-automation coordination. *Cognition, Technology, and Work, 4,* 240–244.

Dismukes, K. (2001). The challenge of managing interruptions, distractions, and deferred tasks. In: *Proceedings of the 11th International Symposium on Aviation Psychology.* Columbus: Ohio State University.

Durso, F. T., Bleckley, M. K., & Dattel, A. R. (2006). Does SA add to the validity of cognitive tests?. *Human Factors, 48,* 721–733.

Durso, F. T., & Dattel, A. R. (2004). SPAM: The real-time assessment of SA. In S. Banbury & S. Tremblay (Eds.), *A cognitive approach to situation awareness: Theory and application* (pp. 137–154). Aldershot, UK: Ashgate.

Durso, F. T., & Gronlund, S. D. (1999). Situation awareness. In F. T. Durso, R. S. Nickerson, R. W. Schvaneveldt, S. T. Dumais, D. S. Lindsay, & M. T. H. Chi (Eds.), *Handbook of applied cognition* (pp. 283–314). New York: Wiley.

Durso, F. T., Hackworth, C. A., Truitt, T. R., Crutchfield, J., Nikolic, D., & Manning, C. A. (1998). Situation awareness as a predictor of performance for en route air traffic controllers. *Air Traffic Control Quarterly, 6,* 1–20.

Durso, F. T., Manning, C. A. (2008). Air traffic control. In: M. Carswell (Ed.), *Reviews of human factors and ergonomics, (Vol 4),* Santa Monica: Human Factors and Ergonomics Society.

Durso, F. T., Rawson, K. A., & Girotto, S. (2007). Comprehension and situation awareness. In F. T. Durso, R. S. Nickerson, S. T. Dumais, S. Lewandowsky, & T. Perfect (Eds.), *Handbook of applied cognition* (2nd ed.) (pp. 164–193). Hoboken, NJ: Wiley.

Durso, F. T., & Sethumadhavan, A. (2008). Situation Awareness: Understanding dynamic environments. *Human Factors, 50,* 442–448.

Endsley, M. R. (1995a). Measurement of situation awareness in dynamic systems. *Human Factors, 37,* 65–84.

Endsley, M. R. (1995b). Toward a theory of situation awareness in dynamic systems. *Human Factors, 37,* 25–64.

Endsley, M. R., Bolte, B., Jones, D. G. (2003). *Designing for situation awareness.* London: Taylor & Francis.

Funk, K. H., II (1991). Cockpit task management: Preliminary definitions, normative theory, error taxonomy, and design recommendations. *International Journal of Aviation Psychology, 1,* 271–286.

Gronlund, S., Ohrt, D. D., Dougherty, M. R. P., Perry, J., & Manning, C. A., (1998). Role of memory in air traffic control. *Journal of Experimental Psychology: Applied, 4,* 263–280.

Hart, S. G., & Wickens, C. D. (1990). WL assessment and prediction. In H. R. Booher (Ed.), *MANPRINT: An approach to systems integration* (pp. 257–296). New York: Van Nostrand Reinhold.

Harwood, K., Barnett, B., Wickens, C. D. (1988). Situational awareness: A conceptual and methodological framework. *Proceedings of the Eleventh Psychology in the Department of Defense Symposium.* USAFA, Colorado.

Hauss, Y., & Eyferth, K. (2003). Securing future ATM-concepts' safety by measuring situation awareness in ATC. *Aerospace Science and Technology, 7,* 417–427.

Helmreich, R. L., Wilhelm, J. A., Klinect, J. R., & Merritt, A. C. (2001). Culture, error, and crew resource management. In E. Salas, C. A. Bowers, & E. Edens (Eds.), *Improving teamwork in organizations: Applications of resource management training* (pp. 305–331). Mahwah, NJ: Erlbaum.

Hilburn, B. (2004). *Cognitive complexity in air traffic control: A literature review* (EEC Note No. 04/04). Brétigny-sur-Orge, France: EUROCONTROL Experimental Centre.

Histon, J. M., Hansman, R. J., Gottlieb, B., Kleinwaks, H., Yenson, S., Delahaye, D., Puechmorel, S. (2002). Structural considerations and cognitive complexity in air traffic control. Paper presented at the *19th IEEE/AIAA Digital Avionics System Conference.*

Jenkins, J. C., & Gallimore, J. J. (2008). Configural features of helmet-mounted displays to enhance pilot situation awareness. *Aviation, Space, and Environmental Medicine, 79,* 397–407.

Kantowitz, B. H. (1986). Mental WL. In P. A. Hancock (Ed.), *Human factors psychology* (pp. 81–122). New York: North-Holland.

Kantowitz, B. H., & Casper, P. A. (1988). Human WL in aviation. In E. L. Wiener & D. C. Nagel (Eds.), *Human factors in aviation.* San Diego, CA: Academic Press.

Kirwan, B., Scaife, R., & Kennedy, R. (2001). Investigating complexity factors in UK air traffic management. *Human Factors and Aerospace Safety, 1,* 125–144.

Kopardekar, P., & Magyarits, S. (2003). Measurement and prediction of dynamic density. Paper presented at the *5th EUROCONTROL/FAA Air Traffic Management Research and Development Conference*, Budapest, Hungary.

Koros, A., Della Rocco, P. S., Panjwani, G., Ingurgio, V., & D'Arcy, J. F. (2003). *Complexity in air traffic control towers: A field study: Part 1. Complexity factors* (DOT/FAA/CT—TN03/14). Atlantic City, NJ: Federal Aviation Administration.

Lee, Y.-H., & Liu, B.-S. (2003). Inflight WL assessment: Comparison of subjective and physiological measurements. *Aviation, Space, and Environmental Medicine, 74*, 1078–1084.

Lee, P. U., Mercer, J., & Callantine, T. J. (2007). Examining the moderating effect of WL on controller task distribution. *Human Computer Interaction, 13*, 339–348.

Loft, S., Sanderson, P., Neal, A., & Mooij, M. (2007). Modeling and predicting mental WL in en route air traffic control: Critical review and broader implications. *Human Factors, 49*, 376–399.

Lorenz, B., Biella, M., Teegen, U., Stelling, D., Wenzel, J., Jakobi, J., & Korn, B. (2006). Performance, situation awareness, and visual scanning of pilots receiving onboard taxi navigation support during simulated airport surface operation. *Human Factors and Aerospace Safety, 6*, 135–154.

Manning, C. A., Mills, S. H., Fox, C., Pfleiderer, E. M., & Mogilka, H. J. (2002). *Using air traffic control taskload measures and communication events to predict subjective WL* (DOT/FAA/AM-02/4). Washington, DC: Federal Aviation Administration.

Morris, C. H., & Leung, Y. K. (2006). Pilot mental WL: How well do pilots really perform? *Ergonomics, 49*, 1581–1596.

Niessen, C., Eyferth, K., & Bierwagen, T. (1999). Modeling cognitive processes of experienced air traffic controllers. *Ergonomics, 42*, 1507–1520.

O'Brien, K. S., & O'Hare, D. (2007). Situational awareness ability and cognitive skills training in a complex real-world task. *Ergonomics, 50*, 1064–1091.

O'Hare, D. (1997). Cognitive ability determinants of elite pilot performance. *Human Factors, 39*, 540–552.

Oman, C. M., Kendra, A. J., Hayashi, M., Stearns, M. J., & Burki-Cohen, J. (2001). Vertical navigation displays: Pilot performance and WL during simulated constant angel of descent GPS approaches. *International Journal of Aviation Psychology, 11*, 15–31.

Parasuraman, R., Sheridan, T. B., & Wickens, C. D. (2008). Situation awareness, mental WL, and trust in automation: Viable, empirically supported cognitive engineering constructs. *Journal of Cognitive Engineering and Decision Making, 2*, 140–160.

Rantanen, E. (2004). *Development and validation of objective performance and WL measures in air traffic control* (AHFD-04–19/FAA-04–7). , Oklahoma City, OK: Federal Aviation Administration Civil Aeromedical Institute.

Rantanen, E. M., Yeakel, S. J., & Steelman, K. S. (2006). *En route controller task prioritization research to support CE-6 human performance modeling Phase II: Analysis of high-fidelity simulation data* (HFD-06–03/MAAD-06–2). Savoy: Human Factors Division, Institute of Aviation, University of Illinois at Urbana-Champaign.

Remington, R. W., Johnston, J. C., Ruthruff, E., Gold, M., & Romera, M. (2000). Visual search in complex displays: Factors affecting conflict detection by air traffic controllers. *Humans Factors, 42*, 349–366.

Roscoe, S. N., Corl, L., & Larouche, J. (2001). *Predicting human performance* (5th ed.). Pierrefonds, Quebec: Helio Press.

Saleem, J. J., & Kleiner, B. M. (2005). The effects of nighttime and deteriorating visual conditions on pilot performance, WL, and situation awareness in general aviation for both VFR and IFR approaches. *International Journal of Applied Aviation Studies, 5,* 107–120.

Schmorrow, D. D., & Kruse, A. A. (2004). Augmented cognition. In W. S. Bainbridge (Ed.), *Berkshire encyclopedia of human-computer interaction* (pp. 54–59). Great Barrington, MA: Berkshire Publishing Group.

Schutte, P. C., & Trujillo, A. C. (1996). Flight crew task management in non-normal situations. In: *Proceedings of the 40th Annual Meeting of the Human Factors and Ergonomics Society.* Santa Monica, CA: HFES (pp. 244–248).

Schvaneveldt, R. W., Beringer, D. B., & Lamonica, J. A. (2001). Priority and organization of information accessed by pilots in various phases of flight. *International Journal of Aviation Psychology, 11,* 253–280.

Seamster, T. L., Redding, R. E., Cannon, J. R., Ryder, J. M., & Purcell, J. A. (1993). Cognitive task analysis of expertise in air traffic control. *International Journal of Aviation Psychology, 3,* 257–283.

Sirevaag, E. J., Kramer, A. F., Wickens, C. D., Reisweber, M., Strayer, D. L., & Grenell, J. F. (1993). Assessment of pilot performance and mental workload in rotary wing aircraft. *Ergonomics, 36,* 1121–1140.

Sohn, Y. W., & Doane, S. M. (2004). Memory processes of flight situation awareness: Interactive roles of working memory capacity, long-term working memory, and expertise. *Human Factors, 46,* 461–475.

Sperandio, J. C. (1971). Variation of operator's strategies and regulating effects on WL. *Ergonomics, 14,* 571–577.

Sperandio, J. C. (1978). The regulation of working methods as a function of WL among air traffic controllers. *Ergonomics, 21,* 195–202.

Spick, M. (1988). *The ace factor: Air combat and the role of situational awareness.* Annapolis, MD: Naval Institute Press.

Stanton, N., Salmon, P. M., Walker, G. H., Baber, C., & Jenkins, D. P. (2005). *Human factors methods: A practical guide for engineering and design.* Burlington, VT: Ashgate Publishing.

Stedman, A. W., Sharples, S., Littlewood, R., Cox, G., Patel, H., & Wilson, J. R. (2007). Datalink in air traffic management: Human factors issues in communications. *Applied Ergonomics, 38,* 473–480.

Tenney, Y. J., & Pew, R. W. (2006). Situation awareness catches on. What? So what? Now what?. In R. C. Williges (Ed.), *Reviews of human factors and ergonomics: Vol. 2* (pp. 89–129). Santa Monica, CA: Human Factors and Ergonomics Society.

Thomas, L. C., & Wickens, C. D. (2006). Display dimensionality, conflict geometry, and time pressure effects on conflict detection and resolution performance using cockpit displays of traffic information. *International Journal of Aviation Psychology, 16,* 321–342.

Tsang, P. S., & Vidulich, M. A. (2006). Mental WL and situation awareness. In G. Salvendy (Ed.), *Handbook of human factors and ergonomics* (3rd ed.) (pp. 243–268). Hoboken, NJ: John Wiley & Sons.

Vogt, J., Hagemann, T., & Kastner, M. (2006). The impact of WL on heart rate and blood pressure in en-route and tower air traffic control. *Journal of Psychophysiology, 20,* 297–314.

Wickens, C. D. (1980). The structure of attentional resources. In R. Nickerson (Ed.), *Attention and performance VIII* (pp. 239–257). Hillsdale, NJ: Erlbaum.

Wickens, C. D. (1984). Processing resources in attention. In R. Parasuraman & R. Davies (Eds.), *Varieties of attention* (pp. 63–101). New York: Academic Press.

Wickens, C. D. (2001). WL and situation awareness. In P. A. Hancock & P. A. Desmond (Eds.), *Stress, WL, and fatigue* (pp. 443–450). Mahwah, NJ: Erlbaum.

Wickens, C. D. (2002). Pilot actions and tasks: Selections, execution, and control. In: P. S. Tsang & M. A. Vidulich (Eds.), *Principles and practice of aviation psychology* (pp. 239–263).

Wickens, C. D. (2008a). Multiple resources and mental WL. *Human Factors, 50,* 449–455.

Wickens, C. D. (2008b). Situation awareness: Review of Mica Endsley's 1995 articles on situation awareness theory and measurement. *Human Factors, 50,* 397–403.

Wilson, G. F. (2002). An analysis of mental WL in pilots during flight using multiple psychophysiological measures. *International Journal of Aviation Psychology, 12,* 3–18.

Wilson, G. F., & Russell, C. A. (2003). Operator functional state classification using multiple psychophysiological features in an air traffic control task. *Human Factors, 45,* 381–389.

Woodhouse, R., & Woodhouse, R. A. (1995). Navigation errors in relation to controlled flight into terrain (CFIT). In: *8th International symposium on Aviation Psychology*. Columbus: Ohio State University (pp. 1403–1406).

Yeh, Y., & Wickens, C. D. (1988). Dissociation of performance and subjective measures of WL. *Human Factors, 30,* 111–120.

Young, M. S., & Stanton, N. A. (2001). Mental WL: Theory, measurement, and application. In W. Karwowski (Ed.), *International encyclopedia of ergonomics and human factors* (pp. 507–509). New York: Taylor and Francis.

9

Team Dynamics at 35,000 Feet

Eduardo Salas, Marissa L. Shuffler,
Deborah DiazGranados

Department of Psychology, and, Institute for Simulation
and Training, University of Central Florida

INTRODUCTION

On January 13, 1982, Air Florida Flight 90, a Boeing 737–222 took off from Reagan National Airport in Washington, DC, amid snowy weather that led to icy flight conditions. Because of a delay in the departure of the aircraft due to inclement weather, the crew was eager to depart in order to ensure that another incoming flight that was supposed to be using the same runway would have a timely arrival. Although conditions were icy, the flight crew did not follow typical engine anti-icing operations and took off with snow and ice on the airfoil surfaces of the aircraft. Furthermore, the captain continued with takeoff even when anomalous engine instrument readings occurred, creating an environment ripe for disaster. Subsequently, following takeoff the aircraft was unable to overcome the ice and snow adhering to it, causing it to crash into the 14th Street bridge that connects Washington, DC, and Virginia while the pilots were attempting to land over the water. Seventy-four passengers perished, as did four motorists in vehicles on the bridge at the time of the crash. An investigation of the accident led to the conclusion that human error was the primary cause of the accident, with the primary factors being the failure of the crew to adequately address the weather conditions before takeoff (http://www.ntsb.gov).

Almost three decades later, on January 15, 2009, U.S. Airways Flight 1549 also experienced problems during takeoff and ended in a watery landing as well, albeit with drastically different results. Leaving LaGuardia International Airport in New York City, the Airbus A320 encountered a flock of birds during takeoff, causing the aircraft to lose power in both engines. While initially the captain and crew planned to either return to LaGuardia or land at a nearby airport in New Jersey, they quickly realized their inability to safely guide the plane to either location. Despite limited control over the aircraft, the flight's captain and crew were able to effectively land the plane in the Hudson River, allowing for all 155 passengers and the five crew members to escape safely (Rivera, 2009). The crew and its captain, Chesley "Sully" Sullenberger, an experienced pilot with his own safety management consulting company, were well coordinated in their efforts to not only land the plane safely, but also to ensure that all passengers were able to escape once the aircraft landed in the water, with Captain Sullenberger rechecking the cabin twice for remaining passengers before leaving the aircraft himself.

Why are the results of these two incidents so drastically different? Both crews received extensive training on how to deal with the technical issues that they were facing, yet one crew was clearly superior in its ability to recognize and avert disastrous circumstances. Perhaps one notable difference has been cited often by Captain Sullenberger is the fact that the crew simply responded with "what we're trained to do": reduce human error in the cockpit (Rivera, 2009). Undoubtedly, these two stories serve to illustrate the fact that drastic changes to the aviation industry and its understanding of aviation crew performance have occurred over the past two decades. Such advances have the potential to avert disaster and overcome the proclivity of human error due to the increased use of crew resource management (CRM) and other effective training techniques.

The prevalence of human error as a major contribution to aviation accidents (e.g., Nullmeyer, Stella, Montijo, & Harden, 2005) is certainly exemplified in the comparison of the previous cases, but so is the ability of well-trained personnel to successfully manage events that could potentially have led to catastrophic results. However, what exactly has changed in the way that crews are trained and interact with one another that has enabled aviation crews to be prepared for incidents such as those mentioned previously? Furthermore, what has changed in the way that we view teams and their inputs and outcomes that impact processes? Understanding these changes and their impacts will provide not only a comprehensive view of where the research currently stands on the management of human error in the aviation industry, but also provide insight towards the future of CRM and other relevant aviation crew performance issues.

Therefore, the purpose of the current chapter is to provide an update to Foushee and Helmreich's (1988) chapter on group interaction and flight crew performance, as much has changed in flight crew performance in the past 20 years. Furthermore, we aim to push the agenda for advancing our understanding of aviation crews, their training, and the factors that influence their performance. Specifically, we will review the need to address performance issues in flight crews, highlight advances in understanding flight crew performance and their training, particularly through the use of CRM, and delineate directions for future research in this area. It is hoped that with this update, we will paint a clear picture of the current state of the art in flight crew performance, as well as address the future human performance concerns for such teams.

WHY DOES AVIATION CREW PERFORMANCE MATTER?

As previously noted, the prevalence of human error in aviation incidents and accidents has been a major propellant for understanding how to manage such errors (Foushee, 1984). A striking statistic based on data gathered from commercial aviation accident statistics from 1959 to 1989 indicate that flight crew actions have been attributed as cause in more than 70% of aviation accidents (Helmreich & Foushee, 1993). Furthermore, a recent analysis of global airline safety by the United Kingdom's Civil Aviation Authority proposed that crew judgment errors are the most consistent causal factor in global catastrophic events (Learmount, 2008).

Statistics such as these explain the advancement of research and the development of training programs aimed at improving crew coordination and the inclusion of crew coordination as a requirement for pilots (United States General Accounting Office, 1997). Of particular interest has been the focus upon understanding how teams can aid in reducing human error. Teams have become widely used in a range of organizations, as they provide a broader pool of cognitive resources and task-relevant knowledge, skills, and abilities (KSAs) than just a single individual. For aviation, teams provide a redundant system in which individuals can monitor and provide back up for one another, reducing the likelihood that errors will go unnoticed (Wilson, Guthrie, Salas, & Howse, 2006).

Perhaps one of the best examples illustrating the benefits of using teams in aviation is the 1989 case of United Airlines Flight 232 (Helmreich, 1997). During the flight from Denver to Chicago, the center engine somehow disintegrated, severing the hydraulic lines that are vital for the functioning of the rudder, ailerons, and other control surfaces. This caused a loss of control over the direction of the plane, yet the crew worked together to assess the situation and develop a plan of action. The crew members worked in tandem to control the aircraft, find a landing site, prepare the cabin crew and passengers for an emergency landing, and abandon tasks that were no longer beneficial for reaching the final goal. Their actions saved the lives of 185 crew members and passengers, including all of the pilots.

The teamwork of this crew enabled members to be aware of the situation as a whole and enhance crucial decision-making regarding their course of action. This is a prime example as to why it is critical to understand the factors that influence aviation crew performance, as being able to replicate the team processes at play in this case study could be beneficial in future situations. Furthermore, assessing the coordination, communication, and other processes that occurred can provide a better foundation for training aviation crews (Prince & Salas, 1999). Foundationally sound training can lead to a reduced error rate, preventing accidents and saving lives—the key to why team performance is so critical to the aviation industry (Salas, Wilson, Burke, & Wightman, 2006).

ADVANCES IN TEAM PERFORMANCE RESEARCH

Team and group research has provided a strong foundation for aviation crew research (Prince & Salas, 1999), yet much has changed in team performance research in the past two decades (Guzzo & Dickson, 1996). These advances in teamwork, team process, and team competencies have enabled a richer understanding of their significance in aviation crew performance. In the following section, we will outline constructs from team research relevant to aviation crew performance and highlight some of the most significant advancements in understanding team processes.

Teamwork

In order for teams to be effective, they must have both teamwork and taskwork skills (Salas, Kosarzycki, Tannenbaum, & Carnegie, 2004). Teamwork is defined as a set of behaviors, cognitions, and attitudes that are enacted in order to achieve mutual goals and meet the demands of the outside environment (Salas, Stagl, Burke, & Goodwin, 2007). Much has been advanced regarding what we know about teamwork in the past several decades, with a large number of input, process, and outcome variables that affect teamwork being identified (Burke, Wilson, & Salas, 2003). From this research, we now have a better understanding of the dynamic and multidimensional nature of teamwork, and the significance of process in determining the relationship between inputs and outcomes (Swezey & Salas, 1992; Weiner et al., 1993).

Several researchers have argued that there are actually two tracks of skills important within teams, teamwork and taskwork (Burke, Wilson, & Salas, 2003). Taskwork involves those skills necessary for team members to perform tasks, whereas teamwork skills focus primarily on the behaviors and attitudes necessary for teams to function in order to accomplish these tasks (Salas et al., 2007). Both of these skill sets are viewed as equally important, although many recognize that taskwork skill should be trained before teamwork. Additionally, several researchers argue that the relationship between taskwork and team effectiveness is mediated by teamwork skills (Hackman & Morris, 1975; Bass, 1990; Burke, Wilson, & Salas, 2003).

In aviation, teams are expected to perform both teamwork and taskwork skills. While being able to efficiently and effectively perform the task at hand is critical to crews in the cockpit, teamwork skills are also crucial for effective coordination and communication among crew members. Merket and colleagues (2000) found that breakdowns in team performance skills played a significant role in aviation errors. Specifically, they found that deficiencies in aircrew coordination skills (e.g., decision making, leadership, adaptability) contributed to 68% of the mishaps examined in the study. Certainly, understanding the role of teamwork skills beyond just taskwork is important to aircrew effectiveness and accident prevention.

Team Competencies

As an understanding of teamwork has progressed, so has knowledge of the skills and competencies necessary to perform such teamwork. Over the past few decades, researchers have worked to identify the core competencies that are necessary for teamwork to occur, including knowledge, skills, and attitudes (Cannon-Bowers et al., 1995). For example, after collecting and synthesizing prior teamwork research, Cannon-Bowers and colleagues (1995) identified a set of eight major teamwork skills, including (1) adaptability, (2) communication, (3) coordination, (4) decision making, (5) interpersonal relations, (6) leadership/team management, (7) performance monitoring/feedback, and (8) shared situational awareness.

Although identifying these competencies has led to great advances in understanding what impacts teamwork, most recently the work of Marks et al. (2001) and Salas et al. (2005) have further contributed to our understanding of the dynamic interdependencies among the components of teamwork. Focusing upon the temporal nature of teams, Marks and colleagues identified a framework of team processes in addition to interpersonal processes that influence team performance. In their framework, Salas and colleagues advanced a "big five" of teamwork, highlighting five core components of teamwork: (1) team leadership, (2) mutual performance monitoring, (3) backup behavior, (4) adaptability, and (5) team orientation. Furthermore, they examined how these core competencies require the support of several coordinating mechanisms, including shared mental models, closed-loop communication, and mutual trust, as well as how these competencies may vary in importance over the lifespan of the team. These competencies certainly have implications for the aviation community, as all are critical in the demanding and complex situations occurring in the cockpit at any given moment.

I-P-O Framework Advancement to IMOI

Traditionally, teamwork and team performance/effectiveness has been studied using an input-process-output (I-P-O) framework (see Figure 9.1), as originally advanced by McGrath (1984). From this perspective, inputs involve antecedent factors that enable and constrain the interactions of team members (Mathieu, Maynard, Rapp, & Gilson, 2008). Inputs can involve individual characteristics, team level factors, and organizational level factors. These factors combine together to drive team processes, or the interactions of members that are directed to accomplishing the team task at hand. Processes are a very important piece to this framework, as

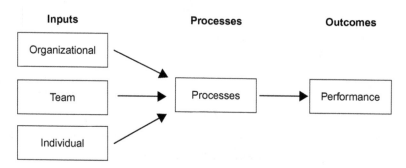

FIGURE 9.1   Input-Process-Outcome (IPO) Team Effectiveness Framework. Reprinted from Mathieu, Maynard, Rapp, and Gilson (2008).

they provide the mechanism by which team inputs are transformed into team outcomes. Outcomes are described as results and by-products of these team processes, and can include factors such as team performance as well as affective reactions (Mathieu, Heffner, Goodwin, Salas, & Cannon-Bowers, 2000).

This model of team effectiveness has been well utilized over the years, with adjustments and modifications being made to some degree in order to better understand team issues (Cohen & Bailey, 1997; Mathieu et al., 2008). For example, some researchers have examined the temporal nature of the model (Marks, Mathieu, & Zaccaro, 2001), while others have looked at the inherently multilevel nature of the individual, team, and organizational inputs that affect processes and outcomes (Klein & Kozlowski, 2000). Most recently, Illgen and colleagues (2005) have advanced a new form of the model which focuses on the cyclical nature of team functioning, the IMOI model (see Figure 9.2).

The IMOI model of team effectiveness adds to the original I-P-O framework by addressing the increased complexity that teams are facing today. Substituting "M," or mediator, for "P" illustrates the broader range of variables that influence teams, their processes,

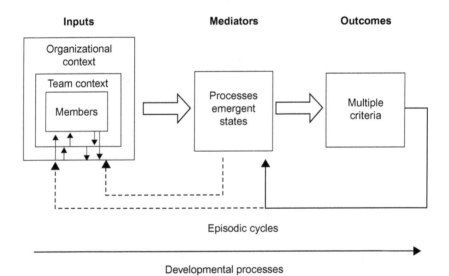

FIGURE 9.2   Input-Mediator-Outcome (IMO) Team Effectiveness Framework. Reprinted from Mathieu, Maynard, Rapp, and Gilson (2008).

and their outcomes (Illgen, et al., 2005). Additionally, the inclusion of another "I" illustrates the fact that the framework is cyclical, with feedback occurring to inform the next iteration. Finally, the removal of hyphens represents that the model is not linear or additive, but is in fact nonlinear or conditional.

For the aviation community, understanding the role of the IMOI process in relation to aircrews is important to the successful operation of such teams. As will be discussed later in this chapter, inputs such as leadership can play a key role in affecting outcomes in aviation crews, particularly if leaders neglect balancing task oriented behaviors (e.g., monitoring progress to goals, providing backup support) and relational behaviors (e.g., developing and motivating team members) (Zaccaro, Rittman, & Marks, 2002). Effective management of such inputs is key, as they affect other processes, such as the development of trust among team members, which can in turn impact additional team processes as well as performance (Burke, Sims, Lazzara, & Salas, 2006). Certainly, the effectiveness of aviation crews cannot be viewed as the outcome of a single, linear process, but instead is impacted dynamically by numerous inputs and mediators/processes which may come into play at different times during a period of performance. Therefore, utilizing models such as those presented here can provide us with a more accurate understanding of the interplay of these variables during aircrew performance.

Summary

In sum, team dynamics research has advanced over the past several decades, which has several implications for understanding aviation crew performance. The advancement of our understanding of the competencies necessary for team functioning certainly has implications for aviation crew training, as does the move from an I-P-O framework to an IMOI model that places greater emphasis on the wide range of factors that may impact team performance and effectiveness. Furthermore, the cyclical nature and iterative feedback that is now incorporated into the framework better captures the complexity of team processes. By understanding team dynamics research, we can more fully comprehend the impacts that such complexities may have on aviation crew performance.

ADVANCES IN AVIATION CREW PERFORMANCE

Foushee and Helmreich (1988) reviewed research on group performance process as it affects crew effectiveness. More specifically, they outlined the factors (i.e., social, organizational and personality) that are relevant for flight safety. They organized their brief review of the literature based on McGrath's (1964) I-P-O model, which explains the factors affecting group performance. The factors include individual, group, and environmental variables as inputs, group interactions as the process variables, and performance variables as outcomes.

The research on flight crew performance before the publication of Foushee and Helmreich's (1988) review took several forms. On the one hand, since a large percentage of aviation accidents were attributed to human error the research focused on performance characteristics of the individual pilot. Furthermore, the majority of early research emphasized the research and development of equipment that would reduce pilot workload and would result in decreased pilot error (i.e., automation). On the other hand, research in aviation focused on personality and individual differences. A great deal of research was conducted on evaluating the stereotypical pilot, as described by Foushee and Helmreich, who is "fearless, self-sufficient, technically qualified, and slightly egotistical" (p. 191). Research focused on identifying characteristics associated with the skills necessary for the job of pilot.

At the time of Foushee and Helmreich's review, group performance research in the realm of aviation crews had not received much attention. Discussions on the factors which affect crew coordination were just beginning. With the development of what is now known to the aviation industry as CRM in 1979 and the implementation of it by U.S. airlines beginning in 1981, the research about aviation teams coordinating effectively was in its infancy. Salas, Prince, Bowers, Stout, Oser and Cannon-Bowers (1999) defined CRM as "being a set of teamwork competencies that allow the crew to cope with situational demands that would overwhelm any individual crew member" (p. 163). From this definition, they consider CRM training to be a "family of instructional strategies designed to improve teamwork in the cockpit by applying well-tested training tools (e.g., performance measures, exercises, feedback mechanisms) and appropriate training methods (e.g., simulators, lectures, videos) targeted at specific content (i.e., teamwork knowledge, skills, and attitudes)" (p. 163).

The early programs of CRM emphasized changing individual styles and correcting deficiencies in individual behavior. This behavior often took the form of juniors lacking the assertiveness when communicating to their captains, or captains being too authoritarian. Early CRM programs were psychological in nature, and focused heavily on psychological testing (Helmreich, Merritt, & Wilhelm, 1999). In later years came the evolution of CRM from a focus on general human relations training to a program which trained the management of errors via teamwork concepts (e.g., situational awareness, feedback, communication, coordination). Specifically, in the late 1980s and early 1990s, a transformation was seen in that CRM was becoming more team-based, and training even extended to others on the "crew" such as flight attendants, dispatchers, and maintenance personnel. Furthermore, cockpit-cabin CRM, by this time called crew resource management rather than cockpit resource management, was being conducted.

In the years since Foushee and Helmreich's (1988) chapter, we have seen a significant change in aviation research. An emphasis on group processes and teamwork has been demonstrated by the evolution and implementation of CRM. Moreover, research in aviation over the last 10 to 15 years has seen a rise in topics such as the effect of culture in the cockpit (see Ooi, 1991; Merritt, 1996, 2000), decision making in the cockpit (see Klein, 2000), and teamwork in the cockpit (see Brannick, Prince, & Salas, 2005; Helmreich & Foushee, 1993; Salas, Fowlkes, Stout, Milanovich, & Prince, 1999). A brief review of these topic areas are provided here.

The Cockpit and National Culture

The crash of Avianca Flight 052 on approach to New York's John F. Kennedy Airport from Bogota, Colombia, on January 25, 1990, is a highly studied case in aviation history. Not only in the examination of the reason for this crash, but in the early 1990s aviation researchers began to consider Hofstede's (1980) work in order to explain how national culture may influence the behaviors of the cockpit crew (i.e., captain, first officer and second officer). As the examination of the details of Flight 052's crash is beyond the scope of this chapter, we direct the interested reader to Helmreich (1994) for more details. However, there are two points regarding the research on national culture, which are illustrated in the assessment of why Flight 052 crashed, which merit discussion. Taken from Hofstede's classification of cultural dimensions, two dimensions were identified as having relevance to the way aviation crews function: power distance and individualism-collectivism.

Power distance defines the nature of the relationship between subordinates and superiors. Those cultures that are considered to have low power distance relate to one another more as equals regardless of job titles. Individuals in cultures that have high power distance accept the hierarchy and acknowledge the power of others who hold higher-ranking jobs. The dimension that describes a culture's individualism versus collectivism refers to the extent to which people are group interdependent or independent. National culture issues like these were relevant in the analysis of the crash of Flight 052 by Helmreich (1994). We now provide a brief explanation of the cultural factors that may have impacted the crew's interactions and thus mitigated the failure to safely land the aircraft.

Helmreich's (1994) assessment of the crash of Flight 052 discussed the national culture dimensions of power distance, individualism-collectivism, and uncertainty avoidance as reasons of the flight's crash. First, Helmreich explained that Colombians are traditionally high in power distance, and given the entire crew was Colombian, the first and second officers may have expected the captain to make the decisions and direct the officers' behaviors. Unfortunately, this did not occur. Second, culture also may have influenced the lack of a declaration of emergency. That is, the Colombian culture is collectivistic and therefore the crew may have chosen to remain as a unit and not disrupt group harmony, in Helmreich's explanation he stated "... [there may have been a] tendency for the crew to remain silent while hoping that the captain would 'save the day'" (p. 282). Third, uncertainty avoidance may have also played a role in the airliner crash. Uncertainty avoidance is the tendency to pre-fer structure and to avoid uncertain situations. Because of this the crew may have maintained their course to land at JFK rather than discussing alternative courses of action. Although none of these explanations as to why Flight 052 resulted in a failed landing can be proven, the possibility that national culture did play a role has not been dismissed.

The recognition that variables such as interpersonal relations, and teamwork processes among the flight crew factor in to a majority of aviation accidents led to the development of CRM. Furthermore, evidence that has suggested that these variables are influenced by national culture (Berry, Poortinga, Segall, & Dasen, 1992; Hofstede, 1991) has advanced the knowledge regarding aircrew performance via research that evaluates the impact of national culture on crew performance. Merritt and colleagues (Helmreich, Merritt, Sherman, Gregorich, & Wiener, 1993; Merritt & Helmreich, 1996, 2007; Merritt, 2000) have made and continue to make great leaps in understanding the influence of national culture on attitudes and processes of flight crew behavior. For example, Merritt (2000) replicated Hofstede's dimensions of culture in a large sample of pilots across 19 countries and confirmed that national culture exerts an influence on cockpit behaviors over and above the professional culture of pilots.

Decision-Making in the Cockpit

Aeronautical decision making (ADM) has been a topic of concern for many years. However, the method that researchers thought pilots used to make decisions (Rational Choice strategy) has since been reassessed. Originally it was believed that pilots made deci-sions by considering a range of options, it was then thought that these options were evaluated and judged on a set of categories, assigning weights to the categories that reflect their level of impor-tance. Once the options were judged it was believed that pilots would then base their decision by choosing the option with the highest score (i.e., the highest weighted score; Klein, 1998).

However, as researchers studied decision making in natural set-tings it was discovered that this method for making decisions had some limitations. This strategy is too data and time intensive for the environment in which pilots often find themselves. Klein (1989) found that in a variety of natural settings experts making decisions rarely compared options. Klein, Calderwood, and Clinton-Cirocco (1986) developed a model called the recognition primed decision (RPD) model, which explains how individuals make decisions in field settings, especially under conditions that make it difficult to generate multiple options and then compare them.

The RPD model states that people use their experience to assess the situation and determine if it is familiar. If a situation is familiar then their method for making a decision is relatively easy, they implement the solution which has been proven to work. If they are unsure of the situation then they develop options and simulate the implementation of those options. When they simulate an option that does not result in something unacceptable, then the decision maker chooses that option.

Recent research on decision making in aviation has not been limited to decisions made by single pilots. Since commercial planes are not flown by individuals and rather crews, crew decision-making has been a critical topic to research and understand. For example, Orasanu (1993) explains that crew decision-making is managed decision making. The captain is responsible for making the decisions but he/she is supported by the input from the entire crew, both those individuals in the cockpit but also on the ground (i.e., air traffic control, dispatch, maintenance). Moreover, crews that work effectively can make better decisions than individuals because the multiple eyes, ears, hands, and minds can increase their cognitive capacity; and crews can offer multiple options, share workload, and often avoid traps to which individuals are often susceptible.

Summary

Research in aviation has changed significantly in the years since Foushee and Helmreich's (1988) original chapter. With the evolution of CRM, teamwork is a critical topic in determining effective aircrew performance. This trend has followed Foushee and

Helmreich's forecast that there would be an increased emphasis on training programs to increase crew coordination and effectiveness.

The transformation in aviation research has also seen an emphasis on variables that may impact team processes. There has been a large avenue of research that has considered the impact of national culture on team effectiveness. Moreover, when CRM training has been implemented outside of the United States, carriers have included national culture as part of the training, as well as customizing the program to ensure training effectiveness.

Another change in aviation research has been the research conducted on decision making. Aviation decision making has been of interest for many years. With the discovery that decisions are most likely made by examining the familiarity of the situation the area of ADM has evolved. With the work by Klein and colleagues, the research on Naturalistic Decision Making, decision making in field settings, is an area that has and will greatly inform aviation and how decisions are made in the cockpit. In order to ensure that aviation remains a high reliability organization, a term used to describe organizations which internally operate at high levels and have succeeded in achieving levels of reliability which are exceptionally high (Weick & Roberts, 1993), research on team coordination and team processes in general, must continue to evolve.

WHAT FACTORS IMPACT CREW PERFORMANCE?

Aviation crews are like any other type of team. For them to be effective the members need to perceive that they are a team, understand each others' roles, and be well trained on teamwork competencies. In this section we discuss some of the critical teamwork processes which influence the effectiveness of a crew. We discuss the captain's role as a leader, shared mental models, situational awareness, adaptability, and communication. We by no means believe this to be an exhaustive list of competencies that impact crew performance, but rather, use these to illustrate some of the key issues prevalent today in understanding aviation crew performance.

Leadership

Legally, the captain is responsible for the safety of the flight. Therefore, no doubt about it, the captain of an aircrew is the leader of the team. Furthermore, the effectiveness of a crew can be traced back to the captain. Ginnett (1993) found this to be the shared perception of crew members after conducting numerous interviews. A crew member stated in one of the interviews:

> Some guys are just the greatest in the world to fly with. I mean they may not have the greatest hands in the world but that doesn't matter. When you fly with them, you feel like you want to do everything you can to work together to get the job done. (p. 84)

So how does the leader impact crew performance? The answer includes those things that a captain does before, during and after the crew are in flight together.

The captain's impact on his/her crew does not begin at takeoff. Rather the captain can help to improve its crew's performance by engaging the crew in preflight discussions or prebriefs to ensure that the crew indeed perceives themselves as a unit and holds a shared mental model (Entin & Serfaty, 1999; Serfaty, Entin, & Deckert, 1993). In commercial aviation, since it is uncommon for the same crew to work together often the moments before a flight can serve as team building. That is, the captain can utilize this time to address expectations and ensure that the crew is prepared to do their job and has the appropriate team related attitude.

During the flight the captain's job is to ensure that he/she is fully aware of the progress of flight. This means, managing the information that he is given from the equipment, those in the cockpit, and those on the ground (i.e., air traffic control). Problems can arise when information is not effectively shared between the flight crew. These breakdowns can occur because of lack of assertiveness (Jentsch & Smith-Jentsch, 2001). A breakdown in the cockpit can result in ambiguity which could lead to disaster. Wickens (1995) found that failures in information sharing accounted for more than 50% of all errors in the cockpit. An effective leader can prevent these errors from occurring.

Action learning is a learning process whereby the individuals study their own actions and experiences in order to improve performance. After a performance session the captain can impact his/her crew future performance by engaging in what is known as an after action review (AAR). The AAR is a method for a team to learn and reflect on their performance. For an AAR to be done effectively a leader must guide the team through the process providing feedback and engaging the members in the process of self-assessment. Captains have the responsibility of managing their crew and can impact crew performance by practicing effective leadership behaviors.

Overbearing captains are not uncommon. Unfortunately, this type of behavior can lead to drastically negative outcomes in times of emergency. Captains are responsible for the flight, however, they are highly dependent on the information provided to them by their first officer and if present second officer. If a captain is over-bearing this can create an environment in which the officers are not comfortable expressing concerns or equally bad, expressing the severity of their concerns. As an illustration, the NTSB ruling on the fatal crash of Air Florida Flight 90 into the Potomac River was that the captain did not react to the co-pilot's concerns. In this case, the co-pilot did express his concerns however, did so subtly and not assertively.

Let us consider the leadership related factors that could have mitigated the tragedy of Flight 90. First, the captain lacked the perception that his co-pilot was there to assist him. Rather, the captain's complete disregard for the co-pilots concerns related to the take off illustrated that the captain in this case believed he knew best. Second, the captain failed to consider the information he was being supplied with from all sources. The co-pilot was the one who felt something was "not right" after brake release. The captain failed to place any relevance to this information. An effective leader would set a team climate that encourages open communication and assertiveness, as well as deferring to expertise (i.e., empowering individuals who are closest to the problem, and deferring to those who have the answer to the problem) regardless of status. If the captain of Flight 90 would have engaged in better leadership processes the results of Flight 90 may have been different.

Shared Mental Models

Shared mental models in teams has been used to explain the effectiveness of team functioning. Effective performance in a dynamic condition has been explained via the use of shared mental models (Cannon-Bowers, Salas, & Converse, 1993). Mental models are organized knowledge structures that allow individuals to predict and explain the behavior of the world around them, to recognize and remember relationships among components of the environment, and to construct expectations for what is likely to occur next (Rouse & Morris, 1986). In sum, mental models aid people in describing, explaining, and predicting events in their environment (Mathieu, Heffner, Goodwin, Salas, & Cannon-Bowers, 2000).

Klimoski and Mohammed (1994) proposed that a team most likely has multiple mental models existing among the members. These models could be models which represent knowledge about the task, the technology, or teamwork (see also Rentsch & Hall, 1994). A task mental model should represent knowledge of how the task is accomplished and the procedures associated with the task. A teamwork mental model should represent how the team interacts; that is, how information is shared, and the interdependencies of the roles. Finally, team members must understand the equipment which they must interact with to accomplish their task. In aviation, understanding the information on the displays and how to utilize the equipment in order to accomplish the task is critical for team functioning. Shared mental models afford the team to anticipate actions, which results in improved team functioning and more effective performance outcomes. Moreover, shared mental models have led to improved processes. That is, teams that hold shared mental models have improved communication, strategy formation, and coordination (Mathieu et al., 2000).

As previously discussed, in January 2009 US Airways Flight 1549 had to engage in an emergency landing into the Hudson River. The successful landing has thus far been attributed to the effectiveness of the crew, how well they were led and their effective coordination. In a recent interview (Klug & Flaum, 2009) with the pilot of

Flight 1549, Captain Sullenberger recalled the few seconds before his crew had to make the emergency landing, stating:

> I made the brace for impact announcement in the cabin and immediately, through the hardened cockpit door, I heard the flight attendants begin shouting their commands in response to my command to brace. "Heads down. Stay down." I could hear them clearly. They were chanting it in unison over and over again to the passengers, to warn them and instruct them. And I felt very comforted by that. I knew immediately that they were on the same page. That if I could land the airplane, that they could get them out safely.

This is an illustration of this crew's shared mental model as well as how their shared mental model impacted the leader's sense of efficacy in their team. Not only in the obvious sense that the cabin crew members knew what they had to do once the captain had made the announcement, but also in the efficacy that the captain had in his cabin crew that once he completed his job, that those in the cabin would know what to do and how to get everyone safely off the plane, which they did with minimal injuries.

Situational Awareness

Situational awareness in aviation has had a well-documented role. Hartel, Smith, and Prince (1991) evaluated over 200 military aviation accidents and determined that the leading causal factor was situation awareness. Furthermore, Endsley (1995) studied accidents among major air carriers and found that 88% of those accidents which involved human errors could be attributed to problems with situation awareness.

Situational awareness is defined by Endsley (1988) as "the perception of the information in the environment within a volume of time and space, the comprehension of their meaning and the projection of their status in the near future" (p. 97). Simply put, a high level of situational awareness involves identifying critical factors in the environment, understanding what those factors mean in relation to the aircrew's goals, and being able to understand what will happen in the near future. This affords the crew to function timely and effectively (Endsley, 1997).

Most of the literature on situational awareness has examined it at the individual level. But as we have discussed commercial planes are not flown by individuals, they are flown by crews. It only seems reasonable that team situational awareness has an impact on crew performance. Researchers who have examined the notion of team situational awareness have discussed what is essential to developing team situation awareness. For example, Bolman (1979) and Wagner and Simon (1990) viewed situation monitoring, cross-checking of information, and coordinating activities as critical to establishing team situational awareness. Schwartz (1990) viewed direction from the captain as crucial to establishing team situational awareness. He argued that the captain's role was to receive person-specific information about the situation and process it into an entire picture. Moreover, Schwartz argued that the level and quality of communication among the crew is critical for developing high levels of team situational assessment.

Salas, Prince, Baker, and Shrestha (1995) presented a framework for building team situational awareness. That is, team members need to have information that will help each individual develop relevant expectations about the team task. Critical to this is communication which can be facilitated by team processes. Salas and colleagues argue that planning and leadership help to build team situational awareness by facilitating effective communication.

To illustrate how the lack of situational awareness can have critical impact on crew performance and aircraft safety we provide the reader with two examples found within archival data assessing flight crashes. On the morning of August 27, 2006, Comair Flight 191 was scheduled to depart from Lexington, Kentucky, to Atlanta, Georgia. The aircraft was assigned to use Runway 22, and in a fatal mistake the crew chose Runway 26, which was too short for a safe takeoff. The crash resulted in the death of 49 of the 50 individuals on board. The NTSB (2007) reported the probable cause of the accident was a loss of positional awareness. This loss of positional awareness was likely due to several causes: (1) the flight crew members' failure to use available cues and aids to identify the airplane's location on the airport surface during taxi, (2) their failure to cross-check and verify that the airplane was on the correct runway before takeoff and (3) the flight crew's irrelevant conversations during taxi.

As another example, in January 1992 Air Inter Flight 148 was on approach to Strasbourg from Lyon, France. The flight crashed into a high ridge near Mount Sainte-Odile in the Vosges Mountains, killing 87 individuals on board. The cause of the crash was found to be a misentered glide slope. Glide slope is the proper path of descent for an aircraft preparing to land. It takes into consideration the angle of distance and altitude from the landing surface. One of the scenarios identified as a possible cause of the crash of flight 148 involved a misentered glide slope (i.e., they entered 3300 feet per minute rather than 3.3 degrees). Based on this scenario, the crew entered a vertical speed of 3300 feet per minute when the correct speed was approximately 800 feet per minute. This means that the plane was descending too quickly and too steeply, which made it impossible to correct and slow the aircraft down. Given this scenario, it is possible that the crew was unaware of the mode that they were in when entering the glide slope data.

Adaptability

In certain contexts the effectiveness of a crew is based on how the team adapts to a multitude of potential unexpected events. These events can be technical (i.e., machine or display failure), external (i.e., bird strikes), and even be person-based (i.e., unexpected incapacitation of an individual on the crew), but how the crews adapt and coordinate to safely fly, land, or embark their aircraft given unexpected events separates those high-performing crews from the rest.

The ability to adapt is an important skill in high-performance teams (Cannon-Bowers, Tannenbaum, Salas, & Volpe, 1995; McIntyre & Salas, 1995). Research has found that groups that engage in adaptive behaviors outperform groups that do not engage in these behaviors. Moreover, studies have demonstrated that the behaviors are not the only factors that impact performance but also the time at which they engaged in adaptive behaviors is critical (Eisenhardt, 1989; Gersick, 1989; Parks & Cowlin, 1995).

In a study by Waller (1999) that examined flight crew performance during unexpected events in a flight simulator found that teams that engaged in adaptive behaviors did have higher ratings of crew performance. Critical to this finding is that the frequency

of engaging in these behaviors did not indicate level of performance, those teams who had a higher frequency of demonstrating
adaptive behaviors did not necessarily do better. Rather, it was the
timing of engaging in these behaviors after an unexpected event
which affected the crew's performance.

Training for adaptability entails a long-term process. This process
includes exposing trainees to extensive experiences. These experiences should address situations that may be frequently encountered
as well as experiences which are novel and vary (Kozlowski, 1998).
However, the circumstances under which air crews are challenged
with are often times unpredictable, and unable to be replicated in
flight simulators. For example, what the air crew of United Airlines
Flight 232 experienced when on a scheduled flight from Stapleton
International Airport, in Denver, Colorado, to O'Hare International
Airport in Chicago on July 19, 1989, has been described by the pilot
as having a-billion-to-one odds of occurring (Haynes, 1991). The
Douglas DC-10 suffered failure of all three of the aircraft's hydraulic
systems (NTSB, 1989).

How the aircrew of Flight 232 responded to the system failures
illustrates how adaptability and good communication, more on
this in the next section, can contribute to effective aircrew performance. The aircrew under the emergency circumstances which
they were experiencing clearly demonstrated that they engaged in
adaptive behaviors. Adaptive behaviors are dictated by the situation, and in this situation the aircrew had to adapt to obstacles
such as not being able to turn left, the failure of the three hydraulic
systems, and the aircraft not responding to flight control inputs.

More specifically, Haynes (1991) reported that the aircrew had no
ailerons—the controlling surface attached to the wing of an airplane—to bank the airplane, had no rudder to turn it, no elevators to
control the pitch (i.e., the up and down movement of the nose of the
plane), no leading-edge flaps or slats to slow the airplane down, no
trailing-edge flaps for landing, no spoilers on the wing to help the
plane down, or help slow down the aircraft once the plane was on
the ground. Furthermore, when on the ground, the aircrew had no
steering, nose wheel or tail, and no brakes. Under those conditions

it was no surprise that the aircrew told air traffic control, "unless we get control of this airplane, we're going to put it down wherever it happens to be" (p. 3, Haynes, 1991). Given the circumstances under which the crew was operating under, and that they were able to save 185 of the passengers on the flight, it is no surprise that this event is often cited as a textbook example of successful CRM.

Communication

Critical to the understanding of team performance is that the processes which we have discussed thus far are not mutually exclusive processes. That is, a team needs to communicate with one another in order to develop a shared mental model, team situational awareness, and adaptability. Therefore the fact that communication has been implicated in 80% of all accidents from 1980 to 2000 (Sexton & Helmreich, 2000) is not surprising. Moreover, 70% of the first 28,000 reports made to the Aviation Safety Reporting System, a system that allows pilots to confidentially report aviation incidents, were related to communication problems (Connell, 1995).

Research conducted in aviation on communication and performance has demonstrated how critical communication is to crew performance. In a full-mission simulator study results showed that crew performance was more closely related with the quality of crew communication than with the technical proficiency of individual pilots (Ruffell Smith, 1979). Moreover, Predmore (1992) discovered that effective flight crews in similar critical situations to Flight 232 made an average of 35 utterances per minute. Predmore (1991) rigorously analyzed the communication of Flight 232 and determined that the communication performance of the aircrew was at a rate of 60 utterances per minute.

However, number of communication utterances is not simply enough. The type of communication is equally important. Research has indicated that teams that engage in proactive communication as well as communication related to planning and situation assessment perform better than teams that do not (Macmillan, Entin, & Serfaty, 2002; Wright & Kaber, 2005). Teams which practice closed-loop communication as well as short and concise communication will also benefit more than those teams who do not. Closed-loop communication consists of three steps: (1) the sender of a message must follow up to ensure that their message was received, (2) the receiver should acknowledge that the message was received, and (3) clarification between the sender and receiver to ensure that the message sent was in fact the intended message (McIntyre & Salas, 1998).

The captain of Flight 232 reported that one of the main factors that contributed to the survival of 185 passengers was the communication experienced in the cockpit as well as the communication with air traffic control and the emergency services on the ground. Haynes (1991) explained that the controller who was assigned to the radar station in Sioux City, where Flight 232 landed, was "calm and concise" (p. 5) in his communication. He provided the crew with information that helped them determine their position in relation to the airport, and information regarding landing options. Moreover the response from emergency services to the call for emergency made by the crew was quick. The 20 minutes that emergency rescue staff and fire-fighters had before United Flight 232 touched ground allowed them to prepare and implement their own disaster plan.

The dynamics in the cockpit are extremely interdependent. Those in the cockpit must communicate with one another but also with those in air traffic control. Communication plays a critical role in issuing and acknowledging commands, conducting briefings, performing callouts, stating intentions, asking questions, and conveying information (Kanki, 1995). Equally as critical is the team climate that is set in the cockpit by the captain. Captains can encourage open lines of communication between the officers and the captain, as well as encouraging assertiveness. Engaging in these processes is critical to facilitate development of team situational awareness as well as engaging in adaptive behaviors.

Summary

A search for a better understanding of what impacts crew performance has been occurring for many years. Aviation researchers have considered critical findings presented in the teams literature regarding team effectiveness and have applied it to the development and assessment of aircrew performance. Team processes is viewed as an indicator of team effectiveness and interventions like CRM training have been implemented in aviation to improve critical processes.

Critical indicators of effective team processes have been identified in the literature (e.g., McIntyre & Salas, 1995; Salas et al., 2005; Stout, Cannon-Bowers, Salas, & Milanovich, 1999). In this chapter we discussed the impact that leadership, shared mental models, situational awareness, adaptability, and communication have on crew performance. This section was not an exhaustive list of processes that impact crew performance rather it was offered as a discussion of critical factors which we highlight as important.

We feel it is key to mention that many of these factors that we discussed as critical should not just be considered as independent factors. Their impact on crew performance may be substantially greater when the factors are considered together. For example, how does the leader impact a crew's communication? How can the leader take poor communication in a team, and improve it? Also, how does a crew's ability to communicate and practice situational awareness help the crew's learning and adaptability? Questions like these are critical to determining the true impact of team processes on aircrew performance and effectiveness.

HOW CAN AVIATION CREW PERFORMANCE BE IMPROVED?

The underlying goal of understanding all of these processes that impact aviation crew performance has consistently been the improvement of such crews. While research on these dynamics has grown, so have the methods and strategies for training aviation crews. As discussed previously, team training promotes teamwork and enhances team performance, and therefore it has become a key part of training for aviation crews (Salas, Cooke, & Rosen, 2008). Additionally, factors such as the increased technological capabilities available have undoubtedly played a large role in the advancement of such training strategies. Thus, in the following section we present some of the key training practices utilized to improve aviation crews, with a strong focus on CRM and its advancement in recent years.

Crew Resource Management (CRM)

Throughout the chapter, we have made reference to CRM as a key training strategy for enhancing aviation crew performance. The concept of CRM training originated from a workshop sponsored by NASA in 1979 (Helmreich, Merritt, & Wilhelm, 1999). This workshop was brought about by the recognition of an increase in human error as the cause of numerous aviation incidents. CRM training has gone through multiple changes and iterations, but the basic principles have remained the same. Salas and colleagues (1999) define CRM training as an instructional strategy "designed to improve teamwork in the cockpit by applying well-tested training tools (e.g., performance measures, exercises, feedback mechanisms) and appropriate training methods (e.g., simulators, lectures, videos) targeted at specific content (i.e., teamwork knowledge, skills, and attitudes)" (p. 163).

From its beginning, CRM training has been an evolving process, going through a series of five major generations (Helmerich, Merritt, & Wilhelm, 1999; Wilson et al., 2006). The first generation of CRM training was implemented by United and KLM airlines in the early 1980s, and was designed primarily upon existing training approaches for management. This program was relatively psychological in nature, as it focused on psychological testing and general interpersonal behaviors, but it lacked an emphasis on team behaviors specific to the aviation community. Thus, the second generation of training was designed to emphasize the team. Part of this involved changing the title from its original "cockpit resource management" to the current "crew resource management." In 1989, the FAA set guidelines by which airlines were to develop, implement, reinforce, and assess CRM training (i.e., Advisory Circular 120–51D).

These guidelines were soon followed by the third generation of CRM training, focused upon an integration of technical and CRM skills training. During this time, CRM also began to expand beyond just flight deck crews to others, including check airmen and cabin crews. The fourth generation of CRM, the Advanced Qualification Program (AQP), was designed to tailor CRM training to specific organizational needs for each airline. Other changes during the fourth generation of CRM involved requiring airlines to train crew members on CRM and line operational flight training (LOFT), a training developed before the introduction of CRM, and nontechnical and technical skills, as well as to use line oriented evaluations (LOEs) for full mission simulators (for more information regarding these programs, see Helmreich et al., 1999; Weiner et al., 1993; Wilson et al., 2006).

A fifth focus shift for CRM training occurred in 1997, when error management became a critical issue. The fifth generation of CRM highlights the idea that human error is inevitable, and introduces three countermeasures to error: avoiding error, trapping incipient errors, and mitigating the consequences of errors that have already occurred. This generation also places emphasis on the importance of feedback as well as the reinforcement of CRM concepts through LOFT, LOE, and in-line checking. This generation is the one currently still in effect in the aviation industry, although previous generations are also still influential (Wilson et al., 2006).

Overall, CRM training has been viewed as highly beneficial to the aviation industry. In a recent meta-analysis, O'Connor and colleagues (2008) examined evaluations of CRM training in order to determine the level of effectiveness that such training has on individual, team, and organizational outcomes. Their results found that CRM training had a significant impact on participants' attitudes and behaviors and a medium effect on their knowledge. Such findings are encouraging in terms of the benefits of CRM training, primarily in terms of its impact at the individual level. However, additional evaluation is necessary in order to ensure its added value to the aviation community, particularly in terms of assessing its impact on organizational and industry level factors such as accident rates.

Tools for Training CRM

Certainly, our understanding of CRM training has changed over the past few decades, as have our training methods. Much of this is due to the fact that the advancement of technology (e.g., introduction of aviation simulators) has impacted the methods by which training is delivered. The use of computer games to train crew members has certainly gained in popularity (Baker, Prince, Shrestha, Oser, & Salas, 1993), as has the use of the rapidly reconfigurable event-set based line-oriented evaluations generator (RRLOE; Wilson, et al., 2006) to determine performance levels and proficiency. Other tools that have been utilized in the application of CRM strategies involve the use of full motion simulators, which provide realistic situations in which crew members can be exposed to a range of events and environmental conditions while in a safe learning environment. However, these can be rather costly and not always necessary for training nontechnical skills. PC-based simulators often serve as an effective alternative to the full motion simulators, providing crew members with a level of physical fidelity that is just as capable of providing a successful learning environment as the more high fidelity environments.

Other Best Practices in Aviation Crew Training

While CRM training is possibly the most widespread instructional strategy used for training aviation crews, it is also important to acknowledge additional training practices that are beneficial for improving aviation crew performance. These include training tools and instructional strategies that can be used in tandem to accent the strengths of CRM. Wilson and colleagues (2006) provide a review of current training strategies utilized in training team processes in aviation, highlighting the benefits of scenario based training, metacognitive training, and assertiveness training as methods that can complement CRM strategies. Scenario based training embeds learning into simulation scenarios derived from critical incidents, which provides a meaningful framework upon which crew members can learn critical skills. Metacognitive training enables crew members to be aware of their decision making process and therefore enables teams to respond appropriately in dynamic environments (Smith, Ford, & Kozlowski, 1997). Assertiveness training is another training method currently used in aviation to complement CRM. This method teaches crew members to clearly address their concerns, ideas, feelings, and needs with other crew members (Smith-Jentsch, Salas, & Baker, 1996). This is an important strategy to learn as junior crew members must be able to feel comfortable speaking up when they notice errors that more senior team members may have not recognized. This type of instruction can be enhanced through role-play, such as through scenario-based training.

Summary

CRM training has been key to improving aviation crew performance, as have the other training methods that complement it. We have come far in our understanding of what aviation crews need to learn in order to reduce human error, and the iterative generations of CRM reflects this advancement. However, this does not mean that CRM training is perfect. Indeed, Salas, Wilson, Burke, and Wightman's (2006) review of CRM training highlights the success of CRM use but also illustrates mixed reviews for the impact that CRM has on organization's bottom line. Understanding the role of measurement in assessing and evaluating the impact of CRM training is critical, as will be discussed in the following section.

WHERE DO WE GO FROM HERE?

Although much advancement has been made in understanding aviation crew performance, there is still much to be discovered. Using the future tasks and directions provided by Foushee and Helmreich (1988) as a framework, in the following section we will briefly highlight how these future directions have been advanced, and discuss how they can continue to be explored. Furthermore, we will highlight additional future research tasks, based on current best practices in aviation crew training., measurement, and performance.

Foushee and Helmreich's Future Directions

Foushee and Helmreich (1988) make several recommendations regarding directions in which aviation crew performance can be explored, with a heavy emphasis on improving training design, instruction, and evaluation. Based on the advances made in both training and understanding team dynamics, we have been able to address their concerns to some extent. However, several are in need of continued research.

First, Foushee and Helmreich identify a need for increased emphasis on training programs aimed at facilitating crew coordination and effectiveness. Undoubtedly, the increased emphasis on CRM training has been an excellent proponent of this particular future direction. However, as will be discussed later, this does not mean that everything is known regarding how to effectively train aviation crews on teamwork. While CRM has provided an excellent starting point for identifying aviation crew needs in terms of human error reduction, more research is needed to ensure that this method is adequately able to prevent human errors from occurring to the fullest extent possible. This has particularly strong implications for the evaluation and assessment of CRM and other aviation crew training strategies. On a related note, Foushee and Helmreich call for evaluation methods to protect individual participants, preventing them from being negatively assessed based on the evaluation results of such training programs. Using appropriate strategies, researchers have for the most part been able to address this issue, such as through the use of sanitized data (Merket, Bergondy, & Salas, 2000). However, researchers must continue to be cognizant of the potential implications of assessment upon individual crew member's willingness and ability to participate in training evaluations.

Another concern of Foushee and Helmreich was the necessity of high-fidelity simulations. While high-fidelity simulations have appeal, the authors questioned their need in terms of providing adequate training for teamwork skills. Research has been conducted to address this issue, and several insights have been gained over the years. While the assumption has typically been that high cognitive and physical fidelity is most effective at ensuring training transfer, this fidelity may not always be necessary to achieve learning goals (Gopher, Weil, & Barket, 1994; Jenstch & Bowers, 1998). Taylor and colleagues (1993) found that when a high-fidelity simulation was used to train pilots, the training actually had minimal effects for on the job performance. Results such as these have led to the argument that lower-fidelity simulations, such as the use of PC-based video games, can be just as effective. The key to this argument is that such methods must have a match between the psychological fidelity and the training needs of the crew members (Bowers & Jenstsch, 2001). Overall, lower cognitive and physical fidelity methods have been found to be successful in the transfer of skills to the cockpit, and thus we should continue to explore their value—as well as the value of high-fidelity simulations—in order to advance our understanding of their benefits.

New Future Research Tasks and Directions

Although there is a need to continue to explore the areas recommended by Foushee and Helmreich (1988), there are additional areas in need of advancement to better understand aviation crew performance. First, research must continue to improve CRM, particularly in terms of accessing data and evaluating programs. Second, there is a need for continued advancement regarding the impact of culture in the cockpit. Third, understanding the role of individual characteristics in crew dynamics is important in comprehending their impact on team effectiveness. Fourth, addressing measurement issues in terms of evaluating team training and team effectiveness is undoubtedly a critical issue. Finally, as CRM has expanded to other industries, it is important that the aviation industry utilizes the advances from these areas to continue improvement of its own training methods. Each of these future research needs will be explored in more detail in the following section.

Continuous Improvement of CRM

We have learned much from the development of CRM training over the past two decades, with numerous advancements made in terms of design, development, implementation, and evaluation of CRM programs (Salas et al., 2006). Table 9.1 provides an overview of the lessons learned from the research and development of CRM training, as presented by Salas and colleagues in their recent review of the CRM literature. However, while CRM has become one of the most successful training methods of its kind, it

TABLE 9.1   Lessons Learned from CRM Training Research & Practice

Themes	Description
Theoretical Drivers	
Lesson 1	Relevant and practical theories abound in CRM research.
Lesson 2	CRM research is solidly guided by meaningful theoretical drivers.
Research Advances	
Lesson 3	Assessment centers and situational judgment tests (SJTs) are strategies for selecting team members with CRM-based skills.
Lesson 4	Culture (organizational and national) matters to CRM-related performance.
Lesson 5	Simulations are a viable alternative to train teams in a safe yet realistic learning environment.
Lesson 6	Low-fidelity simulations offer a cost effective, viable alternative to large-scale, high-fidelity simulations for training CRM-related skills.
Lesson 7	The assessment of team dynamics remains a critical component during training and on the job.
Lesson 8	CRM-related performance is composed of a set of interrelated knowledge, skills, and attitudes.
Lesson 9	There are a number of valid, reliable, and diagnostic measurement tools available to researchers and practitioners.
Training Effectiveness and Evaluation	
Lesson 10	The desired CRM-related learning outcomes must drive which competencies to capture, assess, diagnose, and remediate.
Lesson 11	Simulation-based training is a powerful strategy for acquiring the CRM-related processes and outcomes.
Lesson 12	CRM simulation-based training is effective when CRM-relevant instructional features are embedded in it.

(Continued)

TABLE 9.1 Continued

Themes	Description
Lesson 13	CRM training is only one team training strategy. Additional team-focused strategies are available that should be used in conjunction with CRM training.
Lesson 14	The science of training can (must) guide the design, delivery, and evaluation of CRM training.
Lesson 15	CRM training must be evaluated and transfer encouraged.
Lesson 16	CRM training leads to positive reactions, positive changes in attitudes toward CRM concepts, and the transfer of learned competencies to a simulated or operational environment.
Lesson 17	The impact of CRM training on organizational results (e.g., safety, error reduction) is inconclusive.
Lesson 18	More multilevel training evaluations are needed.
Lesson 19	CRM training may not be having as great an impact on accident rates as once believed; more data are needed.

Adapted from Salas et al., 2006.

is not without flaws. As can be seen by the lessons learned, there is undoubtedly still much to be understood regarding CRM training and its impact on outcomes.

One issue is the fact that there are numerous terms used to describe the same type of training: while commercial aviation refers to it as CRM training, the U.S. military uses a variety of labels (Wilson et al., 2006). Specifically, although the U.S. Navy and Air Force utilize CRM, the U.S. Army refers to its training program as Aircrew Coordination Training. The primary difference between these programs is not content but the developmental environment, it would be beneficial to find a more effective labeling method to bring about uniformity, especially as CRM expands to other industries.

Another area of improvement for CRM is the standardization of competencies included in CRM training (Salas et al., 2006).

Salas and colleagues (2001) identified 36 different skills that have been trained by at least one agency as a part of CRM in aviation. Although the most common competencies are communication, situation awareness/assessment, and decision making, these are not the finite set of competencies that are included in all training (Bowers, Jentsch, & Salas, 2000). Furthermore, this information is derived only from studies that have actually published information regarding their training methodology. Although the FAA does provide some guidance as to how CRM training should be implemented (FAA, 2001), a standardized set of competencies should be selected and implemented in order to make CRM training more consistent (DoD Working Group, 2005).

Improve Knowledge of Cultural Effects

In 1977 two Boeing 747 airliners collided on the runway of Los Rodeos Airport on the Spanish island of Tenerife. This crash resulted in the deaths of 583 people, making it the worst accident in aviation history. The number of fatalities caused by airplane-related crashes was exceeded only by the terrorist attacks on September 11, 2001. Upon investigation, it was found that one of the factors contributing to this crash was the use of nonstandard phrases in communications between one of the pilots and the flight engineer—whose native language was not English, although that was the language being used during the conversation (http://www.airdisaster.com).

As previously discussed, culture is only going to continue to impact aviation crew performance (Merritt & Helmreich, 1996). As can be seen from this incident, understanding how cultural nuances—such as the use of particular words or phrases—can have great impacts on the prevention or escalation of human error (Helmreich & Davies, 2004). This is true in terms of both the national cultural differences that can hinder communication among crew members, as well as aviation industry cultural norms that can enhance or reduce crew members' proclivity for paying attention to errors and promoting a culture of safety (Mjos, 2004; Redding & Ogilvie, 1984). However, we still have much to learn in terms of the specific ways that culture can impact aviation crew performance. For example, it is unknown as to what adjustments may need to be made in order for CRM principles to effectively align with a given culture (Merritt & Helmreich, 1996). Future research should explore this and other cultural issues, including ways in which communication can be improved if crew members are not all from the same culture—or if the people on the ground with whom they are communicating do not speak the same language.

Explore the Role of Individual Characteristics in Teams

Historical perspectives have painted the picture of pilots as being "lone wolves," or individuals who are self-sufficient, technically qualified, and slightly egotistical in the beliefs about their own capabilities (Foushee, 1984). As these individual characteristics are now viewed as factors that could potentially lead to human error since these "lone wolves" may be less likely to listen to their teammates, it is important that we continue to explore how teams are impacted by the individual differences that their group members bring to the table. Driskell and colleagues (2006) recently examined the impacts of personality on team effectiveness, providing a framework of personality facets and how they might interact with team performance. However, further research is needed to empirically test these predictions and link them to specific outcome factors. Additionally, we need to determine what the implications of such research would be for aircrew selection, composition, and training.

Advance the Measurement of Training Evaluation and Team Effectiveness

Being able to accurately assess the impact of training on aviation crew performance is critical to its success in reducing human error. Unfortunately, our current measurement practices of training evaluation and team effectiveness are not as accurate and precise as would be hoped. Although measures such as the nontechnical evaluation skills method (NOTECHS) exist for evaluating performance following training, we still need consistent application of such measures in order for them to be effective (Wilson et al., 2006). For example, in their meta-analysis of CRM training effectiveness studies, O'Connor and colleagues (2008) found that in general, CRM training has led to positive results. However, they noted that many studies evaluating CRM training did not effectively report evaluation results, leading to difficulty in assessing the effectiveness of such programs. As reflected in Table 9.1, Salas and colleagues (2006) found in their review of the CRM training literature that studies investigating the impact of CRM training on organizational outcomes and accident rates are either inconclusive or reveal that CRM may not have as much of an impact as once believed. Therefore, it is vital that future users of CRM training be more rigorous in the reporting of such information, particularly in terms of evaluating the effects of CRM training at the organizational level.

Another issue relevant to the assessment and evaluation of aviation crew performance is the use of automated systems for evaluation purposes. One such system, the Enhancing Performance with Improved Coordination (EPIC) tool, is a computer-based system that has been found to augment the abilities of an instructor when assessing air crew performance (Deaton, Bell, Fowlkes, Bowers, Jentsch, & Bell, 2007). This system allows instructors to use an automated method which complements the instructor's own rating capabilities and provides a more accurate rating of performance. The use of such a system aids in reducing the impact of high workload on team instructor's ability to complete ratings, creating a more powerful and diagnostic rating. The continued advancement and incorporation of such systems should aid in better assessing air crew performance, enabling a more accurate understanding of current and future needs.

Learn from Other Industries

Although CRM and similar training practices began in the aviation industry as a means to reduce human error in flight, these practices have begun to spread to other organizations and industries who are also interested in the reduction of error. Indeed, Lyndon (2006) highlights how high risk domains are beginning to understand the necessary role that teamwork and team performance plays in critical outcomes such as patient survival in healthcare. There has been a major thrust in recent years to better develop teams in these organizations, and aviation may be able to benefit from what is gleaned in these areas, just as they have benefited already from the advances in CRM training in aviation. The nuclear power industry, medicine, and the offshore oil and gas industry are all currently practicing CRM training in some form (Flin, O'Connor, & Mearns, 2002). For example, many hospitals are incorporating CRM into operating rooms, as teamwork is critical in patient safety and care (Salas, DiazGranados, Weaver, & King, 2008). By working with these industries to continue developing CRM and similar training techniques, the aviation industry can maintain growth in understanding human performance and error reduction in complex and stressful environments.

CONCLUDING REMARKS

The understanding of aviation crew performance has come a long way since Foushee and Helmerich's original publication review on flight crew performance research two decades ago. Advances in technology, training design, and the introduction and incorporation of team training and crew resource management principles as basic components of aviation training has done much to reduce the level of human error in the aviation industry. However, there is still much to be uncovered, particularly regarding the evaluation and assessment of such programs in order to ensure their continued effectiveness. Additionally, the globalization of organizations will continue to affect aviation crew culture, both through the connection of differing nationalities and through the adaptation of the aviation industry's own culture. It is hoped that this chapter has given a high-level review—the 35,000-foot view, that is—of how things have changed in the past several decades, as well as give a preview of what is to come in terms of understanding aviation crew performance. While human error can never be reduced to zero, the leaps and bounds made thus far in group dynamics and aviation crew training have certainly started a revolution that will continue to grow and in turn reduce human error as much as possible.

ACKNOWLEDGMENT

The research and development reported here was supported partially by the FAA Office of the Chief Scientific and Technical

Advisor for Human Factors (AAR-100). Dr. Eleana Edens was the contracting officer's technical representative at the FAA. All opinions stated in this article are those of the authors and do not necessarily represent the opinion or position of the University of Central Florida or the Federal Aviation Administration.

References

Baker, D., Prince, C., Shrestha, L., Oser, R., & Salas, E. (1993). Aviation computer games for crew resource management training. *International Journal of Aviation Psychology, 3*(2), 143–156.

Bass, B. M. (1990). From transactional to transformational leadership: Learning to share the vision. *Organizational Dynamics, 18*(3), 19–36.

Berry, J. W., Poortinga, Y. H., Segall, M. H., & Dasen, P. R. (1992). Cross-cultural psychology: Research and applications. Cambridge, UK: Cambridge University Press.

Bolman, L. (1979). Aviation accidents and the "theory of the situation". In G. E. Cooper, M. D. White, & J. K. Lauber (Eds.), *Resource management on the flight deck: Proceedings of a NASA/industry workshop* (pp. 31–58). Moffett Field, CA: NASA Ames Research Center.

Brannick, M. T., Prince, C., & Salas, E. (2005). Can PC-based systems enhance teamwork in the cockpit? *International Journal of Aviation Psychology, 15*, 173–187.

Burke, C. S., Wilson, K. A., & Salas, E. (2003). Teamwork at 35,000 feet: Enhancing safety through team training. *Human Factors and Aerospace Safety, 3*(4), 287–312.

Cannon-Bowers, J. A., Salas, E., & Converse, S. A. (1993). Shared mental models in expert decision making teams. In N. J. Castellan (Ed.), Jr., *Current Issues in individual and group decision making* (pp. 221–246). Hillsdale, NJ: LEA.

Cannon-Bowers, J. A., Tannenbaum, S. I., Salas, E., & Volpe, C. E. (1995). Defining competencies and establishing team training requirements. In R. A. Guzzo, E. Salas, & Associates (Eds.), Team effectiveness and decision making in organizations (pp. 333–380). San Francisco: Jossey-Bass.

Cohen, S. G., & Bailey, D. E. (1997). What makes teams work: Group effectiveness research from the shop floor to the executive suite. *Journal of Management, 23*(3), 239–290.

Connell, L. (1995, November 14–16 1995). Cabin crew incident reporting to the NASA aviation safety reporting system. *Paper presented at the international conference on cabin safety research*. Atlantic City, NJ.

Deaton, J. E., Bell, B., Fowlkes, J., Bowers, C., Jentsch, F., & Bell, M. A. (2007). Enhancing team training and performance with automated performance assesment tools. *International Journal of Aviation Psychology, 17*(4), 317–331.

Driskell, J. E., Goodwin, G. F., Salas, E., & O'Shea, P. G. (2006). What makes a good team player? Personality and team effectiveness. *Group Dynamics, 10*(4), 249–271.

Eisenhardt, K. M. (1989). Making fast strategic decisions in high-velocity environments. *Academy of Management Journal, 32*(3), 543–576.

Endsley, M. R. (1988). Design and evaluation for situation awareness enhancement. *Paper presented at the Proceedings of the Human Factors and Ergonomics Society 32nd Annual Meeting*. Santa Monica, CA.

Endsley, M. R. (1995). A taxonomy of situation awareness errors. In R. Fuller, N. Johnston, & N. McDonald (Eds.), *Human factors in aviation operations* (pp. 287–292). Aldershot, UK: Avebury Aviation Ashgate Publishing Ltd.

Endsley, M. R. (1997). The role of situation awareness in naturalistic decision making. In C. E. Zsambok & G. A. Klein (Eds.), *Naturalistic decision making* (pp. 269–284). Mahwah, NJ: LEA.

Entin, E. E., & Serfaty, D. (1999). Adaptive team coordination. *Human Factors, 41,* 312–325.

Foushee, H. C. (1984). Dyads and triads at 35,000 feet: Factors affecting group process and aircrew performance. *American Psychologist, 39*(8), 885–893.

Foushee, H. C., & Helmreich, R. L. (1988). Group interaction and flight crew performance. In E. L. Weiner & D. C. Nagel (Eds.), *Human factors in aviation* (pp. 189–227). San Diego, CA: Academic Press.

Gersick, C. (1989). Time and transition in work teams: Toward a new model of group development. *Academy of Management Journal, 31*(1), 9–41.

Ginnett, R. C. (1993). Crews as groups: Their formation and their leadership. In E. L. Weiner, B. G. Kanki, & R. L. Helmreich (Eds.), *Cockpit resource management* (pp. 71–98). San Diego, CA: Academic Press.

Gopher, D., Weil, M., & Bareket, T. (1994). Transfer of sill from a computer game trainer to flight. *Human Factors, 36*(3), 387–405.

Guzzo, R. A., & Dickson, M. W. (1996). Teams in organizations: Recent research on performance and effectiveness. *Annual Review of Psychology, 47,* 307–338.

Hackman, J. R., & Morris, C. G. (1975). Group tasks, group interaction process and group performance effectiveness: A review and partial integration. In L. Berkowitz (Ed.), *Advances in experimental social psychology: Vol. 8* (pp. 47–99). New York: Academic Press.

Hartel, C. E., Smith, K., & Prince, C. (1991). Defining aircrew coordination: Searching mishaps for meaning. *Paper presented at the sixth international symposium on aviation psychology.* Columbus, OH.

Haynes, A. (1991). *The crash of United flight 232.* Edwards, CA: NASA Ames Research Center Dryden Flight Research Facility.

Helmreich, R. (1997). Managing human error in aviation. *Scientific American,* 62–67.

Helmreich, R. L. (1994). Anatomy of a system accident: The crash of Avianca flight 052. *International Journal of Aviation Psychology, 4,* 265–284.

Helmreich, R. L., & Foushee, H. C. (1993). Why crew resource management? Empirical and theoretical bases of human factors training in aviation. In E. L. Weiner, B. G. Kanki, & R. L. Helmreich (Eds.), *Cockpit resource management* (pp. 3–45). San Diego, CA: Academic Press.

Helmreich, R. L., Merritt, A. C., Sherman, P. J., Gregorich, S. E., & Weiner, E. L. (1993). The flight management attitudes questionnaire (FMAQ) Austin, TX: The University of Texas. *NASA/UT/FAA Technical Report, 93–94.*

Helmreich, R. L., Merritt, A. C., & Wilhelm, J. A. (1999). The evolution of crew resource management training in commercial aviation. *International Journal of Aviation Psychology, 9*(1), 19–32.

Hofstede, G. (1980). *Culture's consequences: International differences in work-related values.* Beverly Hills: Sage.

Hofstede, G. (1991). Cultures and organizations: Software of the mind. Maidenhead, UK: McGraw-Hill.

Ilgen, D. R., Hollenbeck, J. R., Johnson, M., & Jundt, D. (2005). Teams in organizations: From I-P-O models to IMOI models. *Annual Review of Psychology, 56,* 517–543.

Jentsch, F., & Smith-Jentsch, K. A. (2001). Assertiveness and team performance: More than "just say no". In E. Salas, C. A. Bowers, & E. Edens (Eds.), *Improving teamwork in organizations* (pp. 73–94). Hillsdale, NJ: Erlbaum.

Kanki, B. G. (1995). A training perspective: Enhancing team performance through effective communication. *Paper presented at the proceedings of the methods and metrics of voice communication workshop.*

Klein, B. (2000). Fisher-General motors and the nature of the firm. *The Journal of Law and Economics, 43*(1), 105–142.

Klein, G. (1989). Strategies of decision making. *Military Review,* 56–64.

Klein, G. (1998). Sources of power. Cambridge, MA: MIT Press.

Klein, G., Calderwood, R., & Clinton-Cirocco, A. (1986). Rapid decision making on the fire ground. *Paper presented at the proceedings of the human factors society 30th annual meeting.* Santa Monica, CA.

Klein, G., & Kozlowski, S. (2000). A multilevel approach to theory and research on organizations: Contextual, temporal, and emergent processes. San Francisco, CA: Jossey-Bass.

Klimoski, R. J., & Mohammed, S. (1994). Team mental model: Construct or metaphor? *Journal of Management, 20*(2), 403–437.

Klug, R., Flaum, A. T. (Directors). (February 8, 2009). The amazing pilot and air crew of US Airways flight 1549. In J. Fager(Executive Producer), *60 Minutes.* New York: CBS.

Kozlowski, S. W. J. (1998). Training and developing adaptive teams: Theory, principles, and research. In J. A. Cannon-Bowers & E. Salas (Eds.), *Making decisions under stress: Implications for individual and team training* (pp. 115–153). Washington, DC: American Psychological Association.

Learmount, D. (2008). Crew factors top accident causes. *Flight International, 174*(5165), 16.

MacMillan, J., Entin, E. E., & Serfaty, D. (2002). From team structure to team performance: A framework. In: *Proceedings of the Human Factors and Ergonomics Society 46th Annual Meeting.* Santa Monica, CA: Human Factors and Ergonomics Society (pp. 408–412).

Marks, M. A., Mathieu, J. E., & Zaccaro, S. J. (2001). A temporally based framework and taxonomy of team processes. *The Academy of Management Review, 26*(3), 356–376.

Mathieu, J. E., Heffner, T. S., Goodwin, G. F., Salas, E., & Cannon-Bowers, J. A. (2000). The influence of shared mental models on team process and performance. *Journal of Applied Psychology, 85*(2), 273–283.

Mathieu, J. E., Maynard, M. T., Rapp, T., & Gilson, L. (2008). Team effectiveness 1997–2007: A review of recent advancements and a glimpse Into the future. *Journal of Management, 34*(3), 410–476.

McGrath, J. E. (1984). Groups: Interaction and performance. Englewood Cliffs, NJ: Prentice Hall.

McGrath, J. E. (1964). Social psychology: A brief introduction. New York: Holt, Rinehart & Winston.

McIntyre, M., & Salas, E. (1995). Measuring and managing for team performance: Emerging principles from complex environments. In R. A. Guzzo & E. Salas (Eds.), *Team effectiveness and decision making in organizations* (pp. 149–203). San Francisco, CA: Jossey-Bass.

Merket, D., Bergondy, M., & Salas, E. (1999). Making sense out of team performance errors in military aviation environments. *Transportation Human Factors, 1*(2), 231–242.

Merritt, A. C. (2000). Culture in the cockpit: Do Hofstede's dimensions replicate?. *Journal of Cross-Cultural Psychology, 31*, 283–301.

Merritt, A. C. (1996). *National culture and work attitudes in commercial aviation: A crosscultural investigation.* Unpublished doctoral dissertation, University of Texas at Austin

Merritt, A. C., & Helmreich, R. L. (1996, November 1995). Creating and sustaining a safety culture—some practical strategies (in aviation). *Proceedings of the 3rd australian aviation psychology symposium.* Manly, Australia.

Merritt, A. C., & Helmreich, R. L. (2007). Human factors on the flight deck: The influence of national culture. In F. E. Jandt (Ed.), *Intercultural communication.* Thousand Oaks, CA: Sage Publications.

Mjos, K. (2004). Basic cultural elements affecting the team function on the flight deck. *International Journal of Aviation Psychology, 14*(2), 151–169.

NTSB. (1989). *Aircraft accident report—United Airlines flight 232, McDonnell Douglas DC-10, Sioux Gateway Airport, Sioux City, Iowa (No. NTSB/AAR-90/06).* Washington, DC: National Transportation Safety Board.

Nullmeyer, R. T., Stella, D., Montijo, G. A., & Harden, S. W. (2005). Human factors in air force flight mishaps: Implications for change. *Paper presented at the The Interservice/Industry Training, Simulation & Education Conference (I/ITSEC).* Mesa, AZ.

O'Connor, P., Campbell, J., Newon, J., Melton, J., Salas, E., & Wilson, K. A. (2008). Crew resource management training effectiveness: A meta-analysis and some critical needs. *The International Journal of Aviation Psychology, 18*(4), 353–368.

Ooi, T. S. (1991). Cultural influences in flight operations. *Paper presented at the SAS Flight Academy international Training Conference.* Stockholm, Sweden.

Orasanu, J. (1993). Decision making in the cockpit. In E. L. Weiner, B. G. Kanki, & R. L. Helmreich (Eds.), *Cockpit resource management* (pp. 137–172). San Diego, CA: Academic Press.

Parks, C. D., & Cowlin, R. (1995). Group discussion as affected by number of alternatives and by a time limit. *Organizational Behavior and Human Decision Processes, 62*, 276–285.

Predmore, S. C. (1991). Microcoding of communications in accident investigation: Crew coordination in United 811 and United 232. In R. S. Jenson (Ed.), *Proceedings of the Sixth International Symposium of Aviation Psychology* (pp. 350–355). Columbus: Ohio State University.

Predmore, S. C. (1992). *The dynamics of group performance: A multiple case study of effective and ineffective flight crew performance* Unpublished doctoral dissertation. Austin, TX: University of Texas.

Prince, C., Salas, E. (1999). Team processes and their training in aviation. In D. Garland, J. Wise, D. Hopkins (Eds.), *Handbook of aviation human factors* (pp. 193-213). Mahwah, NJ.

Rentsch, J. R., & Hall, R. J. (1994). Members of great teams think alike: A model of the effectiveness and schema similiarity among team members. *Advances in Interdisciplinary Studies of Work Teams, 1*, 223–261.

Rivera, R. (January 17, 2009). A pilot becomes a hero years in the making. *New York Times*. Retrieved from http://www.nytimes.com/2009/01/17/nyregion/17pilot.html

Rouse, W. B., & Morris, N. M. (1986). On looking into the black box: Prospects and limits in the search for mental models. *Psychological Bulletin, 100*, 349–363.

Ruffell Smith, H. P. (1979). *A simulator study of the interaction of pilot workload with errors, vigilance, and decisions (No. 78482 NASA Technical Memorandum)*. Moffett Field, CA: NASA Ames Research Center.

Salas, E., Cooke, N. J., & Rosen, M. A. (2008). On Teams, teamwork, and team performance: discoveries and developments. *Human Factors, 50*(3), 540–547.

Salas, E., DiazGranados, D., Weaver, S. J., & King, H. (2008). Does team training work? Principles for healthcare. *Academic Emergency Medicine, 15*, 1002–1009.

Salas, E., Fowlkes, J. E., Stout, R. J., Milanovich, D. M., & Prince, C. (1999). Does CRM training improve teamwork skills in the cockpit?: Two evaluation studies. *Human Factors, 41*(2), 326–343.

Salas, E., Kosarzycki, M. P., Tannenbaum, S. I., & Carnegie, D. (2004). Principles and Advice for Understanding and Promoting Effective Teamwork in Organizations. In R. Burke & C. Cooper (Eds.), *Leading in turbulent times: Managing in the new world of work* (pp. 95–120). Malden, MA: Blackwell.

Salas, E., Prince, C., Baker, D., & Shrestha, L. (1995). Situation awareness in team performance: Implications for measurement and training. *Human Factors, 37*(1), 123–136.

Salas, E., Prince, C., Bowers, C. A., Stout, R. J., Oser, R. L., & Cannon-Bowers, J. A. (1999). A methodology for enhancing crew resource management training. *Human Factors, 41*(1), 161–172.

Salas, E., Sims, D. E., & Burke, C. S. (2005). Is there a "big five" in teamwork? *Small Group Research, 36*(5), 555–599.

Salas, E., Stagl, K. C., Burke, C. S., & Goodwin, G. F. (2007). Fostering team effectiveness in organizations: Toward an integrative theoretical framework of team performance. In R. A. Dienstbier (Ed.), *Modeling complex systems: Motivation, cognition and social processes, Nebraska Symposium on Motivation: Vol 51* (pp. 185–243). Lincoln: University of Nebraska Press.

Salas, E., Stagl, K. C., Burke, C. S., & Goodwin, G. F. (2007). Fostering Team Effectiveness in Team Organizations: Toward and Integrative Theoretical Framework. In B. Shuart, W. Spaulding, & J. Poland (Eds.), *Modeling Complex Systems* (pp. 185–243). Lincoln: University of Nebraska Press.

Salas, E., Wilson, K. A., Burke, C. S., & Wightman, D. (2006). Does CRM Training Work? An Update, Extension, and Some Critical Needs. *Human Factors, 48*(2), 392–412.

Schwartz, D. (1990). Coordination and information sharing. In: *Training for situational awareness*. Houston, TX: Flight Safety International.

Sexton, J. B., & Helmreich, R. L. (2000). Analyzing cockpit communication: The links between language, performance, error, and workload. *Paper presented at the Proceedings of the Tenth International Symposium on Aviation Psychology.* Columbus, OH.

Smith, E. M., Ford, J., & Kozlowski, S. W. (1997). Building adaptive expertise: Implications for training design strategies. In M. A. Quinones & A. Ehrenstein (Eds.), *Training for a rapidly changing workplace: Applications of psychological research* (pp. 89–118). Washington, DC: APA Books.

Smith-Jentsch, K. A., Salas, E., & Baker, D. P. (1996). Training team performance-related assertiveness. *Personnel Psychology, 49*(4), 909–936.

Stout, R. J., Cannon-Bowers, J. A., Salas, E., & Milanovich, D. M. (1999). Planning, shared mental models, and coordinated performance: An empirical link is established. *Human Factors, 4*(1), 61–71.

Swezey, R. W., & Salas, E. (1992). Teams: Their training and performance. Norwood, NJ: Ablex.

Taylor, H. L., Lintern, G., & Koonce, J. M. (1993). Quasi-transfer as a predictor of transfer from simulator to airplane. *Journal of General Psychology, 120*(3), 257–276.

United States General Accounting Office. (1997). *Human factors, FAA's guidance and oversight of pilot crew resource management training can be improved (GAO/RCED-98–7).* Washington, DC: GAO Report to Congressional Requesters.

Waller, M. J. (1999). The timing of adaptive group responses to nonroutine events. *Academy of Management Journal, 42*(2), 127–137.

Weick, K. E., & Roberts, K. H. (1993). Collective mind in organizations: Heedful interrelating on flight decks. *Administrative Science Quarterly, 38,* 357–381.

Weiner, E. L., Kanki, B. G., & Helmreich, R. L. (1993). Cockpit resource management. San Diego, CA: Academic Press.

Wickens, C. D. (1995). Aerospace techniques. In J. Weimer (Ed.), *Research techniques in human engineering* (pp. 112–140). Englewood Cliffs, NJ: Prentice Hall.

Wilson, K. A., Guthrie, J. W., Salas, E. (2006). Team processes and their training in aviation: An update. In J. A. Wise, V. D. Hopkin, D. J. Garland (Eds.), *Handbook of aviation human factors* (2nd ed.). Hillsdale, NJ.

Wright, M. C., & Kaber, D. B. (2005). Effects of automation of information-processing functions on teamwork. *Human Factors, 47,* 50–66.

10

Flight Training and Simulation as Safety Generators

John Bent and Dr. Kwok Chan
**Hong Kong Dragon Airlines Ltd.,
Hong Kong International Airport**

'Education is the systematic, purposeful reconstruction of experience'

Chapter 10 reviews the components of airline safety; identifies related crew training factors; and concludes with practical strategies with which to enhance crew training and raise the safety bar. Flight Training and Simulation are vital generators for a safer aviation industry. While huge improvements have been achieved in recent aviation history, continuous process improvement is vital. Two-way connections must exist between safety and crew training, in one continuous process loop. The emphasis today must be to improve the quality of what we do in training, and thus achieve safer operations.

GROWTH

During any phase of rapid growth, average experience levels drop as 'human-ware' struggles to keep pace with expanding fleets.

This industry has experienced unprecedented growth, and symptoms were becoming evident right up to the recession of 2008. The consequences of previous growth may reverberate for some years, and with the resumption of growth the industry may again be unprepared. The following pre-recession commentaries serve as reminders of what may again become valid:

> More planes are flying than ever before, but the number of people who do everything from piloting them to fixing them isn't keeping pace. The growing shortage is raising fresh concerns about air safety. Industry and government experts are worried that a looming dearth of pilots, aircraft inspectors and air-traffic controllers around the world could place new strains on maintaining some of the advances in airline safety of the past two decades. 'We know how to make the system even safer than it is, but we're going to lose ground if we fail to manage growth within the limits of our human resources,' says Bill Voss, president of the Flight Safety Foundation, an international nonprofit organization based in Washington. Behind new worries about safety are massive changes whipsawing the global aviation industry. In mature markets such as the U.S. and Western Europe, soaring fuel prices and rising competition from budget carriers are squeezing airline finances and forcing airlines to do more with less. Smaller work forces and cost-cutting measures like loading and refueling planes in a hurry leave staff less margin for error, industry officials warn. Meanwhile, fast-growing countries such as China, India, and some Middle Eastern states are snapping up jetliners as never before. Airline fleets and passenger numbers are growing so quickly that airports, air-traffic controllers, and safety inspectors can't keep up. Rich new markets like Persian Gulf emirates are handling breakneck growth by hiring staff from other countries, compounding staff shortages in countries from Europe to Africa. Flying is still much less risky than it was two decades ago, even in parts of Africa and Asia long prone to air disasters. Today, crashes in the U.S. are at an all-time low and deaths from air crashes worldwide are also near historical lows, with about one crash for each million departures. ... the industry has yet to grapple with the shortage of personnel. The gap is most pronounced for pilots. 'It's time to ring the warning bell on pilot availability' and devise new solutions because 'this is an issue that will face all of us,' says IATA Director General Giovanni Bisignani. Some experts project a shortage of 42,000 pilots worldwide by 2020. Pilot-union leaders say some U.S. carriers are using special programs allowing co-pilots to fly with as few as 50 hours of cockpit time in big planes—far below the hundreds of hours usually required—because of intense demand. Filling the gap won't be easy because educating pilots takes years. The shortage is raising concerns that some pilots don't have adequate training or experience to deal with adverse conditions, especially in developing countries.

INDUSTRY SAFETY

From the Safety and Quality Management perspective, where does training sit in the broader sense within an organization? Several major accidents exhibited training as a contributory factor, and we can examine how effective training can be used as a mitigating action for accident prevention. In any major accident there are several causal and contributing factors, some of which relate to 'organizational culture' which is that vague, general, indistinct, and often unwritten 'way we do it,' in a particular organization. A key component of this culture is the embedded organizational attitude to system quality and safety.

Organizational Culture

A number of research studies into organizational cultures are now well accepted in the industry. Dr Ron Westrum (1993), of Eastern Michigan University, conducted significant research into organizational culture. From data collected he classified organizational cultures into three broad categories. These categories were refined and published by ICAO as part of the Safety Management Manual (SMM):

1. Positive
   - Hazard information is actively sought
   - Safety messages are trained and encouraged
   - Responsibility for safety is shared
   - Dissemination of safety information is rewarded
   - Failures lead to inquiries and systemic reform
   - New ideas are welcomed
2. Bureaucratic
   - Hazard information is ignored
   - Safety messages are tolerated
   - Responsibility for safety is fragmented
   - Dissemination of safety information is allowed but discouraged
   - Failures lead to local fixes
   - New ideas are new problems (not opportunities)

3. Poor
   - Hazard information is suppressed
   - Safety messages are discouraged or punished
   - Responsibility for safety is avoided
   - Dissemination of safety information is discouraged
   - Failures lead to cover-ups
   - New ideas are crushed

Out of approximately 100 delegates in an international symposium in 1999, only a hand-ful advised that they worked for 'positive' (generative) organizational cultures. One hopes that this sample is not indicative of the broader picture. It would be sobering to think that an aerospace industry which has placed men on the moon and made air transport routine and indispensable employs few people who believe that they work for positive organizational cultures. The 1980s was a decade with a high aviation accident rate, and with reference to Nance's (1986); U.S. Representative James Oberstar[1] had this to say (quote): *'the margins of air safety are diminishing at the very time when they should be increasing. Safety must remain an enforced priority of the FAA.'* John Nance raised many issues three decades ago by pointing (in part) to the unhealthy tensions between cost-reduction and flight crew training.

Columbia and Challenger:

Following the loss on reentry of the *Columbia* space shuttle in February 2003, which followed the 1986 loss of the *Challenger* Space Shuttle, NASA was assessed by the Columbia Accident Investigation Board Report (August 2003), as exhibiting a 'broken' safety culture.

Healthy Unease:

Professor James Reason (1997) uses the term 'chronic unease' to describe a positive feature of an effective safety culture. We could

---

[1]Chairman Subcommittee on Investigations and Oversight, Committee on Public Works and Transportation.

suggest that an 'uncomfortable' airline culture will have the greatest integral safety. Such a culture, which is both outward and inward-looking, seeks to continuously learn and improve in critical training and operational (rather than just commercial and marketing) areas, and is more likely to trap serious problems before they occur. In Dr. Westrum's language, this is a 'generative' (positive) safety culture.

Organisational Complacency

Conversely we have the successful 'comfortable' organization, which can eventually become infected by complacency, and in the worst case, organizational arrogance. There are too many recent examples of long-term 'comfortable' organizational cultures in industry to disregard this issue. The expression 'un-rocked boat' was coined by Perin (1992) to describe an organization infected with a non-generative culture following a long period free of major accidents. Some airlines have even traded on their safety record, but contrary to popular belief, many decades of accident free operations may expose an airline to increased risk, especially if continuous improvement is not embedded in the culture and frequently nurtured.

Regulatory Standards and Quality

Aviation authorities create standards for compliance, or at times guidance, for their commercial operators, and significant differences remain between authorities. This is not to say one State is right and another is wrong, but it does highlight a phenomenon in our industry, that if there is no regulatory standard on a subject, an NAA elects to set its own standard, or looks to see how others do it, seeking industry best practice. A State may even have an operating environment not requiring a specific standard due to local factors. Nevertheless there has to be a basic minimum standard that all States comply with, and for this ICAO sets the standard. States that subscribe to and comply with all the ICAO requirements do so in the belief they meet world-class standards in parallel with all other states. A fundamental requirement for any operator is the attainment of its AOC (Air Operator Certificate); the licence to operate. This is granted by the State's own regulator based on the application of a set of rules and standards that the applicant must or should meet. Furthermore, this licence is re-issued on an annual basis, requiring the operator to maintain those standards and rules in order to continue to operate in a safe and efficient manner. There are also variances in AOC requirements across regulatory groups. FAA requirements set out the hierarchy of each post holder, identifying the importance of each post holder's accountability and responsibility in the overall management of an airline. Two key AOC requirements are Quality System and Crew Training requirements. The EASA/JAA AOC regulation is one major authority that defines a Crew Training requirement. On the other hand, the FAA is the only regulation requiring a Safety function within an AOC. The reality is that each regulatory authority has evolved its own national standards over the years, and some regulations will be tailored to suite each State's own aviation environment. The body of knowledge and experience gained over the defining years of regulation in this rapidly growing aviation industry has led to a certain feeling that 'what we do now' is 'what the industry needs.' Apart from the response to 9/11, the aviation industry has not gone through any really dramatic regulatory changes for many decades. For example, in aircraft design, part 121[2] requirements and standards have served from the beginning of the jet age to the design of the fly-by-wire generation of civil aircraft. It also shows how robust and timeless some of the regulations have become to serve all facets of a growing industry. Regulatory requirements are minimum requirements, but this does not mean compliance with the highest standard of operation, as this cannot be defined, and may indeed not be attainable. What new systems, processes or regulatory standard must be put in place to further enhance an already acceptable standard of operation? Variability can be seen in accident rates across different regions of the world, yet all States must comply with ICAO requirements, so why are there more or fewer accidents in some States than others? If current global regulatory standards

---

[2]**FAA** - Airline Certification 14 CFR **Part 121** Air Carrier Certification.

are not sufficient to ensure a consistently safe operation, what more is needed and where do we need to focus attention for the future? During the evolution of the industry, non-regulatory organizations have stepped in to identify deficiencies and create best practices exceeding basic ICAO standards. The International Air Transport Association (IATA) is one such organization.

Quality Assurance

(QA) processes are by no means new. These have been at the forefront of many of our industrial and manufacturing processes for decades. The QA prime objective has been (1) to determine in a consistent and routine manner the 'quality' of an end product and (2) to evolve process to produce continuous improvement. So how can aviation more effectively use this concept of continuous product improvement? First one has to understand that the safety task is never finished. Once one issue has been resolved, another will emerge, due to ever-changing operating conditions and environment, and constant change demands continuous improvement. A Quality Assurance process can be regarded as one important step toward continuous improvement. It is sometimes difficult to imagine how the airline industry, with its primary business carrying passengers and cargo around the world, can be compared with an industrial process that produces an end product, but it can. An obvious airline product is the satisfied passenger transported from A to B in the most efficient way without incident or injury. This is achieved by meshing a multitude of disciplines together ranging from flight operations to maintenance of complex machinery. Each discipline can be process mapped, after which control measures can be defined to ensure that the process achieves the end product, eg, a flight crew training organisation in the delivery of competent and efficient pilots. So QA is about assuring customers or end users that a system can deliver products and services to the required standard. It can achieve this by documenting the way things are supposed to be done, and then auditing to ensure they are done as intended. The audit process will uncover discrepancies, fed back to the responsible organization for corrective action, thus enabling continuous performance improvement.

Safety and QA Working Together

Safety Management and QA have the common goal of attaining an acceptable level of safety. Safety Management is a reactive process, whereas QA is proactive. In previous paragraphs, we looked at the differing requirements for AOC post holders, where there was a clear emphasis on the Quality System requirement in the EASA standard, and in the Director of Safety in the FAA standard. Knowing the respective roles of Safety and QA, we now know that both have the same end goal, but that the processes employed are different. Imagine therefore that both safety and QA processes are employed jointly, which is already the case in some organisations. This merges powerful toolsets with which to enhance and focus safety management (SM) towards a self-sustaining status, where either the QA or SM discipline becomes dominant in the safety outcome. The QA process is thorough in weeding out discrepancies which are latent threats within the operation. It is unrealistic to assume that every process designed or implemented is perfect, and there remains a constant need to monitor effectiveness in the real environment. The Safety process, in its investigative thoroughness, will inevitably discover process or procedural errors as a causal or contributing factor. The big challenge of any QA process is to ensure that the audit procedure is sufficiently robust in finding discrepancies before these accumulate to cause incidents in the end process. On the other hand accidents and incidents may be caused by factors out of ones own control (e.g., ATC errors putting two aircraft into a collision threat). Safety Management can then unravel causal factors, and new procedures can be created to build an environment of continuous improvement. Safety and Quality Assurance should be complementary processes, working together against safety risks and threats in normal operations.

Training within the QA Process

Training is one of the vital constituents of a Quality System process, delivering the end product of a competent pilot, cabin crew or engineer to meet the demands of an exacting role. The process can be described as *Selection > Training > Monitored Perform ance > Evaluation > Re-training > Monitored Performance,* and so on. This process should ultimately lead us back to the beginning of the *Selection Process.* Yet, can we assume that the criteria set for pilot requirements at the beginning of the jet age are the same today? Just look at the generation of civil aircraft flying today compared with the first generation of large passenger jets. Flying skills required have changed from intense handling and control manipulation to the use of automation designed into the systems to instruct how and where to fly the aircraft. So the quality and relevance of training has had to evolve not only with the technological changes but the changing profile of the traditional airline pilot to be. The challenge for pilot training organisations is to keep pace with changing syllabi and train to sufficient standards in ever-decreasing time. In any commercial environment there are financial constraints on the extent of training, and ensuring the optimum training standard is most challenging when there are restrictions on how much one can train. Hence training must encompass a follow-up process to monitor and measure the effectiveness of each programme, identifying high value priorities to address. The training organization is therefore tasked to deliver just the right amount of training without having to expend greater cost, while achieving the objective of delivering each individual with the optimum knowledge and skills to perform the tasks. Often, when investigating an incident or accident, training has been a causal or contributing factor. So is it correct to assume that the only way to assess the effectiveness of any training regime or programme is that it does not lead to an accident? **This question reveals how critical training is to the overall safety process, as it has a direct impact on safety outcomes of any airline**. The effects may be seen a long way down the line, rather than immediately. Ineffective or negative training is a major latent failure in the overall accident causation chain. Effective training will mitigate weaknesses in procedural or equipment design. Since it has an indeterminate dormancy timescale, it should not be left to the Safety process to uncover its weaknesses. Rather, continuous measurement should be proactively applied to enable improvement.

Learning from Past Accidents

There is no better place to review the safety of our industry than by looking at actual accident statistics. This gives a measure of past performance in managing an ever-challenging operating environment. Safety is a moving target affected by many variables including a growing fleet size and increase in aircraft movements. Imagine an airline that has grown from a fleet of 65 to 120 aircraft in a time of span of 10 years. The level of safety achieved cannot be compared, as the ultimate level has multiplied. Worldwide accident statistics reveal how safe the industry has become, even when considering the limited number of high profile accidents. So the primary role of safety management is to monitor outcomes; the number and type of safety events, to measure the level of risk to which the airline is routinely exposed. Furthermore, when an event or incident occurs, an open non-punitive style of investigation is vital in determining all root cause(s) and contributing factors. If an investigative process just seeks to find a guilty party, then the mitigating action, which normally dismisses the guilty party, will become wholly ineffective because it cannot address the root cause (s). So Safety is a reactive process, where the outcome manifests itself in a safety event or accident. The process of investigation and setting mitigating actions for prevention of future occurrences has been the primary safety management tool for decades. It is still effective today, but to an extent where there are perhaps diminishing returns, especially during rapidly expanding and changing environments.

Any review of accident statistics first reveals that accidents are not caused by a singular event, and human factor errors are not intentional. Past accidents also indicate to us where we need to focus our attention, deploy a fix, or take mitigating action, in order to prevent a recurrence of a similar event. One frequently quoted statistic is that between 70 and 80 percent of all aircraft accidents are due to human error. There are mixed methods employed to tackle human error, ranging from the punitive, to a determined intent to thoroughly understand all causal factors. Effective management of human factor errors has the greatest single potential in accident reduction. There are even some aviation commentators who boldly state that all safety or accident outcomes are 100 percent human factor related, since the machines we operate are designed, piloted, and maintained by humans. A small sample follows of accidents linked to training triggers (AAIB,[3] Tideman, 2006):

- Qantas **B747–400** Bangkok 23 Sept 1999[4] (training & SOP related) The first recommendation from the State's accident investigation authority to the state's regulatory authority is to ensure all operators of high capacity jet aircraft in that State have in place procedures and training to ensure flight crews are adequately equipped for operations in wet/contaminated runways within that State.
- Air Transat **Airbus A330** Azores[5], 2000 (training related) System familiarization, when both crew members were relatively new to type, was cited as one of the causal factors in this remarkable accident where the crew mistakenly opened the aircrafts cross feed valve and depleted all the fuel through a fuel leak one side of the aircraft wing tanks.
- American Airlines **Airbus A300**[6] New York, 2001 (training related)

[3]UK Air Accidents Investigation Branch.

[4]Crew not provided with appropriate procedures and training to properly evaluate the potential effect the Bangkok Airport weather conditions might have had on the stopping performance of the aircraft. In particular, they were not sufficiently aware of the potential for aquaplaning and of the importance of reverse thrust as a stopping force on water-affected runways.

[5]24 Aug 2001 The flight crew did not detect that a fuel problem existed until the Fuel ADV advisory was displayed and the fuel imbalance was noted on the Fuel ECAM page. The crew did not correctly evaluate the situation before taking action. The crew did not recognize that a fuel leak situation existed and carried out the fuel imbalance procedure from memory, which resulted in the fuel from the left tanks being fed to the leak in the right engine. Conducting the FUEL IMBALANCE procedure by memory negated the defence of the Caution note in the FUEL IMBALANCE checklist that may have caused the crew to consider timely actioning of the FUEL LEAK procedure.

[6]PROBABLE CAUSE: 'The in-flight separation of the vertical stabilizer as a result of the loads beyond ultimate design that were created by the first officer's unnecessary and excessive rudder pedal inputs. Contributing to these rudder pedal inputs were characteristics of the A300-600 rudder system design and elements of the American Airlines Advanced Aircraft Maneuvering Program.'

This was perhaps one of the most compelling accidents involving training; where training was determined by the NTSB as a significant causal factor, in which training practices were adopted that were seen as a cure for one concern (improving crews upset recovery techniques), but led to the demise of this flight, when rudder reversals were used which caused the rudder and tail fin to fail.

- Air France **Airbus A340**[7] Toronto 2 August 2005 (training related)

  Training to be instigated to ensure crews use appropriate cues in changing conditions close to touch down. Need for a timely decision to abort a landing even when aircraft is on the runway.

Implications for Training

Many recent accidents can therefore be seen to have their roots in training, through challenges adapting to new technology, or rapid growth exceeding capacity to deploy adequately trained personnel. Training time is limited by the need to reap adequate returns on investment within a reasonable timeframe. But when an accident is caused by the lack of training adequacy the fix is often to add more training, when the quality and relevance of the training may have been the issue, not the amount. So how then can an organization attain the optimum training regime without having to suffer an accident first?

Training as a Mitigant

After a thorough investigation has been conducted on any incident, mitigating action often ordered is crew training. Suddenly training

---

[7] 02 AUG 2005 Although it could not be determined whether the use of the rain repellent system would have improved the forward visibility in the downpour, the crew did not have adequate information about the capabilities and operation of the rain repellent system and did not consider using it. The information available to flight crews on initial approach in convective weather does not optimally assist them in developing a clear idea of the weather that may be encountered later in the approach. During approaches in convective weather, crews may falsely rely on air traffic control (ATC) to provide them with suggestions and directions as to whether to land or not. Some pilots have the impression that ATC will close the airport if weather conditions make landings unsafe; ATC has no such mandate.

appears as an important accident prevention strategy. The blame and train solution is often favoured rather than a deeper look into systemic and dormant factors within an organization. This is not say that training will not deliver an improvement in standards or performance of an individual, but blame may be used after an accident to indicate that the initial training regime did not meet the intent to deliver the valid quality each and every time. Training must be an integral part of an overall quality process encompassing measurable performance markers to determine the standard of training being delivered. In this respect an effective training programme and organisation must have a QA process to continuously evaluate its performance. This process should become part of an overall safety management system where training enters at the forefront in delivering safe operations.

Training in the Safety Management System

ICAO Safety Management Systems (SMS). ICAO mandates SMS (Safety Management Systems) for all airlines in 2009. Extracts from the SMM:

1.3.3: *'The air transport industry's future viability may well be predicated on its ability to sustain the public's perceived safety while travelling. The management of safety is therefore a pre-requisite for a sustainable aviation business.'*

1.4.2: *'An SMS is an organised approach to managing safety, including the necessary organisational structures, accountabilities, policies, and procedures.'*

SMS will become the by-word in safety management. While many organizations claim to have started an SMS approach well before the turn of the decade, it has evolved through industry best practice into a regulatory requirement. The benefits of SMS cannot be disputed and many States will be following the ICAO recommendation to mandate its implementation. SMS may have many definitions but hopefully only one interpretation; that it is a systematic management of all activities in an organization to secure an acceptable level of safety. The industry has had to balance the competing pressures of commercial output or production (generating profit from passengers and cargo) and the maintenance of high levels of safety. In this challenge there can only be two possible outcomes:

1. Dominant attention to commercial outcomes—a risky policy
2. Balancing of commercial outcomes with the needs of the safety system.

Production and safety are opposing forces at work between senior management and the safety specialist. An uneasy truce will exist until the balance is tipped against one or the other objective. To take an extreme view, this decision may become a choice between not staying in business (take no risks), or staying in business with a high level of risk of an accident. So imagine if safety could be brought into every day organizational decision making. SMS is the process which can deliver such a goal, making safety an integral part of the airline business. The lack of investment in safety in some airlines is mainly due to the inability to effectively measure the benefits of a safety programme. Safety and QA programmes are becoming pro-active in nature, aimed at the prevention of unsafe acts, incidents, and accidents. SMS is an investment in the safety and a risk mitigation goal for future operations. An SMS must foster a nonpunitive and just-culture environment which encourages effective incident and hazard reporting. Significant identified hazards today are (1) shortage of experienced personnel, and (2) sub-optimal training. An acceptable level of safety must be affordable, but safety is a profit center if viewed against risk. In the context of safety culture, it is vital that the same executives making the tough calls on cost controls also understand the importance of maintaining resource flow to the vital organs of their safety culture. An essential element of continued resource-flow must include on-going development of effective crew training and safety processes (positive safety culture), with appropriate simulation as a key component. The airline industry must continue to aim for the highest standard of safety in order to further reduce the already low accident rate.

The aviation industry has one of the lowest accident rates in the transport world, but if the accident rate reduces no further, an eventual doubling of the global airliner fleet could result in a significant increase in accidents. So the rate must be further reduced, as a necessary goal to avert passenger concerns and keep the industry safe. But there are an increasing number of threats working against a reduction in accident rate. A 2008 survey[8] found that: 'A majority of aviation professionals do not expect airline safety to improve during the next five years.

The three main factors identified were:

1. A shortage of experienced personnel
2. Fatigue/work practices, and
3. Airline management experience/attitudes/culture.

Furthermore, as if to validate these concerns, since early 2008 a significant increase in fatal airline accidents has occurred. Effective pilot training remains a critical long-term safety generator for airlines. Many factors impact safety, but **human factors** feature in almost all incidents and accidents, and are strongly influenced by training received. Crew Training and Safety have a strong cause and effect relationship, and the industry has clear control of crew training.

SAFETY GENERATORS IN CREW TRAINING

High fidelity[9] simulation is a powerful training tool, with rich potential for improved application. The concept of Threat and Error Management (TEM), forms a useful backdrop for any discussion on pilot training. TEM was developed by Professor Helmreich et al. (2001) of the University of Texas, and embedded in ICAO (International Civil Aviation Organisation) guidance material for pilot training. The airline industry faces almost continuous threats, but the extended cycle of growth prior to Q4 2008 was unprecedented. After the current recession, growth will resume, and the number of airliners in operation is still projected to double from 18,000 to 35,000 in two decades, which translates into 550,000 pilots to fly the fleet; or an increase of 15,000 new pilots per year. Fleet growth is projected to be greatest in Asia. In 2008, fuel costs and global economic crisis put the brakes on growth,

[8]Ascend Survey (2008), London-based aviation consultancy.
[9]fidelity = similarity to the aircraft simulated.

yet some expansion continued in Asia. Half the world's population resides in this developing region; within a circle bounding 5 hours' flight time from Hong Kong. A simple statistical comparison between developing and developed airline systems shows the USA as approximately 6,000 airliners for a population of 0.28b, with China 1,100 airliners for a pop of 1.3b.

The pressures recently felt within an expanding airline system have forced reviews and better practices. As resources became increasingly limited, and the reservoir of experienced pilots shrunk, corrective solutions were urgently sought. Extensions to pilot retirement age were instituted, which helped pilot supply, but only temporarily, as at top of the experience pyramid, there are fewer pilots. As fleets continued to expand, pilot sourcing moved toward ab-initio[10], and as basic training schools ramped-up and delivered, an increasing proportion of new airline pilots entered from ab-initio or lower experience pools, reducing the average experience levels on flight decks. In the following paragraphs, these and other factors are listed as **system 'threats,'** in no particular order of importance.

THREAT 1: Top down: Management perception: From the ICAO Safety Management Manual (SMM), AN/460 (2006), paragraphs 12–13: *'Weak Management may see training as an expense rather than an investment in the future viability of the organisation.'*

THREAT 2: Reduced average pilot experience: As fleets expand, the experienced pilot pool shrinks; and pilot sourcing has to move to ab-initio, reducing average experience levels on airliner flight decks. In China, airline pilot sourcing from general aviation is almost non-existent, as the GA sector is still extremely small, and airline pilots are primarily sourced from ab-initio or the military.

THREAT 3: Reduced Instructor Corps and experience: A secondary effect of expansion is the depletion of the instructor corps. In turn, ICAO MPL places more stringent requirements on instructor training. Even if sufficient recruits could be found, the global instructor shortage restricts the capacity and quality of training which can be delivered. During expansion, ab-initio instructors are poached by

---

[10]Initial pilot training to commercial pilot license.

airlines; licensed simulator instructors return to flying, and training demands cannot be met. Inexperienced pilot instructors may not make effective airline trainers, and sub-optimal initial instruction can inject a latent lurking human factors threat into the system.

THREAT 4: Reduced depth of piloting experience: Ex-military pilots entering civil aviation generally benefited from deeper and broader practical flying experience than available in airline flying. Military pilots often flew to the limits of their aircraft, and mostly in highly skill-challenging environments. However, fewer ex-military pilots are available to civil aviation. Today's airline pilot may therefore have thinner reserves of skill-based experience to call upon if needed. In these circumstances, command experience also becomes more limited, reducing the mentoring expertise available to new pilots in the right seat.

THREAT 5: Improved airliner technology (hidden threat) Airliner technology has progressed continuously adding redundancy, ergonomics, reliability, and self-protection[11]. Improved technology adds significantly to system safety, but there is one downside. Technology protects the system for most of the time, but reduces on-the-job challenges and skill acquisition opportunities for pilots, as they fly increasingly using automation. The need for deep piloting skills and experience has progressively reduced, and major technical failures are thankfully a much lower probability. But superior 'airmanship' is still occasionally needed; for example, 'Souix City DC10[12],' and 'A300 in Baghdad[13]'; the 'BA B777 crash landing at Heathrow[14],' and the 'United Airlines A320 ditching in the Hudson River.' The fact remains that with improved airliner technology, the average pilot has been less skill-challenged, and is less experienced (a foundation of airmanship), and in cases of degraded technology and unusual circumstances, there are thinner reserves of experience to call upon.

THREAT 6: Pilot shortage, increasing pilot mobility, and reduced training quality: In order to meet the demands of growth, the

---

[11]e.g. Airbus flight envelope protection.
[12]United Airlines DC10 – loss of hydraulics & controls - 19 July 1989.
[13]European Air Transport – hit by SAM7 – fire + loss of all hydraulics & controls - 22 Nov 2003.
[14]Loss of thrust on approach to Heathrow Airport - 17 Jan 2008.

airline industry has to select and qualify an unprecedented number of pilots and other specialists. The lead-time needed for training has generated a 'demand surge,' fueling improved pay and conditions for those qualified. This in turn has caused more pilot mobility within the industry from lower to higher paying airlines, placing greater demands on training centers. The more profitable airlines are the last to feel shortages, while feeders at the bottom have to cancel services and ground aircraft. Selection and recruitment today is more challenging, from a pool of more broadly qualified young people, who are generally less keen to join this industry than in the past. A negative factor identified by professionals in the aviation training industry is that the level of 'passion' for aviation (and whole-aviation knowledge) identified in pilot applicants after WW2, is not so evident today. Today a piloting career may only be third or fourth choice on the career wish list. For the training industry, this translates into an increased motivational requirement during induction training. Even if sufficient numbers of recruits can be found to join the industry, the capacity and quality of training systems needs to continue to improve.

THREAT 7: Regulatory lag and minimum regulatory standards
Regulatory 'lag' can be seen in the global pilot licensing standard for the Commercial Pilots Licence by the end of the 20th century. Requirements were laid down in 1944, and no significant changes were implemented until MPL[15] doc 9868 was published by ICAO in 2006. To quote from the Director General of IATA Bisignani (2002) (November 2007):

> the industry is concerned that there are no global standards for training concepts or regulation. Pilot training has not changed in 60 years—we are still ticking boxes with an emphasis on flight hours.

IATA supports the competency-based approach of multi-crew pilot licensing (MPL) training programmes. Unlike traditional pilot training, MPL focuses from the beginning on training for multi-pilot cockpit working conditions. MPL also makes better use of simulator technology. Europe, Australia, and China are moving ahead with implementation of MPL. But legislative change takes time, and

[15]Multi-crew Pilots License.

regulatory requirements may be 'outdated standards' rather than 'current best practice.' Airlines operate in high cost environments, and training managers are often under pressure to ensure that resources applied to training do not exceed regulatory requirements. A regulator may not be able to keep pace with the rapidly changing operating environment faced by an airline. The challenge for the training manager is to produce justification for necessary resources in excess of those prescribed to meet basic regulatory standards. Under severe cost pressures, the only training expenditure sanctioned by an airline may be that which is required to meet minimum regulator standards. But as technology and operational factors change, training and safety managers may see the requirement for new best practice standards first, in defence of their own standards. So in practice, the application of resources should not just be aligned to minimum regulatory standards, but evaluated against operational need as well. This is sometimes forgotten in the quest to continuously cut costs. As if to underline this threat, there is a perception in the airline training world that training standards have gravitated to bare minimum standards.

THREAT 8: Reducing resources for training: Airlines seem to face a continuous cocktail of challenges, such as increasing competition, terrorism, more open skies; emergence of low-cost carriers, fuel price volatility, and economic turmoil. Airline operating costs have been rising continuously, yet on the revenue side, ticket prices have hardly risen in real $-value terms for decades. The only viable policies for survival are therefore continuous cost reduction, combined with increased efficiencies and market share. In this climate, additional dollars for crew training are scarce or non-existent. A further quote from the Director General of IATA in Singapore—February 2008:

> After over US$40 billion in losses 2007 finally saw the industry return to profitability. Airlines made US$5.6 billion on revenues of US$490 billion. That is less than a 2% return, so no investor is going to get too excited. And tough times will continue. Airlines are US$190 billion in debt. Oil is pushing US$100 per barrel and accounting for 30% of operating costs or a total bill of US$149 billion. The revenue cycle peaked in 2006 and the negative impact of the credit crunch is still being calculated. Airlines may be out of intensive care but the industry is still sick.

Training resources tend to become most scarce at the very point in the aviation cycle (expansion) when more, rather than less, should be applied to ensure operational safety. The conundrum is that failure to improve training processes may likely lead to narrowed safety margins and higher operational risk. The pressure to compete and grow market share ensures that airline dollars applied to customer-contact 'products,' such as improved network, and ground and cabin, are protected. Yet other components of airline operations may be required to reduce costs every successive year, with the expectation of securing never ending savings through continuous improvements in efficiency. This may become a paradigm blind to diminishing returns, as we get closer to scraping the barrel. Such strategy justifies itself in the belief that functions such as engineering and training are 'back-office' activities, and as high cost items, are valid targets to prune to offset rising external costs. Cutting resources to a function such as maintenance may not have immediately severe consequences (Professor James Reason's 'Swiss Cheese model'[16]). This is because well-trained pilots catch and trap upstream errors or omissions in the process, as the last line of defense. To those at the operational coal face, the link between effective training and 'safety' is obvious. But to quantify this relationship in 'dollar' terms to the airline commercial manager or CFO[17] is difficult. Arguments for more training resources tend to be nullified by the exceptionally low industry accident 'rate.' The perception that pilot training can be lumped together with other so called 'support functions' is a dangerous one. Failure to train pilots fully and effectively is a high risk policy; *an accident is the greatest inefficiency.* The surge of resources applied to airline training and safety systems after an accident is a matter of industry record. Adequate crew training resources must be protected; prevention is better than cure. ICAO has defined 'safety' in the SMM[18] as: 'the state in which the risk of harm to persons or of property damage is reduced to, and maintained at or below, an acceptable level, through a continuing process of hazard

[16]Depicting a series of slices of Swiss cheese with random holes which when lined up lead to an accident.
[17]Chief Financial Officer.
[18]ICAO Safety Management Manual.

identification and risk management.' The latter phrase is an unambiguous action item for the airline industry. Essential resource flow to vital organs of airline safety culture must be maintained.

THREAT 9: Sluggish global standardization Despite moves toward global harmonization, there is always more to do. As with regulatory standards, a latent barrier to effective training is the variability of training practices and operating procedures around the world, even for the same aircraft types. As pilot mobility increases, this becomes a system safety threat. There are however better signs. Airline managements are slowly recognising the importance of best practice safety initiatives, such as LOSA (Line Operational Safety Audits) and IOSA (IATA Operational Safety Audits). These developments are creating a quiet steady shift toward common global procedures, through the application of best practice to their audit process. The IOSA Standards Manual, used as the prime reference for IOSA audits, is evolving into a useful global reference document for safe global operating procedures. ICAO is underpinning these developments with a raft of recent ICAO training and safety documents:

2002	Doc 9808—Human Factors in Civil Aviation Security Operations;
	Doc 9806—Human Factor Guidelines for Safety Audits;
	Doc 9803—Line Operations Safety Audit (LOSA) Manual
2004	Doc 9835—Manual on Implementation of ICAO Language Proficiency Requirements
2006	Doc 9841—Manual on the Approval of Flight Crew Training Organisations;
	Doc 9868—Training (MPL, competency-based training, embedded TEM;
	Doc 9859—Safety Management Manual (SMM)
2007	Doc 9683—Human Factors Training Manual
2009	Doc 9625—Manual of Criteria for the Qualification of Flight Simulators

Time is needed for implementation of ICAO mandates at National Aviation Authority (NAA) level, and these may not be seen as 'essential' by airline managers, whose main focus is the financial wellbeing of their airline *'Show me the regulation. If it's not an Authority requirement, it can't be that important, and we simply cannot afford it!'*

Summary of System Threats

Rapid growth can exacerbate existing risks, and add new ones. Threats and errors are inevitable in any man-machine system. Risk measurement will estimate the potential consequences of un-trapped threats and errors, and if nothing is done to address these threats, risk will increase. Rather than trying unrealistically to eliminate all risk, the industry must look for ways to cost-effectively minimise consequences through better management processes. 'Cost-effective,' because airlines must be profitable to survive. 'Better management,' as this is the best-practice variable which we can improve on under the concept of 'Threat and Error Management.' Younger airlines have fewer historical paradigms to shift, and contemporary best practice processes can be inserted with relative ease, but more established carriers with long-embedded legacy policies usually find change more challenging to implement; another threat in itself.

TRAINING ENHANCEMENT STRATEGIES

[Was the training completed? Yes, look, all the boxes have been ticked!] From the earliest days commercial aviation demanded strict regulation, and training and testing evolved in a box-ticking environment. Although annual or bi-annual testing of pilots is a policy far ahead of other industries, it is now well understood that 'infrequent snap-shots' cannot provide assurance of continuous performance. A line or route check passed on one day per year provides no comfort of equivalent performance for the remaining 364 days. We would prefer to assure ourselves that continuous performance standards are learnt and sustained by all in the system. Embedded quality process can provide a greater measure of continuous system performance. Much effort needs to be focussed on the known

weakest link: human factors. As advances in airliner technology continue to strengthen the safety system, the human part of the system becomes increasingly exposed In the light of growth-related threats, maintenance of the current remarkably low accident rate will be challenging, and with a doubling of fleet size will result in more accidents. The rate has to be reduced further. Improved pilot training presents a great opportunity to achieve this in the future. Threat & Error Management (TEM) is a useful framework on which to hook better pilot training practices. Pilots need to be taught to detect, avoid, and trap system threats continuously within the training process. Well-trained pilots are produced via well-established educational principles, but to more effectively deliver the necessary Knowledge, Skills, and Attitudes, we need valid tools, people, and process. In the following pages, the tools ('hardware'), people ('software'), and process ('training programming') are examined more closely, and ideas proposed for improvement.

Hardware—FSTDs[19] and Training Tools

The 'Link Instrument Flight Trainer'[20] first appeared in July 1939, dramatically reducing the fuel required for training purposes, and aircraft training was halved. In August 1941, the success of the Link Trainer warranted a separate company in the form of Link Aviation Devices Inc., and the Trainer was in widespread use with many variants, remaining in active service until the late 1950's. These were the beginnings of the remarkable development of flight simulators, which have become mandated training and testing tools for airlines today. How would accident statistics read if we had not been able to use simulators in commercial aviation? Without simulators, pilots would never have been so well prepared for serious in-flight emergencies, which could not be practiced in real airliners. The 'fidelity[21]' of simulators has come a long way in a relatively short time, and will continue to improve, eventually to a point where the 'reality gap' becomes almost imperceptible. But it has not yet been convincingly demonstrated that

[19]Flight Simulation Training Devices.
[20]Built by Link Avionics Ltd., at Binghamton, New York.
[21]fidelity = similarity to the aircraft simulated.

machine-man learning is as effective and man-man learning. The transfer of knowledge skills and attitudes to trainee pilots still depends on experienced motivated instructors who can extract the greatest learning value out of any new training technology, with enthusiasm. Simulators are now essential components of type transition training to new airliner types, and for recurrent training to maintain standards. During transition training, simulators are already used to train pilots to the point where their first exposure to the real aircraft will be with passengers on board[22]. Due to the high level of simulator fidelity available, high cost aircraft training, taking an airliner out of service, has been progressively replaced by simulation.

Governance of simulation and new standards: As simulators became vital ingredients in pilot training, regulators imposed strict guidelines on their use. These guidelines developed regionally and independently, but in 2006 ICAO tasked the Royal Aeronautical Society (RAeS) to research and develop a new set of Global Simulator Standards for simulation, to be implemented as the revised Document 9625—Manual of Criteria for the Qualification of Flight Simulators. This will bring to NAA[23]s more relevant harmonized global guidelines for the application of flight simulation to pilot training processes, with appropriately defined devices better matched to training objectives.

Manufacturers and fidelity: Simulator operators face significant constraints in trying to maintain fidelity for their customers. As simulators are not a traditional part of an aircraft manufacturer's responsibility, the role that the latter plays in modern simulation is a distant one. Aircraft data is sold to simulator operators at significant cost. This cost, plus the time required to re-engineer the data into simulators, have conspired to limit fidelity of simulators with aircraft fleets. However, keeping simulators in harmony with airliner fleets can be critical to operational safety. Aircraft manufacturers increasingly appreciate the importance of simulation

---

[22]zero-flight-time training.
[23]National Aviation Authorities.

to the safe operation of their products. Regulators need to play a stronger role in moving manufacturers and operators toward even greater continuity of simulator fidelity. If manufacturers would grant upgrade data free to simulator operators as a 'component of their business,' more prompt simulator fidelity updates would result. The regulator could then reasonably expect operators to maintain simulator fidelity with airliner fleets, in real time. In Flight Training Organisations (FTOs) and Type Rating Training Organisations (TRTOs) under EASA, FSTD-A requirements already help to ensure that embedded quality processes sustain simulator fidelity.

New FSTDs: There has always been a need for simulator operators to work closely with designers for the development of more user-friendly machines, and in recent years this need has been addressed proactively. Partially mitigating the instructor shortage, improved training tools continue to flow into the industry. The unit cost of high-fidelity simulation is being driven down by higher production rates, competition, and advancing technologies. Electric motion systems are maturing, and enhanced PC-based visual systems are providing dramatic improvements in visual realism, along with lower cost of ownership and operation. New lower fidelity training devices, such as IPTs (Integrated Procedure Trainers) are adding value in terms of effectiveness, combining former classroom/CBT[24]/FMS-FMGS[25]/and Fixed Based simulation[26] into one device, using current aircraft data. In early stages of training, these highly interactive devices, which are designed into course syllabi, reduce the depth of instructor delivery necessary. This technology now enables higher levels of standardization in training delivery, and faster learning of flow-patterns, systems, procedures, and checklists. When applied to advanced transition programs such as APT2 (Airbus Pilot Training 2), the IPT (APTD) takes over approximately 25% of pilot training formerly conducted in a full flight simulator. The pilot trainee emerges from this phase conversant with checklists and procedures, needing much less

[24]Computer-based-training.
[25]Flight Management System/Flight Management Guidance System.
[26]Simulator without the use of motion.

time in the higher cost Full Flight Simulator to acquire handling skills; a significant saving as well as an advance in the learning process for instructors and trainees alike.

Digital analysis in simulation: Although the technology exists to measure and analyze crew performance in simulators, few airlines practically employ this facility to mine data. However, in other areas of operations, airlines are accustomed to tapping and analyzing data; for example, through the flight data analysis programs (FOQA/QAR[27]). The simulator is captive, with 24/7 access, relying on continuous data processing. It is therefore a relatively simple matter to tap this data in real time and analyze results, with appropriate protocols, in a similar way to FOQA. The measurement of crew performance in simulators enhances training effectiveness, and highlights skill and procedural deficiencies in a safe environment. Analysis of simulator data adds a new depth to knowledge and skill acquisition, translating into cost savings through accelerated learning.

Video and data replay: Video-debrief systems allow crews to de-brief with more ownership of the critique, and gain a deeper learning experience. Trainees can self-debrief with minimum instructor input, reviewing specific segments of the flight to observe their own CRM/TEM skills. Synchronized with aircraft animation, flight instrumentation, aircraft systems, flight control deflection, and other parameters, video and data debrief systems are becoming a standard simulator fit. Once instructors learn to facilitate video debrief systems effectively, powerful training outcomes result.

Video clips of correct procedures: 'A picture tells a thousand words.' Some airlines have used the simulator as a tool to video how to 'do it right.' Such video shorts will show a uniformed crew handling an emergency procedure 'as per the SOP.' Production may be resource-intensive, but the final product is a training library of correct practices, accessible by intranet, at optimal learning times.

Air Traffic Control (ATC) simulation: ATC simulation systems using voice recognition software, can now be integrated into Flight Simulation Training Devices (FSTDs). These systems feature:

[27]Flight Operations Quality Assurance/Quick Access Recorder.

- Variable (male/female/accented) voices responding to R/T calls, ground and other communications.
- Reliability rates above 95%, accents understood.
- Correct context of traffic and airport or ATC area, as seen on related visual.
- A requirement to use clear and correct aviation R/T from the start of training.
- Deletion of any instructor requirement to mimic all ATC/ground crew communications.
- PC platform options to train aviation ICAO English[28] in a realistic format, interactively.
- An ICAO MPL requirement for syllabus in Type 1, 2, 3, and 4 devices.
- Such realism that pilots can forget that the responding ATC voice is synthetic.

Simulator instructors know the importance of believable ATC simulation in training, and this capability raises the training bar significantly to address increasing challenges in the ATC environment.

Zero-flight-time simulation: There is an area of simulation where pilots still have had doubts about simulator fidelity-during lateral manoeuvres, particularly near or on-ground. This is manifest when either no motion cueing is perceived, or motion onset is followed a few milliseconds later by an unwanted side force. During taxi and take-off this can lead to nausea, when visual and motion perception do not exactly match. Pilots may feel apparent 'side-slip' over the runway. Correct roll onset can be followed a few seconds later by a spurious reverse lateral side force, and this phenomenon is described as 'leaning; student on the pedals; out of phase.' This phenomenon is mostly noticeable on short finals or during a circling approach visual manoeuvring task (for example when base training in the simulator). In order to remove unwanted side forces, gain settings in many recent simulators are set so low that hardly any motion can be perceived. Motion sense is then restricted to vibrations and special-effects. One training research organisation believes that these shortcomings are due

[28]ICAO standard for aviation English.

to the architecture of the motion drive control laws, as applicable to all current simulators. This area is seen as the weakest point in actual flight simulation. A new 'lateral manoeuvring motion' or Lm² algorithm has been developed for simulator motion systems to correct this limitation. The main features are to match lateral forces as perceived in the pilot's seat during taxi, take-off and landing, and in flight, to provide more natural roll onset with correct side forces. Lm2 promises pilots motion sensations similar to the real aircraft in critical phases of flight. The phases of operations most likely to benefit are taxi, backtrack & line-up, take-off roll, final approach, visual; circuit, circle to land, crosswind landing, and roll-out; critical flight phases for pilot skill development which all come together during base training. Using Lm2, significant cost savings may be possible, if more aircraft base training can be realistically accomplished in a simulator.

Software—the People

The 'people team' in flight simulation consists of simulator designers and manufacturers, aircraft manufacturers (who supply the data to the simulator manufacturers), operators, regulators, certification and maintenance teams, instructors, examiners, and the airline crews who are the prime target of simulation. Together this team delivers regulatory compliance, training and safety value. The responsibility for the delivery of minimum standards of simulation rests with the regulator, but effective and relevant training outcomes are a team effort.

The instructor's art: Effective instructors are essential to optimal training outcomes, compensating for less those effective training tools still in operation. Conversely poor instruction will compromise the effectiveness of the most advanced training devices. Students thrive in a training environment in which dedicated instructors are encouraged and supported in their work. Recognition of the unique importance of each student, combined with the positive learning atmosphere created by a team of experienced, enthusiastic, and empathetic instructors, are most important factors in the generation of cost effective training. But which instructor skill-sets and experience are most appropriate for modern type training? Flight crew operating older technology airliners had to understand aircraft systems at a detailed level, because systems had to be managed manually. Technical demands on such crews required that systems were committed to memory, to ensure correct actions in emergencies, and technical instructors were important at the start of type-training courses. Technical instructors continue to be essential for the training and revalidation of maintenance engineers, and for flight crew basic systems knowledge, but for pilots, the emphasis is shifting. While appropriate technical knowledge remains important, the training process has moved toward a more operational focus due to advances in airliner automation, where pilots have no access to many areas of aircraft systems. New aircraft design has enabled two-pilot flight crews to manage flights in a more holistic way, becoming more responsible for navigation and flight path through more airliner-autonomous systems. Contemporary type training therefore places more emphasis on operationally-based learning, focusing on flight deck operational procedures. The most appropriate instructors for this role would be active instructor pilots, but this has been an option too costly for airlines in terms of lost pilot productivity. New courses today allocate technical instructors to a lesser extent during the initial phase of systems training, placing the main instructional task on simulator instructors who have held airline pilot and instructor positions. As simulator instructors are usually not current line pilots, a high level of instructional continuity is possible, unhindered by flying rosters. Some regular exposure to line operations is nevertheless required in order to maintain operational awareness, and this is achieved by regular observation flights. Turning to the students, even the highest level of simulator fidelity does not automatically generate 'believing trainees.' 'It's OK, I'm only in the simulator' is probably a passing thought in the minds of many pilots under training or test in simulators. The artificial nature of simulator training and testing, in the context of scripted content, works against the 'reality environment.' In recurrent testing, crews often know in advance the sequence of events to be simulated under test. Automatic, mechanical, and robotic simulator sessions can dramatically reduce the real value of these training tools. Scripted testing can cause crews to display pre-emptive, rather than realistic responses to challenge. But once a session is active, the instructor or examiner has the power to modify the script, and create the unexpected. The instructor plays a pivotal role in adding realism to the scene in both type and recurrent training. He or she can create more realism for the crew, by randomly vocalising radio and ground communications, and replicating regional practice and accents as closely as possible. ATC simulation technology will largely remove this requirement, but hundreds of simulators are likely to remain in service without this facility for some years to come. Ad hoc inputs from an experienced instructor, made without regard for the phase of flight or level of crew activity, will raise trainee crew workload in meeting multiple distractions and challenges. Rising workload reduces peripheral awareness, forcing crews to prioritise, task share, as in real operations. Applying the 'instructor's art' effectively requires special skills and experience, acquired over time in operations and training, together with awareness of current industry threats as exposed by safety systems. For this reason, retention of experience within the simulator instructor corps is important. A poorly motivated or inexperienced instructor can severely limit the value of simulator training or testing. A training organization always has the option to accelerate the acquisition of 'the instructor's art' through high impact instructor training programmes, but there is no substitute for experience. Enhancement of the instructor's art is a high value initiative to improve training effectiveness and system safety.

Teamwork in simulation: In normal operations, the tasks and work-style of simulator instructors and technicians are necessarily different. Engineers try to maintain simulators at high levels of reliability, against the clock, 24/7. Technicians may spend considerable periods of time searching for the cause of a failure. On the other hand, instructors use simulators as the training tools with which to apply their art, juggling with trainee learning rates, also against the clock. Technicians are not normally trained to understand the significance of a component failure from an instructor's perspective, and the instructor may not appreciate the technical work required to rectify faults. Mutual understanding between mechanics and instructors is bound to lead to less simulator down time, and greater quality and efficiency in simulation.

Crew performance markers: Poor skills, procedures and handling contribute to accidents, but inappropriate crew behaviors are frequently at the heart of these events. Valid crew behaviors do not come naturally. An effective crew-member has to adapt to the unique environment of the cockpit, and most pilots need to learn and re-learn how. Since the early 1990s, from research undertaken by Proffessor Helmreich (1998) and his Crew Research Project Team at the University of Texas, airlines and authorities progressively introduced behavioural (Klampter et al., 2001) performance markers for use during crew checks. These markers were enhanced using data from audit programmes, and are now at the heart of crew human factors assessments. Improved application of appropriate performance markers in simulator training has great potential to improve crew performance.

Crew fidelity—uniformed crews in simulators: Simulator training is designed to create the most realistic possible environment outside actual flight, at great hardware cost. Yet many airlines permit pilot trainees to enter simulators wearing a wide variety of clothing, none of which would be worn in commercial service. A pilot dressed in slacks and tea-shirt would be most unlikely to gain access to a real cockpit; especially in the current security environment. Accepting that a uniform provides a strong visual impression of ranking, it is surprising that crew procedures and skills are tested in simulators without uniforms. This would suggest that understanding of human factors in team performance remains immature in the training industry despite overwhelming clues from safety systems pointing to the part human factors play in accidents. Much research has been conducted on the way teams interact in same-cultural and cross-cultural settings. The term 'power-distance,' coined by Professor Hofstede (1986), describes an aspect of team hierarchy, particularly relevant to airliner crews. From Hofstede's research, it can be deduced that the uniform is likely to be a significant component of 'power distance.' It therefore follows that crew response to an emergency may be influenced by attire. In all cultures, some more than others, age and seniority are already 'power-distance' factors, and uniforms and gold-braid add emphasis to this. Cross-cultural issues on flight decks assume greater importance as the mobility of the pilot workforce increases across the world. To evaluate a crew in a simulator, dressed in casual clothing, seems bizarre, especially as it would be prohibited in the real environment which simulation attempts to replicate. While the technical fidelity of simulation continuously improves, the absence of uniforms in simulation is an increasingly exposed anomaly. In those airlines in which uniforms are used in simulation, the training atmosphere tends to be more professional. In recognition of the part that uniforms play, an uncomplicated company policy would not be difficult for an operator to enforce at zero cost, as all airline pilots have uniforms. Uniformed crews in simulation match crew fidelity with simulator fidelity, and instill a stronger sense of relevance, purpose, and professionalism, to each session.

The Training Process

The Multi-Crew Pilot License (MPL) triggered healthy reviews of process across the airline industry. MPL is at the leading edge of process improvement. However, in legacy training environments, recognizing the great advances in simulator technology, do we really maximize the use of simulation? Since the 1960s, the method of using flight simulators has altered very little. We still tend to apply inflexible precanned syllabi, the basis of which was developed decades ago, administered via a somewhat robotic process. Recurrent training has contracted toward minimum regulatory standards, while other more topical training needs are often omitted. We could ask if current training outcomes are proportional to the capabilities of modern simulators? Do we use simulators efficiently, considering the capital cost of USD 10–15 M per unit and cost recovery in excess of US 1,500 per session? Operators have limited influence over design, but can affect the way simulators are used, and have control over best practice application.

Measurement and assessment: An important step in any course design is to integrate measurement capability from the outset in order to provide for continuous improvement. There are numerous examples of data collection where the data is collected but has never been analyzed, despite the fact that analysis would produce powerful signposts to drive training objectives with. The practical objective of any measurement process must be to obtain actionable information for improvement. It's pointless measuring for the sake of measuring. Measuring should be organized for a purpose. Measurement programs are often mandated from above because *'we don't trust what you're doing, so we want you to measure the impact,'* or *'we really need a measurement system which enables us to advise the executive how much 'value' we are getting out of the training.'* The driver for training measurement should be process improvement aimed at operational competence and safety, which must lead to enhanced efficiency.

The assessment of trainee performance has not always been acceptable as a concept in the airline industry, as it carried overtones of 'testing.' The key to developing such systems is the terminology used, and the objective described. Airline crews already accept almost continuous assessment in the workplace. The cockpit voice recorder (CVR) tapes every word of the operating crew (although only retrievable via strictly controlled protocols), and video systems may follow. Used exclusively to improve training effectiveness, assessment can be effectively applied to simulator training, and the simulator will be able to play an even more useful role. Data emerging from such assessments will provide the rationale for training programme and instructor improvement; what works best, and what does not. Data showing real learning rates, matched against trainee background and experience, will help to enhance selection and recruitment, leading to further cost savings.

Training syllabus: Simulators are part of the syllabus for all courses designed to qualify pilots to operate new airliner types. The Airbus 'ADOPT' (Airbus Operational Pilot Training) course was the product of task analysis and instructional design, now replaced by APT (Airbus Pilot Training), which emerged from an even more intense period of research and development. APT matches training equipment to training objectives. Boeing (Alteon) have similar programs planned for B787 type training. Decades ago, aircraft manufacturers tended to provide basic training programmes, sometimes limited in scope and relevance to airline operations. So airlines developed in-house training programmes based on their operational experience, and extended in-house development into the area of Standard Operating Procedures (SOPs); usually modified from those supplied by the manufacturer. Some operators went further, hoping to generate internal efficiencies through commonality across fleets, designing 'common SOPs' and 'training footprints' to be used for type training on any aircraft type operated. While a logical goal, design differences and operating approaches between the major manufacturers can get in the way of this objective. Merging different manufacturer SOPs and training programmes into a common format can compromises training quality and relevance for both types. Such a compromise may also transfer liability from the manufacturer to the airline, adding more risk to the operation. Cost savings from the perceived synergies of this approach, and the argument that aircraft manufacturers do not conduct airline operations, strengthens the conviction that the airline knows best, even in areas where this is patently not true. For example, one airline-developed training module was executed without manufacturer endorsement, contributing to a tragic outcome. At the time, the industry had experienced a spate of jet-upset accidents, and it was not unreasonable to conclude that this area of training needed attention. Airline pilots who had not been trained in the military had usually not seen extreme aircraft attitudes. Neither had they learned appropriate recovery techniques from airliner upsets. The establishment of jet-upset training was seen as beneficial to all. With this laudable motive, an airline set out to produce a jet-upset training program in-house; in the process modifying simulators (aerodynamically) to match the objective. *Simulators can only simulate the real aircraft with any accuracy when the simulation is based on actual data collected during flight test. But no data existed, because the actual aircraft had not been test flown into the extreme attitude which were to be simulated, and accurate simulation could not be created.* This training programme was not endorsed by the manufacturer. After a subsequent fatal accident, this policy came under scrutiny, and the training was withdrawn. Seeing liability as an increasing business risk ahead, manufacturers have poured resources into best training practice research and development. Training has now become an important business component, and is far more mature than a decade ago. Perceived past limitations in manufacturer training are therefore no longer so arguable. Modern manufacturer training programmes have been developed with resources beyond those readily available to airline training departments. Manufacturer design philosophies are firmly entrenched in the task analysis and instructional design of their training programmes, leading to relevant, accurate, and safe syllabi, and such courses are being adopted by airlines. A major argument for more globally standardized SOPs is that as pilots move from operator to operator, they would easily slip into the new operation on the same aircraft type, using essentially the same SOPS. Currently, pilots may move to a new employer, operating the same aircraft on which they are type rated, who uses different SOPs. IATA has taken an interest in encouraging a harmonized approach to training and SOPs, recognizing that the manufacturer's SOPs and training now provide more relevant core type training, which can still be enhanced with add-on airline-specific modules.

Programme relevance: Early in the pilot training process, the commercial pilot license training needed upgrading, and MPL started this process. It has been normal for military pilot training to be ruthlessly focused on the in-service task. Military pilots are often needed in a hurry, and lengthy sequences of training may not be acceptable. The track record speaks for itself. Large multi-million dollar 4-engine jet aircraft have been safely commanded by military pilots sometimes in their early to mid-twenties, with experience levels unacceptable to airlines for left-seat qualification. Many military pilots in the sixties learnt to fly from day one on jet aircraft, as they were no longer likely to operate older types in their service careers. Yet 45 years later, the airline industry was still teaching pilots to fly light propeller aircraft, to qualify for a Commercial Pilots License (CPL); inappropriate preparation for modern airliner operations. This process was inefficient, consumed more time than necessary, requiring an element of 'unlearning' later. Later reversion under pressure to 'first learnt principles' also created a 'lurking threat in the system.' In an airline industry focused on growth, researchers have been looking for more efficient and relevant training methods for some time, and this is how the ICAO MPL (Multi-crew Pilot License) emerged. Contrary to some commentary MPL was not a cost-driven programme. It was developed over six years to make airline pilot training more relevant. The MPL adds simulation to the training syllabus, and reduces aircraft flight training on less appropriate aircraft. MPL training promotes the early use of simulation, embedding crew-coordination and TEM principles into the training process. The use of more effective tools such as ATC simulation, combined with configuration of training aircraft more closely aligned with airliner cockpits (flat screen displays), drives training purposefully towards the airliner flight deck; a very different objective to the standard CPL. The use of cockpit video replay of each flight, takes the initial learning process to a new level. MPL training leads from selection through to airliner type rating.

Training beyond regulatory requirements: The opportunity for Flight Operations or Training Departments to add more recurrent simulation for their crews is often denied on grounds such as: 'we meet legal requirements, don't we?' For this reason, airlines must capture and analyze data more effectively to provide factual information to drive training needs (*'Without data you are just another person with an opinion'*). Until data collection and analysis systems mature, the volume of simulation applied to flight crew is likely to remain related more to cost and regulatory requirements than to actual training needs.

Recurrent training: In airline operations, the ability of the crew to react correctly to emergency situations is a major function of simulation, and must be continuously demonstrated. The key ingredients of an effective recurrent training and testing process must be quality, relevance, frequency, and time of exposure. Regulations generally require a professional pilot to demonstrate competency in skills and procedures on a regular basis, usually every six months. Regulators mandate that for license retention candidates must demonstrate competency in a list of simulated emergencies and abnormalities over a test cycle of one to three years, and in most jurisdictions, crews are required to undergo a minimum of 16 hours recurrent training and testing in simulators per crew per year. But airline operations change, and a closer look at the relevance of recurrent simulation will always yield opportunities for improvement. Recurrent training should be proactive in preparing crews for new and developing threats. Effective data analysis programmes may signal a need for more than the minimum regulatory requirement of simulator training for crews. For example, facing fleet growth of over 9% per year, the Civil Aviation Administration of China (CAAC), applied 28 hours of recurrent simulation per crew per year; a 75% above standard Western practice.

Recurrent training to type of operation: For long- and short-haul operations, there is a 'base volume' of simulation needed to ensure competency in handling abnormal procedures, regardless of the type of operation. But in the area of skill retention, short-haul pilots accomplish takeoffs and landings at a much higher frequency, allowing skills and critical competencies to be maintained in actual service. Therefore for short haul operations, current volumes of recurrent simulator training may be optimal. However, on long haul fleets, pilots are exposed to a low frequency of take-offs and landings, which may drop as low as one take off and landing in three months. It is therefore necessary to recognize the specific needs of long haul pilots in protecting them against skill decay. Recurrent simulation could therefore be tailored more to the type of operation, allocating additional time to long haul crews.

Recurrent training—cabin crew observation: Cabin crew observations of specified simulator sessions, enhances understanding across the flightdeck door, adding to system safety, especially in the post-911 environment of locked cockpit doors.

Recurrent training—airline system feedback to simulator training: An extract from a Flight International editorial on runway excursions and stabilized approaches (Flight International, 2008):

> Data is a wonderful tool, and until the digital age got into full swing it was not used to anything like the extent it is now. It has the power to provide understanding of an issue or problem, and to illuminate the options for improvement. Gathering information often provides surprises, not every unstabilised approach ends up as a runway excursion, but almost every runway excursion starts as an unstable approach. There is no requirement to train pilots to manage aircraft on contaminated runways. It's quite weird to discover that these glaring gaps in safety management exist in such a sophisticated industry.

Traditional recurrent simulation is a 'prescriptive' and regulated process, based on operational needs perceived some time ago. But accident investigators use simulators to replay event data, and simulators could be applied more to line pilots to review 'recent event scenarios' (suitably de-identified), or issues of importance arising from industry data. Unfortunately, such components are rarely seen in modern recurrent training programs, as they are not yet widely mandated. Airlines now tap multiple data sources to monitor their operations, including FOQA (Flight Data Analysis using quick access recorders-QARs), crew performance data (from line and simulator checks), LOSA (Line Operational Safety Audits), LOAS (Line Operations System), and Confidential reporting data. These data feeds form an important part of the ICAO mandated Safety Management System (SMS). But very few airlines consolidate this data into an integrated centralized 'airline health check,' although this is technically feasible. Most data required is mined, but often in different IT platforms and departments, and is traditionally reviewed at infrequent meetings. A continuous 'health check screen' would be far better tool for training and safety managers, providing real time awareness of all training and safety-relevant information. More informed decision making would result, and preventative measures could be inserted quickly into recurrent training programmes.

Recurrent training—simulator programming: Emergencies and abnormalities are not scripted in the real world. Abnormalities occur randomly and unexpectedly, and mental preparation time is often denied in commercial operations. If a rigid annual programme includes no time for topical incident-driven training, then the programme becomes a routine anticipated task for both instructor and trainee, and 'just another chore' in the life of an airline pilot. Ideally, most recurrent training session content should not be transparent to crews. There should be a unstructured component of the annual simulator programme which provides for topical, remedial, and 'free-trainee' training time to practice self-recognized deficiencies.

Human factors training: In any discussion of airline safety, the significance of human factors in accidents is beyond dispute, and a high value area for simulator training development. Crew Resource Management (CRM) training has been applied for many years, often integrated with cabin crew. Under the European Aviation Safety Authority (EASA) CRM is also referred to as MCC (Multi-Crew Cooperation) training which must be completed before Type Rating issue. Modern CRM training continues to provide key guidance on effective communication, task sharing, team-building, and teamwork; employing appropriate cockpit behaviors for safe operations. TEM training promotes pre-emptive strategies of threat recognition, avoidance, and management. Both CRM and TEM should be driven by data from incidents, accidents, Flight Operations Quality Assurance (FOQA) programmes, and Line Operational Safety Audits (LOSA).

Line Oriented Flight Training (LOFT): While CRM and TEM are already applied in many training areas, the most effective training platform yet devised is Line Oriented Flight Training (LOFT), which sets up a simulated A to B flight scenario. LOFT continues to offer high potential returns in training and safety, creating in simulators the most realistic flight scenarios possible, where crew performance will more closely match that seen in the operational environment. LOFT has been poorly and sporadically applied, and only recently mandated by some regulators. Simulator video equipment used in LOFT is becoming standard fit, together with crew performance and analysis systems. In most airline training departments there are those instructors who are passionate about LOFT, and those who have less time for this process. Some regulators have advised that LOFT should be a requirement of recurrent simulator training programmes, but such guidance is allocated a lower priority than other components of the 1–3 year recurrent programmes, and LOFT may be sacrificed when rostering disruptions occur. Despite the critical relationship between human factors and safety, the proportion of airline simulator time currently dedicated to genuine recurrent LOFT is low, except in regulatory regions where LOFT is mandated. LOFT is dedicated specifically to training, but a broader term is used to describe all line oriented simulator training and testing: LOS (Line Operational Simulation). LOS and LOFT facilitators recognize that a properly constituted uniformed crew, not an individual pilot, delivers an airliner safely to destination. LOFT and LOS sessions therefore provide the most effective platform yet for the development of safer crews. Well-positioned cameras, realistic scenarios, reliable hardware, and the instructor's art are all essential to the desired outcome. And *'help them to reach their full potential—catch them doing something right!'*

SUMMARY

In the face of current industry threats the 'no-change path' leads nowhere. Legacy standards may protect for a while, but the deliberate avoidance of fresh defensive strategies today may generate tragic consequences tomorrow. Less forgiving liability laws will result in closer scrutiny of earlier policies. Toughening laws on corporate manslaughter are worth further consideration. But the change-challenge also creates opportunity. Risk reduction can create competitive advantage. Improved process will lead to greater efficiency; cost reduction, and enhanced system safety. New airliner acquisition is relatively straightforward, but the production of safe well-trained pilots is a more complex task. Resources must be allocated with care to this fragile part of the system. Minimum training standards prescribed by the regulator may not adequately protect airlines. Wiser management teams will seek out and apply best practice. **Training is a controllable variable in the airline safety system**. Airlines intent on maintaining or improving safety margins can simply elect to enhance the quality and relevance of crew training at little additional cost. Approximately 1,000 airline simulators support the training operation for more than 18,000 airliners in service. Simulators are powerful training tools for the application of best training practices, but are far from fully exploited. There is always more that can be done to improve instructional skills. Differing views may exist on best practices, but global harmonization is being driven organically through the expanding pool of transparent safety data, ICAO, FAA, EASA, IATA, RAeS, and Flight Safety Foundation (FSF) initiatives, and from other influential industry groups. The standardisation of global best practice is underway. IATA, through the commercial inducement of retained IATA membership, has made huge strides in harmonizing safety standards through IOSA. However, at the highest regulatory level, ICAO recommendations may take time to filter into common practice, and time is a precious commodity in a rapidly expanding industry. Latent resistors in the system remain, especially where airlines have experienced decades of commercial success. Here change may be blocked by those who may not understand the risks; until there is a serious wake-up call. The 'status quo' may be vigorously defended from the apparent security of corporate 'silos.' So operators must look more openly at themselves, within a just safety culture, responding to the performance data they mine, implementing appropriate change without delay. Safety Management Systems will increasingly drive airlines into training enhancement programmes, creating **embedded continuous improvement** loops within the process. The traditional crew training model has already been replaced by the MPL (ICAO mandated in 2006 after 6 years of development). MPL requirements combine the concepts of airline relevance, CRM, competency, and embedded TEM, with high levels of dedicated training technology. Change is the only certainty we have, but **'the only thing harder than getting a new idea into the head, is getting the old one out'** (General Patton). Training programs must now **be designed for continuous change** based on industry feedback from the expanding **reservoir of data** available. As much as individuals display inappropriate paradigms, so do organisations. But there is one airline paradigm which will always be true: *'an accident is the ultimate inefficiency.'* Top management must know enough to be able to inform every airline employee how to recognize and balance the healthy tension between commercial and safety objectives. During periods of growth and recession pressure to shrink crew training resources increases, yet the accident rate must be driven down further. Without adding significant cost, the QA tools now exist to apply more effective and relevant simulation, and protect and expand safety margins. **The link between crew training and safety is now widely recognised, but the conviction to act is essential. Flight crew training requires renewed industry effort to embed quality and continuous improvement, and generate lower accident rates in the future**.

References

Westrum, R. (1993). Cultures with requisite imagination. In J. A. Wise, V. D. Hopkin, & P. Stager (Eds.), *Verification and validation of complex systems: Human factors issues*. New York: Springer.

John J Nance Blind Trust (1986). Safety Margins in 1980s.

Columbia Accident Investigation Board Report (2003). NASA. Chap. 7. A broken Safety Culture. Chap. 8.4 organisation, culture, and unintended consequences.

Reason, J. T. (1997). *Managing the risks of organizational accidents*. Ashgate, Aldershot.

Perin, C. (1992). British Rail – the case of the unrocked boat.

Tideman, R., (Asia Pacific Airline Symposium – Singapore 2006) – Principle Inspector of Air Accidents - UK Air Accidents Investigation Branch (AAIB). Topic: Accidents and Training.

Helmreich, R. L., Klinect, J. R., & Wilhelm, J. A. (2001). System safety and threat and error management: The line operations safety audit (LOSA). In: *Proceedings of the Eleventh International Symposium on Aviation Psychology*. Columbus, OH: The Ohio State University (pp. 1–6).

Bisignani, G., IATA Director General and CEO (2002), Hong Kong presentation – A Call for New Global Solutions, & Hong Kong presentation (22 Sept 2006) – Leadership Challenges & Opportunities. & speeches in November 2007, and at Singapore IATA Conference in February 2008.

ICAO documents: 9808, 9806, 9803, 9806, 9835, 9841, 9868, 9859, 9683.

Helmreich, R. L., & Merritt, A. C. (1998). *Culture at Work in Aviation and Medicine: National*. Organizational and Professional Influences.

Klampfer, B., Flin, R., Helmreich, R. L., Hausler, R., Sexton, B., Fletcher, G., Field, P., Staender, St., Lauche, K., Dieckmann, P., & Amacher, A. (2001). *Enhancing performance in high risk environments: Recommendations for the use of behavioural markers. Presented at the Behavioural Markers Workshop sponsored by the Daimler-Benz Stiftung GIHRE-Kolleg*. Zurich: Swissair Training Center July 5-6, 2001 (UTHFRP Pub26).

Hofstede, G. (1986). Cultural differences in teaching and learning and related: (1980). Culture's consequences: International differences in work-related values. Newbury Park, CA: Sage. (1991). Cultures and organizations: Software of the mind. London: McGraw-Hill. *International Journal of Intercultural Relations, 10*, 301–320.

Flight International - editorial on runway excursions and stabilized approaches, dated March 2008.

General George S Patton. Remembered for his fierce determination and ability to lead soldiers, Patton is now considered one of the greatest military figures in history.

11

Understanding and Analyzing Human Error in Real-World Operations

R. Key Dismukes

Chief Scientist for Human Factors NASA Ames Research
Center Mail Stop 262–2, Moffett Field, CA

"Giant Airways 123, cleared for takeoff runway 25R." With that clearance, the captain of flight 123 taxies into position on the runway as the first officer completes final preparation for takeoff. Both pilots are highly experienced in their large jet, carrying nearly 300 passengers, and between them have over 20,000 hours of flying for major airlines. The captain advances the throttles for takeoff and the aircraft begins to roll forward normally, but the pilots are startled by a loud, insistent warning horn. In their surprise it takes a moment for the pilots to comprehend what is happening, but then they both recognize that the configuration warning system is telling them that something about the aircraft is not properly set, and the captain retards the throttles while the aircraft is still moving slowly. The first officer radios the control tower that they have rejected the takeoff, and the controller instructs them to taxi off the runway at the first turnoff. Scanning the cockpit as they taxi, the pilots discover that the warning horn sounded because the flaps were not set to the proper takeoff position. Somehow the crew forgot to set the flaps and overlooked the omission when they ran the taxi checklist, which directed them to verify the flap setting, in spite of having set flaps and run checklists thousands of times on previous flights.

An unlikely scenario? Not at all. The Aviation Safety Reporting System (ASRS) receives reports of such incidents almost monthly, and these reports are probably a small fraction of the actual occurrences. Well, perhaps one can argue "no harm, no foul"; the system protections worked as intended, and no untoward consequences resulted from these incidents? But consider this: At least two major airline accidents have occurred in the United States when the takeoff configuration warning system failed, and the crew attempted to take off with flaps inadvertently not set (NTSB, 1988, 1989). Beyond the threat to safety from the error of failing to set flaps, rejected takeoffs impose significant costs to airlines. If the aircraft is required to exit the runway and move back into the conga line of departing aircraft, the delay and the extra fuel burned affect the airline financially, not to mention upsetting passengers. And this is only one example of the many kinds of errors that affect flight operations.

Most aviation accidents are attributed to human error—predominantly pilot error, although maintenance error and controller error are also sometimes cited. Among commercial jet airplane operations worldwide, 66% of hull-loss accidents were attributed to flight crew in the period between 1992 and 2001 (Boeing, 2006). In general aviation the percentage is even higher; for example, around 79% of U.S. fatal accidents in 2006 were attributed to pilot error (Krey, 2007).

Beyond safety issues, human error imposes considerable costs on the airline industry in the form of equipment damaged, flights delayed, and fuel costs, although the total amount is not known. The magnitude of these costs is illustrated by a single problem area: ramp damage—collisions between aircraft and ground vehicles or structures, excluding taxiway operations. These collisions, 92% of which are attributed to human error, cost the airline industry about $10 billion each year worldwide (Lacagnina, 2007).

Human error is by no means unique to aviation; it plays a comparable role in accidents and financial losses in almost every industry. However, aviation offers an advantage for studying the nature and causes of human error and for finding countermeasures. Airline accidents are investigated in great detail, generating a substantial body of information about the errors made by human operators, the context in which those errors are made, and the consequences of errors. Further, in contrast to many industries, airline operations use highly scripted procedures, and reasonable consensus exists on whether operator actions are appropriate or inappropriate in the context of those procedures. Thus research on human error in aviation generates methods and knowledge that can be used to address error in other critical domains, such as medicine.

Given that human error plays such a central role in aviation accidents, it is crucial to understand the causes of error and the ways errors contribute to accidents. Furthering this understanding should be the foundation of any safety research program, and any operational safety program should be built on this foundation. Finding ways to improve safety and efficiency and to reduce costs are central aspects of the work of human factors experts, thus these experts require a deep and veridical understanding of the nature and context of human error. This chapter attempts to provide a foundation for that understanding and a framework for practical application.

The way in which errors are analyzed and interpreted hinges on one's perspective on what causes error. In this chapter I will present a perspective on the nature of errors made by skilled professionals such as pilots and the roles those errors play in accidents. I will discuss the types of information we need in order to understand why error-related accidents occur, examine sources of this information, and describe diverse approaches to analyzing errors and accidents. I will illustrate some analytical approaches by describing studies my colleagues and I have completed recently, and will conclude by examining the implications of these studies for developing countermeasures for reducing vulnerability to error.

This chapter focuses on the errors made by skilled, conscientious professionals performing their tasks in real-world operations, rather than the errors of novices and the errors made by participants in laboratory experiments designed to explore the fundamental nature of perceptual and cognitive processes. (However, our ability to analyze the skilled performance of professionals draws heavily on the foundation of fundamental knowledge generated by laboratory studies.) The examples in this chapter focus on errors attributed to pilots because the majority of research and of accident investigations focuses on pilot error. However, the errors of controllers, mechanics, dispatchers, and other aviation personnel are equally important, though under investigated. Fortunately, the principles set forth in this chapter apply equally to the work of all aviation personnel—indeed, the work of personnel in all domains of skilled performance.

A PHILOSOPHICAL PERSPECTIVE ON ERRORS AND ACCIDENTS

How organizations interpret and respond to human-operator errors is driven largely by their perspective on why humans make errors. For many years the dominant perspective was that professionals who make errors performing tasks appropriate to their skill-level evince some sort of personal shortcoming: They must lack skill or diligence. Consequently, organizational response was to admonish, punish, re-train, or fire the individual whose errors adversely affected the organization, an approach that may be characterized as "blame and punish." Although blame and punish still holds sway in many quarters, in recent years a strong consensus has emerged among scientists in several disciplines that this approach is misinformed and impairs efforts to improve safety (Reason, 1997; Dekker, 2002; Dismukes, Berman, and Loukopoulos, 2007).

A modern, scientific perspective holds that the errors made by skilled experts such as professional pilots, controllers, mechanics, and dispatchers are not root causes of accidents but symptoms of the flaws and inherent limitations of the overall sociotechnical system in which these experts work (Bogner, 1994; Dismukes, 2009; Perrow, 1999; Rasmussen, 1990; Reason, 1990, 1997; Woods, Johannesen, Cook, and Sarter, 1994).

This perspective also holds that the causes of errors are intimately related to the cognitive mechanisms that make skilled performance possible. Further, both correct performance and problematic performance must be understood in the context of the experience, training, and goals of the individual; characteristics of the tasks performed; human-machine interfaces; events—routine and unanticipated; interactions with other humans in the system; and organizational aspects. These aspects include both the explicit and the implicit manifestations of the organization's goals, reward structures, policies, and procedures. All of these factors interact in a highly dynamic fashion with the inherent characteristics and limitations of human cognitive processes.

Accident investigators and designers of equipment systems, operating procedures and policies sometimes erroneously assume that, if skilled experts can normally perform some task without difficulty, then they should be able to always perform that task correctly. But there is an enormous difference between being able to perform a task correctly the vast majority of time and being able to perform with 100% reliability. Even the most conscientious and skilled individual occasionally makes errors, and this is true in every domain, from aviation to medicine to musical performance. Experts make errors not because humans are intrinsically inadequate but because some of the tasks they must perform do not allow 100% reliability. True, some tasks can be performed by computers with almost perfect reliability, and it may be better to assign those tasks to computers. But some tasks demand the unique expertise of humans; for example, deciding what course of action to take in novel situations or situations in which information is incomplete, ambiguous or conflicting or in which value judgments are required.

Pilots (and other personnel) must balance competing organizational goals, often in situations in which not enough information is available to be certain of the outcome of their actions. For example, consider a cockpit crew that has just been cleared to depart from an airport in the vicinity of thunderstorms. The controller's clearance implies that it is safe to takeoff, but the decision rests with the airplane's captain. On-board radar and wind reports give some information about the storms' location and strength, but sometimes the decision is a close call, and the captain must balance the certain costs of delaying departure against the highly uncertain level of risk in departing immediately. Perhaps two other airliners immediately in front have departed without difficulty, suggesting to our captain that other experts consider the level of risk acceptable. Also, all three captains may have taken off previously under similar conditions without great difficulty, reinforcing the perception that the risk is not great. However, on this occasion, when our captain commences a takeoff he encounters windshear and crashes.

In this situation the accident captain would almost certainly be judged by investigators to have made an error in judgment that "caused" the accident. Ironically, rarely is anything said about the decisions of the captains of the aircraft departing immediately in front of the accident aircraft. But consider that these non-accident captains had the same information as the accident captain, made the same decision, but—because of the not entirely predictable nature of thunderstorm dynamics—escaped mishap and blame. In these types of situation, is it really appropriate to say that the captain made an error that caused an accident, or would it be better to conclude that the aviation industry operates in conditions in which it is not always possible to evaluate the level of risk precisely and to balance the goals of safety and efficiency with perfect certainty? The latter conclusion suggests that the industry should focus on providing better information to pilots, providing better guidance for dealing with uncertain situations, and accepting responsibility for (thankfully) rare accidents rather than blaming human operators.

In other situations the nature of the tasks human operators are required to perform, and the circumstances in which they are performed, are not well matched to the characteristics and limitations of human information processing. Consider situations in which the operator must continuously switch attention between two or more tasks to keep track of the status of the tasks and manage them, often called "multitasking" (Loukopoulos, Dismukes, and Barshi, 2009). For example, when a crew receives an amended departure clearance while the aircraft is taxiing, the first officer must put the new information in the flight management computer, interleaving steps with other tasks, such as monitoring the course of the taxi and responding to radio calls. Most of the time experienced pilots can switch attention back and forth in this way without great difficulty, but when one task suddenly becomes more difficult (e.g., the computer will not accept the new data), pilots are vulnerable to having their attention absorbed by the demanding task and losing track of the other tasks.

Many individuals believe they can multitask efficiently (as in using a cell phone while driving), but the research evidence is that human ability to manage multiple tasks at the same time is severely limited (Loukopoulos et al., 2009). Even when sufficient time is available to perform several tasks concurrently, error rates—especially errors of omission—go up. To provide the level of reliability required for cockpit operations, we must design procedures to minimize multitasking demands, design equipment to help pilots keep track of multiple tasks, and train pilots about the nature of vulnerability to error when multitasking. This vulnerability should be thought of as system vulnerability rather than a shortcoming of pilots.

The issues we have been discussing have strong implications for how one should think about the causes of accidents and ways to prevent accidents. We must be careful to avoid hindsight bias (Dekker, 2002; Dismukes et al., 2007): *after* an accident involving problematic judgment and decision making, it is easy to identify things the crew could have done differently to avoid the accident. But *the crew did not know* the outcome of their flight; as far as they knew up until things started to go wrong, they were conducting the flight and responding to circumstances appropriately. If we want to understand error-related accidents we must put ourselves in the place of the crew and ask why it seemed appropriate to do what they did at each stage of the flight, knowing only what they did at each moment.

Hindsight bias leads us to define error around the outcome of human operator actions, but it is far more useful to ask whether the operator's goals, knowledge, and skills were appropriate to the situation at hand, whether adequate information was available, and what level of reliability can be expected of skilled operators performing the particular combination of tasks required under the conditions encountered.

Another problem blocking true understanding of error-related accidents is that looking only at the performance of operators involved in accidents, and not examining the performance of other operators in similar situations, constitutes a sampling bias that can drastically distort perception. For example, several airline disasters have occurred when crews attempted to land from an unstabilized approach. (Parameters such as airspeed, descent rate, or engine settings were not on target, or the aircraft was not configured for landing early enough). Airlines provide guidance in flight operating manuals for conducting stabilized approaches.[1] Thus one might interpret an accident resulting from attempting to land from an unstabilized approach simply as a case of a rogue crew violating formal standard procedures. But if one took a sample of nonaccident flights a different picture might emerge. For example, Chidester's (2004) analysis of data from flight data busses revealed that landing from unstabilized approaches occurs with disturbing frequency—usually without mishap, although the margins of safety are compromised. Unstabilized approaches are most likely to occur when air traffic controllers vector an aircraft in such a way that it ends up with too much energy and not enough time to configure the aircraft to dissipate the energy before it reaches the runway ("slam-dunk" approach clearances).

Although landing from an unstabilized approach is an error (and a violation of written procedures), the fact that it occurs fairly frequently under certain circumstances suggests that an accident resulting from landing from an unstabilized approach may be the manifestation of a systemic problem rather than of a deviant crew.

[1]This was not always the case, and when guidance was initially provided it was often couched as suggestion rather than mandatory policy and may not have specified exact parameters. However, in recent years, the industry has been moving to provide detailed, mandatory guidance.

In order to assess the crew's decision to land we need information on the norms for what other professional crews typically do in similar situations. The concept of "norms" does not imply that every pilot always does the same thing in a given situation, but rather encompasses the range of behaviors commonly exhibited.

Pilot errors occur in even the most routine of flights, and on challenging flights it is common for several errors to be made (Klinect, Wilhelm, and Helmreich, 1999). In a cockpit observation study focusing just on checklist use and monitoring during routine flights, Berman and Dismukes (in preparation) found multiple errors occurring on most flights, although the number of errors varied substantially from one flight to the next. (To put this in the proper context, the opportunities for error on a given flight probably run into many thousands, so the error rate is not high, given complex and dynamic task demands.) Fortunately, most errors are either caught before causing serious problems or their outcome is not consequential. (An example of an inconsequential error is a pilot missing an item on a checklist in a situation in which the item checked was already properly set.) These and other studies, as well as theoretical considerations, reveal that the occurrence of error is probabilistic, the product of many factors (both proximal and latent) interacting in ways that are only partly systematic and are to a considerable degree random.

Similarly, whether an error leads to an accident is highly probabilistic rather than deterministic. Through years of experience, safeguards have been developed to counter most single-point equipment failures and pilot errors. For example, if a crew forgets to set flaps to the takeoff position, two safeguards are in place to prevent them from attempting to take off improperly configured: the Before Takeoff Checklist and the Takeoff Warning Configuration Warning System. Thus, when an accident does occur, it is usually because multiple factors occur randomly and interact to defeat multiple defenses. In one case, in which an airliner crashed with the flaps not set, the crew unexpectedly received clearance to take-off ahead of other waiting aircraft, and in rushing to comply the crew inadvertently failed to perform the Before Takeoff Checklist. Unfortunately, the configuration warning system failed to alert the crew because of dirty switch contacts, and they attempted to take off, with disastrous consequences.

The probability of any particular set of factors occurring and interacting randomly to produce an accident is extremely low, yet the number of potential combinations of all possible factors is very large, which makes predicting and countering potential accident scenarios in reliable industries, that have developed many layers of protection against equipment failures and human errors, extraordinarily difficult. Further, large dynamic systems with extensive interactions among multiple human and nonhuman agents are prone to unpredictable emergent behavior that is almost always problematic and sometimes destructive (Perrow, 1999).

This probabilistic character of accidents renders traditional concepts of accident causality at best simplistic. Perhaps, because the crew members of an accident flight are the people who have the last chance to prevent the accident, their errors are most often identified as the "cause" of the accident. Yet to understand those errors and develop effective accident prevention strategies we must focus on the full interplay of multiple factors that is the true progenitor of accidents. Consider the interplay of factors in the CFIT (controlled flight into terrain) accident depicted in Figure 11.1. Many factors interacted in this accident, some of them quite unusual (e.g., the airport control tower closing because its windows blew out). Eliminate any of these factors and the accident would have been much less likely to happen.

This discussion has focused exclusively on the errors of professional pilots. Admittedly, the accidents of private pilots often reveal lack of skill—losing control of an aircraft landing in a moderate crosswind—or egregious judgment—buzzing a friend's house. Undoubtedly, these errors reflect the fact that private pilots receive far less training and scrutiny than professional pilots, and typically have far less experience. And, of course, most private pilots lack many of the safeguards available in airline operations, such as triple-redundant equipment systems, dispatchers to assist in making decisions, and highly structured operating procedures. Nevertheless, to understand the errors and accidents of private

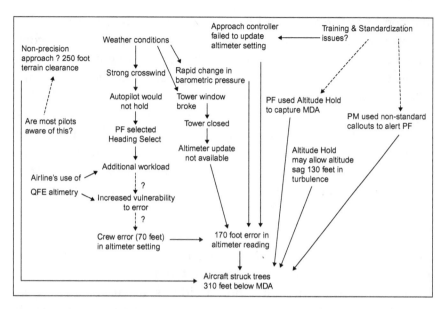

FIGURE 11-1   Confluence of factors in controlled flight into terrain accident.

pilots we must still consider the constellation of factors that affect pilot performance. For example, one of the most common forms of fatal accident in light aircraft operations occurs when private pilots inadvertently fly into weather beyond their skill to manage or the capabilities of their aircraft (Wiegmann, Goh, and O'Hare, 2002). In these accidents it would be well to ask whether adequate information is available for these pilots to evaluate risk adequately and whether training is sufficient to develop robust skill in risk evaluation.

SOURCES OF ERROR DATA

Diverse kinds of data are available to researchers studying the nature and context of errors made by aviation professionals performing work tasks.

Accident Reports

Accidents causing fatalities or large financial losses are investigated in great detail. The ability of investigation teams to reconstruct the conditions, events, and human actions leading to an accident is impressive. Cockpit voice recordings and flight data recordings (and, too seldom, survivor testimony) provide considerable information on what the accident crew did and did not do and allow limited insight into what the crew members were thinking. Investigation teams also examine training records and procedures, operating procedures, and company policies, which helps understand crew performance to some degree. Unfortunately, the level of detail retained in training records is quite limited, and normative data on what a population of pilots might do in the situation of the accident pilots is almost entirely missing.

The biggest limitation of airline accident reports is that they are a very small sample of the many ways in which a flight crew might get into trouble, and this sample cannot represent the full spectrum of possibilities. (General aviation accidents are, unfortunately, far more frequent, however these accidents are rarely investigated in the depth of airline accidents; nevertheless, some statistical data can be gleaned from these reports.) Given the probabilistic nature of errors and accidents, the airline accident that did happen may have been no more probable than many that did not happen.

Incident Reports

Voluntary reporting systems, such as NASA's Aviation Safety Reporting System (ASRS) and the airlines' Aviation Safety Action Partnership (ASAP), provide a much larger sample than accident reports. These programs collect voluntary reports from aviation personnel who experience or observe a situation that threatens safety, though without serious mishap. ASRS, in operation for over a quarter of a century, has collected more than a half million reports, mainly from pilots, though reports are now coming in from mechanics, flight attendants, and controllers. These reports, which can be searched electronically, provide an excellent means to sample the universe of possible ways in which things can go wrong. ASRS analysts have been able to identify problem hot spots, such as an airport arrival procedure that leads to traffic conflicts, and alert appropriate authorities before an accident occurs. When accidents do occur the National Transportation Safety Board

(NTSB) often asks ASRS to search for reports of incidents occurring under conditions similar to those of the accidents.

These reporting systems have several limitations. Because reports are voluntary and submitting them is subject to diverse motivations, the database cannot be assumed to represent the statistical distribution of actual occurrences. The report narratives are by their nature subjective and reflect whatever biases and limited understanding the reporters may have. Further, the reports, although often providing invaluable insight from the reporters, are typically no more than a page or so long and lack the detail of accident reports. However, it is possible for researchers to collaborate with ASRS analysts, who have the capability to call reporters soon after a report is received to obtain more detailed information about the events.

Systematic Surveys of Large Populations

Some of the inherent limitations of voluntary reporting systems could be bypassed by conducting surveys of large samples of pilot, controllers, mechanics and other key aviation personnel. This approach would provide a representative sample of the distribution of problems occurring in flight operations and would allow analysis of trends over time. This method would be particularly beneficial in assessing changes in the overall airspace system as new technologies and procedures are introduced. NASA developed the methods for such surveys using telephone interviews with a sample of more than 25,000 pilots with air transport pilot certificates. Careful attention was paid to using appropriate scientific methodology for sampling, survey organization, and interviewing participants, and to protecting confidentiality of participants. Ideally, with the methods for collecting and analyzing data established, the research program would have led to a continuing operational program run by the aviation industry in collaboration with researchers. Unfortunately, as of the date of this writing, that does not seem to be happening.

In-depth Interviews and Questionnaires

Experienced aviation personnel are an invaluable—essential, actually—source of information about the nature of aviation operations, how skilled operators go about performing their work tasks, the errors they make, and the consequences of those errors. Interviewing experts informally is often the best first step for researchers starting to study a human performance issue. Structured interviews provide a more rigorous approach than informal interviews and can help guard against interviewer biases and insure systematic probing of the interviewee's knowledge. Questions are identified in advance; however, it is useful to also allow follow-up questions to explore unanticipated aspects revealed in the course of the interview.

Similarly, written questionnaire surveys can be used to probe domain experts' knowledge and experience in specific work settings. Although these surveys have most commonly been used to study workplace culture, they can also be used to collect data on the types of problems and errors human operators experience and to glean insight into factors associated with those problems and errors. A good technique is to use informal interviews to gain a general sense of the issues in an area, design question items to probe specific topics, and ask domain experts to comment on the items before administering the questionnaire to the target population. An experienced survey researcher may go through many iterations of a survey, using the results of one survey to delve deeper into issues and explore new issues uncovered (Parke, Orasanu, Castle, and Hanley, 2005).

Surveys have the advantage of garnering information from moderate size populations relatively quickly. Surveys and interviews are both limited, in the kind of information they can provide, by the fact that domain experts have limited explicit declarative knowledge (as opposed to procedural skills) about the cognitive processes that underlie their skilled performance and are vulnerable to organizing their explanatory schema for errors around pop psychology. Thus surveys and interviews are generally more useful for exploring the conditions under which operators work and the errors they make than for determining the cognitive processes that underlie error.

Related to surveys and interviews are knowledge elicitation techniques, described later in this chapter, which can be both a source of error data and a method for understanding error.

Laboratory Experiments

Studies conducted in laboratory settings are of course the main-stay of research on human perceptual and cognitive processes. Laboratory paradigms are designed to provide exquisite control of variables in order to tease apart fundamental processes. But it is very difficult to use this approach to study the dynamic interaction of the many factors that drive correct performance and errors of skilled experts performing real-world tasks. Further, to experimentally investigate all of the many combinations of factors occurring in a complex system would be prohibitively expensive. It is also a challenge to ensure that laboratory paradigms evoke the skills of domain experts in the same way that real-world tasks do and thus produce the same kinds of error. Nevertheless, the knowledge generated by laboratory studies is the foundation for any effort to understand skilled performance in real-world tasks.

Direct Observation of Crew Performance

Under carefully controlled conditions it is sometimes possible for observers to collect data from the cockpit jumpseat during actual flight operations. The airlines themselves do this, in a program called line operations safety audits (LOSA), to gain insight into the challenges facing flight crews and how the crews respond to those challenges (ICAO, 2002; Chapter 14, this volume). Observers use a standardized observation and recording form and are trained with tapes of crew performance in an attempt to achieve reliability among observers. The airlines use the data, which needless to say are highly confidential, to identify problem areas, develop remedies, and monitor trends.

Airlines on some occasions allow researchers to ride in the cockpit jumpseat to collect observational data on particular topics. For example, later in this chapter I describe an observational study on concurrent task demands (multitasking) and inadvertent errors of omission. This study took an ethnographic approach, which goes beyond simply observing events and crew actions, to interviewing pilots, and analyzing training, operating procedures and other factors that shape pilot performance. The ethnographic approach requires observers to have substantial domain knowledge of the work being observed.

Observational studies have the advantage of sampling a cross-section of normal operations and of coupling crew performance data with information about cockpit tasks and real-world operating conditions. It is possible to obtain moderate sample sizes. This is potentially one of the richest sources of information about flight crew performance in real-world operations. Unfortunately, since 9/11, the United States has rarely allowed cockpit observers unless they are employees of the airline, the Federal Aviation Administration (FAA) or the NTSB.

Simulation Studies

Modern flight simulators provide impressive fidelity to actual cockpit displays and controls, flight dynamics, and handling characteristics. (See Chapter 10 in this volume and Chapter 8 in the first edition.) Full-field visual displays depict the world outside the cockpit realistically. In the airline world almost all training is conducted in these simulators, including maneuvers, procedures, and full-mission training.

Around the world, several simulation facilities have been developed for research on pilot and crew performance, using realistic full-mission simulators instrumented to collect data on control inputs and aircraft response. Video and audio recordings, sometimes supplemented by direct observation within the simulator cab and postflight interviews, provide very detailed data. Simulator studies combine the advantages of highly realistic flight operations and of experimental control, which is not possible with observation of actual line operations. Experimenters can manipulate events and conditions systematically and can use control groups, a hallmark of laboratory experimentation. Some of the most important studies of flight crew performance have been conducted in these research simulation facilities; studies, for example on fatigue, crew familiarity, and crew resource management (see Foushee and Helmreich, 1988, for references).

The main limitation on this research approach is the high cost and limited availability of such research simulators and the high cost of operating them. Consequently, the number of published studies using this approach is not large, and the number of study participants is typically fairly small, limiting statistical power.

Some researchers have collaborated with the airline industry to collect data from crews performing recurrent full-mission training (line operational flight training, LOFT) and evaluation (line operational evaluation, LOE). Studies using training simulators more closely resemble observations of actual flight operations than experimental studies because of the lack of control groups and the inability to manipulate conditions beyond those employed for training. In principle, large numbers of participants could be used, but the constraints of arranging the studies and collecting and analyzing observational data have typically limited sample size.

Fortunately, much can be learned about pilot performance using lower-fidelity and part-task simulators that are much cheaper, and studies can include control groups and experimental manipulations. For example, the effects of age on pilot performance have been studied extensively using a moderate fidelity general aviation simulator (Taylor, Kennedy, and Yesavage, 2007), the team performance of military pilots has been studied in low fidelity simulations (Salas, Bowers, and Rhodenizer, 1998), and pilots use of automation has been studied in part-task simulators (Sarter and Woods, 1994). The critical issue in using lower-fidelity and part-task simulators is determining that the tasks and situations used in these simulators tap into the skilled performance of study participants in the same ways that real-world flight tasks and situations do.

Combined Approaches to Obtaining Error Data

Each of the approaches to obtaining error data has advantages and limitations. The most powerful approaches to studying the errors of skilled human operators is to combine several approaches iteratively. One might start with accident and incident reports and interviews with domain experts to identify phenomena of interest and the surrounding context. This analysis can guide collection of data from observational studies and research simulator studies, which in turn can generate hypotheses testable in well-controlled laboratory experiments. However, it is crucial not to stop here, for the conclusions of laboratory experiments should be tested under real-world conditions, or their best approximation, in research simulators. Computational modeling and fast-time simulation also allow ways to explore complex interactions of system components. An example of a study combining some of these approaches is provided later in this chapter

Ways to Gain Insight from Error Data

I will not discuss the many statistical techniques used to analyze research data, which are well known and are described in great detail in many texts. Instead, it may be useful to examine ways of using error data to identify and analyze operational issues. The list that follows illustrates some approaches that have been taken, but is far from exhaustive.

Epidemiological analysis of demographic data. Some researchers have used large accident databases to examine correlations of accident rates with demographic variables such as age, gender, flight experience, and previous accident history of the pilot (Lubner, Markowitz, and Isherwood, 1991; Li, Baker, Grabowski, Qiang, McCarthy, and Rebok, 2003). Careful statistical analysis is required to separate the effects of co-varying factors, such as age, flight experience, and type of aircraft flown. Similar techniques can be used to examine correlation of accident rates with weather, season, time of day, type of aircraft, and other aspects of accident flights. In principle, epidemiological techniques could be used to examine the correlations of different types of error, the conditions surrounding errors, and demographic variables, but in practice this has not been done extensively, in part because accident reports do not use a standardized system for classifying errors. (See Li, Grabowski, Baker, and Rebok, 2006 for a study including error data.) Further, only the few major airline crashes involving loss of life and substantial aircraft damage receive the in-depth investigation required to generate detailed information about how errors occurred.

Taxonomic classification. Errors can be classified into categories using various taxonomies, many of which have been proposed and which offer quite diverse perspectives. Because of space limitations I will illustrate this approach with a single taxonomy that has recently received considerable attention and has been applied to diverse error data sets: the Human Factors Analysis and Classification System (HFACS), developed by Shappell and Wiegmann (2001). These researchers drew upon Reason's (1990) theoretical perspective on the ways in which errors, situations, and organizational factors interact to produce accidents. HFACS describes human error at four levels: organizational, unsafe supervision (middle management), preconditions for unsafe acts, and unsafe acts by operators. Each of these four levels is further divided into several categories; for example, unsafe acts are divided into errors and violations and errors in turn are divided into three types: decision, skill-based, and perceptual.

In a study of commercial aviation accidents Shappell, Detwiler, Holcomb, andHackworth (2007) reported that coders could be trained to categorize error data reliably. These authors argue that HFACS is a comprehensive diagnostic system that allows the relative occurrence of different categories of error to be ascertained and allows tracking of trends over time. Shappell and Wiegmann (2001) give the example of an HFACS analysis of military pilot accident data that found an increase in skill-based errors in accidents and associated this with cutbacks in flying time. A standardized system for classifying errors, such as HFACS, and the context in which those errors occur might make epidemiological analysis more tractable.

Some caveats apply to taxonomies. Shappell et al. (2007) reported organizational factors to be present in less than six percent of the accidents, a finding at variance with the studies by Reason (1990) and others and with the perspective presented earlier in this chapter identifying a pervasive influence of organizational factors. Very probably, organization factors were found in such few accidents in this HFACS study simply because the accident reports analyzed in the study did not often identify organizational factors as causal or contributing. Historically, accident investigators have focused on human operator error much more than on organizational factors—in part because it is difficult to be certain of the role of the latter in any given accident. Obviously, any error taxonomy can be no better than the data fed into it.

Another concern is that taxonomies can enumerate the several types of factor reported to occur in accidents, but do not capture the complicated interplay of multiple factors that I and others argue lies at the heart of most accidents. However, conceivably one might expand a taxonomy to include information about this interplay. Also, there is the issue of what level of granularity of error categorization is needed to appropriately characterize what went wrong in an accident. In the HFACS study of military accidents the level of granularity of skill-based error was reported to be sufficient to understand and remedy the problem. However, it is not clear whether this level of categorization would be sufficient in most cases. For example, errors involving difficulty in managing multiple tasks concurrently, errors involving poor crosswind landing techniques, and errors involving failure to monitor alphanumeric indications of automation-mode changes would all be categorized as skill-based, yet these three problems have very different cognitive underpinnings and require very different countermeasures.

Finally, there is concern that error taxonomies may encourage a false dichotomy between correct performance and incorrect performance (Dekker, 2002; Woods and Cook, 2004). Errors must be understood not as an aberration but as the product of the same factors that drive effective performance. Most errors occur when competing goals, task conditions, inadequate information, problematic equipment design, or some unanticipated combination of these factors do not allow skilled performance to be sufficiently reliable (Perrow, 1999). Taxonomies do not explain human error, but perhaps their value is in providing a way to keep track of what errors occur and the context of those errors, allowing comparisons across different types of operation and monitoring of trends.

Case studies. At the other end of the spectrum from epidemiological analysis and taxonomic classification of large sets of accident data are case studies in which human operator performance and the conditions surrounding performance in a particular accident are analyzed in great detail. Accident investigations conducted by the NTSB and its counterparts abroad are similar to case studies; however, since the purpose of accident investigations is to determine the "cause" of a specific accident (as a means of preventing similar accidents in the future), these investigations tend not to fully explore the interactions of factors whose contribution to the outcome of the accident cannot be determined with confidence. However, accident investigation reports can be used as the starting point for more-extensive case studies by scientists. (See, for example the analysis of voice recordings in the Avianca accident by Helmreich, 1994.) Ideally, case studies should focus on the interplay of the many layers of factors influencing operator performance and on the influence of operator performance on events, an approach taken by the Commercial Aviation Safety Team (http://www.cast-safety.org/). In this approach it is sometimes useful to evaluate how a representative population of skilled operators might react in the situation of the accident operators, an approach illustrated later in this chapter.

The case study approach can also be applied to incident reports, if they contain detailed information, and to data from simulation studies.

Knowledge elicitation, cognitive task analysis, and work analysis. Understanding skilled performance—both correct and problematic—requires analysis of the tasks operators perform, the demands those tasks make on cognitive processes, the affordances and limitations of the equipment used, the information available to operators, the goals of the organization and the degree to which those goals are consistent or conflicting, the policies and procedures of the organization, the training provided, the norms for how operators actually accomplish tasks, and the degree to which norms are consistent with policies and procedures.

Techniques exist for performing some of this analysis. Beyond questionnaires and structured interviews, formal knowledge elicitation techniques provide a way to probe the nature of expertise more deeply. Reviewing the literature on knowledge elicitation is beyond the scope of this chapter, however several excellent reviews are available (see, for example, Cooke, 1994; Hoffman, 2008). An illustrative example of knowledge elicitation is the critical decision method (Klein, Calderwood, and MacGregor, 1989; Hoffman, Crandall, and Shadbolt, 1998). Interviewers use this method to have domain experts describe in increasingly detailed iterations an actual incident in which the experts had to evaluate the situation and develop an appropriate plan of action. Although this method has been used mainly to study how experts recognize and respond to critical situations, it could well be used to study how errors occur.

Knowledge elicitation can be thought of as a core aspect of cognitive task analysis, which goes further to develop models of the mental representations experts have of their tasks. In general, cognitive task analysis draws upon five methods: cognitive interviewing of domain experts, analysis of team communications, graphing domain experts' conceptual maps, systematic collection of verbal reports, and scaling analysis (e.g., cluster analysis) to identify links between items of interest (Seamster, Redding, and Kaempf, 1997). Cognitive task analysis is not limited to domain experts' knowledge, it can also be used to examine their procedural skills and decision-making skills. (However, I question how fully such analysis can capture procedural skills, which, by definition, domain experts acquire by performing tasks. Domain experts have limited ability to describe exactly what they do. For example, cognitive task analysis of the work of a virtuoso pianist might yield useful insight but would fall far short of fully characterizing all that skilled pianists actually do.)

Cognitive task analysis focuses on the operator's mental representation of the knowledge and skills required to perform tasks and is also a tool for exploring how operators' cognitive processes come into play during task performance. Additional types of analysis are required to fully understand operator performance. So, for example, one might use eye-tracking methods to assess how operators dynamically allocate attention during performance of a task. (The actual allocation may differ from the operators' concepts of how they allocate attention, and individual operators may vary substantially in their allocation from moment to moment.)

In addition to operators' mental representations of tasks, it is important to also examine the external character of tasks. For example, operators' work is often interrupted and often requires juggling multiple tasks concurrently and adapting procedures to manage dynamically changing conditions. Ethnographic observation helps characterize the nature of the tasks themselves, the affordances and limitations of the equipment and the overall sociotechnical system in which operators work. Burian and Martin (2008) term this expanded approach to cognitive task analysis cognitive and operational demand analysis (CODA).

Beyond cognitive task analysis, it is also crucial to examine organizational factors, as illustrated by the insightful analyses of NASA's two shuttle disasters from a social and organizational psychology perspective, which helped explain the flawed decision-making processes leading to the disasters (Starbuck and Farjoun, 2005; Vaughan, 1996). Aviation accidents have also been analyzed from an organizational perspective (Maurino, Reason, Johnston, and Lee, 1999). Organizational factors include goals, policies, procedures, practices, training, and institutional rewards. The term "practices" includes the norms for how tasks are typically conducted in line operations, which is not always identical to formally prescribed procedures. Reason (1997) makes a crucial distinction between the active errors of operators and the latent errors made upstream by system designers, policy makers, and managers, and he argues that addressing latent error is essential to understanding and managing the risk of accidents. Scientists now recognize that organizational factors almost always play a central role in determining the effectiveness of pilots' performance, and accident investigation bodies are slowly beginning to pay more attention to this role.

Cognitive work analysis is an approach that overlaps with cognitive task analysis, but with a more ecological emphasis (Roth, 2008). Work analysis starts with external aspects of tasks (rather than characterization of expertise) and emphasizes goals, means and constraints of the system in which the operator works. Thus it is a systems approach, emphasizing interactions within the overall work system, rather than focusing solely on the individual operator.

It seems unlikely that any one approach can generate information on all the diverse factors that drive operator performance. The crucial point we should take away from considering the issues discussed in this section is the need to situate operator performance in the concentric layers of factors that shape what the operator attempts to do and how well he or she succeeds.

Modeling. In basic laboratory research, computational modeling is used to explore the complex interactions of processes theorized to underlie perception and cognition and to force theories to be detailed and explicit. This type of modeling has generally focused on fundamental aspects of perceptual and cognitive processes in which considerable information is available about the critical parameters and constants. Computational modeling is also being extended to the complex behavior of human experts performing real-world tasks, although this poses considerable challenges (Foyle and Hooey, 2008; Pew and Mavor, 1998). Human cognition is vastly complex; as yet, we understand only the broad themes, and scientists hotly debate competing theoretical interpretations of data. Humans are also so exquisitely adaptive that performance varies substantially with subtle aspects of tasks and operators' goals. Modelers deal with these challenges in several ways; for example, in some cases, rather than attempting to capture undetermined cognitive processes, modelers may insert in their model output functions that mimic empirical performance data over some range of conditions, as a substitute for a model of the underlying cognitive processes.

Currently, computational modeling of expert performance of real-world tasks is primarily a research tool that helps us explore the demands tasks place on cognitive processes and how well human operators might manage those demands. Although one can create models of error per se, it seems more useful in most cases to model overall performance, with error as one of the manifestations of performance, as a function of task conditions.

The potential of modeling for addressing aviation operations is well illustrated by a NASA project in which five research teams used quite different models to address to pilots' performance in two tasks: Taxiing in low-visibility and approach to landing using a synthetic vision system (Foyle and Hooey, 2008). Human-in-the-loop performance of actual pilots in a flight simulator provided a common set of data for the modelers to address. This allowed direct comparison of the approach of different types of model, the power of each to capture expert performance in operational settings, and the limitations of each approach.

Although still primarily a research tool, computational modeling has great potential for helping evaluate how human performance will be affected by the design of displays, controls, and procedures under consideration. Ideally, this evaluation will become part of the early design process. Currently, the greatest potential seems to be in comparing the relative advantages and disadvantages of competing designs and in identifying frailties in proposed systems. Modeling might also be used in accident reconstruction and analysis. However, the challenge is ensure that models capture the dynamic nature of task demands and the highly adaptive (and thus variable) cognitive processes human operators use in managing the flow of task demands to achieve diverse goals.

EXAMPLES OF ERROR STUDIES

Many of the themes of this chapter can be illustrated by two studies, which I chose from many possibilities.

Case Study of Airline Accidents Attributed to Crew Error

Dismukes et al. (2007) reanalyzed the 19 major U.S. accidents from 1991 through 2000 that the NTSB attributed primarily to crew error. We used the NTSB's operational and human performance analyses as the starting point for our own analysis, and we did not second-guess the NTSB as to what errors were made or how those errors contributed to the accident. Rather, we worked our way through the sequence of events of each accident and, at each step of the sequence, asked: Why might any highly experienced crew in the position of the accident crew, and knowing only what the accident crew knew at that moment, have been vulnerable to performing in ways similar to the accident crew? This approach seemed justified because nothing in the accident reports portrayed the accident pilots as differing from their peers in experience, training, or competence.

At each step of the accident sequence, we attempted to characterize the multiple task demands confronting the crew, their interactions with other human agents, the goals embedded in their work,
the formal procedures guiding that work, and the norms for how
tasks are typically carried out in line operations. Drawing upon
extensive literatures in cognitive psychology, perception, human
factors, sociology, and organizational psychology, we were able to
identify plausible reasons any skilled pilots might take the same
actions and be vulnerable to the same errors as the accident pilots.
This is not to say that all pilots or even most pilots would have
performed the same way in each situation. Skilled performance
is variable when the information available to human operators
is incomplete or ambiguous, when limited time is available, and
when the operator must balance competing goals.

As an example of variability of pilot performance in these situations, consider an accident in which, just as the flying pilot was
rotating the aircraft during takeoff, the stall warning system gave
a false warning, an extremely rare occurrence for which pilots are
not trained (Dismukes et al., 2007, Chapter 11). The warning, consisting of stick-shaker activation and designed to demand attention, was quite startling, and the crew had at most a second or two
to evaluate the situation and decide whether to continue the takeoff or to abort it, even though they were past the normal point for
aborting. In either case, the wrong decision could be fatal. Having
no way to quickly ascertain whether the stall warning was true or
false, the captain aborted the takeoff at high speed, and the aircraft
touched down hard and ran off the end of the runway, destroying
the aircraft. Fortunately, all injuries were relatively minor.

The NTSB discovered that sometime previously another airline
crew had also experienced a false stall warning at rotation, but
in this incident the captain decided to continue the takeoff. From
this the NTSB concluded that the accident crew should have made
the same decision and identified the decision to abort as causal.
But it seems unreasonable to assume that pilot decision making in
these extremely demanding conditions is likely to be reliable. We
have data from only two cases; the crew of one took one course of
action, the crew of the other took the other course. The responses

of these two crews demonstrate that we must distinguish between what pilots can in principle do and what they can be expected to do reliably.

Looking across the 19 accidents in this study reveals several common themes, only a few of which can be discussed in this space. A totally unexpected finding was that 12 of the accidents involved situations, like that of the false stall warning, in which crews had only seconds to respond to an anomaly and lacked adequate information to analyze the situation so quickly. This finding was startling because most abnormal situations allow adequate time to assess and decide what to do, and wise instructors often counsel pilots not to rush. We concluded that the number of accidents in our sample requiring rapid response was disproportionately large because, even though these situations are quite rare, pilots are especially vulnerable to error when they do occur.

Another important theme that emerged involved pilot judgment in ambiguous situations that hindsight proved wrong—for example, deciding to land or to takeoff from an airport in the vicinity of thunderstorms, discussed earlier in this chapter. Although airline crews manage to balance the competing goals of production (on-time arrival at the intended airport) and safety the great majority of the time, it is simply not possible to always pick the best course of action when the information available is incomplete or ambiguous, or arrives piecemeal.

We found the problem of inadequate information to be compounded by various cognitive biases to which all humans are subject. For example, in at least nine accidents, crews appeared to have been influenced by plan continuation bias, which is a proclivity to continue a planned or habitual course of action past the point when changing conditions require altering the plan. (Pilots often call this "get-there-it is.") The cognitive underpinnings of plan continuation bias are not fully understood, but we suggest this bias results from multiple interacting factors, including other types of bias. For example: Expectancy and confirmation biases strongly influence individuals' mental models of their current situation. Especially when a flight starts out resembling the thousands of previous flights, pilots are slow to notice subtle cues indicating that the situation is not what is expected or has gradually drifted into a more threatening mode. Further, when looking for information to evaluate their situation, individuals tend to notice cues consistent with their current mental model more readily than cues that are inconsistent.

Individuals are also influenced by a "sunk costs" proclivity that makes them reluctant to abandon an effort in which they have invested, even when a more pragmatic approach would be to cut one's losses. For pilots, this might play out as a reluctance to abandon an approach to landing at an intended destination because so much has been invested in getting almost to the successful conclusion of the flight. Further, airline pilots are quite aware that the viability of their company in an extremely competitive economic climate is hurt by delays and diversions to alternate airports. Although professional pilots are not likely to consciously gamble with safety for economic reasons, these concerns may influence decision-making more subtly at a subconscious level. Finally, pilots' previous experience with potentially threatening situations may distort their understanding of the degree of inherent risk. Air transport operations are designed to provide wide margins of safety. Consequently, pilots may repeatedly respond in a certain way to some type of threatening situation without experiencing a bad outcome, for example landing from an unstabilized approach or in the vicinity of thunderstorms. Their repeated successes in these landings may lead to a false mental model of the level of risk involved, and they may underestimate the potential for subtle variations from one incident to the next to increase risk.

Another theme cutting across most of the 19 accidents was the challenge of managing multiple tasks simultaneously competing for attention. In some of the accidents that started out routinely, workload grew very heavy as the situation became problematic, partly as a result of the crew's decision to continue an approach rather than diverting. The problem with high workload is not just finding enough time to perform all tasks, but, more insidiously, the vulnerability of attention being captured by one challenging task, causing the pilot to lose track of other tasks. The interaction of plan continuation bias with the snowballing workload it can lead to greatly increases vulnerability to error (Berman and Dismukes, 2006). We found that crews overwhelmed with workload often responded by abandoning strategic thinking (which places considerable demand on mental resources) in favor of simply reacting to events as they occur, which of course made the crew less likely to find a safe way out of their predicament. Monitoring, a crucial defense against threats and errors, also seemed to falter in favor of active control tasks, further impairing the ability of crews to assess and manage risk.

Taken together, the 19 accidents in this study challenge conventional notions of causality in airline accidents in which human error plays a role. None of these accidents would have occurred without convergence of many factors, none of which by itself would have led to an accident. The co-occurrence of multiple factors is partly random, corresponding to spinning of the safeguard wheels in Reason's (1997) famous "Swiss cheese" model in which holes in the safeguards sometimes line up. However, the confluence of factors is not entirely random. In some accidents, one event or error triggered another after another downstream. For example, plan continuation bias led crews into situations of snowballing workload, which in turn led to inadvertent errors of omission such as failing to arm spoilers and failure to complete checklists that would have caught the errors of omission.

A final conclusion of this study was that both accident investigation and scientific analysis of errors and accidents is greatly impeded by lack of normative data about the range of behavior of the overall population of pilots in situations similar to those confronting accident pilots.

Ethnographic Study of Concurrent Task Demands in the Cockpit

Several of the methods of investigating error described earlier were used in a series of studies my colleagues and I conducted. Both the accident case study just described and a review of ASRS incident reports (Dismukes, Young, and Sumwalt, 1998) revealed that concurrent task demands frequently impaired crew performance, often manifested as forgetting to perform some intended action. This led to an ethnographic study combining jumpseat observations, analysis of formal operating procedures and pilot training, analysis of incident reports, and analysis of the cognitive demands involved in prototypical multitasking situations (Loukopoulos et al., 2009). Since this type of study requires investigators to have substantial domain expertise, two of the authors who are rated pilots obtained airline transport pilot type ratings in the particular aircraft studied and attended an airline's training in the aircraft for new-hire pilots. The other author, who was trained as a Navy aviation psychologist, attended training for the aircraft at another airline.

Analysis of flight operations manuals (FOMs) revealed that operating procedures were depicted as linear, predictable, and controllable. That is, tasks were described as occurring one at a time in fixed and predictable sequence, and crews could control the tempo of operations. Jumpseat observations revealed a much more complicated situation. Crews were frequently interrupted, especially when getting the aircraft ready for departure; circumstances sometimes forced procedures to be performed out of the prescribed sequence, and often several tasks had to be performed in parallel. Our analysis focused on the cognitive demands imposed by these real-world perturbations of the ideal depicted by FOMs and on the errors associated with those perturbations.

The observed perturbations were used to generate terms for searching the ASRS database, which led to a large collection of problematic situations and associated crew errors. These were sorted into four prototypical situations: Interruptions and distractions, tasks that cannot be executed in their normal, practiced sequence, unanticipated new task demands, and multiple tasks that must be interleaved concurrently. Three substantial bodies of research literature were found to provide a framework for understanding the vulnerability to error created by these four prototypical situations: Prospective memory (remembering to perform actions intended for a later time), automatic processing of habitual tasks, and attention switching. These descriptions of cognitive processes in these literatures were used to generate plausible, albeit somewhat speculative, accounts of why pilots are vulnerable in the prototypical situations.

Prospective memory is a relatively new research domain, but a fair degree of consensus exists on its essential features (Kliegel, McDaniel, and Einstein, 2008; McDaniel and Einstein, 2007). Intended actions that must be deferred can be thought of as deferred goals stored in long-term memory, along with the conditions under which the goal should be activated and accomplished. (It is probably not practical to maintain the deferred goal continuously in attention, while performing other real-world tasks, such as entering data in a flight management computer, as these other tasks make heavy demands on the limited cognitive resources of attention and working memory.) Unless the individual undertakes a deliberate search of memory at the crucial moment for what to do next, retrieval of the deferred intention from memory requires the individual to notice some cue associated in memory with the deferred intention. The pilot's environment may contain many such cues, but they are also associated with other memory items and with other goals, so retrieval is probabilistic, as a function of multiple factors such as the strength of the association between cue and intention, saliency of the cue, and attention demands of the ongoing task.

An aviation example of prospective memory failure is forgetting to set wing flaps to takeoff position before attempting to take off, illustrated at the beginning of this chapter. Fortunately, this oversight is usually caught by the aircraft's configuration warning system, but this has on occasion failed. Our investigation of these incidents and accidents found that often the crew was interrupted during times they normally would set the flaps. In other cases the crew was forced by weather conditions to defer setting the flaps, removing the procedural step from the normal procedural sequence of actions

Much of the work skilled human operators perform relies on highly practiced skills that become automatic over time. Automatic processing has the advantage of being fast, efficient, and making minimal demands on limited cognitive resources. Normally highly reliable, it becomes far less reliable if the cues that normally automatically prompt the individual to initiate a task are removed or if the individual's attention is diverted from the cues. Also, when individuals intend to diverge from a normal step in a habitual procedure, they are vulnerable to habit capture if they do not pay careful attention—reverting to the habitual step rather than performing the intended one. For example, pilots who are given a modified departure clearance, differing in one step from the standard clearance they typically receive, sometimes revert to the standard departure procedure when their attention is consumed by heavy task demands during the departure climb.

We found many situations in which pilots must interleave two (or more tasks) concurrently, switching attention back and forth frequently to manage both tasks. For example, when a crew is given a revised departure clearance during taxi, the first officer must enter the new data in the flight management computer, but at the same time must continue to monitor the progress of the taxi—especially when in low visibility conditions—to make sure the captain does not make a wrong turn or cross an active runway without permission from the air traffic controller. Usually, experienced first officers can handle this by looking down at the computer to make a few key strokes, then looking up to check the progress of the taxi, and then back down again, back and forth, until data entry is complete. But if an unexpected glitch in entering data occurs, solving this problem tends to engage all of the first officer's attention, and monitoring the taxi may fall by the wayside. Our cognitive analysis of this situation suggested that part of the problem is that the required frequency for switching attention and the duration of dwell of attention up and down is highly variable as a function of the specific conditions; thus it is difficult to develop a reliable habit of attention switching that works across variable conditions.

This ethnographic study suggested a way to follow up one aspect of concurrent task demands in a laboratory study. Our observations and cognitive analysis suggested that individuals often forget to resume interrupted tasks because the interruption is so abrupt and salient that the individual does not fully encode an intention to resume the interrupted task. Also, after the interruption is over, new tasks demands arise, further diverting attention from cues that might remind the individual that the interrupted task has not been completed. An experimental paradigm was designed to test these hypotheses, both of which were supported by the experimental data (Dodhia and Dismukes, 2008).

This example illustrates the power of combining several approaches to analyzing human error in real-world situations—in this case, cognitive analysis of incident data, an ethnographic field study, and an experimental laboratory study. Although each approach has appreciable limitations, the combination can circumvent some of those limitations and provide richer insight.

Implications for Reducing Vulnerability to Errors and Accidents

The findings of the two studies described above were used to develop suggestions for ways to reduce vulnerability to error and to accidents. Several of those suggestions are summarized here.

The "blame and punish" mentality toward error—especially the errors made by skilled professionals highly motivated to avoid accidents—is inimical to safety, deflecting attention from identifying and reducing system vulnerability. An appropriate system safety perspective holds that the design of equipment, operating procedures, organizational policies and practices, and training should start with the understanding that errors will occur but can be minimized with proper design. Systems must be designed to be resilient to error by providing ways to detect and correct errors before they jeopardize system operation.

Organizations—operators, regulators, and investigators—should explicitly recognize that human operators are subject to competing organizational pressures, in particular, the inherent tension between production and safety. Only by acknowledging that tension can it be managed appropriately. Operators quickly recognize the difference between organizational rhetoric ("Safety is our highest priority") and actual support for safe operations.

Organizations should systematically analyze the messages sent to operators and the incentives provided them—both implicit and explicit—to determine whether these messages and incentives provide a realistic and appropriate balance between production and safety goals. Operators must balance competing organizational goals in the way they work and the daily decisions they make. Although this balancing is often not at a conscious level, it is shaped by organizations in ways subtle and not-so-subtle. For example, airline pilots are acutely aware that the survival of their company depends on on-time arrivals and controlling fuel costs. Company emphasis on quick-turns (minimizing the time an aircraft spends on the ground between flight legs), operating procedures that have crews conducting checklists while taxiing to the runway, and publishing information on the on-time arrival rates of competing airlines, each of which may be reasonable approaches to encouraging efficiency, can in their aggregate, if not carefully worded and placed in appropriate context, subtly bias the actions and decisions of pilots to favor production over safety.

Similarly, recognizing that attempting to land from an unstabilized approach compromises safety, airlines may establish clear-cut criteria for stabilized approaches, but the effectiveness of these criteria is undercut if they are not emphasized in training and checking, or if crews are called in to the chief pilot's office to justify going around from an unstabilized approach.

Well-designed training is crucial for preventing and managing error, although it cannot make up for poorly designed equipment and procedures. In recent years, many airlines have added training in Threat and Error Management to their curricula (Klinect et al, 1999). This is a very positive development, but much more research is required to provide pilots with specific techniques for identifying and assessing threats, catching errors, and managing the consequences of errors.

Training could be enhanced by explaining to pilots the ways in which cognitive biases, such as plan continuation bias, subtly affect their decision-making and by explaining how concurrent tasks and prospective memory tasks increase vulnerability to error. This training should include specific techniques for countering biases and vulnerabilities. For example, pilots might be trained in "debiasing" techniques, such as imaging how a planned course of action might fail, before committing to that plan.

Checklists and monitoring are crucial defenses against equipment failures and errors, but these two defenses sometimes fail because of the way performance of habitual tasks becomes automatic and because of competing demands for attention from concurrent tasks. These two defenses could be strengthened by explaining to pilots why these defenses sometimes fail and by providing techniques to better manage attention—for example, slowing down the process of executing checklists, pointing to or touching items checked and voicing the response in a deliberate manner.

Rushing is a normal tendency when humans are under time pressure. But, in the operational world, rushing saves only seconds at best and greatly increases probability of error. In fact, the likelihood of having to redo a procedure, if an error is caught, nullifies whatever time was gained by rushing. Training, policies, and procedures should be explicitly designed to counter the lure of rushing.

A major recommendation of the ethnographic study described here was that every organization should periodically review all operating procedures to determine whether those procedures are well designed to help pilots deal with interruptions, concurrent task demands, deferred tasks, and unanticipated new task demands. Organizations should also examine whether existing procedures themselves create situations in which interruptions, concurrent tasks, and deferred tasks are more likely to arise. This review cannot be done in an office—it requires systematic observation of actual flight operations, such as LOSA.

The case study of 19 accidents attributed to crew error suggests that we cannot understand the causes of errors and their role in accidents by looking only at flights in which accidents occur. The aviation industry urgently needs data on what happens in apparently normal flights in a wide range of operational situations in order to identify problem areas and to characterize how pilots perform in diverse situations. Much of this data could be provided by existing programs such as:

- Flight operations quality assurance (FOQA)—a voluntary program in which air carriers routinely collect data from aircraft data busses and analyze how flights are being conducted.

- LOSA (described earlier)—provides observational data on threats and errors crews encounter and how they respond in normal line operations.
- ASAP (described earlier)—provides individual airlines voluntary reports from their personnel about incidents threatening safe operation.
- Line oriented evaluation (LOE)—similar to LOFT full-mission flight simulations, but in LOE the performance of crews is evaluated and graded.

Although these programs generate enormously useful data, so far these data have not been fully exploited. Sophisticated data mining tools, technical expertise, and considerable personnel time are required to extract all the information from the voluminous data generated by programs like FOQA. It is difficult for cash-strapped airlines to dedicate the resources needed for extensive data mining. Further, what is really needed is a picture of operations across the industry, which requires means of combining data from airlines and making the database available to all the parties who could analyze the data and publish the results for the common good. Although airlines that participate in the FAA's Advanced Qualification Program do submit their training performance data to the FAA, so far the industry does not have a shared database that is widely accessible and which would allow systematic exploration of issues cutting across the industry and of questions such as: How often do aircraft land from unstabilized approaches? At what airports and under what antecedent conditions?

The same issue of developing a industry-wide accessible database applies to the other existing programs. LOSA data is collected and analyzed by the LOSA Collective, an independent organization, but so far the analysis of that data is directed primarily to providing feedback to the individual airlines, feedback that is of necessity confidential.

CONCLUDING THOUGHTS

The cost of "human error" is enormous—in fatalities, injuries, equipment damage, operating costs, and loss of productivity.

The consequences of error in aviation are certain to increase as traffic volume rises, system complexity increases, and economic conditions drive cost-cutting measures. Although most aviation accidents are attributed to operator error, it is facile and counter-productive to interpret this simply as evidence of human inadequacy. It is far more useful to understand operator errors as symptoms of system weaknesses, and to recognize that human operators are performing tasks under conditions in which computers would be woefully inadequate. We should recognize that for every accident attributed to human error, vast numbers of accidents are averted by the skilled intervention of human operators. These averted accidents receive little attention, though we could learn much from studying them if the data were available.

Human error and the ways in which errors contribute to accidents and to inefficient operations are as complex as the human brain itself; thus no single approach to studying these phenomena will suffice. In this chapter I have described diverse sources of error data and various ways error and the conditions leading to error can be analyzed systematically. Each of these sources and analytical approaches has advantages and limitations, thus combining the insight from different approaches is essential. Vulnerability to human error (more accurately, system error) and accidents can be reduced with practical measures described in the previous section of this chapter. Even more powerful countermeasures can be achieved by further research on the nature of error and its role in accidents.

This review has focused on the errors attributed to pilots because that is the most common operator error identified by the NTSB and because most research has focused on pilot error. Yet it is certain that the errors of controllers, mechanics, and other operators (not to mention managers!) play a crucial role, though understudied.

It is ironic that, although most accidents are attributed to operator error, the amount of research conducted on it is minuscule in comparison to the investment spent on developing new technologies. Safe and efficient operation of these technologies requires that all stages of development be firmly grounded in deep and veridical understanding of human performance. That understanding is as yet fragmentary and will remain that way without long-term commitment to funding the necessary research.

ACKNOWLEDGMENTS

This research was funded by NASA's Aviation Safety Program (IIFD Project).

Kim Jobe contributed substantially to this chapter with literature searches and manuscript preparation.

References

Berman, B. A., & Dismukes, R. K. (2006). Pressing the approach: A NASA study of 19 recent accidents yields new perspective on pilot error. *Aviation Safety World, 206*, 28–33.

Berman, B. A., & Dismukes, R. K. (2008). Checklists and monitoring: Why two crucial defenses against threats and errors sometimes fail. Manuscript in preparation.

Boeing Commercial Airplanes. (2006). *2006 Statistical Summary of Commercial Jet Airplane Accidents, Worldwide Operations 1992–2001*. Seattle: Boeing.

Bogner, M. S. (Ed.). (1994). *Human error in medicine*. Hillsdale, NJ: Erlbaum.

Chidester, T. R. (2004). Progress on advanced tools for flight data analysis: Strategy for national FOQA data aggregations. *Shared Vision of Flight Safety Conference*. San Diego, CA.

Cooke, N. J. (1994). Varieties of knowledge elicitation techniques. *International Journal of Human-Computer Studies, 41*, 801–849.

Dekker, S. (2002). The Field Guide to Human Error Investigations. Hampshire, UK: Ashgate Publishing Limited.

Dismukes, R. K. (2009). Overview of Human Error. In R. K. Dismukes (Ed.), *Human Error in Aviation*. Aldershot, UK: Ashgate Publishing Limited.

Dismukes, R. K., Berman, B. A., & Loukopoulos, L. D. (2007). The Limits of Expertise: Rethinking pilot error and the causes of airline accidents. Aldershot, UK: Ashgate Publishing Limited.

Dismukes, K., Young, G., & Sumwalt, R. (1998). Cockpit interruptions and distractions: Effective management requires a careful balancing act. *ASRS Directline, 10*, 4–9.

Dodhia, R. M. & Dismukes, R. K. (2009). Interruptions create prospective memory tasks. *Applied Cognitive Psychology*.

Foushee, H. C., & Helmreich, R. L. (1988). Group interaction and flight crew performance. In: *Human factors in aviation*. San Diego: Academic Press (pp. 189–227).

Foyle, D. C., & Hooey, B. L. (Eds.), (2008). *Human performance modeling in aviation*. Boca Raton, FL: CRC Press.

Helmreich, R. L. (1994). Anatomy of a system accident: The crash of Avianca flight 052. *International Journal of Aviation Psychology, 4*(3), 265–284.

Hoffman, R. R. (2008). Human factors contributions to knowledge elicitation. *Human Factors, 50*(3), 481–488.

Hoffman, R., Crandall, B., & Shadbolt, N. (1998). Use of the critical decision method to elicit expert knowledge: A case study in the methodology of cognitive task analysis. *Human Factors, 40*, 254–276.

International Civil Aviation Organization. (2002). The LOSA experience: Safety audits on the flight deck. *ICAO Journal, 57*(4), 5–15.

Kliegel, M., McDaniel, M. A., & Einstein, G. O. (Eds.), (2008). *Prospective memory: Cognitive, neuroscience, developmental, and applied perspectives.* New York: Erlbaum.

Klein, G., Calderwood, R., & MacGregor, D. (1989). Critical decision method for eliciting knowledge. *IEEE Transactions on Systems, Man, and Cybernetics, 19*, 462–472.

Klinect, J. R., Wilhelm, J. A., & Helmreich, R. L. (1999). Threat and error management: Data from Line Operations Safety Audits. *Proceedings of the 10th International Symposium on Aviation Psychology* (pp. 683–688). Columbus: Ohio State University.

Krey, N. (2007). *The Nall Report 2007: Accident Trends and Factors for 2006.* Frederick, MD: AOPA Air Safety Foundation.

Lacagnina, M. (2007). Defusing the ramp. *AeroSafetyWorld* (May), 20–24. Retrieved February 21, 2008, from http://www.flightsafety.org/asw/may07/asw_may07_p20–24.pdf.

Li, G., Baker, S. P., Grabowski, J. G., Quiang, Y., McCarthy, M. L., & Rebok, G. W. (2003). Age, flight experience, and risk of crash involvement in a cohort of professional pilots. *American Journal of Epidemiology, 157*(10), 874–880.

Li, G., Grabowski, J. G., Baker, S. P., & Rebok, G. W. (2006). Pilot error in air carrier accidents: Does age matter? *Aviation, Space and Environmental Medicine, 77*, 737–741.

Loukopoulos, L. D., Dismukes, R. K., Barshi, I. (2009). *Multitasking in Real World Operations: Myths and Realities.* Aldershot, UK: Ashgate

Lubner, M. E., Markowitz, J. S., & Isherwood, D. A. (1991). Rates and risk factors for accidents and incidents versus violations for U.S. airmen. *The International Journal of Aviation Psychology, 1*(3), 231–243.

Maurino, D. E., Reason, J., Johnston, N., & Lee, R. B. (Eds.), (1999). *Beyond aviation human factors.* Aldershot: Ashgate.

McDaniel, M. A., & Einsein, G. O. (Eds.), (2007). *Prospective memory: An overview and synthesis of an emerging field.* Los Angeles: Sage.

NTSB. (1988). *Aircraft accident report: Northwest Airlines, McDonnell Douglas DC-9–82, N312RC, Detroit Metropolitan Wayne Country Airport, Romulus, Michigan, August 16, 1987* (NTSB/AAR-88/05). Washington, DC: National Transportation Safety Board.

NTSB. (1989). *Aircraft accident report: Delta Air Lines, Boeing 727–232, N473DA, Dallas-Fort Worth International Airport, Texas, August 31, 1988* (NTSB/AAR-89/04). Washington, DC: National Transportation Safety Board.

Parke, B., Orasanu, J., Castle, R., & Hanley, J. (2005). Identifying organizational vulnerabilities in space operations with collaborative, tailored, anonymous surveys Available from http://www.esa.int/esapub/pi/proceedingsPI.htm (SP-599). In H. Lacoste (Ed.), *Proceedings of the International association for the advancement of space safety conference: Space safety, a new beginning* (pp. 577–583). The Netherlands: ESA Publications.

Perrow, C. (1999). Normal accidents: Living with high-risk technologies. Princeton NJ: Princeton University Press Originally published 1984.

Pew, R. W., & Mavor, A. S. (Eds.), (1998). *Modeling human and organizational behavior: Application to military simulations.* Washington, DC: National Academy Press.

Rasmussen, J. (1990). Human error and the problem of causality in analysis of accidents. *Phil. Trans. R. Soc. Lond. B, 327,* 449–462.

Reason, J. (1990). *Human Error.* New York: Cambridge University Press.

Reason, J. (1997). *Managing the Risks of Organizational Accidents.* Aldershot, England: Ashgate.

Roth, E. M. (2008). Uncovering the requirements of cognitive work. *Human Factors, 50*(3), 475–480.

Salas, E., Bowers, C. A., & Rhodenizer, L. (1998). It is not how much you have but how you use it: Toward a rational use of simulation to support aviation training. *International Journal of Aviation Psychology, 8*(3), 197–208.

Sarter, N., & Woods, D. (1994). Pilot interaction with cockpit automation II: An experimental study of pilots' model and awareness of the flight management system. *International Journal of Aviation Psychology, 4*(1), 1–28.

Seamster, T. L., Redding, R. E., & Kaempf, G. L. (1997). *Applied Cognitive Task Analysis in Aviation.* Aldershot, UK: Avebury.

Shappell, S., & Wiegmann, D. (2001). Applying reason: The Human Factors Analysis and Classification System (HFACS). *Human Factors and Aerospace Safety, 1,* 59–86.

Shappell, S., Detwiler, C., Holcomb, K., & Hackworth, C. (2007). Human error and commercial aviation accidents: Analysis using the human factors analysis system. *Human Factors, 49*(2), 227–242.

Starbuck, W. H., & Farjoun, M. (Eds.), (2005). *Organization at the limit: Lessons from the Columbia disaster.* Malden, MA: Blackwell.

Taylor, J. L., Kennedy, Q., Noda, A., & Yesavage, J. A. (2007). Pilot age and expertise predict flight simulation performance: A 3-year longitudinal study. *Neurology, 68,* 648–654.

Vaughan, D. (1996). The Challenger launch decision: Risky technology, culture, and deviance at NASA. Chicago: University of Chicago.

Wiegmann, D., Goh, J., & O'Hare, D. (2002). The role of situation assessment and flight experience in pilots' decisions to continue visual flight rules flight into adverse weather. *Human Factors, 44,* 189–197.

Woods, D. B., & Cook, R. I. (2004). Mistaking error. In B. J. Youngberg & M. J. Hatlie (Eds.), *The patient safety handbook* (pp. 95–108). Sudbury, MA: Jones & Bartlett.

Woods, D. B., Johannesen, L. J., Cook, R. I., & Sarter, N. B. (1994). *Behind human error: Cognitive systems, computers, and hindsight.* Wright-Patterson AFB, OH: CSERIAC Program Office.

12

Cognitive Architectures for Human Factors in Aviation

Kevin Gluck
Air Force Research Laboratory

INTRODUCTION, MOTIVATION, AND ORGANIZATION

There has been a great deal of investment and resultant progress in the development and evaluation of, improvements to, and comparisons of cognitive architectures over the last several decades. Not all—however, certainly the majority—of that work has taken place since the publication of Weiner and Nagel's (1988) first volume on *Human Factors in Aviation,* so it is understandable both that there was no mention of computational cognitive modeling or cognitive architectures in that first edition, and also that the editors of the second edition are interested in expanding coverage of the text to include this relevant development in the scientific study of human performance and learning.

The overarching interest and motivation for the existence of the aviation human factors community is improving the operational safety of current and future aviation systems. The people serving in the roles of pilots, navigators, maintainers, controllers, or other user-operator positions in all aviation-related sociotechnical systems are both enabled and constrained by their cognitive architectures. By improving our understanding of the human cognitive architecture, we improve our understanding of an important component of the larger system of systems in which those people are performing. Through better understanding of aviation systems, to include the human components of those systems, we can improve their overall performance standards and safety levels. This is why the aviation human factors community should care about basic and applied research on the human cognitive architecture.

The purpose of this chapter is to provide an introduction to cognitive architectures that will be useful to anyone who wants to understand what a cognitive architecture is and who wants some pointers regarding what to read and consider in order to use an architecture effectively in their research. Given the broad range of educational and professional backgrounds among the scientists, engineers, acquisition professionals, instructors, and students that one might reasonably consider to be in the target readership for a volume such as this, I have positioned the chapter as a relatively high-level introduction to the topic. The first section of the chapter provides definitions and characteristics of cognitive architectures to help the reader develop an understanding of what cognitive architectures are and what they are intended to be. The second section identifies recently published key reference materials for those interested in a deeper investigation of the topic than is possible in this single chapter. The third section of the chapter describes current efforts by my colleagues and me at the Air Force Research Laboratory to improve on an existing cognitive architecture. The fourth and final section summarizes challenges and vectors for those interested in evolving cognitive architectures from research programs to applied products that are useful in aviation and elsewhere.

WHAT ARE COGNITIVE ARCHITECTURES?

The concept of *cognitive architecture* and its formal study in humans has its scientific origins in calls for an information-processing psychology and its use in the development of computational process models as means for improving our understanding of human cognition (Simon, 1956; Newell, Shaw, & Simon, 1958, Newell & Simon, 1972). At their core all attempts to implement computational theories of the human cognitive architecture are inherently consistent with the theoretical and methodological position first expressed in Newell, Shaw, and Simon's (1958) *Psychological Review* paper that, ". . . an explanation of an observed behavior of the organism is provided by a program of primitive information processes that generates this behavior." Thus it was that the research agenda was set in motion for cognitive architects, who demonstrate an improving understanding of the mind with computational information processing models that replicate human behavior in simulation.

Anderson (2007) traces the introduction of the actual term *cognitive architecture* into the lexicon of cognitive science, crediting Allen Newell as its progenitor. Anderson's (1983) and Newell's (1990) seminal books provided what may still be the best-known and most widely cited proposals for specific cognitive architectures and their important role as engines of theoretical integration and unification within cognitive science.

Much has been written about cognitive architectures since their inception as a scientific endeavor, and naturally some of that writing has involved attempts at (more or less) concise definitions of what they are. Here is a sampling of such definitions, organized chronologically:

Anderson (1983)—" . . . a theory of the basic principles of operation built into the cognitive system."

Pylyshyn (1984)—" . . . those basic information-processing mechanisms of the system for which a nonrepresentational or nonsemantic account is sufficient."

Newell (1990)—". . . the fixed structure that forms the framework for the immediate processes of cognitive performance and learning."

VanLehn (1991)—"In general, the architecture of a computing system leaves out details about the implementation of the system and includes only a description of its basic operations and capabilities. An architecture for the mind would describe the way memory and attention operate but it would not describe how they are implemented biologically."

Byrne (2003)—"A cognitive architecture is a broad theory of human cognition based on a wide selection of human experimental data, and implemented as a running computer simulation program."

Sun (2004)—". . . a cognitive architecture is the overall essential structure and process of a domain-generic computational cognitive model, used for a broad, multiple-level, multiple-domain analysis of cognition and behavior."

Kieras (2007)—"Cognitive architectures are the current form of the traditional computer metaphor in which human cognition is analyzed in information-processing terms."

Anderson (2007)—"A cognitive architecture is a specification of the structure of the brain at a level of abstraction that explains how it achieves the function of the mind."

Laird (2008)—"Cognitive architectures must embody strong hypotheses about the building blocks of cognition that are shared by all tasks, and how different types of knowledge are learned, encoded, and used, making a cognitive architecture a software implementation of a general theory of intelligence."

The details of these definitions vary considerably with respect to how they position cognitive architectures in classic and ongoing debates regarding levels of abstraction, metaphors for mind, and the complicated intersections of neuro-bio-cogno-silico methods and theories. For instance, Anderson (1983), Pylyshyn, Newell, and Sun all associated cognitive architectures with the invariant "basic principles," "mechanisms," and "fixed" or "essential" structure. Byrne, Sun, and Laird approach it from the other side of that coin, explicitly emphasizing domain generality in their definitions, while others left this characteristic implicit. VanLehn stated that cognitive architectures leave out details regarding biological implementation, whereas Anderson (2007) associated cognitive architectures with abstract specifications of the structure of the brain. Note the interesting historical contrast with Anderson's 1983 definition, which does not mention brain structure. Of course, that earlier definition was written well before the Decade of the Brain and more than 20 years before he explicitly adopted the goal of answering Newell's question, "How can the human mind occur in the physical universe?" Finally, both Byrne and Laird included implementation in software as a feature of cognitive architectures, while there is no mention of this feature in the other definitions.

Despite the coverage provided in the set of definitions here, there are a couple of characteristics of cognitive architectures that are missing from that set, and that I think are important and worth emphasizing. One characteristic is that cognitive architectures that are being developed and used in ongoing research efforts are *evolving*. This often is not clear to those outside the cognitive architecture community. Although there are some core architectural features that are similar or identical, the Soar described in Newell's (1990) *Unified Theories of Cognition* is different than the Soar architecture that exists today. Similarly, the ACT theory described in Anderson's (1976) *Language, Memory, and Thought* is different than the ACT-R architecture that exists today. Indeed, both of these architectures' websites reference architectural evolution. The Soar Web site[1] states, "It has been in use since 1983, evolving through many different versions to where it is now Soar, Version 8.6." The ACT-R Web site[2] states, "As the research continues, ACT-R evolves ever closer into a system which can perform the full range of human cognitive tasks: capturing in great detail the way we perceive, think about, and act on the world." Of course, this characteristic holds not just for ACT-R and Soar but also for any cognitive architecture that is being used in an ongoing research program. Their evolutionary nature is a consequence of the intention to continually expand the functional and explanatory breadth and depth of the architectures.

None of the architectures that exist today is considered complete. They all have weaknesses, deficiencies, and idiosyncrasies that serve as pointers toward the next evolutionary adaptation. Rather than being cast aside they are modified and extended. In this manner, cognitive architectures serve as a formal instantiation of progress in cognitive science (Cooper, 2007; Lakatos, 1970; Newell, 1990).

A second characteristic of cognitive architectures that is not represented in the set of definitions above is that their development is often motivated by an interest in *application*. That is, an objective of cognitive architects is often that the architectures have some applied utility. Note the careful inclusion of the adverb "often" here, to reflect the fact that this characteristic is not universally present, or at least there is not explicit evidence of an interest in

[1](http://sitemaker.umich.edu/soar/home)
[2](http://act-r.psy.cmu.edu/)

application in the publications and other available materials on every existing architecture. However, the evidence for this is clear in some cases. For example, Anderson (1976) actually ends his book introducing the ACT theory with a statement of the importance of application for his research program by saying, "I would like to conclude this chapter with a remark about one of the ultimate goals I have set for my research efforts . . . that is, that it produce a theory capable of practical applications" (p. 535). Newell's (1990) position was that, "Applications are an important part of the frontier of any theory. . . . A unified theory of cognition is the key to successful applied cognitive science" (p. 498). A third example is found in the EPIC architecture, for which some of the earliest publications (Kieras & Meyer, 1997; Kieras, Wood, & Meyer, 1997) make it clear that applications in system design served an important motivational role in its creation. The fact that this motivation persists is clear on the EPIC Web site[3] which states that EPIC is, ". . . for constructing models of human-system interaction that are accurate and detailed enough to be useful for practical design purposes."

Regardless of which definition and combination of characteristics one prefers as a description of cognitive architectures, there has been enough progress in the creation of cognitive architectures over the course of the last half century, and especially in the last two decades, that lately there have been published a number of reports, articles, comparisons, and collections on this topic. These publications are the focus of the next section, for the benefit of any readers who are interested in more detailed and complete coverage of the state of the science and practice of cognitive architectures than could possibly be achieved in a single chapter on the topic.

RELEVANT RECENT PUBLICATIONS

Review Reports

There have been three noteworthy review reports that focused on, or at least included the topic of, cognitive architectures. In all

[3]http://www.eecs.umich.edu/~kieras/epic.html

three cases the reports were requested and/or funded by defense-related organizations. This is no surprise, given that financial support for the development of and improvements to cognitive architectures has originated primarily (though not exclusively) from the military services.

The first of these reviews was a report by Pew and Mavor (1998), published as a book summarizing the conclusions of their Panel on Modeling Human Behavior and Command Decision Making: Representations for Military Simulations. The panel was organized by the U.S. National Research Council, at the request of an organization then known as the U.S. Defense Modeling and Simulation Office (DMSO). Pew and Mavor used the term "integrative architecture," rather than cognitive architecture, in order to accommodate the fact that complete processing systems of this sort include functionalities beyond exclusively the cognitive. Their report reviewed the state of theory and modeling across a variety of important aspects of human information processing from the individual to military unit levels of analysis, with chapters on attention, multitasking, learning, memory, decision making, situation awareness, planning, and behavior moderators (e.g., physiological stressors, intelligence, personality, emotions). Most directly relevant to this chapter is Pew and Mavor's Chapter 3, in which they describe and compare 10 cognitive architectures across these dimensions: purpose and use, underlying assumptions, architecture and functionality, operation, current implementation, support environment, extent of validation, and applicability for military simulations. Their general conclusions were that substantial progress has been made in the formal computational modeling of human behavior, no single one of the architectures provides all that is needed to address the range of the military's interests and requirements in human modeling and simulation, and significant ongoing investment in this area is warranted and advised. Pew and Mavor also made recommendations regarding challenges and future directions. I return to the topic of recommendations later in the chapter, so will not elaborate here.

Ritter, Shadbolt, Elliman, Young, Gobet, and Baxter (2003) responded to and extended the Pew and Mavor (1998) report. Their supplementary review identified several additional architectures either overlooked by or unavailable to the Pew and Mavor report and also emphasized some important challenges and issues that Pew and Mavor did not, such as integration (both across architectures and between architectures and simulation environments) and usability. It is interesting to note, at least to those who follow such things as sources of research support, that preparation of the Ritter et al. report was supported by a combination of the U.K. Defence Evaluation and Research Agency (DERA), the Australia Defence Science and Technology Organization (DSTO), and the U.S. Office of Naval Research (ONR). It was eventually published through the Human Systems Information Analysis Center at Wright-Patterson Air Force Base, making this the most ecumenically supported publication on cognitive architectures and related technologies to date.

Another National Research Council report was published recently which includes material on cognitive architectures (Zacharias, MacMillan, & van Hemel, 2008). This review was requested by the U.S. Air Force Research Laboratory's Human Effectiveness Directorate, with a focus on the state of the science and practice in societal modeling. To their credit, the assembled experts on the NRC review panel recognized the relevance of individual and organizational modeling approaches, given AFRL's interests, and expanded the scope of their study accordingly. Thus, the Zacharias et al. report encompasses individual, organizational, *and* societal (IOS) modeling and simulation accomplishments, challenges, and prospects, oriented around the military's application interests. They propose a four-part framework for IOS models: (1) micro, (2) meso, (3) macro, and (4) integrated, multilevel models. Cognitive architectures explicitly fall into their micro-level modeling category, along with affective models and cognitive-affective architectures, expert systems, and decision theory and game theory. Within the material on cognitive architectures, Zacharias et al. provide brief descriptions of 12 architectures and discuss current trends and persistent issues.

Articles

There are two "must read" articles for those in the aviation human factors community interested in knowing more about cognitive architecture. Conveniently, these articles were written with the

Human-Computer Interaction (HCI) and the Human Factors and Ergonomics communities in mind. The first is Byrne's (2003) article in the Jacko and Sears (2003) *Human-Computer Interaction Handbook.* The paper begins with background information on the character-istics, strengths, and limitations of cognitive architectures and their relevance to HCI, then explains the relationship of cognitive architectures to three systems and theories that have had an impact in HCI research, including the Model Human Processor (MHP; Card, Moran, & Newell, 1983), Cognitive Complexity Theory (CCT; Bovair, Kieras, & Polson, 1990; Kieras & Polson, 1985), and Collaborative Activation-based Production System (CAPS; Just & Carpenter, 1992). Byrne then describes four cognitive architectures that were under active, ongoing development and application within HCI at the time he wrote that article. Those four architec-tures were LICAI/CoLiDeS (Kitajima & Polson, 1997; Kitajima, Blackmon, & Polson, 2000), Soar (Newell, 1990), EPIC (Kieras & Meyer, 1997), and ACT-R/PM (Byrne & Anderson, 1998). The arti-cle ends with some prospective comments on issues, challenges, and the future of cognitive architectures.

A second helpful article for those interested in a historical perspec-tive on developments of and in cognitive architectures is Gray's (2008) article in the 50th anniversary special issue of the journal *Human Factors.* Gray emphasized the use of cognitive architectures in the context of cognitive engineering, making the point that accomplishing this involves acknowledging and taking advantage of the tight coupling between mind and environment. Indeed, the article is actually subtitled "Choreographing the Dance of Mental Operations with the Task Environment." An interesting contribu-tion of Gray's article is his proposed cognitive architecture tax-onomy. The taxonomy is intended to help organize and explain the relationships among the 50 or so architectures that exist in the world today. The top-level branch in Gray's taxonomy divides the space of existing cognitive architectures into "Architectures for Developing Cognitive Science Theory" and "Hybrid Architectures for Cognitive Engineering." This is a quite reasonable initial dis-criminator among the architectures available today, and some of them do fall cleanly into one branch and not the other. Yet we find the situation gets a bit more complicated another level into the taxonomy, where some of the theory-oriented architectures are "Occasionally Used for Cognitive Engineering" and where some of the engineering architectures include "Cognitive Theories of Control of Cognitive Systems." This reflects the fact that many of these architectures exist today at the productive, exciting, bidirectional interplay of basic and applied research.

Comparisons

The Air Force Research Laboratory and NASA both sponsored research efforts in model and architecture comparison within the last decade. These are of relevance to this chapter not only because they involved using and improving on cognitive architectures, but also because both projects focused on task contexts, such as air traffic control and piloting, that are central to the interests of those working in aviation human factors.

AFRL's AMBR Model Comparison Project. The Gluck and Pew (2005) book on modeling human behavior describes the human data, models, and lessons learned from the U.S. Air Force Research Laboratory's Agent-based Modeling and Behavior Representation (AMBR) Model Comparison. This research effort involved four modeling teams using different architecture-based modeling systems (ACT-R, COGNET/iGEN, DCOG, and EASE), a moderator team (BBN Technologies), and several related architecture-based model development, evaluation, and validation efforts over a period of more than four years. The processes and performance levels of computational cognitive process models are compared to each other and to human operators performing the identical tasks. The tasks are variations on a simplified en route air traffic control hand-off task and emphasize multitasking, workload, and category learning. The book is divided into three sections. The first section of the book is background material, including: an overview of the effort, followed by a description of the method and results from the human experiments, the rationale for the choice of tasks, a detailed description of the task software and its dynamics, the human operator requirements, how the software was set up to allow seamless introduction of either a human operator or a computational process model that simulates the human operator, and the way in which the models were connected into the simulation. The second section of the book includes a separate chapter for each of the participating architectures (ACT-R, COGNET, DCOG, and EASE) and the models that were developed with those architectures. At the end of each of these chapters the authors answered a set of summary questions about their models. The last third of the book presents a discussion of the practical and scientific considerations that arise in the course of attempting this kind of model development and validation effort. It starts with a discussion of how the architectures and models were similar and different and how they performed the target tasks as compared with human data. Included are comments on how the results of the models' performances were related to and derived from the architectures and assumptions that went into the models. The last three chapters are of general interest to those working in the area of cognitive modeling, including a chapter that relates the AMBR models of category learning to other models of category learning in the contemporary psychological literature (Love, 2005), a chapter on a variety of important issues associated with the validation of computational process models (Campbell and Bolton, 2005), and the final chapter, which includes reflections on the results of the project and a proposed research agenda to carry the field of human behavior modeling forward (Pew, Gluck, & Deutsch, 2005).

NASA's HPM Project. NASA recently sponsored a Human Performance Modeling (HPM) project within its Aviation Safety and Security Program. The results are published in Foyle and Hooey (2008). There are many high-level similarities between this research project and the AFRL research described previously, primary among them being that the NASA project also involved quantitative and qualitative comparisons among models, in this case developed with four cognitive architectures (ACT-R, IMPRINT/ACT-R, Air MIDAS, and D-OMAR) and a model of situation awareness called A-SA. There also are a variety of differences between the AFRL and NASA efforts. For instance, the tasks used in the NASA HPM project were piloting tasks (navigating around airport taxiways, approach and landing with synthetic vision systems), whereas AFRL's AMBR project used a simplified en route air traffic control task. Another difference is that the NASA project was more applied in orientation. They list as goals of the project: (1) investigating and informing specific solutions to actual aviation safety problems, and (2) exploring methods for integrating human performance modeling into the design process in aviation. The first section of the Foyle and Hooey text includes a chapter that introduces the NASA HPM project, a background chapter on human performance modeling in aviation, and a chapter that describes the simulators and human subjects studies used in their project. The second section of the book provides details regarding the participating architectures, with a full chapter on each. The third section includes a cross-model comparison chapter, a "virtual roundtable" chapter in which the model developers all respond to each of 13 questions, ranging from the more general (e.g., "Why model?") to the relatively specific (e.g., "In terms of supporting the aviation research and development community, what issues and research questions are HPMs best able to address? What issues are HPMs not yet capable of addressing? What will it take to address those issues?"), and a final chapter that includes comments on the achievements of the NASA HPM research project and ongoing challenges for the science and application of human performance models.

Collections

Polk and Seifert's (2002) book, titled *Cognitive Modeling,* is a collection of previously published journal articles pulled together by the editors into a single reference book. Part I of the book covers modeling architectures, which are divided into *symbolic* and *neural network* categories. The "symbolic architectures" include Construction-Integration, ACT, Soar, EPIC, and CAPS. The "neural network architectures" (also sometimes referred to as "approaches" or "paradigms" in their preface) include back-propagation networks, recurrent networks, Hopfield networks, Adaptive Resonance Theory, and Optimality Theory). Part II of the book is a collection of papers on specific use cases of these architectures and approaches, and Part III includes articles on issues and controversies in cognitive modeling.

Gray's (2007a) book on *Integrated Models of Cognitive Systems* is also an edited collection of writings, but it is new material, rather than previously published articles. It is the first book in Oxford University Press' new series on Cognitive Models and Architectures. Gray (2007b) draws a distinction between *single-focus models of cognitive functions* (e.g., visual attention, categorization, situation awareness, working memory) and *integrated models of cognitive systems* (e.g., ACT-R, CLARION, EPIC, Polyscheme, Soar) and emphasizes the complementary and preferably congenial relationships that should exist among people working in those two areas. Gray identifies, and contributors to the book elaborate on, three theory "types" that must be involved in any integrative cognitive modeling effort. Type 1 theories involve an implementation of the control mechanisms among functional components of the system. If one is going to propose a cognitive architecture that integrates multiple functional components into a full, end-to-end, simulation system, it is necessary to specify in detail the relationships among those functional components. How, if at all, do the visual, motor, and knowledge components of the architecture interact? Which contents, if any, of the various functional components are accessible by the other components? What are the roles of the different components in learning new knowledge and skill, in adapting based on performance feedback, and in prioritizing across different possible courses of action? Within Gray's tripartite typology, the implementation details that address these sorts of questions are Type 1 theories of the cognitive architecture. Type 2 theories are implementations of the internal processes within the functional components. What are the representations and processes that are produced and used by vision? By audition? What kinds of memories are available to the system, and how are they created, maintained, strengthened, and/or lost? What are the representations and processes that enable the motor system to take action in the environment? Type 3 theories are implementations of knowledge that grounds the control mechanisms and functional components within a particular context. All by themselves, the theories that are the functional components of the architecture and the theory of how they interrelate can not produce performance in any particular context. There must be at least some knowledge in the system in order to get it moving, and the Type 3 theory addresses the structure and content of that knowledge.

The set of articles, reviews, comparisons, and collections described in this section provide a thorough overview of what has been done and where things stand in computational cognitive modeling, and especially in the development of cognitive architectures. It is an active, vibrant, ongoing research area. The next section of the chapter describes the ways in which my colleagues and I at the Air Force Research Laboratory are working to advance the theory and application of cognitive architectures in ways that are relevant to aviation and other complex contexts.

IMPROVING HUMAN PERFORMANCE AND LEARNING MODELS FOR WARFIGHTER READINESS

The role of the Air Force Research Laboratory (AFRL), like the other service laboratories, is to conduct the basic and applied research and advanced technology development necessary to create future technology options for the Department of Defense. At the Warfighter Readiness Research Division of AFRL's Human Effectiveness Directorate we have a research program focused on mathematical and computational cognitive process modeling for replicating, understanding, and predicting human performance and learning. This research will lead to new technology options in the form of human-level synthetic teammates, simulation-based cognitive readiness analysis tools, and performance tracking and prediction algorithms. Creating a future in which these objectives become realities requires tightly coupled, multidisciplinary, collaborative interaction among scientists and engineers dedicated to overcoming the myriad challenges standing between the reality of the present and our vision for the future.

The Performance and Learning Models (PALM) team was formed to pursue this agenda. Our research approach is organized around a set of methodological strategies with associated benefits. First, we are using and improving on the ACT-R (Adaptive Control of

Thought—Rational) cognitive architecture (Anderson et al., 2004; Anderson, 2007) because it provides important, well-validated theoretical constraints on the models we develop, facilitates model reuse among members of the ACT-R research community, and serves the integrating, unifying role intended for cognitive architectures. Second, we use the architecture, or equations and algorithms inspired by it, to produce precise, quantitative forecasts about the latencies and probabilities of human performance and learning in order to facilitate eventual transition to applications that require such capabilities. Third, we develop models in both abstract, simplified laboratory tasks and in more realistic, complex synthetic task environments in order to bridge the gap between the laboratory and the real world. Fourth, we compare the predictions of our models to human data, in order to evaluate the necessity and sufficiency of the computational mechanisms and parameters that are driving those predictions and in order to evaluate the validity of the models. We are pursuing this research strategy in several lines of research, which are briefly described next.

We have one research line that is entirely mathematical modeling and does not involve a computational simulation component. Progress to date involves an extension and (we think) improvement to the general performance equation proposed by Anderson and Schunn (2000) that allows us to make performance predictions or prescribe the timing and frequency of training, both in aviation-related and other domains (Jastrzembski, Gluck, & Gunzelmann, 2006; Jastrzembski, Portrey, Schreiber, & Gluck, submitted). On the computational modeling side we have research underway in all of the following areas: (1) natural language communication in knowledge-rich, time-pressured team performance environments similar to those encountered in real-world situations, such as unmanned air vehicle reconnaissance missions (Ball, 2008; Ball, Heiberg, & Silber, 2007); (2) a neurofunctional and architectural view of how spatial competence is realized in the brain and the mind (Gunzelmann & Lyon, 2008; Lyon, Gunzelmann, & Gluck, 2008) and how spatial cognition interacts with vision and language to produce situated action (Douglass, 2007; Douglass & Anderson, 2008); (3) implementing new architectural mechanisms and processes that allow us to replicate the effects of sleepiness on the cognitive system, in order to predict what the precise effects of sleep deprivation or long-term sleep restriction will be in a given performance context (Gunzelmann, Gluck, Kershner, Van Dongen, & Dinges, 2007; Gunzelmann, Gross, Gluck, & Dinges (2009)); (4) the interactive dynamics of cognitive coordination for development of a synthetic teammate (Myers, Cooke, Ball, Heiberg, Gluck, & Robinson, submitted); (5) the creation of a meta-computing software infrastructure for faster, broader, and deeper progress in computational cognitive modeling (Gluck & Harris, 2008; Harris & Gluck, 2008; http://www.mindmodeling.org); and (6) a new initiative at the intersection of cognitive modeling and ultra-large-scale software engineering and systems simulation that will create new methods and capabilities that enable the development, exploration, understanding, and validation of computational cognitive process models and software agents (whether in standard ACT-R, some modified version of ACT-R, or some other formalism) on an unprecedented scale.

These ambitious lines of research were carefully selected on the basis of scientific merit and relevance to the U.S. Air Force mission, aviation-related and otherwise. They represent the range of basic and applied research efforts we chose to pursue with the resources made available to date. It is easy to make the case for the relevance of these lines of research to civilian aviation contexts, as well, where capabilities such as natural language communication, spatial reasoning, and vision all are required, where performance and learning take place in complex, time-pressured, dynamic situations, where better performance often requires that people work effectively as teammates, and where stressors like sleepiness may lead to undesirable or even catastrophic degradations in performance. Even this ambitious range of carefully considered and relevant research lines is only a small sampling of the possible investments that could and should be made in improving on the state of the science and practice in cognitive architectures. In the final section of this chapter I review and elaborate on opinions regarding significant scientific and technical gaps in cognitive architectures and recommendations for future investments in this area.

CONSIDERATIONS, CHALLENGES, AND RECOMMENDATIONS FOR THE FUTURE OF COGNITIVE ARCHITECTURE RESEARCH

Many of the published reports and comparisons described earlier included material in which the authors recorded their advice regarding recommendations for future research. In some cases, such as the Pew and Mavor (1998) and Zacharias et al. (2008) National Research Council reports, generating such recommendations was explicitly a motivation for preparing the report in the first place. In all cases, the recommendations are helpful and worthy of attention, especially by those who are in positions of influence or authority at organizations that control basic and applied research and development investments, such as the FAA, for instance. Cognitive architectures are ambitious, incomplete, research programs in progress, with an emphasis on *in progress*. For all of the impressive progress that already has occurred, there is important work yet to be done. Given that the space of possible work that could be done is infinitely large, it seems advisable to review what some of those with experience in cognitive architectures have said about what should be done, both from the perspective of improving on their completeness as unifying theories of human cognition and also from the perspective of improving on their application potential.

Pew and Mavor (1998) summarize their conclusions and recommendations in the form of a research and development framework, with near-, mid-, and far-term objectives, with investment toward objectives at all three timeframes advised to begin immediately and run concurrently. Their framework, also explicitly described as a research program plan, addresses a variety of key issues and challenges faced by the human behavior modeling community, military or otherwise. The Pew and Mavor program plan for the development of models of human behavior is:

- Collect and disseminate human performance data
- Create accreditation procedures (including verification and validation)
- Support sustained model development in focused domains
- Develop task analysis and structure

- Establish model purposes
- Support focused modeling efforts
- Promote interoperability
- Employ substantial resources
- Support theory development and basic research in relevant areas

Ritter et al. (2003) propose 22 specific project-level research activities as important ways to contribute to the science and technology of architecture-based human performance modeling. I point the interested reader to the Ritter et al. report for details regarding each of the 22 projects, choosing here to mention the three higher-level issues which the full set of projects is intended to address. These issues, which Ritter et al. clearly propose as important ongoing or future directions for research in this area, are:

- More Complete Performance (extending the ability/ functionality of cognitive architectures)
- Integration (within architectural modules and also between architectures and external simulations of task contexts)
- Usability

Byrne (2003) describes major ongoing challenges for cognitive architectures in the context of HCI. Although his focus is on challenges and limitations, Byrne's description of these seems written with an eye toward suggested research directions. Indeed, he even elaborates on a few existing research efforts addressing some of the limitations. Byrne mentions:

- Subjectivity (e.g., preference, boredom, aesthetics, fun)
- Social interaction
- The knowledge engineering bottleneck
- Usability, especially for larger-scale models
- Interfacing architectures with simulation environments

The final chapter of Gluck and Pew (2005) describes challenges for and guidance to those who may be interested in conducting a model comparison, followed by a list of improvements that are needed in the theory and practice of computational human modeling

(Pew, Gluck, & Deutsch, 2005). That list of needed improvements included:

- Robustness
- Integrative behavior
- Validation
- Establishing the necessity of architectural and model characteristics
- Inspectability and interpretability
- Cost-effectiveness

The final chapter of Foyle and Hooey (2008) begins with an explanation of the synergistic manner in which they used a combination of human-in-the-loop simulations and human performance model runs to advance their objectives (Hooey & Foyle, 2008). They then describe key modeling advances that were achieved in the context of the NASA project, including modeling the human-environment interaction, visual attention, situation awareness, and human error. They end the chapter by considering important challenges for modeling complex aviation tasks:

- Selecting an architecture (matching architectural strengths to intended application)
- Developing models (knowledge engineering; strategic variability)
- Interpreting model output (Transparency of tasks, procedures, architectures, and models)
- Verification, Validation, and Credibility (matching method, to model, to intended use).

Zacharias et al (2008) listed 10 suggested future directions for research and development with cognitive architectures. Their report was requested by the Air Force Research Laboratory and was written primarily with the military behavioral modeling and simulation communities as the intended audience, but their suggestions regarding research directions are just as relevant when considering the current challenges and prospects for using cognitive architectures to explain and predict human cognition and performance in non-military contexts, such as commercial aviation. Zacharias et al.'s 10 suggested directions are:

- Facilitate architecture development (standardization, interoperability, IDEs)
- Facilitate architecture instantiation (domain ontologies and data repositories)
- Facilitate knowledge base development (address knowledge engineering bottleneck)
- Enhance model explanation capabilities (inspection and visualization tools)
- Address the brittleness problem (larger knowledge bases and learning for robustness)
- Enhance realism (embodiment, emotion, personality, and cultural factors)
- Validation (develop common methods, metrics, and test suites)
- Explore new modeling formalisms (e.g., chaos theory, genetic algorithms)
- Models of groups and teams (via abstraction to group/team-level processes)
- Context and task models (formal estimates of generalizability of architecture-based cognitive models via task and context taxonomies)

This collection of suggested research directions for cognitive architectures is not exhaustive, but certainly is comprehensive. There is plenty of scientific and technical justification for pursuing any one or more of these and other avenues of research in cognitive architecture. In closing this section I will elaborate on a practical issue that must be addressed through additional scientific and technical progress: cost. At Wayne Gray's Integrated Models of Cognitive Systems workshop in 2005, I polled the attendees regarding their estimates of person-years and dollars required to implement and validate an architecture-based model that interacts with a simulation of a moderately complex task. "Moderately complex" in this case would be something like the simplified air traffic control simulation used in AFRL's AMBR project (Gluck & Pew, 2005). The average of the estimates offered by the 13 respondents (all people with some firsthand experience with the development of such models) was 3.4 years and $400,000. That still strikes me as a reasonable estimate today. If it is, in fact, a reasonable estimate, then it is an indication of how far we have left to go as a research community.

That sort of time and money for a single model is prohibitive and clearly stands in the way of transitioning the use of cognitive architectures to the aviation community or any other application context. Cognitive architectures will never have the kind of applied impact that many of us would like as long as model development is measured in a timescale of years and a budget of hundreds of thousands of dollars. Cognitive architectures are still very much *research* programs. They are research programs with a great deal of potential for revolutionizing the way we evaluate and improve on our system designs and training programs, but that revolution has not yet been realized. Perhaps progress toward those revolutionary new capabilities can be measured, at least partly, in terms of the money and time required to develop and validate models with cognitive architectures.

SUMMARY AND CONCLUSION

This chapter has been an introduction to cognitive architectures, intentionally written at a relatively high level of description and review in the hope that the chapter would be an approachable and useful reference for people in the aviation human factors community. As mentioned in the beginning of the chapter, the topic of human cognitive architecture is central to the interests of researchers and practitioners in aviation human factors. This is true whether or not they actually think of themselves as working on or within cognitive architecture theory or application. It is in the theory and application of the human cognitive architecture that an understanding of component processes, phenomena, and stressors, such as sensing and perception, information processing, situation awareness, group and team dynamics, and fatigue (to draw some important examples from the contents of other chapters in this volume), must come together in an integrated, generative, action-producing system. No analysis, representation, or simulation of a planned or operational aviation system is complete without the inclusion of the human component. Cognitive architectures are both theoretical claims about the fundamental nature, strengths, and weaknesses of that human component, and also modeling and simulation research tools for formally exploring the performance implications of changes in next generation system design or training regimen. Progress to date in the development of cognitive architectures has been impressive, yet scientific gaps, technical challenges, and practical issues remain. The research and application contexts in aviation human factors are fertile ground for continuing the evolution of cognitive architectures from promising research programs to useful products for improving the operational safety of current and future aviation systems.

ACKNOWLEDGMENTS

The views expressed in this chapter are those of the author and do not reflect the official policy or position of the Department of Defense or the U.S. government. Many thanks to Don Lyon, Bob Patterson, Ed Salas, and Dan Maurino for their helpful suggestions regarding how to improve on a previous version of this chapter.

References

Anderson, J. R. (1983). The architecture of cognition. Cambridge, MA: Harvard University Press.

Anderson, J. R. (2007). How can the human mind occur in the physical universe? New York: Oxford University Press.

Anderson, J. R., Bothell, D., Byrne, M. D., Douglass, S., Lebiere, C., & Qin, Y. (2004). An integrated theory of the mind. *Psychological Review, 111*, 1036–1060.

Anderson, J. R., & Schunn, C. D. (2000). Implications of the ACT-R learning theory: No magic bullets. In R. Glaser (Ed.)*Advances in instructional psychology: Educational design and cognitive science: Vol. 5* . Mahwah, NJ: Erlbaum.

Ball, J. (2008). A naturalistic, functional approach to modeling language comprehension. *Paper presented at the AAAI fall symposium on naturally inspired AI*. Arlington, VA.

Ball, J., Heiberg, A., & Silber, R. (2007). Toward a large-scale model of language comprehension in ACT-R 6. *Proceedings of the 8th international conference on cognitive modeling.*

Byrne, M. D. (2003). Cognitive architecture. In J. Jacko & A. Sears (Eds.), *The human-computer interaction handbook: Fundamentals, evolving technologies and emerging applications* (pp. 97–117). Mahwah, NJ: Lawrence Erlbaum.

Cooper, R. P. (2007). The role of falsification in the development of cognitive architectures: Insights from a Lakatosian analysis. *Cognitive Science, 31*(3), 509–533.

Douglass, S. A. (2007). *A computational model of situated action* Unpublished doctoral dissertation. Carnegie Mellon University.

Douglass, S. A., & Anderson, J. R. (2008). A model of language processing and spatial reasoning using skill acquisition to situate action. In B. C. Love, K. McRae, & V. M. Sloutsky (Eds.), *Proceedings of the 30th annual conference of the cognitive science society* (pp. 2281–2286). Austin, TX: Cognitive Science Society.

Foyle, D. C., & Hooey, B. L. (Eds.), (2008). *Human performance modeling in aviation.* Boca Raton, FL: CRC Press.

Gluck, K. A., & Pew, R. W. (Eds.), (2005). *Modeling human behavior with integrated cognitive architectures: Comparison, evaluation, and validation.* Mahwah, NJ: Lawrence Erlbaum Associates.

Gluck, K. A., & Harris, J. (2008). MindModeling@Home [Abstract]. In B. C. Love, K. McRae, & V. M. Sloutsky (Eds.), *Proceedings of the 30th annual conference of the cognitive science society,* p. 1422. Austin, TX: Cognitive Science Society.

Gray, W. D. (Ed.). (2007a). *Integrated Models of Cognitive Systems.* New York: Oxford University Press.

Gray, W. D. (2007b). Composition and control of integrated cognitive systems. In W. D. Gray (Ed.), *Integrated models of cognitive systems* (pp. 3–12). New York: Oxford University Press.

Gray, W. D. (2008). Cognitive architectures: Choreographing the dance of mental operations with the task environment. *Human Factors, 50*(3), 497–505.

Gunzelmann, G., Gluck, K. A., Kershner, J., Van Dongen, H. P. A., & Dinges, D. F. (2007). Understanding decrements in knowledge access resulting from increased fatigue. In D. S. McNamara & J. G. Trafton (Eds.), *Proceedings of the 29th annual meeting of the cognitive science society* (pp. 329–334). Austin, TX: Cognitive Science Society.

Gunzelmann, G., & Lyon, D. R. (2008). Mechanisms of human spatial competence. In T. Barkowsky, C. Freksa, M. Knauff, B. Krieg-Bruckner, & B. Nebel (Eds.), *Lecture notes in artificial intelligence #4387* (pp. 288–307). Berlin, Germany: Springer-Verlag.

Gunzelmann, G., Gross, J. B., Gluck, K. A., & Dinges, D. F. (2009). Sleep deprivation and sustained attention performance: Integrating mathematical and cognitive modeling. *Cognitive Science, 33*(5), 880–910.

Harris, J., & Gluck, K. A. (2008, July). MindModeling@Home. *Poster presented at the 30th annual conference of the cognitive science society.* Washington, DC.

Hooey, B. L., & Foyle, D. C. (2008). Advancing the state of the art of human performance models to improve aviation safety. In D. C. Foyle & B. L. Hooey (Eds.), *Human performance modeling in aviation* (pp. 321–349). Boca Raton, FL: CRC Press.

Jacko, J. A., & Sears, A. (Eds.), (2003). *The human-computer interaction handbook: Fundamentals, evolving technologies, and emerging applications.* Mahwah, NJ: Lawrence Erlbaum Associates.

Jastrzembski, T. S., Gluck, K. A., & Gunzelmann, G. (2006). Knowledge tracing and prediction of future trainee performance. In: *The proceedings of the interservice/industry training, simulation, and education conference.* Orlando, FL: National Training Systems Association (pp. 1498–1508).

Kieras, D. (2007). Control of cognition. In W. Gray (Ed.), *Integrated models of cognitive systems* (pp. 327–355). New York: Oxford University Press.

Kieras, D., & Meyer, D. E. (1997). An overview of the EPIC architecture for cognition and performance with application to human-computer interaction. *Human-Computer Interaction, 12,* 391–438.

Kieras, D. E., Wood, S. D., & Meyer, D. E. (1997). Predictive engineering models based on the EPIC architecture for a multimodal high-performance human-computer interaction task. *ACM Transactions on Computer-Human Interaction, 4,* 230–275.

Laird, J. E. (2008). Extending the Soar cognitive architecture. In *Proceedings of the 1st artificial general intelligence conference,* Memphis, TN.

Lakatos, I. (1970). Falsification and the methodoligy of sceintific research program. In I. Lakatos & A. Musgrave (Eds.), *Criticism and the Growth of Knowledge* (pp. 91–196). Cambridge, UK: Cambridge University Press.

Lyon, D., Gunzelmann, G., & Gluck, K. A. (2008). A computational model of spatial visualization capacity. *Cognitive Psychology, 57*(2), 122–152.

Newell, A. (1990). Unified theories of cognition. Cambridge, MA: Harvard University Press.

Newell, A., Shaw, J. C., & Simon, H. A. (1958). Elements of a theory of human problem solving. *Psychological Review, 65,* 151–166.

Newell, A., & Simon, H. A. (1972). Human problem solving. Englewood Cliffs, NH: Prentice-Hall.

Pew, R. W., Gluck, K. A., & Deutsch, S. (2005). Accomplishments, challenges, and future directions for human behavior representation. In K. A. Gluck & R. W. Pew (Eds.), *Modeling human behavior with integrated cognitive architectures: Comparison, evaluation, and validation* (pp. 397–414). Mahwah, NJ: Lawrence Erlbaum Associates.

Pew, R. W., & Mavor, A. S. (Eds.), (1998). *Modeling human and organizational behavior: Applications to military simulations.* Washington, DC: National Academy Press.

Polk, T. A., & Siefert, C. M. (2002). Cognitive Modeling. Cambridge, MA: MIT Press.

Pylyshyn, Z. W. (1984). Computation and Cognition: Towards a Foundation for Cognitive Science. Cambridge, MA: MIT Press.

Ritter, F. E., Shadbolt, N. R., Elliman, D., Young, R. M., Gobet, F., & Baxter, G. D. (2003). *Techniques for modeling human performance in synthetic environments: A supplementary review (HSIAC-SOAR-2003–01).* Wright-Patterson Air Force Base, OH: Human Systems Information Analysis Center.

Simon, H. A. (1956). Rational choice and the structure of the environment. *Psychological Review, 63,* 129–138.

Sun, R. (2004). Desiderata for cognitive architectures. *Philosophical Psychology,* *17*(3), 341–373.

VanLehn, K. (Ed.). (1991). *Architectures for Intelligence.* Hillsdale, NJ: Lawrence Erlbaum Associates.

Wiener, E. L., & Nagel, D. C. (Eds.), (1988). *Human factors in aviation.* New York: Academic Press.

Zacharias, G. L., MacMillan, J., & Van Hemel, S. B. (Eds.), (2008). *Behavioral modeling and simulation: From individuals to societies.* Washington, DC: National Academies Press.

13

Aircrew Fatigue, Sleep Need and Circadian Rhythmicity

Melissa M. Mallis[a], Siobhan Banks[b] and David F. Dinges[c]

[a]Chief Scientist for Operational & Fatigue Research
Institutes for Behavior Resources, Inc., 2104 Maryland
Avenue, Baltimore, Maryland
[b]Research Assistant Professor of Sleep in Psychiatry
Unit for Experimental Psychiatry, Division of Sleep and
Chronobiology, Department of Psychiatry, University of
Pennsylvania School of Medicine, 1013 Blockley Hall, 423
Guardian Drive, Philadelphia, Pennsylvania
[c]Professor, Chief, Division of Sleep and Chronobiology, Director,
Unit for Experimental Psychiatry, Department of Psychiatry
University of Pennsylvania School of Medicine, 1013 Blockley
Hall, 423 Guardian Drive, Philadelphia, Pennsylvania

INTRODUCTION

Fatigue is classically defined as a decrease in performance or performance capability as a function of time on task. This definition forms the basis for the now 70-year-old federal duty time regulations in U.S. commercial aviation. However, a vast amount of scientific work in the past half century has established that human fatigue is dynamically influenced by now well-described neu-robiological regulation of sleep need and endogenous circadian rhythms, which interact nonlinearly to produce changes in human alertness and cognitive performance over time.

With human alertness and performance modulated by sleep need and circadian rhythms, it is evident that humans were simply not designed to operate effectively under the pressure of 24/7 schedules. Fatigue cannot be eliminated from aviation operations because of the inherent schedule requirements for transmeridian travel, irregular and unpredictable schedules, long duty days, early report times, night flights, and reduced sleep opportunities. Many commercial aviation industry practices induce fatigue via sleep loss and circadian misalignment in flight crews. Therefore, there is a need to develop scientifically valid fatigue-management approaches to mitigate sleep loss, enhance alertness during extended duty periods and cope with circadian factors that are primary contributors to fatigue-related aviation incidents and accidents.

This chapter begins by reviewing the homeostatic and the biological processes that regulate sleep, fatigue and alertness. Building on this foundational material, evidence is presented to demonstrate how sleep loss and circadian misalignment contribute greatly to fatigue and performance risks in short haul, long haul and ultra long range flight operations. The chapter concludes with a discussion of scheduling approaches, countermeasure application and fatigue management systems and technologies that can be used, in combination, as part of a comprehensive approach for ensuring that fatigue is effectively minimized in human centered aviation operations.

BIOLOGICAL REGULATION OF SLEEP, ALERTNESS, AND PERFORMANCE

Circadian System

Most biological organisms show daily alterations in their behavior and physiology that are linked to Earth's 24-hour rotation on its axis. These rhythms in mammals are known as circadian rhythms and are primarily controlled by an internal biological clock in an area of the brain called the suprachiasmatic nucleus (SCN), which it located in the hypothalamus (Moore, 1999). Circadian rhythms can be synchronized to external time signals (e.g., the light/dark cycle) but they also continue in the absence of such signals or when environmental signals are outside the entrainment capacity of the endogenous oscillator. For example, in the absence of time cues the SCN shows an average "free running" intrinsic period of 24.18 hours (Czeisler et al., 1999). This internal circadian pacemaker modulates the daily high and low impacts in many physiological and neurobehavioral functions including core body temperature, plasma cortisol, plasma melatonin, alertness, subjective fatigue, cognitive performance, and sleep patterns. The lowest point of the circadian rhythm, which varies between individuals but occurs approximately between 03:00 and 06:00 when entrained to a stable light-dark cycle of 24 hours, is associated with an increased sleep propensity (Dijk & Czeisler, 1995).

Homeostatic Sleep Drive

A second fundamental neurobiological process involved in the timing of alertness and quality of optimal cognitive performance is the homeostatic sleep drive, which interacts with the endogenous circadian pacemaker. All mammals require periods of sleep. Biological pressure for sleep builds over the wake period until the ability to resist sleep becomes very difficult or impossible. For most adults the continuous wake duration at which the sleep homeostatic drive begins to manifest itself in behavior as sleepiness and performance deficits is approximately 16 hours (Van Dongen et al., 2003). The biological increase in homeostatic sleep drive is not only responsive to time awake, but also to sleep quantity and sleep consolidation (Van Dongen et al., 2003). Consequently, three factors can result in elevated homeostatic sleep drive: (1) increasing time continuously awake; (2) inadequate sleep duration for one or more consecutive days; and (3) sleep that is physiologically disrupted (fragmented) due to medical conditions (e.g., untreated sleep disorder such as obstructive sleep apnea) or environmental factors (e.g., attempting to sleep upright or in an uncomfortable environment).

The neurobiological mechanisms underlying homeostatic sleep pressure are beginning to be identified (Saper, et al., 2001; Mignot et al., 2002) and the promotion of both wakefulness and sleep appears to involve a number of neurotransmitters and nuclei located in the human forebrain, midbrain, and brainstem.

Interaction of Sleep and Circadian Influences

The increase in homeostatic sleep drive with time awake (or inadequate sleep) and the circadian cycle interact to produce dynamic nonlinear changes in human fatigue and functional capability. For example, the sleep homeostatic and circadian interaction determines the extent and timing of daytime sleepiness, fatigue and neurobehavioral performance (Czeisler & Khalsa, 2000; Durmer & Dinges, 2005; Van Dongen & Dinges, 2005) . During the day, an individual's homeostatic sleep drive and circadian rhythm act in opposition to promote wakefulness, while during biological night (i.e., habitual nocturnal sleep time) these system are in synergy such that both sleep drive and reduced circadian promotion of waking keeps the sleeper asleep for 7–8 hours. Figure 13-1 illustrates the dynamic interplay over time of sleep homeostatic drive and circadian oscillation. This interaction has a fundamental influence on ensuring that sleep occurs during biological night and wakefulness is stable across the day.

For most mammals, the timing of sleep and wake under natural conditions is in synchrony with the circadian control of the sleep cycle. For example, after awakening from a sleep period, the homeostatic pressure for sleep is low, and the circadian drive for wakefulness is increasing. As wakefulness continues across the day, homeostatic pressure for sleep builds, while at the same time, the circadian pressure for wakefulness increases (this is the opposition that produces the effect of being most endogenously alert in the late afternoon or early evening). The effect of the opposition of the two processes is to maintain a relatively even level of alertness across approximately 16 hours of waking. At night, however,

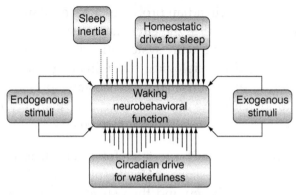

FIGURE 13-1 Schematic representation of the (speculative) oppositional interplay of circadian and homeostatic drives in the regulation of alertness, performance, and related neurobehavioral functions. As illustrated in the upper part of the figure, wakefulness typically begins with rapidly dissipating sleep inertia, which suppresses neurobehavioral functioning for a brief period after awakening. The homeostatic drive for sleep accumulates throughout wakefulness, and progressively down regulates neurobehavioral performance and alertness (while increasing subjective sleepiness). Unlike the circadian system, which is limited by its amplitude, the homeostatic drive for sleep can accumulate far beyond the levels typically encountered in a 24h day (illustrated by the increasing density of downward arrows). In opposition to these downward influences on performance and alertness is the endogenous circadian rhythmicity of the biological clock. Through its promotion of wakefulness, the circadian system modulates the enhancement of performance and alertness. The improvement in waking neurobehavioral functions by the circadian drive is an oscillatory output which periodically involves robust opposition to the homeostatic drive, alternated with periods of withdrawal of the circadian drive for wakefulness. Critical modulators of neurobehavioral functions other than the sleep and circadian drives are subsumed in the schematic under the broad categories of *endogenous* and *exogenous* stimulation. Although common in the real world, these factors are considered masking factors in most laboratory experiments. They can include endogenous (e.g., anxiety) or exogenous (e.g., environmental light) wake-promoting processes that oppose the homeostatic drive for sleep. Alternatively, they can include endogenous (e.g., narcolepsy) or exogenous (e.g., rhythmic motion) sleep-promoting processes that oppose the circadian drive for wakefulness either directly, or indirectly by exposing the previously countered sleep drive. The neurobiological underpinnings of these exogenous and endogenous processes are undoubtedly diverse, and few of their interactions with the circadian and homeostatic systems have been studied systematically. (Adapted from and reprinted with permission from Van Dongen, H. P. A. and Dinges, D. F. (2005). Circadian Rhythms in Sleepiness, Alertness and Performance. Principles and Practice of Sleep Medicine (4th ed.). M. H. Kryger, T. Roth and W. C. Dement, W.B. Saunders Company.)

the homeostatic and circadian processes act simultaneously to promote sleep. In the evening the circadian pressure for wakefulness gradually lowers along with systematic changes in many other circadian mediated physiological functions (e.g., decrease in core body temperature), whereas the homeostatic pressure for sleep continues to increase. As a result, there is a relatively rapid shift from alertness to sleepiness from 21:00 to 00:00.

During sleep, homeostatic pressure for sleep steadily dissipates while the circadian pressure for wakefulness declines further to the a.m. nadir in midsleep and then steadily increases thereafter. Thus, there is little waking pressure throughout the night, which, in healthy individuals, results in a consolidated period of sleep. In the morning just before waking, the circadian pressure for wakefulness increases again, exceeding the dissipated homeostatic pressure for sleep, resulting in the termination of sleep (although most people now enforce an earlier awakening than might occur naturally by using an alarm clock).

Sleep homeostatic and circadian neurobiology have been studied extensively and the two physiological processes have been incorporated into contemporary theoretical (Achermann, 2004) and biomathematical models for predicting human alertness and performance (Jewett et al., 1999; Neri, 2004). Even though the alternation between sleep and wakefulness is regulated fairly precisely, humans frequently choose to ignore the homeostatic and circadian-mediated signals for sleep. This commonly occurs when the sleep-wake cycle is out of phase with the internal rhythms that are controlled by the circadian clock (e.g., during night-shift work or jet lag), causing adverse effects for health and safety (Dijk & Czeisler, 1995). The synchrony of an organism with both its external and internal environments is critical to the organism's well-being and behavioral efficacy. Disruption of this synchrony can result in a range of difficulties, including fatigue, deficits in cognitive functions, sleepiness, reduced vigilance, altered hormonal functions, and gastrointestinal complaints.

WORK FACTORS IN RELATION TO BIOLOGICAL CONTROL OF FATIGUE

Continuous Hours of Work

Many industries, including aviation, seek to manage fatigue relative to the demands of 24-hour operations. Any occupation that requires individuals to maintain high levels of alertness over extended periods of time (> 16 hours) or at a circadian phase that permits sleep to occur (biological night) is open to the neurobehavioral and cognitive consequences of sleep loss and circadian disruption. The resulting effects on performance (e.g., slowed reactions times, difficulties in problem solving and perseveration) can mean an increase in the risk of errors and accidents (Dinges, 1995). Unfortunately, the hazards associated with extended duty schedules are not always obvious and there is a general lack of awareness regarding the importance of obtaining adequate amounts of sleep (Walsh et al., 2005). Chronic sleep restriction is a frequent consequence of schedules in which individuals must work at night or for durations in excess of 12 hours.

Workload Interaction with Sleep Loss

Although fatigue can result from sleep loss and circadian disturbances, it can also occur due to excess cognitive or physical workload. The risks of accidents and injuries increase as workload increases—especially after more than 12 hours of work a day or more than 70 hours of work a week (Caruso et al., 2004, 2006; Knauth, 2007; Nachreiner, 2001), which is why federal statutes and regulations have historically limited duty-hours in all transportation modes and other safety-sensitive industries. Excessive workload can also erode time available for recovery and sleep (Macdonald & Bendak, 2000).

Higher cognitive workload has been reported to increase the adverse effects of sleep loss on cognitive performance (Stakofsky et al., 2005). Significantly greater deficits were found for vigilant attention, cognitive speed and perceived effort and sleepiness for the high cognitive workload condition. These data suggest that higher cognitive workload interacts with sleep loss to potentiate the effects.

Cumulative Effects of Reduced Sleep Opportunities

One of the most difficult issues in work-rest regulations is the question of how much rest or time off work should be provided to avoid the accumulation of fatigue. Rooted in many federal work rules is the assumption that 8 hours for rest between work periods will result in adequate recovery. However, people rarely use (and/ or physiologically cannot use) every minute of non–work time for sleep. As a result, physiological sleep appears to account for about 50% to 75% of rest time in a daily off-duty period, which means that 8 hours off duty allows for only about 4–6 hours of sleep in most individuals.

Sleep debts from repeated days of reduced sleep time have been documented to occur in long-haul commercial airline pilots (e.g., Rosekind et al., 1994) but until recently it has not been certain whether sleep debt results in significant increases in fatigue and risk of errors. Major experiments on healthy adults ages 22–55 years of age have now established that behavioral alertness and a range of cognitive performance functions involving sustained attention, working memory, and cognitive throughput, deteriorate system-atically and cumulatively across days when nightly sleep dura-tion is less than 7 hours (Belenky et al., 2003; Van Dongen et al., 2003). In contrast, when time in bed for sleep is 8–9 hours, no cumulative cognitive performance deficits were seen. Figure 13-2 shows data from these studies. In one study, commercial truck drivers were randomized to 3, 5, 7, or 9 hours time in bed per night for 7 nights (Belenky et al., 2003). In the other study (Van Dongen et al., 2003) participants were restricted to 4, 6, or 8 hours time in bed per night for 14 consecutive nights. Cumulative daytime defi-cits in cognitive performance were observed for vigilant attention, memory and for cognitive throughput in subjects randomized to the 4- and 6-hour sleep periods relative to participants in the

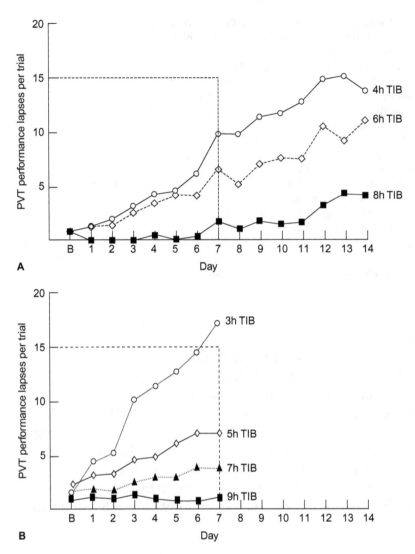

FIGURE 13-2 Findings from two major studies (A, B) on the performance effects of chronic sleep restriction, showing lapses of attention during vigilance performance as a function hours time in bed (TIB) for sleep across 14 consecutive days (panel A) and 7 consecutive days (panel B). Panel A was produced from an experiment in which sleep was restricted for 14 consecutive nights in 36 healthy adults (Van Dongen et al., 2003). Subjects were randomized to 4-hour (N = 13), 6-hour (N = 13) or 8-hour (N = 9) time in bed (TIB) each night. Performance was assessed every 2 hours (9 times each day) from 07:30 to 23:30. The graph shows cumulative increases in lapses of sustained attention performance on the Psychomotor Vigilance Test (PVT) per 10-minute test bout across days within the

4-hr and 6-hr groups (p = 0.001), with sleep-dose differences between groups (p = 0.036). The horizontal dotted line shows the level of lapsing found in a separate experiment when subjects had been awake continuously for 64–88 hours. In Panel A, subjects in the 6-hour TIB condition averaged 54 lapses that day (6 per test bout), while those in the 4-hr TIB averaged 70 lapses that day (8–9 per test bout). Panel B shows comparable data from Belenky et al. (2003). In this study sleep was restricted for 7 consecutive nights in 66 healthy adults. Subjects were randomized to 3-hour (N = 13), 5-hour (N = 13), 7-hour (N = 13) or 9-hour (N = 16) time in bed (TIB) each night. Performance was assessed 4 times each day from 09:00 to 21:00. The graph shows cumulative increases in lapses of sustained attention performance (PVT) per test bout across days within the 3-hour and 5-hour groups (p = 0.001). The horizontal dotted line shows the level of lapsing found in a separate experiment by Van Dongen et al. (2003) when subjects had been awake continuously for 64–88 hours. For example, by day 7, subjects in the 5-hr TIB averaged 24 lapses that day, while those in the 3-hr TIB averaged 68 lapses per day. (Adapted from and reprinted with permission from Institute of Medicine (2006–07), Pubic Health Significance of Sleep Deprivation and Disorders, The National Academies, Washington, DC).

8-hour condition. In order to determine the magnitude of cognitive deficits experienced during 14 days of restricted sleep, the findings were compared with cognitive effects after 1, 2, and 3 nights of total sleep deprivation (Van Dongen et al., 2003). This comparison found that both 4- and 6-hour sleep conditions resulted in cognitive impairments that quickly increased to levels found after 1, 2, and even 3 nights of total sleep deprivation. These data suggest that 8 hours for sleep between work bouts is inadequate unless the individual obtains physiological sleep for 90% or more of the time (i.e., 7+ hours).

Individual Differences in Response to Fatigue

Although restriction of sleep periods to below 7 hours results in neurobehavioral deficits in healthy adults, interindividual differences exist in the degree to which sleep loss produces cognitive deficits (Doran et al., 2001). While the majority of people suffer neurobehavioral deficits when sleep deprived—some very severely—there is a minority that is less affected. Recently, it has been shown that these different responses are stable across subjects and over time (Van Dongen et al., 2004). The biological basis of this differential vulnerability to sleep loss is not known and may be genetic. Interestingly, individuals were not really aware of their differential responses to sleep loss, and more importantly, subjective reports of fatigue and sleepiness often underestimate actual performance deficits from fatigue (Van Dongen et al., 2003).

Working at an Adverse Circadian Phase

Overnight operations pose a challenge because circadian biology promotes sleepiness and sleep at night (Figure 13-1). Since the circadian effect on cognitive functions magnifies with increasing sleep pressure, neurobehavioral and cognitive deficits associated with night work are likely to be most acute with extended wakefulness. There is a nocturnal decline in the cognitive processes of attention, vigilance and alertness during laboratory-controlled simulated night shifts (Santhi et al., 2007). These nocturnal deficits in attention and alertness highlight that occupational errors, accidents, and injuries are more pronounced during night work compared to day work. The consequences of fatigue for performance when remaining awake into the nocturnal circadian phase for sleep can be as serious as other risks. For example, a classic study by Dawson and Reid (1997) examined the effects of working through the night on a hand–eye coordination test compared to the effects on the same test of doses of alcohol up to and beyond 0.05% BAC. The results showed that between 03:00 and 08:00, performance on the task was equivalent to the effects found with a BAC of between 0.05% and 0.10% (Dawson & Reid, 1997). Because there are established legal limits for alcohol intake while performing certain safety-sensitive activities (e.g., driving), this study illustrates the extent to which being awake all day and then working at night (during the circadian nadir) can pose a risk to performance.

Jet Lag: Circadian Placement of Sleep

When an individual travels across a number of time zones, the endogenous circadian rhythm is out of synchrony with the destination time. This is technically referred to as circadian misalignment (but better known as jet lag). The misalignment of the biological timing of the body with local (social) clock time results in the physiological and behavioral effects of jet lag (Waterhouse et al., 2007). Jet lag is a chronic problem because the circadian biological clock in the human brain is slow to adapt to a new time zone. Circadian rhythms in cognitive functions, eating, sleeping, hormone regulation, body temperature, and many others require days to reentrain to the new time zone (Cajochen et al., 2000).

As a result of modern commercial flight operations, transmeridian flight crews are rarely in a nondomicile time zone long enough to adapt biologically. For example, as Figure 13-3 shows, flight crews can continue to sleep for shorter durations in the new time zone when the sleep period is in the diurnal portion of their home times, even if the local time for sleep is at local night and the layover is 24 hours. Such inadequate circadian adjustment to time zones can continue even after crews have been flying consecutive routes for 3–7 days. The reduced sleep duration crews can experience when sleeping at local night (but domicile day) in the new time zone is a result of the brain's circadian clock neurobiologically promoting wakefulness during local night. In addition to reduced sleep duration, circadian misalignment can cause problems with the ability to initiate and consolidate sleep, all of which contribute to reduced ability to sleep at the new destination (Dijk & Czeisler, 1994).

Because jet lag reduces the ability of the individual to sleep effectively at the destination, the resulting effects on cognitive performance and fatigue are similar to that of total sleep deprivation, and if occurring frequently enough, can cause a cumulative sleep debt and a resulting cumulative buildup of cognitive deficits (e.g., Rosekind et al., 1994). Jet lag can result in safety issues when flight crews are unable to sleep effectively when required to do so in flight or during a layover rest period. The risk of performance impairment and fatigue caused by lack of sleep and working during circadian misalignment can then be further compounded if the individual has to work long duty hours or in an environment of heavy workload.

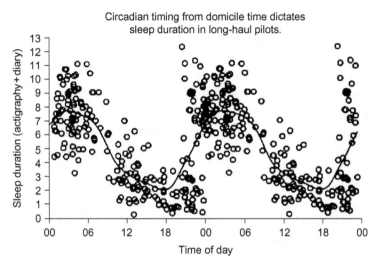

FIGURE 13-3   Sleep durations in N = 21 long-haul commercial flight crew members flying the middle four consecutive transmeridian Pacific flight legs (out of 8 legs) from the USA. Each point is a single sleep episode for a single crew member on layover. All four layover periods were between 21 and 29 hours. Sleep duration is double-plotted as a function of clock time relative to each crew-member's home (domicile) time. The data show that layover sleep duration was longer (> 6 hours) when sleep in the layover city occurred between 00:00 and 09:00 at the crewmember's (domicile) home. In contrast, shorter sleep durations (<6 hours) occurred in the layover city when sleep was taken between 10:00 and 22:00 at the crewmember's domicile. These data suggest that long-haul crews do not make a substantial circadian adjustment to the time zones they fly into, but instead experience sleep durations that more closely reflect circadian entrainment in their domicile. This is one major reason why fatigue management is important in long-haul commercial aviation. (Data from a study by Rosekind, M.R., Graeber, R.C., Dinges, D.F., Connell, L.J., Rountree, M., Gillen, K.A.: Crew factors in flight operations: IX. Effects of cockpit rest on crew performance and alertness in long-haul operations. NASA Technical Memorandum Report No. 103884, 252, 1994.)

FATIGUE CHALLENGES IN OPERATIONAL ENVIRONMENTS

Operational Relevance

Although aviation technology has advanced significantly over the years—permitting exponential growth in the air transit industry—sleep and circadian biology and their effects on human alertness and performance have not changed. Although there is the expectation that motivated individuals can maintain optimum neurocognitive functioning and alertness levels throughout the 24 hours day, there is no substitute for adequate sleep. The demands of aviation operations continue to challenge the brain's sleep and circadian systems, thus contributing to fatigue and increased performance risks in flight crew. This is further complicated by highly automated cockpits that require minimal interaction with aviation systems, which results in a high requirement for relatively passive vigilance in flight crews. Unfortunately, sustained attention and prolonged vigilance are among the most difficult tasks to perform reliably when people are sleep deprived or working at night (Dorrian et al., 2005; Durmer & Dinges, 2005; Lim & Dinges, 2008).

The Code of Federal Regulations (CFRs), previously referred to as the Federal Aviation Regulations (FARs), have not adequately incorporated the most current research in sleep and circadian science (Dinges et al., 1996). Therefore, the challenge of pilot fatigue continues to increase and is demonstrated by the occurrence of fatigue related accidents and incidents and objective and subjective data collected from flight crew in short haul, long haul and ultra-long range operations. Independent of the type of operation (short haul, long haul or ultra-long range), each is faced with its own unique fatigue issues. Pilots are faced with fatigue related operational challenges including irregular and unpredictable schedules, long duty days, early report times, night flights, reduced sleep opportunities and circadian disruption, and the extent to which these various factors are problems can vary as a function of the type of operation (e.g., short haul, long haul, ultra-long range).

Short Haul Operations

Research examining fatigue in commercial aviation has primarily been conducted in long haul operations and is therefore limited in short haul environments. This is partially due to the fact that fatigue research has focused on the effects of extended duty days and jet lag relative to fatigue in aviation. Although these are not commonly associated with short haul operations, short haul pilots are not immune to extended duty days. One operational evaluation showed that short haul pilots, on average, worked longer than long haul pilots, 10.6 hours versus 9.8 hours, respectively (Gander et al., 1998a).

Recent surveys have shown that the challenge of fatigue in short haul pilots is continuing to worsen with new operational requirements in both scheduled and low-cost carriers (Jackson & Earl, 2006). Due to economic demands, short haul operations are often scheduled with short turn-around times between flights, resulting in an increase in take-offs and landings and additional time constraints contributing to an increased workload. Short haul pilots have reported that schedules consisting of 4–5 legs of flight are one of the more fatiguing schedules to fly (Bourgeois-Bougrine et al., 2003). Short haul pilots also report schedules that require early morning report times across multiple days and extended duty days as contributing to fatigue levels (Bourgeois-Bougrine et al., 2003; Powell et al., 2007).

Pilots beginning their duty during the early morning hours are required to awaken during, or very close to, their early morning circadian low, when sleep propensity is high. As shown by Powell and colleagues, it is not uncommon for pilots to have start times prior to 07:00 (Powell et al., 2007). Such an early morning awakening can truncate the sleep period because of the circadian challenge of trying to initiate sleep earlier than the "usual bedtime" on the evening prior. Sleep log and actigraphy data collected in a sample of pilots over a 15-day period have shown that this sleep prior to a duty day can be reduced by almost 2 hours (Roach et al., 2002). Consequently, pilots are beginning their flight with a sleep debt, and if sleep is truncated across consecutive days, the debt can be cumulative. The sleep debt can also be exacerbated by extended duty days that involve work hours that continue into biological night. Thus, short haul schedules that require combinations of early start times and late evening (or early morning) end times make it difficult to maintain a regular sleep-wake cycle and reduce the opportunities available for recovery sleep. Chronically reducing sleep is associated with cumulative performance decrements (Figure 13-2).

Long Haul Operations

Fatigue in long haul flight crews has been extensively studied using both subjective and objective measures. The occurrence of microsleeps in cockpit crews during flight has been documented (Rosekind et al., 1994; Samel et al., 1997; Wright & McGown, 2001; Bourgeois-Bougrine et al., 2003) and multiple factors have been shown to impair alertness and performance during long haul operations. Long haul pilots frequently report jet lag from circadian misalignment and flying during nighttime periods as the main contributors to their fatigue (Bourgeois-Bougrine et al., 2003).

Crossing multiple time zones is a common occurrence in long haul operations, resulting in jet lag even in seasoned flight crews (Klein & Wegmann, 1980). Long haul operations also require pilots to fly on the backside of the clock, performing during times when they would normally be asleep. The misalignment of the biological clock relative to the new environment can be similar to the circadian disruption and sleep loss experienced by night shift workers. Pilots report the majority of their flying to occur between the hours of midnight and 6:00 a.m. (Bourgeois-Bougrine et al., 2003). Schedules that involve outbound and inbound flights occurring during nighttime hours with daytime layover periods were reported to be particularly fatiguing. When flights begin at night, continuous hours of wakefulness are increased, especially if the pilot is not able to take an afternoon nap. They may awaken in the morning and remain awake all day before their duty day starts that evening for a night flight. Studies have found that it is not uncommon for long-haul pilots to be awake longer than 20 hours, especially on outbound flight segments (Samel et al., 1997; Gander et al., 1998b). Therefore, it is important to consider the hours since awakening in scheduling and not rely on duty time hours.

The rate at which one adapts to transmeridian travel has been shown to depend on the number of time zones crossed and the direction of travel. Adjustment to eastward travel tends to be slower than adjustment to westward travel, due to the former requiring shortening of the circadian period (i.e., reducing the day to less than 24 hours) while the latter requires some lengthening of the circadian period (Boulos et al., 1995; Sack et al., 2007). Because circadian adaptation is not likely to occur during long haul trips of less than 3 days, pilots can minimize circadian disruption by keeping their sleep-wake schedule as close to their home time zone as possible (Lowden & Akerstedt, 1998; Lowden & Åkerstedt, 1999). Studies have shown that the exact timing of the circadian clock and adaptation rate with multiple time zone crossings cannot be easily predicted (Gander et al., 1998b), therefore, it is difficult to apply a prescriptive formula that calculates the exact number of days needed for circadian adjustment in long haul crews without considering multiple factors (physiological, environmental, and operational).

Ultra Long Range Operations

Ultra Long Range (ULR) Flights, which are greater than 16 hours block-to-block, can pose special challenges to performance relative to sleep need and circadian biology. Duty times (preflight, flight, and postflight activities) approach 20 hours during many ULR operations, which makes some circadian misalignment and extended wakefulness unavoidable. It may seem logical that the best way to mitigate fatigue during ULR operations is by increasing the duration of time flight crew members have available for sleep in the airplane bunk, since longer duration flights potentially offer more opportunities for consolidated sleep periods. However, studies conducted during ULR flights, in which bunk sleep was assessed by electroencephalographic measures, have shown that providing longer or more frequent bunk sleep opportunities during flight does not guarantee an increase in sleep duration (Ho et al., 2005). Not surprisingly, when given a 7-hour crew rest period in the bunk, sleep times averaged only 3.27 hours (Signal et al., 2003). It appears therefore that URL pilots are sleeping less than 50% of the time during scheduled rest periods, which is equivalent to a long nap more so than a consolidated daily sleep period.

Time since awakening when beginning the duty day, occurrence of nighttime flying and circadian disruption associated with ULR operations are all factors sensitive to the precise scheduling of

ULR flights. Preliminary data collected during a simulated ULR flight showed that nighttime departures are associated with significantly more performance lapses and slowed reaction times than morning departures (Caldwell et al., 2006). The nighttime group, which departed after having already been awake for at least 13.5 hours, was especially impaired during the first half of the flight, during times at which homeostatic and circadian factors combined to produce particularly low levels of alertness. The daytime group, which departed after having been awake only for about 3.5 hours, evidenced much less impairment throughout the flight, although it was clear that towards the latter portions, as continuous hours of wakefulness increased, performance decrements emerged. While departing during biological night may allow an opportunity for a nap during the afternoon period when napping is biologically possible (Carskadon, 1989; Dinges, 1989), there is no guarantee that flight crews will take advantage of this napping opportunity or that they will be able to obtain sleep (many people cannot nap). Sleep opportunities at the ULR city destination are likely to be in direct opposition to the biological clock, thus reducing the ability to obtain restorative recovery sleep post-flight or before another flight leg. Therefore, it is important that scheduled layovers have some built-in sleep time redundancy to ensure multiple sleep opportunities to maximize total sleep time prior to the return flight. Studies have shown that total sleep time per 24 hours, even when it is obtained as a combination of a major sleep period and a nap (i.e., a split-sleep schedule) is the dominant determinant of performance capability (Mollicone et al., 2008). The effects of the circadian disruption also need to be considered in the timing of take-off and landing, both critical phases of flight.

The advancement of technology that provides aircraft with the capability of flight beyond 16 hours has exceeded the work that has occurred in updating the CFRs. Current CFRs do not address the increase in flight duration of ULR flights and the FAA is currently using a case-by-case approach to approve carrier specific city pairs by issuing a non-standard operations specification paragraph (OpSpec A332) "Ultra Long Range (ULR) Flag Operations in Excess of 16 hours Block-to-Block Time." As part of OpSpec A332, carriers are required to "optimize opportunities for flight crew members to obtain adequate rest to safely perform their operational duties during all phases of this ULR trip sequence." ULR flights have only been flying since 2006 and are considered relatively new in the industry, therefore operational evaluations are necessary to determine how fatigue levels are influenced by actual ULR schedules, allowing for the development of the most effective and appropriate fatigue management approaches specific to ULR operations.

INNOVATIVE APPROACHES TO OPERATIONAL MANAGEMENT OF FATIGUE

Scheduling Approaches

Prescriptive and Nonprescriptive Regulations

Research documenting the fatigue risks associated with sleep restriction, sleep loss, and circadian disruption and the effects of the interaction between the sleep and circadian systems was nonexistent in the 1930s when duty and flight time regulations were originally developed. In fact, the dominant model of fatigue at the time the regulations were created was a time-on-task theory that attributed all fatigue to prolonged periods of work, rather than to sleep loss and circadian misalignment, which are now known to be the primary causes of fatigue in commercial aviation. Although a large body of scientific research in the past 40 years has established that the interaction of sleep need and circadian dynamics determines fatigue levels in otherwise healthy individuals such as pilots, the prescriptive scheduling regulations from the 1930s remain in place today with very few changes. In other words, the federal regulations and practices of the industry have not utilized the large body of science on causes, consequences and prevention of fatigue that has been generated by public funds (through DOT, NASA, NIH, and DOD). The prevalence of fatigue in today's aviation operations, as shown by numerous operational evaluations and research studies, clearly demonstrates that the current prescriptive approaches do not adequately address sleep and circadian challenges associated with 24/7 operations, nor do they provide operational flexibility. This has resulted in increasing interest in evidence-based nonprescriptive approaches for the management of fatigue in aviation operations. Two examples of this approach are discussed below.

Fatigue Risk Management Systems

Fatigue Risk Management Systems (FRMS) potentially offer non-prescriptive approaches for addressing the complexity of aviation operations and fatigue challenges associated with aviation operations. FRMS strives to be evidence-based and it includes a combination of processes and procedures that can be employed within an existing Safety Management System (Signal et al., 2006) for the measurement, mitigation, and management of fatigue risk. FRMS programs provide an interactive way to address performance and safety levels of operations on a case-by-case basis—in that sense they are more idiosyncratic than general systems.

Aviation carriers, regulators and groups worldwide are addressing the benefits of incorporating FRMS programs into current aviation operations (recent examples include Easyjet, Air New Zealand, United Airlines, Continental Airlines, Australia Civil Aviation Safety Authority, Flight Safety Foundation and ICAO, etc.). Preliminary data from Easyjet's FRMS program have demonstrated its effectiveness in reducing fatigue (Stewart 2006). Air New Zealand has been a leader in demonstrating the operational benefits of an FRMS as presented at the recent FAA Symposium, "Aviation Fatigue Management Symposium: Partnerships for Solutions" on June 17–19, 2008, in Vienna, Virginia. Based on the available research and demonstrated success of such programs, the FAA is also exploring ways to implement effective FRMS based programs in operations where fatigue has been identified as an inherent risk. An FRMS offers a way to more safely conduct flights beyond existing regulatory limits and is one alternative to prescriptive flight and duty time and rest period regulations.

Biomathematical Scheduling Tools

The need for predictive modeling algorithms as part of scheduling tools that focus on minimizing fatigue related errors, incidents and accidents during flight operations is becoming more apparent. These scheduling tools are being developed with the goal of quantifying the impact of underlying interaction of sleep and circadian physiology on performance levels. The foundation of most of these models is based on the two-process model of sleep regulation (Achermann, 2004). A number of major efforts are underway internationally that focus on the application of biomathematical modeling in a software package in order to: (1) predict the times that neurobehavioral functions and performance will be maintained; (2) establish time periods for maximal recovery sleep; and (3) determine the cumulative effects of different work-rest schedules on overall performance (Mallis et al., 2004). These models can be very useful in the prediction of performance or effectiveness levels when comparing different operational schedules, particularly in evaluating the placement and timing of critical phases of flight and the occurrence of in-flight rest periods and opportunities for layover sleep. While most of the biomathematical modeling software shows promise in the prediction of performance, there is still a critical need to establish their validity and reliability, as well as their sensitivity and specificity based on real world data (Dinges, 2004).

Appropriate Countermeasure Applications

Cockpit Napping

Numerous studies have shown that prophylactic naps (also called "power" naps) can significantly improve alertness and performance when taken during long periods of otherwise continuous wakefulness (Dinges et al., 1987; Dinges, 1989; Rosa 1993; Driskell and Mullen, 2005). Based on the widely known NASA cockpit napping study (Rosekind et al., 1994), which provided direct evidence that a 40-min cockpit nap opportunity (resulting in an average 26-min nap) was associated with significant improvements in subsequent pilot physiological alertness and psychomotor performance, multiple international carriers have implemented cockpit napping into their everyday operations (e.g., Air Canada, Air New Zealand, British Airways, Emirates, Finnair, Lufthansa, Swissair, Qantas). However, cockpit napping is not currently sanctioned by the FAA for use by U.S. carriers, despite the fact that the United States paid for the research to demonstrate its efficacy (Rosekind et al., 1994). This paradox highlights one of the barriers to implementing practical fatigue countermeasures—a reluctance to acknowledge that fatigue affects airline pilots.

Although cockpit napping is not currently sanctioned by the FAA, survey results suggest that flight crew are already using cockpit napping as a strategy when fatigue countermeasures options are limited (Rosekind et al., 2000) and this has been observed during long haul operational research (Gander et al., 1991). Naps can help flight crew to maintain alertness and performance levels, particularly during unexpected delays, which can result in delays of the next rest opportunity. However, cockpit napping should not be used in place of bunk sleep and is not a substitute for the responsibility of the flight crew to try to obtain sufficient amounts of recovery sleep in preparation for flight operations.

Activity Breaks

Activity breaks, especially when taken during early morning periods around the circadian nadir, have limited effectiveness as a countermeasure for maintaining alertness levels. Research has shown that pilots, when studied in a high fidelity flight simulator during a nighttime flight, showed significantly less physiological sleepiness up to 15 min after a 7-min activity break and significantly greater subjective alertness up to 25 min after the break (Neri et al., 2002). Short activity breaks can improve alertness by relieving the pilot temporarily from the vigilance task and monotony of monitoring a highly automated cockpit. However, they can also be used to reduce accumulating workload by distributing the performance tasks throughout the work period. Activity breaks allow for a disengagement from the operational task, potentially including some physical activity and providing an opportunity to increase social interaction. Research has shown that sleepiness ratings are significantly greater when frequent breaks are not taken (Neri et al., 2002). Whether engaging in a break involving activity or simply taking advantage of the opportunity to disengage from the current task, the available data suggest that the resulting increase in social interaction often accompanying a work break can play a role in maintaining alertness, especially during the early morning hours around the circadian nadir (Dijkman et al., 1997).

Caffeine as a Countermeasure

There has been extensive research on use of caffeine as a fatigue countermeasure to increase alertness and improve performance levels (Van Dongen et al., 2001; Bonnet et al., 2005). After consumption of caffeine, it takes 15–30 min to enter the bloodstream. Therefore, the alertness effects are not seen instantaneously, but can persist up to 4 to 5 hours after ingestion. Caffeine (contained in common beverages such as soft drinks, coffee, and tea) is commonly used by pilots as a countermeasure to help maintain alertness levels. Short-term use of caffeine will prevent undesirable effects (e.g., tolerance, stomach difficulties, and increased blood pressure). Individual differences exist in how people respond to caffeine (both on performance and the degree to which it affects sleep structure when taken too close to bedtime) and should be explored prior to use as an operational countermeasure during flight operations.

Future Technologies for Fatigue Management in Aviation

Unobtrusive, objective ways to detect fatigue in human operators have begun to be the focus of considerable research on technologies that validly and reliably predict, detect, and/or prevent performance risks due to fatigue. The realization of this technology in commercial aviation is inevitable. The generation coming into power over the next 10–20 years grew up immersed in technology. They accept human-machine interaction in nearly all aspects of life. In their minds, the computer should be sentient-like, in that it should read human intentions, anticipate human actions, and do other things that enhance human capability. Those expectations will bring the emergence of ever-more sophisticated human-machine interfaces, which will undoubtedly change the nature of human work in all transportation modes, including commercial aviation. Fatigue is an area where such human-machine interfaces can have a profound effect by preventing, predicting, detecting and mitigating fatigue-related risks (Figure 13-4).

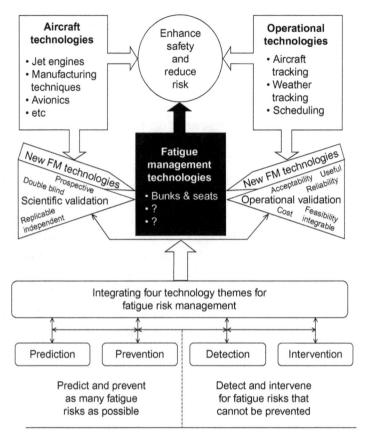

FIGURE 13-4 Key elements of a technology-based Dynamic Fatigue Risk Management System. Fatigue management (FM) technologies have lagged behind aircraft technologies and operational technologies in efforts to enhance safety and reduce risk in commercial aviation. It is now possible to develop the former to aid in prevention of fatigue, prediction of when fatigue is most likely, detection of fatigue during operations, and interventions to reduce fatigue or its risks when it occurs.

The following three concepts provide a framework for thinking about how technologies for fatigue management might be integrated into commercial aviation.

1. The integration and validation of new scientific information on human fatigue and its mitigation, and on technologies that predict and detect fatigue could form the basis for a dynamic fatigue risk management system that can be adapted to changing operational needs and idiographic factors that contribute to fatigue risk.

2. A system approach based on integrated components that are scientifically valid and operationally practical might emphasize prevention, prediction, detection, and intervention to dynamically manage fatigue and risk (Figure 13-4).
3. The empirical development and validation of system components requires both evidence of positive benefits that exceed current practices, and evidence that there are no unwanted consequences to safety, costs, or personnel. It should include the best available information on the biology of fatigue and its risk mitigation with novel technologies.

Perhaps the most compelling argument for the development of fatigue management technologies in commercial aviation is the fact that no matter what scheduling limits are placed on commercial aviation, the circadian and sleep-dependent nature of fatigue ensures it will occur in some operations—such as night flights and transmeridian flight schedules. However, this fact opens up the possibility of predicting when fatigue will occur, using mathematical models validated on sleep and circadian dynamics relative to performance (Dinges, 2004; Mallis et al., 2004). While advances in aviation technology (e.g., avionics, jet engines) and operational technology (e.g., tracking of aircraft and weather) have given air travel a good safety record, people (flight crews, maintenance personnel, air traffic controllers) remain at the heart of a safe air transit system. In this sense, the safety of commercial aviation remains human-centered. Fatigue management is designed to prevent, detect, and reduce fatigue as a risk factor in a human-centered, safety-sensitive industry. However, fatigue management technology should be more than quality of seats and bunks for crew rest in airplanes (Figure 13-4), which along with regulated duty-hour limits have been low-tech approaches to managing fatigue in flight crews.

Criteria for identifying human-centered technologies that predict and/or detect fatigue in flight operations have been detailed, but first and foremost is the requirement that they meet systematic scientific validity (Dinges & Mallis, 1998). This should include double-blind testing of the accuracy of a given technology relative to a gold-standard performance-based measure of fatigue, and assurance that it is accurate when used in every person. This necessitates swift elimination of invalid approaches. One cost-effective strategy to getting to the most valid technologies is to leverage what has already been discovered by research supported through other federal agencies. The biology of fatigue is common to all occupations, and some discoveries in other transportation modalities can be applied to aviation.

A scientifically valid fatigue management technology should then undergo operational validation, which refers to the extent to which a technology is feasible, reliable, and acceptable by operators in an operational environment. For example, it is obvious that a scientifically valid fatigue-detection technology must be deployable in an aircraft cockpit if it is intended to be used by pilots. Pilots must also perceive the feedback from the technology to be useful to them in managing their fatigue. The technology must work reliably, and have both high sensitivity (i.e., detect fatigue) and high specificity (i.e., detect primarily fatigue). Finally, to be used, it must be as unobtrusive as possible.

Fatigue management technologies that have potential for use in commercial aviation include the following: (1) technologies that predict the occurrence and severity of fatigue and as such can be used to create schedules that are more fatigue-management friendly; (2) technologies that help deliver education on fatigue management and optimal countermeasure use to individuals; (3) technologies in the workplace (on the operator or embedded in the work system) that detect when an individual is showing signs of fatigue; and (4) intervention technologies that help people to be more alert and free of fatigue. We now discuss two of the more promising areas for fatigue management technologies—those that predict fatigue and those that detect fatigue.

Fatigue Prediction Technologies

Human performance (e.g., alertness, attention, working memory, problem solving, reaction time, situational awareness, risk taking, etc.) is dynamically controlled by the interaction of waking biological processes sensitive to time awake, sleep quantity, and circadian phase (Durmer & Dinges, 2005; Van Dongen & Dinges, 2005). Although the effects of time awake and sleep duration can be modeled as near-linear processes within and between days, the circadian interaction with these processes makes the prediction of performance nonlinear. For example, when remaining awake for 40 hours, it is a counterintuitive fact that fatigue and performance deficits are worse at 24 hours than at 40 hours awake. The circadian system also influences the duration of recovery sleep that is possible to achieve, and the circadian system is slow to adapt to sleep in new time zones (see Figure 13-3). It is this nonlinearity that makes inadequate and imprecise many work-hour limits based solely on a linear model of fatigue (i.e., the longer one works the more fatigued one will become). This nonlinearity in the brain's performance capability over time is the reason that developing mathematical models that predict performance is increasingly regarded as essential.

Mathematical models of fatigue prediction are the fatigue management technologies that have received the most attention in the past 15 years thanks to interest and support from DOD (Jewett et al., 1999), NASA and DOT (Neri, 2004). Based on the dynamic interaction of the human sleep homeostatic drive and circadian rhythms, some of these mathematical models have advanced to the critical point of integrating individual differences into the modeling predictions for a more accurate estimate of the timing and magnitude of fatigue effects on individuals (Van Dongen et al., 2007), which should facilitate more precise use of countermeasures (e.g., naps, recovery sleep, caffeine intake).

Fatigue Detection Technologies

There are three scientifically based reasons why objective fatigue detection technologies are needed in safety-sensitive operations such as commercial aviation: (1) Humans are often unable to accurately estimate how variable or uneven their alertness and performance have become due to inadequate sleep or working at night. When fatigued they tend to estimate their alertness based by their best responses and ignore their worse responses. (2) Performance deficits from fatigue accumulate over days to high levels when recovery sleep is chronically inadequate (Belenky et al., 2003; Van Dongen et al., 2003). Awareness of these cumulative deficits appears to be less accurate as performance declines (Van Dongen et al., 2003). (3) While everyone eventually develops performance deficits from fatigue, some people do so very rapidly while others take much longer, and these differences appear to be stable characteristics of people (Van Dongen et al., 2004) and therefore they may reflect biological differences among them (Viola et al., 2007). There are currently no reliable biomarkers for one's performance vulnerability to fatigue, making detection of fatigue a primary goal—although the discovery of such may be useful in fatigue prevention strategies.

Fatigue detection technologies have been of interest to the DOT for some time. A decade ago, the National Highway Traffic Safety Administration (NHTSA) and Federal Motor Carrier Safety Administration (FMCSA) funded a laboratory study that systematically evaluate the validity of the "most promising" fatigue detection technologies, which included brain wave (EEG) measures, eye blink devices, a measure of slow eyelid closures (called PERCLOS), and a head position sensor. In a number of highly controlled, double-blind experiments, the technologies were evaluated to determine the extent to which each technology detected the alertness of subjects over a 40-hour period, as measured by lapses of attention on the Psychomotor Vigilance Test (PVT)—a well validated and highly sensitive measure of the effects of fatigue on neurobehavioral alertness (Dorrian et al., 2005). Only PERCLOS reliably and accurately tracked PVT lapses of attention in all subjects, outperforming not only all the other technologies, but also subjects' own ratings of their fatigue and alertness (Dinges et al., 1998; Dinges et al., 2002).

Subsequently, a group of technologies that included an infrared-based PERCLOS monitor, were evaluated in an over-the-road study of commercial drivers, to determine whether feedback from fatigue detection technologies would help truck drivers maintain their alertness in actual working conditions (Dinges et al., 2005b). The details of this study are extensive and need not be reviewed here, but suffice it to say that the infrared PERCLOS monitor did not perform well due to environmental factors (ambient light) and operator behavior (head turning to view mirrors). However, a technique is now being developed for NASA that involves optical computer recognition (machine vision) of the human face to identify expressions of stress and fatigue (Dinges et al., 2005b; Dinges et al., 2007). This system has a number of advantages. It requires no sensor or conspicuous technology, it can track the face as it moves in 3-dimensional space, and it can process information online in real time.

In the over-the-road study of the effects of feedback from fatigue-detection technologies on commercial truck drivers, it was expected that when the technologies signaled a driver was drowsy it would result in the driver taking countermeasures, including stopping to rest or nap. However this rarely happened. On the other hand, it was found that the drivers felt the fatigue detection devices (and the PVT test they performed in the middle and at the end of each trip) informed them of their fatigue levels and prompted them to acquire more sleep on their days off duty. Both the debrief interviews with the drivers, as well as the actiwatch data acquired on them confirmed that they increased their sleep by an average of 45 minutes on days off duty (Dinges et al., 2005a). This is a remarkable and unexpected outcome, and it suggests another purpose for fatigue detection technologies in the workplace—namely, to urge operators to sleep more during off-duty periods. Recent research efforts underway for NIH and NASA on recovery sleep following a period of sleep restriction reveal that getting extra sleep during off-duty periods and days off work is one of the most important fatigue countermeasures—but it will only be effective if sufficient time is permitted for sleep off duty. If a fatigue management technology could be used to teach people to use their downtime to sleep more, the risk of fatigue could be reduced substantially, for it is known that in the U.S. population as a whole, work duration is the primary activity that is reciprocally related to sleep duration (Basner et al., 2007).

CONCLUSIONS

Although aviation technology that allows travel over multiple time zones in a single day is a major advance, it poses substantial challenges to sleep and circadian physiology, which can result in flight crew fatigue. Operational demands resulting in extended work days, increased workload levels, reduced sleep opportunities and circadian disruption continue to pose significant challenges during aviation operations. The homeostatic drive for sleep and the circadian clock interact in a complex manner that is nonlinear, rather than the simplistic linear models that dominate prescriptive duty time regulations for flight crews.

The research summarized in this chapter reveals that many commercial aviation industry practices induce fatigue via sleep loss and circadian misalignment in flight crews, and that aspects of the current federal prescriptive approaches to fatigue prevention in commercial aviation appear to be archaic. Current prescriptive approaches do not adequately address sleep and circadian challenges associated with 24/7 operations, nor do they provide operational flexibility. This has resulted in increasing interest in evidence-based nonprescriptive approaches for the management of fatigue in aviation operations. Multicomponent approaches that are based in science and provide interactive ways to address performance and safety levels during flight and allow for flexibility will contribute to the overall safety of aviation operations.

There is a need to develop approaches that exceed current fatigue management practices, by implementation of scientifically valid and operationally feasible technologies that adjust, in real-time, to an individual's fatigue level and provide an intervention to help manage the fatigue in flight crew members. While it may not be possible at this time to know which fatigue management technologies will be most useful and acceptable in commercial aviation, it is fairly certain that in order for valid technologies to be used, they must not violate the privacy rights of individuals. It is for this reason that fatigue management technologies in commercial aviation should first be developed as personal aids. Moreover, these technologies should be used responsibly—they are not a substitute for reasonable working conditions. If these principles are followed, information from fatigue management technologies can help people involved in commercial aviation be less fatigued and more alert, and that this is an achievable goal worthy of our best efforts.

ACKNOWLEDGMENT

The substantive evaluation on which this chapter is based was supported by NIH grant NR04281; by the National Space Biomedical Research Institute through NASA NCC 9–58; and by the Institute for Experimental Psychiatry Research Foundation.

References

Achermann, P. (2004). The two-process model of sleep regulation revisited. *Aviation, Space and Environmental Medicine, 75*(3 Suppl), A37–A43.

Basner, M., Fomberstein, K. M., Razavi, F. M., Banks, S., William, J. H., Rosa, R. R., & Dinges, D. F. (2007). American time use survey: sleep time and its relationship to waking activities. *Sleep, 30*(9), 1085–1095.

Belenky, G., Wesensten, N. J., Thorne, D. R., Thomas, M. L., Sing, H. C., Redmond, D. P., Russo, M. B., & Balkin, T. J. (2003). Patterns of performance degradation and restoration during sleep restriction and subsequent recovery: a sleep dose-response study. *Journal of Sleep Research, 12*, 1–12.

Bonnet, M. H., Balkin, T. J., Dinges, D. F., Roehrs, T., Rogers, N. L., & Wesensten, N. J. (2005). The use of stimulants to modify performance during sleep loss: A review by the Sleep Deprivation and Stimulant Task Force of the American Academy of Sleep Medicine. *Sleep, 28*(9), 1163–1187.

Boulos, Z., Campbell, S. S., Lewy, A. J., Terman, M., Dijk, D. J., & Eastman, C. I. (1995). Light treatment for sleep disorders: consensus report. VII. Jet lag. *Journal of Biological Rhythms, 10*(2), 167–176.

Bourgeois-Bougrine, S., Carbon, P., Gounelle, C., Mollard, R., & Coblentz, A. (2003). Perceived fatigue for short- and long-haul flights: a survey of 739 airline pilots. *Aviation, Space, and Environmental Medicine, 74*(10), 1072–1077.

Cajochen, C., Zeitzer, J. M., Czeisler, C. A., & Dijk, D.-J. (2000). Dose-response relationship for light intensity and ocular and electroencephalographic correlates of human alertness. *Behavioral Brain Research, 115*, 75–83.

Caldwell, J. A., M., M. M., Colletti, L. M., Oyung, R. L., Brandt, S. L., Arsintescu, L., DeRoshia, C. W., Reduta-Rojas, D. D., & Chapman, P. M. (2006). *The effects of ultra-long-range flights on the alertness and performance of aviators (NASA Technical Memorandum 2006-213484).* Moffett Field, CA: NASA Ames Research Center.

Carskadon, M. A. (1989). Ontogeny of human sleepiness as measured by sleep latency. In D. F. Dinges & R. J. Broughton (Eds.), *Sleep and alertness: Chronobiological, behavioral, and medical aspects of napping* (pp. 53–69). New York: Raven Press.

Caruso, C. C., Bushnell, T., Eggerth, D., Heitmann, A., Kojola, B., Newman, K., Rosa, R. R., Sauter, S. L., & Vila, B. (2006). Long working hours, safety, and health: Toward a national research agenda. *American Journal of Industrial Medicine, 49*(11), 930–942.

Caruso, C. C., Hitchcock, E. M., Dick, R. B., Russo, J. M., & Schmit, J. M. (2004). *Overtime and extended work shifts: Recent findings on illnesses, injuries, and health behaviors*. National Institute for Occupational Safety and Health, Centers for Disease Control and Prevention, U. S. Department of Health and Human Services.

Czeisler, C. A., Duffy, J. F., Shanahan, T. L., Brown, E. N., Mitchell, J. F., Rimmer, D. W., Ronda, J. M., Silva, E. J., Allan, J. S., Emens, J. S., Dijk, D. J., & Kronauer, R. E. (1999). Stability, precision, and near-24-hour period of the human circadian pacemaker.. *Science, 284*(5423), 2177–2181.

Czeisler, C. A., & Khalsa, S. B. S. (2000). The human circadian timing system and sleep-wake regulation. In M. H. Kryger, T. Roth, & W. C. Dement (Eds.), *Principles and practice of sleep medicine* (pp. 353–376). Philadelphia: W. B. Saunders Company.

Dawson, D., & Reid, K. (1997). Fatigue, alcohol and performance impairment. *Nature, 388*(6639), 235.

Dijk, D. J., & Czeisler, C. A. (1994). Paradoxical timing of the circadian rhythm of sleep propensity serves to consolidate sleep and wakefulness in humans. *Neuroscience Letters, 166*(1), 63–68.

Dijk, D. J., & Czeisler, C. A. (1995). Contribution of the circadian pacemaker and the sleep homeostat to sleep propensity, sleep structure, electroencephalographic slow waves, and sleep spindle activity in humans. *Journal of Neuroscience, 15,* 3526–3538.

Dijkman, M., Sachs, N., Levine, E., Mallis, M., Carlin, M. M., Gillen, K. A., Powell, J. W., Samuel, S., Mullington, J., Rosekind, M. R., & Dinges, D. F. (1997). Effects of reduced stimulation on neurobehavioral alertness depend on circadian phase during human sleep deprivation.. *Sleep Research, 26,* 265.

Dinges, D. F. (1989). Napping patterns and effects in human adults. In D. F. Dinges & R. J. Broughton (Eds.), *Sleep and alertness: Chronobiological, behavioral, and medical aspects of napping* (pp. 171–204). New York: Raven Press.

Dinges, D. F. (1995). An overview of sleepiness and accidents. *Journal of Sleep Research, 4*(Suppl. 2), 4–14.

Dinges, D. F. (2004). Critical research issues in development of biomathematical models of fatigue and performance. *Aviation, Space and Environmental Medicine, 75*(3), A181–A191.

Dinges, D. F., Graeber, R. C., Rosekind, M. R., Samel, A., & Wegmann, H. M. (1996). *Principles and guidelines for duty and rest scheduling in commercial aviation*. Ames Research center, Moffett Field, CA, National Aeronautics and Space Administration 1–10.

Dinges, D. F., Maislin, G., Brewster, R. M., Krueger, G. P., & Carroll, R. J. (2005a). Pilot test of fatigue management technologies. *Safety: Older Drivers; Traffic Law Enforcement; Management; School Transportation; Emergency Evacuation; Truck and Bus; and Motorcycles*(1922), *6,* 175–182.

Dinges, D. F., & Mallis, M. M. (1998). Managing fatigue by drowsiness detection: Can technological promises be realized?. In L. Hartley (Ed.), *Managing fatigue in transportation* (pp. 209–229). Oxford, Pergamon.

Dinges, D. F., Mallis, M. M., Maislin, G., Powell, J. W. (1998). Evaluation of techniques for ocular measurement as an index of fatigue and the basis for alertness management. U.S. Department of Transportation, National Highway Traffic Safety Administration, Contract No. DTNH22-93-D-07007.

Dinges, D. F., Orne, M. T., Whitehouse, W. G., & Orne, E. C. (1987). Temporal placement of a nap for alertness: Contributions of circadian phase and prior wakefulness. *Sleep*, *10*(4), 313–329.

Dinges, D. F., Price, N. J., Maislin, G., Powell, J. W., Ecker, A. J., Mallis, M. M. and Szuba, M. P. (2002). Prospective Laboratory Re-Validation of Ocular-Based Drowsiness Detection Technologies and Countermeasures. Subtask A in Report. W. W. Wierwille, R. J. Hanowski, R. L. Olson, D. F. Dinges, N. J. Price, G. Maislin, J. W. Powell, A. J. Ecker, M. M. Mallis, M. P. Szuba, E. Ayoob, R. Grace and A. Steinfeld, NHTSA Drowsy Driver Detection And Interface Project, DTNH 22-00-D-07007; Task Order No. 7.

Dinges, D. F., Rider, R. L., Dorrian, J., McGlinchey, E. L., Rogers, N. L., Cizman, Z., Goldenstein, S. K., Vogler, C., Venkataraman, S., & Metaxas, D. N. (2005b). Optical computer recognition of facial expressions associated with stress induced by performance demands. *Aviation Space and Environmental Medicine*, *76*(6), B172–B182.

Dinges, D. F., Venkataraman, S., McGlinchey, E. L., & Metaxas, D. N. (2007). Monitoring of facial stress during space flight: Optical computer recognition combining discriminative and generative methods. *Acta Astronautica*, *60*(4-7), 341–350.

Doran, S. M., Van Dongen, H. P., & Dinges, D. F. (2001). Sustained attention performance during sleep deprivation: Evidence of state instability. *Archives Italiennes de Biologie*, *139*(3), 253–267.

Dorrian, J., Rogers, N. L., & Dinges, D. F. (2005). Psychomotor Vigilance Performance: Neurocognitive Assay Sensitive to Sleep Loss. In C. Kushida (Ed.), *Sleep Deprivation: Clinical Issues, pharmacology and sleep loss effects* (pp. 39–70). New York, NY: Marcel Dekker, Inc.

Driskell, J. E., & Mullen, B. (2005). The efficacy of naps as a fatigue countermeasure: A meta-analytic integration. *Human Factors*, *47*(2), 360–377.

Durmer, J. S., & Dinges, D. F. (2005). Neurocognitive consequences of sleep deprivation. *Seminars in Neurology*, *25*(1).

Gander, P. H., Graeber, R. C., Connell, L. J., et al. (1991). Crew factors in flight operations VIII: Factors influencing sleep timing and subjecive sleep quality in commercial long-haul flight crews. *Moffett Field, CA, NASA Ames Research Center*, *6*.

Gander, P. H., Gregory, K. B., Graeber, R. C., Connell, L. J., Miller, D. L., & Rosekind, M. R. (1998a). Flight crew fatigue II: short-haul fixed-wing air transport operations. *Aviation Space & Environmental Medicine*, *69*(9 Suppl), B8–B15.

Gander, P. H., Gregory, K. B., Miller, D. L., Graeber, R. C., Connell, L. J., & Rosekind, M. R. (1998b). Flight crew fatigue V: long-haul air transport operations. *Aviation Space & Environmental Medicine*, *69*(9 Suppl), B37–B48.

Ho, P., Landsberger, S., Signal, L., Singh, J., & Stone, B. (2005). The Singapore experience: Task force studies scientific data to assess flights. *Flight Safety Digest*, 20–40.

Jackson, C. A., & Earl, L. (2006). Prevalence of fatigue among commercial pilots. *Occupational medicine (Oxford, England)*, *56*(4), 263–268.

Jewett, M. E., Borbely, A. A., & Czeisler, C. A. (1999). Proceedings of the workshop on biomathematical models of circadian rhythmicity, sleep regulation, and neurobehavioral function in humans. *Journal of Biological Rhythms, 14*(6), 429–630.

Klein, K. E., & Wegmann, H. M. (1980). Significance of circadian rhythms in aerospace operations. *France, Neuilly-Sur-Seine*.

Knauth, P. (2007). Extended work periods. *Industrial Health, 45*(1), 125–136.

Lim, J., & Dinges, D. F. (2008). Sleep deprivation and vigilant attention1129. In: *Molecular and biophysical mechanisms of arousal, alertness, and attention*. Annals of the New York Academy of Sciences (pp. 305–322). 1129

Lowden, A., & Akerstedt, T. (1998). Retaining home-base sleep hours to prevent jet lag in connection with a westward flight across nine time zones. *Chronobiology International, 15*(4), 365–376.

Lowden, A., & Åkerstedt, T. (1999). Eastward long distance flights, sleep and wake patterns in air crews in connection with a two-day layover. *Journal of Sleep Research, 8,* 15–24.

Mallis, M. M., Mejdal, S., Nguyen, T. T., & Dinges, D. F. (2004). Summary of the key features of seven biomathematical models of human fatigue and performance. *Aviation, Space and Environmenatl Medicine, 75*(3 Suppl), A4–A14.

Macdonald, W., & Bendak, S. (2000). Effects of workload level and 8-versus 12-h workday duration on test battery performance. *International Journal of Industrial Ergonomics, 26*(3), 399–416.

Mignot, E., Taheri, S., & Nishino, S. (2002). Sleeping with the hypothalamus: emerging therapeutic targets for sleep disorders. *Nature Neuroscience, 5*(Suppl), 1071–1075.

Mollicone, D. J., Van Dongen, H. P. A., Rogers, N. L., & Dinges, D. F. (2008). Response surface mapping of neurobehavioral performance: testing the feasibility of split sleep schedules for space operations. *Acta Astronautica, 63*(7-10), 833–840.

Moore, R. Y. (1999). A clock for the ages.. *Science, 284,* 2102–2103.

Nachreiner, F. (2001). Time on task effects on safety. *Journal of Human Ergology, 30*(1-2), 97–102.

Neri, D. (2004). Preface: Fatigue and performance modeling workshop June 13-14. *Aviation, Space & Environmental Medicine, 75*(3), A1–A3.

Neri, D. F., Oyung, R. L., Colletti, L. M., Mallis, M. M., Tam, P. Y., & Dinges, D. F. (2002). Controlled breaks as a fatigue countermeasure on the flight deck. *Aviation, Space and Environmental Medicine, 73*(7), 654–664.

Powell, D. M. C., Spencer, M. B., Holland, D., Broadbent, E., & Petrie, K. J. (2007). Pilot fatigue in short-haul operations: Effects of number of sectors, duty length, and time of day. *Aviation Space and Environmental Medicine, 78*(7), 698–701.

Roach, G. D., Rodgers, M., & Dawson, D. (2002). Circadian adaptation of aircrew to transmeridian flight.. *Aviation Space & Environmental Medicine, 73*(12), 1153–1160.

Rosa, R. R. (1993). Napping at home and alertness on the job in rotating shift workers. *Sleep, 16*(8), 727–735.

Rosekind, M. R., Co, E. L., Graeber, K. B., & Miller, D. L. (2000). Crew factors in flight operations XIII: A survey of fatigue factors in corporate/executive aviation operations. *Moffett Field, CA, NASA Ames Research Center,* 1–83.

Rosekind, M. R., Graeber, R. C., Dinges, D. F., Connell, L. J., Rountree, M. S., Spinweber, C. L., & Gillen, K. A. (1994). Crew factors in flight operations IX: Effects of planned cockpit rest on crew performance and alertness in long-haul operations. *Moffett Field, CA, NASA Ames Research Center*, 1–82.

Sack, R. L., Auckley, D., Auger, R. R., Carskadon, M. A., Wright, K. P., Jr, Vitiello, M. V., & Zhdanova, I. V. (2007). Circadian rhythm sleep disorders: part I, basic principles, shift work and jet lag disorders. An American Academy of Sleep Medicine review. *Sleep, 30*(11), 1460–1483.

Samel, A., Wegmann, H. M., & Vejvoda, M. (1997). Aircrew fatigue in long-haul operations. *Accident Analysis and Prevention, 29*(4), 439–452.

Santhi, N., Horowitz, T. S., Duffy, J. F., & Czeisler, C. A. (2007). Acute sleep deprivation and circadian misalignment associated with transition onto the first night of work impairs visual selective attention. *PLoS ONE, 2*(11), e1233.

Saper, C. B., Chou, T. C., & Scammell, T. E. (2001). The sleep switch: hypothalamic control of sleep and wakefulness. *TRENDS in Neurosciences, 24*(12), 726–731.

Signal, L., Gander, P., & van den Berg, M. (2003). *Sleep during ultra-long range flights: A study of sleep on board the 777-200 ER during rest opportunities of 7 hours.* Wellington, New Zealand: Published for the Boeing Commercial Airplane Group Sleep/Wake Research Centre, Massey University.

Signal, L., Ratieta, D. and Gander, P. (2006). Fatigue management in the New Zealand aviation industry, Australian Transport Safety Bureau Research and Analysis Report.

Stakofsky, A. B., Levin, A. L., Vitellaro, K. M., Dinges, D. F., & Van Dongen, H. P. A. (2005). Effect of cognitive workload on neurobehavioral deficits during total sleep deprivation. *Sleep, 28*, A128–A129.

Stewart, S. (2006). *An integrated system for managing fatigue risk within a low cost carrier.* Paris, France: annual international safety seminar (IASS).

Van Dongen, H. P., Baynard, M. D., Maislin, G., & Dinges, D. F. (2004). Systematic interindividual differences in neurobehavioral impairment from sleep loss: evidence of trait-like differential vulnerability. *Sleep, 27*(3), 423–433.

Van Dongen, H. P., Mott, C. G., Huang, J. K., Mollicone, D. J., McKenzie, F. D., & Dinges, D. F. (2007). Optimization of biomathematical model predictions for cognitive performance impairment in individuals: accounting for unknown traits and uncertain states in homeostatic and circadian processes. *Sleep, 30*(9), 1129–1143.

Van Dongen, H. P. A., & Dinges, D. F. (2005). Circadian Rhythms in Sleepiness, Alertness and Performance. In M. H. Kryger, T. Roth, & W. C. Dement (Eds.), *Principles and practice of sleep medicine fourth edition.* W.B. Saunders Company.

Van Dongen, H. P. A., Maislin, G., Mullington, J. M., & Dinges, D. F. (2003). The cumulative cost of additional wakefulness: Dose-response effects on neurobehavioral functions and sleep physiology from chronic sleep restriction and total sleep deprivation. *Sleep, 26*(2), 117–126.

Van Dongen, H. P. A., Price, N. J., Mullington, J. M., Szuba, M. P., Kapoor, S. C., & Dinges, D. F. (2001). Caffeine eliminates psychomotor vigilance deficits from sleep inertia. *Sleep, 24*(7), 813–819.

Viola, A. U., Archer, S. N., James, L. M., Groeger, J. A., Lo, J. C. Y., Skene, D. J., von Schantz, M., & Dijk, D. J. (2007). PER3 polymorphism predicts sleep structure and waking performance. *Current Biology, 17*(7), 613–618.

Walsh, J. K., Dement, W. C., & Dinges, D. F. (2005). Sleep Medicine, Public Policy, and Public Health. In: *Principles and practice of sleep medicine* (4th ed.). W.B. Saunders. (4th ed.).

Waterhouse, J., Reilly, T., Atkinson, G., & Edwards, B. (2007). Jet lag: trends and coping strategies. *Lancet, 369*(9567), 1117–1129.

Wright, N., & McGown, A. (2001). Vigilance on the civil flight deck: incidence of sleepiness and sleep during long-haul flights and associated changes in physiological parameters. *Ergonomics, 44*(1), 82–106.

PART IV

HUMAN FACTORS IN
AIRCRAFT DESIGN

14

Aviation Displays

Michael T. Curtis and Florian Jentsch
University of Central Florida
3100 Technology Pkwy Suite 100 Orlando

John A. Wise
The Wise Group, LLC; Glendale AZ

INTRODUCTION

In aviation, the transition of cockpit display technology is continuing. In the past, especially from the 1960s to 1980s, the cockpit was a marvel of large, crowded instrument panels filled with an array of interrelated gauges that would make Gestalt theorists shudder. The sheer number of sources of information that necessitated separate display space imposed a vast visual environment resulting in an increased potential for visual and cognitive overload. In stark contrast, today's display technology invokes relief from the crowded instrument panels of the past, and the cockpits of the third millennium look clean and uncluttered.

The implementation of current display technology has reduced the expanse of flight critical information into a minimal number of electronic displays more closely resembling a computer monitor than a mechanical gyroscope. These changes to the cockpit were intended to help to address the overload problems that pilots were encountering.

Yet, despite these improvements, the information load on pilots has not decreased. In fact, if anything, the amount of information available has increased causing the potential for "information glut" to overburden pilots even more (Wiener, 1987). As a result, the source of the overload that pilots are experiencing is changing; high visual workload has changed to high cognitive workload.

In the previous edition of this book, Stokes and Wickens (1989) effectively addressed the issues confronting the then-current status of cockpit displays. They focused primarily on issues related to visual overload. Specifically, they discussed display issues of navigational and flight path displays and design characteristics that can impact information retrieval (e.g., color and sound). Their discussion effectively presented both strengths and weaknesses of design techniques. Whereas display efficiency and visual overload were the dominant topics of concern when Stokes and Wickens reviewed the literature and state-of-the-art of cockpit displays in the 1980s, improvements in computer processing and digital integration have solved many of the problems they discussed, while creating new ones, and, in fact, in some cases reintroducing old problems in new ways.

Within this chapter, we will discuss how advances in display technology have altered the role of human information processing in the cockpit. We will describe both, how these advances have improved pilot performance, and the emerging challenges that resulted. Finally, we will discuss how recent aviation industry initiatives influence current and future display technologies.

FROM VISUAL TO COGNITIVE

In the late 1970s, a revolution in cockpit displays began with the development of the first "glass cockpits." At the time, the state-of-the-art in aviation instrumentation comprised complex configurations of individual displays and input devices that required in excess of 100 individual components which pilots were charged with managing. With each new aircraft, the challenge for cockpit display designers was somehow to find more cockpit display real estate, and to optimize the amount of information displayed in that limited space (Reising, Liggett, & Munns, 1999). Because of technological limitations, each gauge had to represent an individual piece of information. All of which, had to be attended and integrated with other related pieces of information.

From a human-factors perspective, maximizing display efficiency hinged on ensuring that the most critical information was positioned in a manner that required minimal visual scan, and was configured so that related information was proximally spaced. At its best, the cluster of gauges provided a continuously visible reading of aircraft states. More frequently, the cockpit represented a growing overload of visual information in increasingly complex configurations of gauges which also happened to be decreasing in size due to limited space. All in all, the visual complexity in the cockpit had evolved to a state in which it was increasingly difficult to display, and subsequently monitor, everything effectively.

Just when instrument panel real estate was becoming a truly limited resource, advances in computer technology opened the door to digital displays (Reising, Liggett, & Munns, 1999). Since multiple pieces of digitally processed information could be integrated into a much smaller display area, many of the constraints of limited cockpit space were relieved. Over the course of the next three decades, advancements in display presentation and information processing technology altered the nature of information retrieval in the cockpit (Figure 14-1). The previous task of integrating readings from individual gauges, in order to derive aircraft states, has been replaced with the integration of multiple sources of information into a few displays. Instead of visually scanning the environment, pilots can now select the appropriate information to access from a menu, on a screen that resembles a laptop computer—the implications of which represent a shift from presenting large amounts of information with high visual demand to presentation of that same information in a potentially cognitively demanding way.

Despite what one may believe at first, fundamental cognitive processes, like working memory and attention, that were important in the vast visual display cockpit layouts of the past, continue to be central to human interaction with increased automation and computerized instrumentation. Awareness of when information

FIGURE 14-1 Illustration of difference between traditional cockpit and "glass cockpit." Note the representation change from many gauges to a handful of screens. Top image courtesy of NASA and bottom image courtesy of FAA.

should be attended (Podczerwinski, Wickens, & Alexander, 2002), how to navigate menus to access it (Paap & Cooke, 1997), and correct interpretation of what is displayed all contribute to increased cognitive demand. Considering displays as an integral part of information processing in the cockpit (Hutchins, 1995), suggest that display design can be utilized to offset some of the human information processing limitations. Computer-based displays make it easier than ever to integrate information in the cockpit. Consequently, display designers now should not neglect the display design research of the past, but also look to human computer interaction (HCI) research to guide effective aviation display design.

HUMAN FACTORS IMPLICATIONS FOR COCKPIT DISPLAYS

The shift of issues associated with cockpit display is best understood in the context of human information processing. Displays, in any context, are intended to provide information to users. In complex systems, like the cockpit, the display provides the operational link to system state and environmental information. In the cockpit, this link involves both observational and input interfaces from which the pilot interacts with the aircraft. The success of the pilot/aircraft system is determined by the efficiency which information is presented, observed, and acted on. Perception, attention, and memory are all key aspects of information processing that have implications in display design.

Perception

At the most basic level, any information displayed must be perceived for action to commence. In essence pilots are surrounded by a wealth of sensory information from which they must extract critical information to the execution of the flight. Displays provide a mostly visual experience from which visual perception plays a critical role. After all, if displays are crammed with too much information, spread out over a wide viewing area or poorly organized, it will be more costly to scan the visual environment to locate critical information. Provided that aviation is rich with visual stimuli and that additional perceptual inputs (e.g., auditory, tactile) also influence pilot interaction, display design can have a major impact on the perception of critical information.

Attention

Most of the tasks involved in aviation are contingent on the ability to attend to multiple sources of information efficiently. This includes balancing between tasks that require focus on specific flight critical information to complete a task and monitoring multiple different sources of information of varying relation. Tasks that require focused attention should emphasize display of task relevant information while minimizing unnecessary information. As in most complex systems, there are numerous tasks that may change from monitoring to problem solving in a matter of seconds. As a result, display design should not neglect the fact that pilots must also be able to divide their attention between multiple sources of information to maintain overall system stability. There is no easy answer to finding the proper balance between focused and divided attention in the cockpit. Understanding the attentional limitations in human information processing can be helpful, though, in deciding how information is displayed.

Memory

Storage of information for later retrieval and use is also critical to information processing. In many cases the information that is displayed in the cockpit needs to be integrated with other sources of information to complete a task. By presenting information in a semipermanent fashion, displays can serve as an external memory source. Instead of overtaxing temporary memory stores (i.e., working memory) displayed information can help offload the burden on memory, for quick reference without having to commit and integrate information mentally. By keeping relevant information grouped together, processing delays caused by working memory overload can be mitigated.

Summary

Individually each of these concepts can be affected by how information is displayed in the environment. Due to the interconnectedness of these in information processing, each must be considered individually and in combination. Designers are largely constrained by current technology and these human limitations when developing display tools. Unfortunately, there is no optimal or "best" way to address human information processing limitations in design, but by understanding these concepts, a designer is better equipped to make design decisions that can optimize human performance and limit features that may induce error.

In the following section, we will briefly describe how the integration of information has changed the face of cockpit displays. We will then provide a description of each of four separate methods of display integration that shapes current display design. While the glass cockpit, in many instances, improved upon the flow of information in the human cockpit system, in some cases it introduced a new set of human factors challenges.

COCKPIT INTEGRATION

The combination of electronic display units (EDU) that integrates display information into a more condensed display format is often referred to as the glass cockpit. The glass cockpit contains software based displays that provide a platform for multiple methods of integration of information. By now, most pilots in military, commercial, corporate and general aviation have encountered glass cockpits. In fact, an entire industry has spawned, dedicated to retrofitting cockpits with glass cockpit displays. Just because the content of the display can change does not mean that display placement and configuration is not of concern. For example, different sized aircraft call for a variety of different configurations for glass displays. In large commercial airliners, such as the Airbus A320, there is ample space for multiple LCD display screens. However, instrument panel space is at a premium in general aviation cockpits. Consequently, the number of screens that will fit is limited, and therefore, the amount of information obtained from each, by necessity, is greater. Despite the differences across glass cockpits, the technology itself can be generally characterized by the integration of information into meaningful and functional groupings.

For cockpit displays, integration represents a combined representation in multiple source, type, function, and environmental configurations (Table 14-1). The human tendency to strive for parsimony would suggest that the fewest displays with the most information available are most desirable. Unfortunately, human

TABLE 14-1   Integration Methods Utilized in the Cockpit

Integration Method	Description	Human Factors Issues to Consider
Source	Combination of multiple pieces of information gathered from separately measured sources into one EDU	– Menu navigation – Attentional tunneling
Type	Simultaneous overlay of similar types of information onto display space	– Clutter – Focused attention tasks – Multidimensional display
Function	Integration of flight relevant actions and behaviors into a reduced number of input requirements.	– Automation monitoring – Multimodal sensory overload – False Alarm
Environment	Presentation of display information onto the visual environment outside of the cockpit	– Clutter – Attentional tunneling – Limited sensor range for EVS – Lack of real time imagery for SVS

information processing does not fall directly in line with this. Going to the minimalist extreme for display design, will require more menu depth and breadth, and less space to simultaneously display related information. Limitations associated with information processing such as working memory, workload and attention can result in increased processing time, insufficient attention to critical information, and a potential increase in pilot error. As a result, display design should reflect the limitations in human information processing. Designers should mirror the human computer interaction design goals of intuitive design. This is needed to prevent exponentially growing training time and monetary cost as technology grows at a similar rate. Poor implementation of these principles of design was the cause of initial skepticism with glass cockpit displays (Lintern, Waite, & Talleur, 1999).

In an environment rich with complex and extensive information, the careful integration of displays becomes critical. In addition, displayed information represents the link between pilots and increasingly automated aviation systems. In high workload activities

such as approach and landing, quick, salient access to appropriate information can spell the difference between routine flight and
catastrophe. If executed well, display integration can prove to be
a valuable aid. The following sections provide a description of
the different methods of integrating information made possible
through glass cockpits and overall technological advancement.
Where applicable, examples from current display design will be
used to illustrate each type of integration. Additionally, we will
focus on describing how each type of integration presents its own
unique set of design issues to consider.

SOURCE INTEGRATION

The cockpit is a rich supply of information that comes from a vast
array of sources. These include weather updates, measures of aircraft status, and environmental factors. Source integration is the
combination of multiple pieces of information gathered from separately measured sources into one EDU. The visual demand that
individual gauge displays of yore presented, were predominately
caused by each source of information having a separate physical location in the cockpit. Execution of flight maneuvers, such as
choosing the appropriate approach angle, might require reference to
altitude, flight speed, navigational charts and power settings. All of
this would require visual reference from separate display sources.
The attention needed for this series of visual scans and internal calculations would then need to be stored in working memory long
enough to mentally integrate the information for processing. Any
forgotten information calls for re-reference, costing additional scan
time. Now that digital processing is prevalent, the combination of
multiple pieces of information in one source is now reality.

Source Integration in the Cockpit

The multifunction display (MFD; Figure 14-2) is a good example of
source integration. The MFD utilizes computer generated imagery
to display navigational, weather, and aircraft state information at
the pilot's discretion. The MFD, in effect, acts as the storage center
for information that needs to be referenced on occasion, but unlike
primary flight display (PFD) items, does not necessitate constant

FIGURE 14-2 Source integration illustration. A photograph of a multifunction display (MFD). Image courtesy of FAA.

display. Generally, navigation between functions on the MFD is conducted using keys that border the display, or in newer versions, even touch screen capability. The MFD display format presents a growing potential for storage at roughly the same pace that computer information storage is progressing. Provided the amount of information that is now being housed in a MFD, the interface continues to evolve into a menu driven display that bears strong resemblance to personal computing. When used effectively, the MFD serves as an excellent savings on visual scan time and provides numerous sources of information to support awareness of the aircraft and the surrounding environment in which it operates. Despite the improvements that come with source integration, several issues are associated. Two of the most salient issues are the depth of menus and the display of dynamic information (Wickens, 2003).

Issues Associated with Source Integration

Display menu navigation

As displays begin to more closely resemble computers, information acquisition in the cockpit is also following a similar evolutionary track. Instead of having a physical location for information, important information is organized hierarchically in a series of menus accessible at the pilot's discretion. Many of the issues now associated with retrieval of information in aviation displays mirror that which is studied in human computer interaction research. Increased volume of information is stored in digital instead of physical space. If it is not possible to display all relevant information at the same time on the display screen, pilots run the risk of what is known as the keyhole property (Woods & Watts, 1997). The keyhole property occurs when all relevant information to a task cannot be viewed in parallel (Watts-Perotti & Woods, 1999). If menus are not carefully constructed users might get lost or forget where pertinent information is located within the network of menus.

A number of human computer interaction studies investigating how the depth and breadth of menu structure influence response performance suggest that there are tradeoffs to consider for both (Paap & Roske-Hofstrand, 1986; Lee & MacGregor, 1985; MacGregor, Lee, & Lam, 1986). Taken further, Jacko and Salvendy (1996) investigated how the breadth and depth of a menu's structure were affected by task complexity. Although these findings suggest that the use of less depth would be more helpful for menus, taking a more practical look reveals that menu depth can, in some instances, still produce optimal user interaction (Paap & Cooke, 1997). Mejdal, McCauley, and Beringer (2001) suggest that hierarchical structures that contain familiar items will benefit from additional menu breadth, while less familiar information is better suited for deep menu structure. The fact that pilots are generally highly experienced supports use of broader menu structure. Provided that menus are structured as such, the potential for novice pilots to experience resulting menu navigation delays, should not be ignored.

Out of Sight Out of Mind

The decision to condense the display space, given the vast amount of information available to pilots, can also influence how pilots react to changes that occur during flight. Given the numerous dynamic sources of information in the cockpit, there is a chance that changes relevant to flight planning will occur while not being displayed, and thus not attended to by pilots. When optimally aware, pilots can balance between multiple display selections to sample all aircraft and external environment information. Since pilots occasionally operate under conditions of high workload, this can impact a pilot's attention. In fact, conditions of especially elevated workload can result in what is known as "cognitive" or "attentional" tunneling (Dirkin, 1983; Wickens, 2005). This may result in a pilot ignoring or simply forgetting to attend to information presented outside of their current focus. Tunneling behavior becomes especially concerning given the possibility of sudden changes in weather, traffic or aircraft state that could dramatically alter the appropriate course of flight action. In context of source integration, this phenomenon has been called the "out of sight, out of mind" phenomenon (Podczerwinski, Wickens, & Alexander, 2002). Source integration itself cannot completely account for this phenomenon. In the following sections, we will describe how other methods of integration can help offset the issues associated with source integration.

Summary

Source integration provides a method of combining information from separately measured sources into one display. It serves as a method for reducing the processing demands that hinge on visual scan. By integrating multiple sources, displays are no longer constrained by the physical display space in the cockpit. Simply combining multiple sources into one display, however, does not guarantee performance gains. Given the large amount of information that can be combined using source integration, display design must account for information processing constraints associated with retrieval and monitor of information that is not currently visible.

TYPE INTEGRATION

When information is integrated based on source alone, it provides a reduction in visual scan, but does little to address the workload caused when multiple sources of information need to be referenced to make flight decisions. To account for the cost associated with referencing multiple sources of information to derive a flight plan, most glass cockpits have adopted displays that utilize type integration. Type integration involves the overlay of similar types of information onto display space at the same time. By simultaneously displaying related information, type integration reduces the imposed constraints on working memory. This is especially true when several individual sources of information need to be considered, that previously would have been presented in isolation. In many cases, on a display where numerous pieces of information need to be attended, distinguishing which types of information should be attended at a time can be taxing. Type integration can serve to defuse some of this difficulty. Given that glass cockpits can now display high quality representations of maps, aircraft states, and weather information, the dominating design question has changed from what sources of information can we fit into one display to what information should be displayed together. Moreover, designers are faced with the dilemma of determining if there are times that combining information is counterproductive to flight tasks.

Type Integration in the Cockpit

Navigational displays provide a good example of the advantages and disadvantages of type integration. Navigational information such as weather, terrain, and air traffic are dynamic pieces of information that are influenced by the aircraft's location in space. Any one of these pieces of information is relevant and in many cases the combination of information will give the most complete picture of the flight environment. When utilized appropriately, an overlay of multiple pieces of information can enhance a pilot's awareness of the surroundings. Using the proximity compatibility principle, in classic cockpits, served to lighten visual scan load, and it is no less important in the glass cockpit (Wickens & Carswell, 1995). The proximity compatibility principle is characterized by the notion that similar task operations should be located close in display space. At its best, type integration techniques involve displaying related information near each other, while separating disparate information that might serve to interfere with attentional resources. Type integration overlay can relieve workload associated with taking in information from different screens. When type integration is not carefully implemented, however, displays can impart more cognitive effort on pilots. Given that multiple pieces of information are displayed at once, it can be easy to overclutter the screen.

Another way in which type integration has been implemented in the aircraft is the use of 3D displays. The argument against traditional 2D displays is that in a three-dimensional plane of movement, it is difficult to accurately represent movement on all planes in one display. Instead additional processing time and working memory resource are needed to alternate from separate displays on the vertical and horizontal plane. 3D displays account for this by combining vertical and horizontal information into one display. This can be especially helpful in time critical tasks such as traffic monitoring and collision avoidance. In fact, the 3D display has been found to improve reaction time in collision avoidance and also elicit an increase in vertical avoidance maneuvers as opposed to the 2D map in which individuals were more apt to make only horizontal corrections (Garland, Wise, Guide, Jentsch, & Koning, 1991). Although there are many possible viewpoints from which the environment can be displayed, the predominant distinction is whether displays are from an egocentric or exocentric perspective (Wickens, 2003). Egocentric 3D displays provide display of the world view as it would appear from in the cockpit looking out to the world. Egocentric viewpoint displays are more useful for flight path guidance tasks (Haskell & Wickens, 1993). Exocentric 3D displays, on the other hand, provide a view of the aircraft from outside the cockpit and are generally more beneficial for navigational comparison judgments. Unlike egocentric viewpoint displays, the exocentric display gives a broader worldview, increasing the visibility of important environmental features surrounding the aircraft (Wickens, 2003). Given that time is a critical element of flight, additional displays such as quickened displays (Neal, Wise, Deaton, & Garland, 1997) and 4D displays (Krishnan, Wise, Bibb, Winkler, 1999; Krishnan, Kertesz, Wise, 2000) have been incorporated to provide predictive behaviors of the aircraft. Although there are advantages to utilization of type integration for multiple aviation display needs, in the next section we will provide a discussion of some of the implementation issues associated with type integration.

Issues Associated with Type Integration

Clutter

More information is capable of being displayed in a limited space by utilizing type integration. Making sure that the display space is unaffected by clutter of too much information is as important as it was when displays were sprawled across the entire cockpit dash. Wickens (2003) defines clutter as the number of marks, or visually contrasting regions, that are within a given display area. By this definition clutter itself is not detrimental, but the display of too much clutter is where the problem occurs. Display of unrelated information in the same area can serve to distract and slow the information seeker from obtaining information that they need. Based on this premise, it would be more beneficial to separate navigation and flight control types of information. Although in the bigger picture these pieces of information are both flight critical, displays combining different types of data such as altitude information on a weather map could generate unnecessary clutter of information. Often, excessive use of symbols can contribute to display clutter.

Symbology is used in displays as an artificial representation of real-world information. The use of symbols when carefully considered can serve to enhance displays. Symbolic representations can signify environmental feature (e.g., other aircraft, terrain features, etc.), engine status, or presentation of additional flight critical information on the environment that may not otherwise be visible. In order to reduce clutter, any symbology used should, be clearly distinguishable from other symbols (Wood & Wood, 1987); otherwise, cluttered display space or difficult distinction between dissimilar information could produce detrimental effects.

Less experienced users have been found to have more difficulty interpreting symbology on a display (Cushman & Rosenberg, 1991). Growth in the aviation display design industry, promotes design of unique products that stand out from the competition. Unfortunately, this lends itself to more variation in displays and subsequent variations in symbology used. As a result, it would be helpful to establish symbology standards that provide consistency across industry design (Harris, 2004). Without industry-wide consistency, pilot transition to a new aircraft will necessarily involve relearning a new set of symbols that could conflict with what they previously used.

We feel that this chapter would not be complete without a brief mention of the potential impact of the aesthetic of the interface on system performance. Gannon (2010) demonstrated that the more aesthetically pleasing a cockpit is to pilots, the better they think they and the system perform. While one can make arguments as to why this may potentially be a concern, it could have a positive effect in times of high stress. The impact of aesthetics, in any case, is an area that needs further investigation.

Focused Attentional Tasks

Unfortunately maximization of related information and minimization of unrelated information is not the one-stop solution to the problem. In a dynamic complex environment like the cockpit, at any one time display of information that does not serve to further a pilot's task can impose additional clutter to the task at hand. That is, when pilots need to focus their attention on a specific piece of information in a display, type integration can burden information processing. Varying techniques such as selective decluttering (Mykityshyn, Kuchar, & Hansman, 1994) and intensity coding (Wickens, Kroft, & Yeh, 2000) have been implemented to address the issues of clutter and focused attention. Selective decluttering involves providing the pilot with the ability to select information that they want to view. Any deselected items will not be visible on the display without further selection. The clear benefit here is that the pilot can self filter type integrated information to minimize the effects of clutter. Because pilots can narrow the displayed information, selective de-cluttering has the downside of reintroducing the source integration issue of "out of sight, out of mind" (Podczerwinski, Wickens, & Alexander, 2002) in addition to the processing time cost associated with the necessary display interaction. Intensity coding, on the other hand, maintains the full representation of type integrated information. Instead of filtering by subtraction, intensity coding involves highlighting the information of importance at any given time. Essentially, this display technique eliminates the "out of sight, out of mind" issue at the cost of keeping muted clutter on the display screen.

3D Display Issues

3D displays offer a number of type integration advantages that are beneficial in the cockpit. Despite this, there are a few limitations that restrict any recommendations for full replacement of 2D displays with 3D. As mentioned above, the egocentric viewpoint provides individuals with the restriction of only forward facing viewpoints. This provides a limited representation of the surrounding environment, which is not particularly useful in the event of activity outside of the forward facing viewpoint that may affect flight. The exocentric viewpoint can be used to offset the drawbacks of the egocentric view, but they are not particularly well suited for precise positional information relevant to traffic, weather, and terrain judgments (Boyer & Wickens, 1994; May, Campbell, & Wickens, 1995). Tasks requiring pilots to make line of sight judgments, such as trajectory for flight path adjustments, will not be able to achieve a level of precision that would be considered safe (Wickens, 2003).

Summary

Whereas source integration merely involves integrating access to information into a compacted display system, type integration involves combining information onto the visible display screen. Effective use of type integration provides a more complete picture of related information in a combined display space. By combining related information, displays can reduce the amount of working memory and mental integration load that occurs when related pieces of information are displayed separately. Implementation of type integration should be done with caution. Unrelated information in the visible display can clutter the screen and impact user ability to locate and focus on critical information.

FUNCTION INTEGRATION

To this point, the focus of discussion has been on how much information can be integrated into displays. The third type of integration that we will discuss, functional integration, is a slight departure from this theme. Functional integration actually does not specifically refer to display integration. Rather, functional integration is the integration of flight relevant actions and behaviors into a reduced number of input requirements. Functional integration stems from the increased amount of automation that is involved in flight. However, seeing that this is a chapter on displays, functional integration itself is not the focus of our conversation. We should emphasize that automation problems are not necessarily display problems. However, automation issues can be attributed to problems due to information management and communication, of which, the display contributes (Wiener, 1987). For pilots, displays serve as the contact point for any functionally integrated aircraft actions and therefore should be considered a tool for monitoring automated systems. Advanced automated flight systems dictate that functional integration occurs out of sight from the pilot. With a risk similar to the "out of sight, out of mind" condition described earlier, displays should supplement the pilots understanding of aircraft behaviors and effectively direct attention when necessary. Because functional integration information is not generally the primary focus of pilots attention, displays of this information should use attention-grabbing techniques such as motion, color, or multimodal display, when information should be attended to. Functional integration is a beneficial way of reducing pilot workload; however, removing the pilot from the functional loop can have detrimental effects in the event of system failure.

Functional Integration in the Cockpit

The flight management system (FMS) is the pilot's primary means of navigating. Using a series of waypoints, pilots can input desired flight paths from which the FMS system adjusts heading, pitch, and roll automatically to execute the desired path. In most cases, all of this is done without the need of constant pilot supervision. This functional integration of flight path action frees up the pilot to focus attention on other flight relevant tasks. In a majority of cases, the actions triggered by the FMS are not closely monitored by pilots. Otherwise, the workload gains of automation would be lost on tracking every action of the flight.

Display needs for the FMS comes into play when a pilot needs to manually take control of the flight or to make changes in the flight plan. In some cases, the perceived benefits of reduced workload, by pilots, disappear when adjustments to the FMS are required in already high workload phases of flight (Wiener, 1985; Wiener, 1989). In addition to this, actions of the FMS have not always been clear to pilots (Wiener & Curry, 1980; Sherry & Polson, 1999). In the extraordinary circumstance that the FMS is not functioning properly or if the pilot needs to take control, the displays must provide pertinent functionally integrated information to keep the pilot informed and afford for efficient interface interaction, especially in instances of high workload. Display techniques like color coding and motion (i.e., flashing) are both ways that designers use to draw pilot's attention to important visual display information. If focused attention is being given to another display, visual displays can be supplemented with the use of other sensory modalities.

Collision prevention displays are an example of the use of multimodal displays to convey functional integration information. The Ground Proximity Warning System (GPWS) and Traffic alert and Collision Avoidance System (TCAS) are both display mechanisms intended to prevent collision with terrain or other air traffic. Although each system provides slightly different information, the principle of the information conveyed is essentially the same. When an aircraft comes within a predetermined distance of making contact with another object these displays are engaged. These are intended to provide sufficient time for the flight crew to evaluate and react to the alert (Wiener, 1985). Not only are there visual display notifications generated by the system, but there are also auditory signals that are intended to alert pilots to this potential problem and depending on the severity of the alarm, provide avoidance recommendations on how to maneuver to avoid collision (i.e. pull up). By providing redundant alerts in both the visual and auditory modality, the alert information is more likely to be detected (Boff & Lincoln, 1988).

Unfortunately, multimodal displays that use visual and auditory alerts have a limited value if there are multiple displays utilizing similar attention grabbing techniques. For each additional source of visual and/or auditory stimulation, a more cluttering effect will occur. In an effort to investigate other modalities there have been promising findings in the domain of tactile display (van Erp, Veltman, & van Veen, 2003). Tactile displays have shown promise to enhance navigation (van Erp, van Veen, Jansen, & Dobbins, 2005) and directional attention (Van Erp, 2005).

Issues Associated with Functional Integration

Mode Errors

As cockpits get more sophisticated the number of automated features directly affecting flight management increase. A pilot's ability to recognize and understand automated states of the aircraft, also known as mode awareness (Sarter & Woods, 1995; Sarter & Woods, 1992), is important especially when a pilot has to intervene (i.e., take control of the aircraft). Pilot unawareness of automated states can lead to mode errors. Mode errors are a result of pilot action based on the state that they believe the aircraft is in, even though their belief does not match the state of the aircraft (Sarter, Mumaw, & Wickens, 2007). This, in turn, can lead pilots to experience automation surprise when the aircraft actions do not match their mental model (Sarter, Woods, & Billings, 1997). Since displays serve as the connection point between pilot and machine, displays provide a platform for addressing issues of mode awareness. Simply displaying a notification of what mode the autopilot is in, does not guarantee understanding. In addition to simply drawing attention to changes in automation, displays should provide information on the available automated functions and inform pilots of how current automation modes affect the behavior of the aircraft (Sarter, 2008). By providing additional feedback on automation states, displays can serve to enhance pilot mental models of automation modes. By improving pilot mental models, displays can offset confusion caused by unexpected actions of the automated system.

Sensory Overload

Displays are intended to provide pilots with relevant and timely information regarding the functionally integrated aspects of an aircraft. In recent findings the color and motion coding that is intended to attract attention has been found to not always have the intended effect. Nikolic, Orr, and Sarter (2004) found that flashing notifications do not necessarily have the presumed attention drawing capability in dynamic, multicolor, data rich display environments (2004). As a result, multimodal displays seem promising, but as alluded to in the previous section, there are a limited number of alternate modalities that have proven viable for conveying information. Auditory display, like that which the TCAS and GPWS utilize, is common, but the possibility for multiple sounds producing a masking effect, increases as more sounds are used. As mentioned earlier, there are a number of studies that support the use of displays that utilize tactile sensors to convey messages to the pilot. Although tactile displays show promise, Ferris and Sarter (2008) caution that multimodal tactile displays may induce slower response times for visual-tactile display depending on the location of each cue.

False Alarms

TCAS and GPWS alerts are useful tools for maintaining safe distances from terrain and other aircraft. For these systems to be effective they must be designed so that there is no instance where the system "misses" detection of unsafe flying distance. If this were not the case, pilots would have little reason to trust these alerts. Since no system is 100% reliable, the trade-off for making sure that the system gets no misses is that the system will have/must have some false alarms. If there are too many false alarms, pilots will begin to mistrust the display alert and thus begin to disregard its message. In the case of collision avoidance displays, if pilots receive warnings even when they are making purposeful and routine flight maneuvers, it is likely that they will begin to ignore the warning as a nuisance. Thus, display alert systems should be calibrated to minimize the number of false alarms while at the same time not missing any potential hazards. For a system that uses just one alarm the pilot must focus attention on the cause of the alarm, assess the severity of the condition, and decide the proper course of action. All of this may divert attention from other important information in the cockpit. One way that this has been addressed is by using alarms that have multiple levels of alert instead of one just one single alert to signify potential danger (Sorkin, Kantowitz, & Kantowitz, 1988). By providing different levels of alert, the system can serve to notify pilots of situations as they develop.

Summary

In high end aviation, many of the functions that were previously manually operated by pilots have been replaced by automated systems. Functional integration involves assimilation of notifications and updates of automated system states into the display to keep pilots aware of actions triggered by the automation. Since the information associated with functional integration, is mostly associated with monitoring for abnormal states, this information must effectively draw attention without burdening normal operation when unusual states are not present. In order to adequately implement functional integration, display designers should consider automation issues, including mode error and false alarm, and the potential for sensory overload in order to make sure that display information is not disregarded at critical moments.

ENVIRONMENTAL INTEGRATION

The final type of integration discussed is environmental integration. Environmental integration involves the display of information in a manner which overlays onto the visual environment outside of the cockpit. In effect, displays utilizing environmental integration are intended to reduce the need to alter gaze from outside of the windscreen by superimposing critical flight information onto the external environment. Implementation of environmental integration can help to reduce visual scan time, prevent re-accommodation delays associated with near and far sight transitions, and provide imagery modifications to aid in attending to the environment (Wickens, Ververs, & Fadden, 2004). Two technologies, the head-up display (HUD) and the near-to-eye display (NTE), display cockpit relevant information in a way that pilots do not have to alter their gaze from the view out the window.

Environmental Integration in the Cockpit

HUD and NTE

The concept of projecting flight information into the field of view out the cockpit is not new. HUD technology has been around since prior to the availability of computer generated displays, however to this point it has remained largely a military application. In the last 20 years, however, the use of HUD technology has slowly crept into commercial aviation. Alaska Air served as the de facto pioneer in commercial aviation, using HUD technology to aid pilots on challenging approaches (Lopez, 1991). HUD technology today consists of computer generated imagery that is projected forward on a fixed location of the wind screen (Figure 14-3). The projected image fuses near and far visual information so that displayed imagery appears at the same distance as the external surroundings. This collimation eliminates time to alter gaze from near to far which provides savings in the time required to refocus (Martin-Emerson & Wickens, 1997). A number of studies have investigated whether the glass HUD display screen or NTE affect

FIGURE 14-3   HUD image using conformal runway imagery. Image courtesy of NASA.

an individual's ability to accommodate effectively to the outside visible environment. A number of early studies suggested that even after collimating imagery, individuals still had accommodation issues (Hull, Gill, & Roscoe, 1982; Roscoe, 1985). Improving HUD and NTE display technology may be responsible for more recent findings to the contrary (Wise & Sherwin, 1989; Valimont, Wise, Nichols, Best, Suddreth, & Cupero, 2009), suggesting that there was little accommodation difficulty from the display to the outside world. An additional benefit of the collimated display is the use of conformal symbology, an image superimposed over the visible scene to create more salient representations. Conformal symbology can be used to make specific environmental features such as the horizon or runway, appear more salient. In military applications, HUDs provide everything from acronym PFD information to target acquisition information. Due to the reduction in visual scan, studies have shown improved traffic detection, and slight improvements in display detection (Wickens, Ververs, & Fadden, 2004; Beringer & Ball, 2001).

Similar to HUD technology, NTE displays (a.k.a. head- or helmet-mounted display, HMD) capitalize on the display space available by overlay on the environment. As an advantage to the HUD, the information in an NTE is not limited to forward-facing display. Instead, NTE imagery can be continuously overlaid regardless of head position. This increased range of viewing can be especially beneficial in conditions that require pilots to utilize a wider range of the visual field (i.e., tactical combat and terminal airspace). Although NTE has been developed using both binocular and monocular visual cues, research findings have not agreed which provides the most performance gain (Lippert, 1990; Patterson, Winterbottom, & Pierce, 2006; Valimont, Wise, Nichols, Best, Suddreth, & Cupero, 2009)

EVS and SVS

There are two technologies that have emerged in environmental integration that aim to further improve pilot's perception of the external environment. These technologies are the enhanced vision system (EVS) and the synthetic vision system (SVS). Both are intended to convey visual information where degraded or unavailable visual conditions prevail.

An EVS is a display that, through the use of sensors, provides enhanced imagery of the real environment. Forward-looking infrared (FLIR) is the most common example of EVS (Sweetman, 1991; Aarons, 1992). FLIR uses thermal image sensors at wavelengths that produce visible images in degraded visual environments (Krebs, Scribner, Miller, Ogawa, & Schuler, 1998). Aircraft equipped with modern FLIR technology can provide images that pierce some types of cloud cover (Figure 14-4). In addition to the obvious vision enhancement, EVS also has the added benefit of being a true representation of the outside world. Environmental sensors equipped on the aircraft can serve to "see" things that are not possible with the naked eye. Although representations are sampled from the real world, in some cases this imagery can still appear much different from how we would naturally perceive the world. In this case, EVS sensor generated imagery has the added cost of training time, and could still result in increased interpretation time for pilots.

FIGURE 14-4 Sample of EVS imagery. Taken using FLIR vision sensor technology. Image Courtesy of NASA.

In contrast to the EVS, SVS produces an artificial representation overlaid onto the visual scene (Schnell, Kwon, Merchant, & Etherington, 2004). SVS generates symbols and computer generated imagery to provide a visual depiction of environmental information. In one sense, PFD data that is displayed on HUD and NTE is an example of SVS display imagery. More complex SVS imagery, including full computer-generated representations of the outside world, can create an image that expands beyond the limitations of the windscreen (Figure 14-5). As a result of improved graphic rendering software, display prototypes such as the synthetic vision integrated primary flight display (SV IPFD; He et al., 2007), combining standard PFD display information with computer-generated ground proximity imagery, are available. The advantage of this type of imagery is in being able to provide a representation of external sources of information in a manner that remains consistent, regardless of weather and visibility issues. The use of synthetic displays may help with depth cues and time to contact judgments (Delucia, Kaiser, Bush, Meyer, & Sweet, 2003).

While EVS and SVS displays each bring a unique strength to the table, display researchers have turned their attention to combining the two. Both of these in combination can provide additional

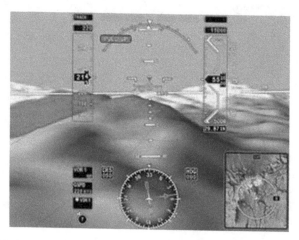

FIGURE 14-5 Synthetic vision system display (Feyereisen, He, Conner, Wyatt, Engels, Gannon, Wilson, & Wise, 2007). Image courtesy of Honeywell.

display improvements (Jennings, Alter, Barrows, Bernier, & Guell, 2002). Combining the two capitalizes on the terrain and static obstacle awareness of the SVS and the real world reliabilty of the EVS that synthetic representations do not provide. Bailey, Kramer, and Prinzel (2007) examined the performance effects of a combined EVS/SVS display system. Their results indicate no workload decrements as a result of the combined system and improvements some improvement on situation awareness. They did, however, find conflicting results on runway incursion detection, indicating the EVS/SVS display development is still a burgeoning concept. Provided that the EVS/SVS system can be tuned to complement human performance, the improved vision capabilities could dramatically reduce issues associated with low visibility.

Issues with Environmental Integration

HUD/NTE Issues

Many of the issues associated with environmental integration are similar to those described in the other integration sections. In some instances, the issues previously associated with other methods of integration are more pronounced when environmental integration is utilized. The effect of cognitive tunneling, for instance, has been reported to impact performance with HUD/NTE technology. A number of studies have found decrement in response time and detection of objects on the runway in an approach task in which HUD is used (Hofer, Braune, Boucek, & Pfaff, 2000; Fischer, Haines, & Price, 1980; Fadden, Ververs, & Wickens, 2001; Weintraub, Haines, & Randall, 1985). The effects of cognitive tunneling can in turn be magnified by the presence of too much clutter in the heads up display (May & Wickens, 1995). In HUD/NTE display clutter not only affects retrieval of projected information, but, given its overlay on the external environment, it could negatively influence a pilot's capability to attend to important external visual cues. Any important information that is not located in the HUD/NTE will necessarily have to be located on the dash. The transition of focus between cockpit instrumentation and out the window viewpoints has been associated with the occurrence of temporary spatial disorientation

(Patterson, Cacioppo, Gallimore, Hinman, & Nalepka, 1997). Appropriate use of symbology and display location can both contribute to a reduction in the effect of clutter. Conformal symbology is one method that has been found to reduce clutter (Boston & Braun, 1996). Although there have been attempts to provide standardization recommendation on HUD displays (Weinstein, Ercoline, McKenzie, Bitton, & Gillingham, 1993), designers are not tethered to symbology design standards. Display design for HUD/NTE, as a result, should consider the implications of all previously mentioned integration techniques.

EVS/SVS Issues

Given that EVS and SVS technologies are still evolving, it is important to be aware of where each may not provide optimal performance. EVS technology is dependent on gathering information from the external environment. Given that sensors have a limited range from which they gather information, there is still the possibility that conditions will fall outside the window of operable performance. Since technologies such as the FLIR are intended to make it possible to lower the minimum altitude required to commit to a landing, there will be less time if something unexpected occurs to adjust. The EVS is also not equipped to aid in interpreting the displayed information. Given that EVS imagery produces a degraded image similar to that which one might view from a low-grade security camera, there may be instances where additional symbolic representations can help to clarify unfamiliar scenes.

Provided that SVS displays are artificial representations of real-world information, the issues of symbology apply. Now that SVS technology is being used to construct full computer generated representations of the external environment, the issue becomes one of currency of the database of imagery information. A computer generated image of an airport, if not updated in real time, will not present accurate depictions of traffic or new obstructions. Given the multitude of airports that this information would have to be continually updated for, the support required to maintain this technology can be extensive.

Summary

A large portion of flight continues to be associated with visual contact with the world outside the cockpit. Environmental integration deals with the presentation of flight relevant information combined with out-of-the-cockpit viewpoints. This method of integration can be helpful to reduce processing time associated with changing gaze from inside to outside the cockpit. A number of issues associated with the other integration methods are mirrored when displaying information on the visible environment. Additionally, one must consider the limitations associated with new environmental integration technologies such as EVS and SVS.

INTEGRATION SUMMARY AND BLACKOUTS

Current cockpit display technology involves using a number of integration techniques to obtain the most optimal utilization of human information processing and system-generated flight critical information. Based on the discussion of each, it should be evident that no one type of integration provides the solution to optimal design, and that there are a number of trade-offs and processing issues that can result from each. Instead, display designers should spend more time considering how best to balance these display integrations for the given aviation task. The differences between a combat fighter pilot, a commercial flight crew, and single pilot general aviation aircraft produces a unique set of requirements that should be considered in the development of displays that optimize the task requirements of each. Despite the number of advantages that integration provides in displays, there is another concern with the glass cockpit, the loss of power to the display screens.

One of the biggest strengths of the glass cockpit, the ability to integrate information in a less visually demanding display, can also, in extraordinary circumstances, represent one of its biggest weaknesses. An NTSB investigation revealed that there have been somewhere in the neighborhood of 50 incidents in which a loss of electrical power to one or more displays occurred (Katz, 2008). In several of these instances, there was a complete electrical failure in which all of the glass cockpit displays failed. Given that circumstances involving the loss of flight critical information will cause a workload increase on the flight crew, the consequences of this occurring in an increased workload phase of flight could result in something far more catastrophic, than what has been reported thus far.

In the event of a complete power failure the glass cockpit goes from a rich source of flight information to a black screen providing no information. In the past, cockpit understanding comprised learning a lengthy list of gauges that provided different and important information. As with all systems, there were occasions in which gauges would malfunction or provide faulty readings. In the event of a gauge failure, the pilot would have to recognize that a particular piece of information was no longer available. Since, in the past, the gauges were displayed separately, one could still operate using readings from other related gauges still available. In the glass aircraft, a much bigger problem arises. Since all of the flight critical information is displayed on LCD screens, a failure can lead to a more substantial loss of flight-critical information. Electrical failure to the glass cockpit display can leave pilots with less and in some cases almost none of the information they are *accustomed* to referencing.

Although a full electrical failure is extremely rare, partial electrical failures, while still very uncommon, occur with more frequency. Blackouts expose an issue that can have serious implications for flight. Pilots have to be equipped to address the potential for losing access to a large portion of the information in the cockpit. Fortunately, cockpits are fitted with backup displays for the most critical flight information. The systems are designed and certified to allow a pilot to land an aircraft with the available information from backup displays. Due to the level of heightened stress that no doubt will accompany a failure of this magnitude, the backup information should be organized such that pilots have easy visual access to them. Since the information is intended to backup to the main system it has to be located in a place where it does not distract from the glass display, and contains only critical information or else it runs the risk of over-cluttering the visual environment again.

DISPLAYS IN THE NEXT GENERATION

Based on the widespread use of glass cockpit technology, researchers should continue to address current issues to optimize displays. Some suggest that any display advances will evolve slowly as a result of organizational demands in the commercial and corporate aviation industry (Harris, 2004). Under this line of reasoning, if not done properly the cost of installing new technology and training pilots on new displays will limit the quick update of modern display technology. After all, it has taken roughly 30 years from the outset of the glass cockpit to fully realize its potential in a high percentage of aircraft. However, taking into account that displays are now software driven, this suggests the opposite may be true outside of commercial flight. Instead of having to remove and replace gauges, new display updates can be installed, much like a personal computer update, by uploading new databases or software to the already existing display computer. These quicker update capabilities are especially important given the future vision of aviation. The following section will discuss how future industry mandates and developing aircraft technology will shape the design of displays in future generations of aviation.

NextGen

Looking toward the future, the Federal Aviation Administration (FAA) and Eurocontrol have identified increased air traffic as one of the primary concerns. NextGen and Single European Sky Air Traffic Management Research (SESAR) are industry-changing programs geared toward evolution from a ground-based air traffic control to a satellite-based air traffic management system (European Commision, 2005; FAA, 2009a). Ultimately, the system will increase pilot involvement in flight path management and traffic separation. The primary goal of NextGen/SESAR is to create more efficient traffic management through improved systems. Reducing the additional burden that additional traffic could have on air traffic control (ATC) has been the focus of attention for the NextGen/SESAR program. The idea is to place more of the navigation responsibility on the pilots. Doing this will, in effect, open up the skies to a wider variety of flight paths, and at the same time place more of the burden of separation of traffic on pilots.

Ultimately, the reduction in ATC involvement in flight path selection will have major implications for display design. Using data communication link technology, what was previously a voice command from ATC, will be displayed more permanently (FAA, 2009b). While this will relieve the working memory load of remembering and acting on ATC voice commands, at the same time this will produce additional attentional demands for pilots. Given that pilots will have more control over their cruise path, display technology has to be designed to effectively notify flight path deviations and provide recommended corrections. Absence of these notifications could lead to more deviations into restricted airspace, or, worse, midair collisions in the overcrowded airspace.

A second resulting outcome of NextGen/SESAR is to make aviation a more accessible mode of transport. Provided that there is a growing industry of low cost light sport aircraft, the number of general aviation pilots will assuredly increase. One of the consequences of making flight more easily available is that there will be an increase in less experienced pilots taking to the skies. In one sense, NextGen/SESAR systems are designed to improve commercial flight especially in busy traffic areas. But in the sense of increased availability, the NextGen/SESAR system also has to account for this new generation of pilots. It is already common for light sport aircraft to come equipped with modern displays, but near-future additions of a NextGen/SESAR system will also have to accommodate growth in general aviation as well.

Unmanned Aerial Vehicles

Although displays for use in piloted flight has been the predominant focus of discussion in this chapter, the industry of nonpiloted or unmanned aerial vehicles (UAV) is also worth mentioning. Utilization of UAVs are dedicated primarily to military operation, but much like HUD technology in the 1980s and 1990s, it is likely to infiltrate civilian airspace in the foreseeable future. In military operation, UAVs are generally intended for surveillance and ordinance delivery in combat scenarios. On the battlefield, the perceived cost for loss of control of a UAV is mostly monetary. Unfortunately this perceived cost does not take into account the damage that occurs on the ground. When a UAV goes down, destroyed structures and in some cases loss of life can result. Before UAVs can successfully make the transition into civilian airspace, UAVs will have to come equipped to interact with piloted air traffic with improved operator controls to reduce risk of accidents resulting in ground incursions (Berglund, Tyseng, Westin, Nisser & Hilburn, 2007). In light of this, UAV systems development has to focus on balancing between the resilience and stability of the sociotechnical system formed by the aircraft and the operator (Johansson & Lundberg, 2010). By striking a balance, the system will be able to adapt to unexpected occurrence in an ever changing environment while at the same time providing some consistent action in the face of more commonly occurring disturbances.

UAV operators face a number of additional challenges unique to uninhabited flight. Pilots in inhabited aircraft have the benefit of referencing the feel of the aircraft while in flight. The vestibular forces on the pilot can help to indicate changes in aircraft speed and direction, and also provide initial feel indications if something goes wrong. If a pilot experiences loss of engine power, the feedback from the shift in aircraft orientation will immediately indicate that something has gone wrong. In UAVs, the operator, or pilot, is remotely controlling the craft from the ground (Figure 14-6). The lack of motion cues may cause a lag in reaction. In the case of engine loss, this can result in significant altitude loss or worse a crash. Due to sensory isolation (McCarley & Wickens, 2005), UAV displays are the only sensory link between the human operator and the aircraft. This becomes especially critical in the transition from automatic to human operator control (Williams, 2006). Given that current UAVs suffer from limited field of view, growing research in SVS and EVS technology could produce richer visual representations of the operational environment.

FIGURE 14-6 UAV system control station display panel for the Predator drone. Image courtesy of United States Air Force.

CONCLUDING REMARKS

In the mid-1980s, the concept of developing a "Super Cockpit" was introduced (Furness, 1986; Nordwall, 1987). At the time, current technology advances could not adequately support many of the futuristic technologies discussed and were considered a "radical" departure from existing displays. The foresight of these technologies are remarkable considering much of what was discussed is either commonly implemented or currently researched for near term implementation. At the time, these improvements were seen as solutions, but what should be clearer today than ever is that despite continuous improvements, new display technologies are not free from the constraints of human information processing. With limited standardization on newer display components, it is more important to understand the processing implications that increased integration have on the pilot-cockpit system.

The information age has enabled the development and advancement of modern aviation displays. Provided that the human information processing limitations are considered when integrating different forms of information, the glass cockpit now serves to perform this integration automatically. Although there are obvious benefits associated with these display advances, poor consideration of the costs associated with different methods of integration will result in displays that tax individuals, the same if not more, than the traditional gauge displays of the past. Constant growth and shifting industry goals for the future, dictate that integration of flight critical information in displays, capitalize on the strengths and weaknesses of both the human and computer in the aviation system. It is important to realize that even when keeping human factors principles in mind for display design, errors will likely not cease to occur. Instead keeping human factors principles in mind for design will help to design systems that, through a process of addressing cost and benefit of design recommendations, will produce more optimally designed displays.

ACKNOWLEDGMENTS

Preparation of this chapter was in part supported by Research Grant FAA 99-G-047 from the Federal Aviation Administration (FAA) to the University of Central Florida (UCF). Dr. Eleana Edens was the technical grant monitor. The opinions stated in this chapter, however, are those of the authors alone and do not necessarily represent the position of the FAA, or any of the organizations with whom the authors are affiliated.

References

Aarons, R. (1992). Safety: Looking ahead. *Business and Commercial Aviation*, 54–57.

Bailey, R. E., Kramer, L. J., & Prinzel, L. J. (2007). Fusion of synthetic and enhanced vision for all-weather commercial aviation operations. *NATO-OTAN*, 11.11–11.26.

Berglund, A., Tyseng, M., Westin, C., Nisser, T., & Hilburn, B. (2007). UAVs in civil airspace: Toward a human-centered concept of autonomous flight operations. *Presentation at the 2007 Human Factors of UAVs Workshop*, Mesa, AZ.

Beringer, D., & Ball, J. (2001). General aviation pilot visual perfromance using conformal and non-conformal head-up and head-down highway -in-the-sky displays. *Proceedings of the international symposium on aviation psychology*. Columbus, OH: The Department of Aerospace Engineering, Applied Mechanics and Aviation, The Ohio State University.

Boff, K., & Lincoln, J. (1988). Guidelines for designing auditory signals. In: *Engineering data compendium: Human perception and performance*. Wright Patterson AFB, OH.

Boston, B. N., & Braun, C. C. (1996). Clutter and display conformality: Changes in cognitive capture. *Proceedings of the Human Factors and Ergonomics Society 40th Annual Meeting* (pp. 57–61), Santa Monica, CA.

Boyer, B. S., & Wickens, C. D. (1994). *3D weather displays for aircraft cockpits* (Tech. Rep. No. ARL-94–11/NASA-94–4). Savoy, IL: University of Illinois, Aviation Research Laboratory.

Cushman, W. H., & Rosenberg, D. J. (1991). Human factors in product design. Amsterdam: Elsevier Inc..

Delucia, P., Kaiser, M., Bush, J., Meyer, L., & Sweet, B. (2003). Information integration in judgements of time to contact. *The Quarterly Journal of Experimental Psychology, 56A*(7), 1165–1189.

Dirkin, G. (1983). Cognitive tunneling: Use of visual information under stress. *Perceptual and Motor Skill, 56*(1), 191–198.

European Commision. (2005). The SESAR program: Making air travel safer, cheaper and more efficient. Electronic memo retrieved May 10, 2009, from <http://ec.europa.eu/transport/air_portal/sesame/doc/2005_11_memo_sesar_en.pdf/>.

Fadden, S., Ververs, P., & Wickens, C. (2001). Pathway HUDS: Are they viable?. *Human Factors, 43*(2), 173–193.

Federal Aviation Administration. (2009a, February 12). What is NextGen? Retrieved March 16, 2009, from <http://www.faa.gov/about/initiatives/nextgen/defined/what/>.

Federal Aviation Administration. (2009b, March 12). Data communications. Retrieved March 16, 2009, from <http://www.faa.gov/about/office_org/headquarters_offices/ato/service_units/techops/atc_comms_services/data-comm/>.

Ferris, T. K., & Sarter, N. B. (2008). Cross-modal links among vision, audition, and touch in complex environments. *Human Factors, 50,* 17–26.

Feyereisen, T., He, G. Conner, K., Wyatt, S., Engels, J., Gannon, A., Wilson, B., & Wise, J.A. (2007). EGPWS on Synthetic Vision Primary Flight Display. 6th EUROCONTROL Innovative Research Workshops & Exhibition. Eurocontrol Experimental Centre, December 4–6.

Fischer, E., Haines, R., & Price, T. (1980). *Cognitive issues in head-up displays.* Moffett Field, CA: NASA Ames Research Center.

Furness, T. A. (1986). The super cockpit and its human factors challenges. *Proceedings of the Human Factors and Ergonomics Society 30th Annual Meeting* (pp. 48–52), Santa Monica, CA.

Gannon, A. J. (in press) Flightdeck aesthetics and pilot performance new uncharted seas. In J. A. Wise, V. D. Hopkin, & D. J. Garland (Eds.), *Handbook of aviation human factors* (2nd ed.). Boca Raton, FL: CRC Press LLC.

Garland, D. J., Wise, J. A., Guide, P. C., Jentsch, F. G., & Koning, Y. P. (1991). *Pseudo three dimensional cockpit navigation display, format and decision-making performance* (Rep. No. CAAR-15410–91–1). Daytona Beach, FL: Embry-Riddle Aeronautical University.

Harris, D. (2004). Head-down flight deck display design. In D. Harris (Ed.), *Human factors for civil flight deck design* (pp. 69–102). Burlington, VT: Ashgate.

Haskell, I. D., & Wickens, C. D. (1993). Two- and three-dimensional displays for aviation: A theoretical and empirical comparison. *The International Journal of Aviation Psychology, 3*, 87–109.

He, G., Feyerseisen, T., Conner, K., Wyatt, S., Engels, J., Gannon, A., & Wilson, B. (2007). EGPWS on synthetic vision primary flight display. *Proceedings for SPIE, 6559*, Orlando, FL.

Hofer, E., Braune, R., Boucek, G., & Pfaff, T. (2000). *Attention switching between near and far domains: An exploratory study of pilots' attention switching with head-up and head-down tactical displays in simulated flight operations.* Seattle, WA: The Boeing Commercial Airplane Co.

Hull, J. C., Gill, R. T., & Roscoe, S. N. (1982). Locus of the stimulus to visual accommodation: Where in the world, or where in the eye?. *Human Factors, 24*, 311–319.

Hutchins, E. (1995). How a cockpit remembers its speeds. *Cognitive Science, 19*, 265–288.

Jacko, J., & Salvendy, G. (1996). Hierarchical menu design: Breadth, depth, and task complexity. *Perceptual and Motor Skills, 82*, 1187–1201.

Jennings, C., Alter, K., Barrows, A., Bernier, K., and Guell, J. (2002). Synthetic vision as an integrated element of an enhanced vision system. *Proceedings for SPIE, 4713*.

Johansson, B., & Lundberg, J. (in press). Balancing resilience and stability. In J. A. Wise, V. D. Hopkin, & D. J. Garland (Eds.), *Handbook of aviation human factors* (2nd ed.). Boca Raton, FL: CRC Press LLC.

Katz, P. (2008). Glass-cockpit blackout. Retrieved January 27, 2009, from *Plane and Pilot Magazine*: <http://www.planeandpilotmag.com/pilot-talk/ntsb-debriefer/glass-cockpit-blackout.html/>.

Krebs, W., Scribner, D., Miller, G., Ogawa, J., & Schuler, J. (1998). Beyond third generation: A sensor fusion targeting FLIR pod for the F/A-18. In B. Dasarasthy (Ed.), *Proceedings of the SPIE-sensor fusion: Architectures, algo-righms, and applications II* (pp. 129–140). Bellingham, WA: SPIE-International Society for Optical Engineering.

Krishnan, K., Kertesz, S. Jr., & Wise, J.A. (2000). Putting four dimensions in "perspective" for the pilot. *Proceedings of the IEA 2000/HFES 2000 congress, 3 – 81-3-84*.

Krishnan, K., Wise, J. A., Gibb, G. D., & Winkler, E. R. (1999). Cockpit display of traffic information in free flight: A concept based on an analogy with the Minkowski space-time diagram. *Proceedings of the 10th International Symposium on Aviation Psychology*, Columbus, OH.

Lee, E., & MacGregor, J. (1985). Minimizing user search time in menu retrieval systems. *Human Factors, 27*, 157–162.

Lintern, G., Waite, T., & Talleur, D. (1999). Functional interface design for the modern aircraft cockpit. *The International Journal of Aviation Psychology, 9*(3), 225–240.

Lippert, T. M. (1990). Fundamental monocular/binocular HMD human factors. *SPIE Helmet-Mounted Displays II, 1290*, 185–191.

Lopez, R. (1991). Thick Alaskan soup stirs visual review. *Interavia Aerospace Review, 67*.

MacGregor, J., Lee, E., & Lam, N. (1986). Optimizing the structure of data-base menu indexes: A decision model of menu search. *Human Factors, 28*(4), 387–399.

Martin-Emerson, R., & Wickens, C. (1997). Superimposition, symbology, visual attention, and the head-up display. *Human Factors, 39*(4), 581–601.

May, P. A., Campbell, M., & Wickens, C. D. (1995). Perspective displays for air traffic control: Display of terrain and weather. *Air Traffic Quarterly, 3,* 1–17.

May, P. A., & Wickens, C. D. (1995).The role of visual attention in head-up displays: Design implications for varying symbol intensity. *Proceedings of the Human Factors and Ergonomics Society 39th Annual Meeting* (pp. 50–54), Santa Monica, CA.

McCarley, J., & Wickens, C. (2005). *Human factors implications of UAVs in the national airspace.* Savoy, IL: University of Illinois Institute of Aviation, Aviation Human Factors Division.

Mejdal, S., McCauley, M., & Beringer, D. (2001). *Human factors design guidelines for multifunction displays.* Washington, DC: Office of Aerospace Medicine.

Mykityshyn, M., Kuchar, J., & Hansman, R. (1994). Experimental study of electronically based intrument approach plates. *International Journal of Aviation Psychology, 4*(2), 141–166.

Neal, M. J., Wise, J. A., Deaton, J. E., & Garland, D. J. (1997). Evaluation of a display with both quickened and status information for control of a second order system. *Proceedings of the 9th International Symposium on Aviation Psychology,* (pp.335–340), Columbus, OH.

Nikolic, M., Orr, J., & Sarter, N. (2004). Why pilots miss the green box: How display context undermines attention capture.. *The International Journal of Aviation Psychology, 14*(1), 39–52.

Nordwall, B. D. (1987 20). Advanced cockpit development effort signals wide industry involvement. *Aviation Week and Space Technology,* 72–77.

Paap, K., & Cooke, N. (1997). Design of menus. In M. Helander, T. Landauer, & P. Prabhu (Eds.), *Handbook of human-computer interaction* (pp. 533–572). New York: Elsevier.

Paap, K., & Roske-Hofstrand, R. (1986). The optimal number of menu options per panel. *Human Factors, 28*(4), 377–385.

Patterson, F. R., Cacioppo, A. J., Gallimore, J. J., Hinman, G. E., & Nalepka, J. P. (1997). Aviation spatial orientation in relationship to head position and attitude interpretation. *Aviation, Space, and Environmental Medicine, 68*(6), 463–471.

Patterson, R., Winterbottom, M. D., & Pierce, B. J. (2006). Perceptual issues in the use of head-mounted visual displays. *Human Factors, 48,* 555–573.

Podczerwinski, E., Wickens, C., & Alexander, A. (2002). *Exploring the "out of mind out of sight" phenomenonin dynamic settings across electronic map displays* Technical Report ARL-01-8/NASA-01-4. Moffett Field, CA: NASA Ames Research Center.

Reising, J. M., Liggett, K. K., & Munns, R. C. (1999). Controls, displays, and workplace design. In D. J. Garland, J. A. Wise, & V. D. Hopkin (Eds.), *Handbook of aviation human factors* (pp. 327–357). Mahwah, NJ: Lawrence Erlbaum.

Roscoe, S. N. (1985). Bigness is in the eye of the beholder. *Human Factors, 27,* 615–636.

Sarter, N. (2008). Investigating mode errors on automated flight decks: Illustrating the problem-driven, cumulative, and interdisciplinary nature of human factors research. *Human Factors, 50*(3), 506–510.

Sarter, N., & Woods, D. (1995). How in the world did we ever get into that mode? Mode error and awareness in supervisory control. *Human Factors, 37*(1), 5–19.

Sarter, N., & Woods, D. (1992). Pilot interaction with cockpit automation: Operational experiences with the flight management system. *International Journal of Aviation Psychology, 2*(4), 303–321.

Sarter, N., Mumaw, R., & Wickens, C. (2007). Pilots' montioring strategies and performance on automated flight decks: An empirical study combining behavioral and eye-tracking data. *Human Factors, 49*(3), 347–357.

Sarter, N., Woods, D., & Billings, C. (1997). Automation surprises. In G. Salvency (Ed.), *Handbook of human factors/ergonomics* (2nd ed.) (pp. 1926–1943). New York: Wiley.

Schnell, T., Kwon, Y., Merchant, S., & Etherington, T. (2004). Improved flight technical performance in flight decks equipped with synthetic vision information system displays. *The International Journal of Aviation Psychology, 14*(1), 79–102.

Sherry, L., & Polson, P. (1999). Shared models of flight managmenet system vertical guidance. *International Journal of Aviation Psychology, 9*(2), 139–153.

Sorkin, R. D., Kantowitz, B. H., & Kantowitz, S. C. (1988). Likelihood alarm displays. *Human Factors, 30*, 445–459.

Stokes, A., & Wickens, C. (1989). Aviation displays. In E. Wiener & D. Nagel (Eds.), *Human factors in aviation* (pp. 387–431). San Diego, CA: Academic Press.

Sweetman, B. (1991). Infra-red offers new landing aid competition. *Interavia Aerospace Review*, 62–6.

Valimont, B., Wise, J. A., Nichols, T., Best, C., Suddreth, J., & Cupero, F. (2009). When the wheels touch the earth and the flight is through, pilots find one eye is better than two. *Proceedings of SPIE Defense, Security, and Sensing*, Orlando, FL.

Van Erp, J. (2005). Presenting directions with a vibro-tactile torso display. *Ergonomics, 48*, 302–313.

Van Erp, J., van Veen, H., Jansen, C., & Dobbins, T. (2005). Waypoint navigation with a vibrotactile waist belt. *ACM Transactions on Applied Perception, 2*, 106–117.

Van Erp, J., Veltman, J., & van Veen, H. (2003). A tactile cockpit instrument to support altitude control. *Proceedings of the Human Factors and Ergonomics Society 47th Annual Meeting* (pp. 114–118). Denver, CO.

Watts-Perotti, J., & Woods, D. (1999). How experienced users avoid getting lost in large display networks. *International Journal of Human-Computer Interaction, 11*(4), 269–299.

Weinstein, L., Ercoline, W., McKenzie, I., Bitton, D., & Gillingham, K. (1993). *Standardization of aircraft control and performance symbology on the USAF head-up display*. Brooks Air Force Base, TX: Air Force Material Command.

Weintraub, D. J., Haines, R. F., & Randle, R. J. (1985). Head-up display (HUD) utility: Runway to HUD transition monitoring eye focus and decision times. *Proceedings of the Human Factors and Ergonomics Society 29th Annual Meeting* (pp. 615–619), Santa Monica, CA.

Wickens, C. (2003). Aviation Displays. In P. Tsang & M. Vidulich (Eds.), *Principles and practice of aviation psychology* (pp. 147–199). Mahwah, NJ: Lawrence Erlbaum.

Wickens, C., & Carswell, C. (1995). The proximity compatibility principle: Its psychological foundation and relevance to display design. *Human Factors, 37*(3), 473–494.

Wickens, C., Kroft, P., & Yeh, M. (2000). Database overlay in electronic map design: Testing a computational model. *Proceedings of the Human Factors and Ergonomics Society 44th Annual Meeting* (pp. 451–454), San Diego, CA.

Wickens, C., Ververs, P., & Fadden, S. (2004). Head-up displays. In D. Harris (Ed.), *Human factors for civil flight deck design* (pp. 103–140). Burlington, VT: Ashgate.

Wiener, E. L. (1985). Beyond the sterile cockpit. *Human Factors, 27,* 75–90.

Wiener, E. L. (1987). Fallible humans and vulnerable systems: lessons learned from aviationNATO ASI Series. In J. A. Wise, & A. Debons, (Eds.)*Information systems: Failure analysis: Vol. F32* (pp. 165–181). Heidelberg: Springer-Verlag.

Wiener, E. L. (1989). *Human factors of advanced technology (glass cockpit) transport aircraft* (Rep 177528). Moffett Field, CA: NASA Ames Research Center.

Wiener, E. L., & Curry, R. E. (1980). Flight deck automation: Promises and problems. *Ergonomics, 23,* 995–1011.

Williams, K. (2006). *Human factors implications of unmanned aircraft accidents: Flight-control problems.* Oklahoma City, OK: FAA Civil Aerospace Medical Institute.

Wise, J. A., & Sherwin, G. W. (1989). An empirical investigation of the effect of virtual collimated displays on visual performance. *Proceedings of the 5th International Symposium on Aviation Psychology,* Columbus, OH.

Wood, W. T., & Wood, S. K. (1987). Icons in everyday life. In G. Salvendy, S. L. Sauter, & J. J. Hurrell (Eds.), *Social, ergonomic and stress aspects of working with computers* (pp. 97–104). Amsterdam: Elsevier.

Woods, D., & Watts, J. (1997). How not to have to navigate through too many displays. In M. Helander, T. Landauer, & P. Prabhu (Eds.), *Handbook of human-computer cooperation* (2nd ed.) (pp. 617–650). New York: Elsevier.

15

Cockpit Automation: Still Struggling to Catch Up…

Thomas Ferris and Nadine Sarter

Center for Ergonomics, Department of Industrial
and Operations Engineering, University of Michigan

Christopher D. Wickens

Alion Science and Technology

INTRODUCTION

In 1988, for the first edition of this book, Earl Wiener wrote in his chapter titled "Cockpit Automation" that "The rapid introduction of computer-based devices into the cockpit has outstripped the ability of designers, pilots, and operators to formulate an overall strategy for their use and implementation. The human factors profession is struggling to catch up." He ended by saying that a harmonious crew-automation interface needs to be developed, guided by an overall design philosophy.

Over the past 20 years, progress has been made toward this goal but surprisingly many issues are still unresolved. New ones are emerging as a result of the increasing complexity and volume of air traffic operations and the introduction of yet more automated systems that are not well integrated. The human factors profession has "caught up" in the sense that, since the first edition of this book, a large body of research has improved our understanding of (breakdowns in) the interaction between pilots and automated flight deck systems, such as the Flight Management System (FMS) or the Traffic Alert and Collision Avoidance System (TCAS). Also, promising solutions to some known problems in the form of improved design, training, and procedures have been proposed and tested. The main goal of this chapter is to summarize this work and provide an update of the existing knowledge base on issues related to the design and use of cockpit technologies. Different levels and capabilities of automated systems will be reviewed. Next, breakdowns in pilot-automation interaction will be discussed, both in terms of research methods that were used to identify and analyze problems and with respect to the nature of, and contributing factors to, observed difficulties with mode awareness, trust, and coordination.

While undoubtedly progress has been made with respect to under-standing and addressing past problems, it appears that, at the same time, "the human factors profession is [yet again] struggling to catch up." The envisioned Next Generation Air Traffic System (NEXTGEN; JPDO, 2008)—a completely transformed aviation system intended to accommodate the expected increase in traffic volume and complexity by 2015—is likely to create new oppor-tunities but also new challenges for airborne and ground-based operators as well as designers of support technologies. It will lead to the addition and reallocation of tasks and functions, require high levels of coordination between all players in the system, and call for the introduction of more sophisticated automated systems and procedures. Currently, NEXTGEN operations are still rather ill-defined, and the human factors community thus lacks a clear target to be sufficiently proactive and thus anticipate and avoid likely new demands and difficulties. A second goal of this chap-ter is therefore to lay out the current vision of the future aviation system and highlight, in the last part, some of the likely challenges and implications of moving to this more flexible and distributed mode of traffic operations.

AUTOMATION LEVELS, PROPERTIES, AND CAPABILITIES

The introduction of automation to the flight deck can be justified on many grounds, including increases in safety, economy, and precision, as well as workload reduction. The extent to which these goals are achieved, however, depends ultimately on the particular implementation of the automation and its ability to communicate and coordinate activities with its human operators. The question is no longer whether or not to automate and whether or not pilots should turn the automation on or off—an oversimplification often invoked in the early days of cockpit automation. Rather, a more complex view has emerged that distinguishes various stages, levels, and forms of automation and is concerned with how to support pilots in dynamically configuring and collaborating with their machine counterparts.

For example, Parasuraman, Sheridan, and Wickens (2000) have developed a taxonomy of stages and levels in which four stages of automation correspond to the stages of human information processing that are assisted by the automation:

1. sensation and selective attention through automated information acquisition and filtering;
2. perception, working memory, and situation assessment through automation information integration and intelligent inference;
3. action choice through automated decision aiding;
4. action execution through automated control.

Clearly, these stages of information processing are not independent of one another and are not necessarily completed in sequence. However, for the purpose of classifying, selecting, and combining automated systems, this taxonomy can be a useful tool.

At each stage, the automation can assume more or less authority and autonomy. For example, consider an automation system that supports pilots in traffic conflict detection and avoidance. At a low level of stage 1, all traffic around the own aircraft may be presented on a CDTI (cockpit display of traffic information). At the highest level of stage 1, only the aircraft representing the highest threat (closest temporal proximity) is presented (Wickens, Helleberg, & Xu, 2002). At stage 2, the system portrays only predicted tracks (low level stage 2), or, portrays the predicted tracks coupled with an estimate of the time until loss of separation and the likelihood of this event (high automation level at stage 2). At stage 3, the system may recommend that an avoidance maneuver is needed (low level) or may recommend a particular maneuver choice (e.g., "turn right"; high level). As another example at stage 3, the automation may implement a chosen maneuver provided pilot approval has been obtained (low level—also referred to as "management-by-consent"), or it may proceed with the maneuver if the pilot does not veto it within a certain time frame (highest level—an approach called "management-by-exception").

Finally, at stage 4, increasing levels of automation may reduce the amount of physical work the pilot must carry out, from "hand flying" the aircraft (continuous control of ailerons, thrust, and elevators; low level) to controlling trajectories of heading, vertical speed, and airspeed via the mode control panel (higher level) to directly inputting waypoints in the FMS (highest level). In control theory terms, this may be described as progressively decreasing the amount of *inner loop control* (Wickens, 2003b).

The above taxonomy underlies some of the options that designers of aviation systems have considered in their research. For example, high levels of stage 1 automation are invoked when "decluttering" tools are used in displays (Mykityshyn, Kuchar & Hansman, 1994). The distinction between status and command displays (Hicks & DiBrito, 1998; Sarter & Schroeder, 2001; Wickens, 2003a) is inherent in contrasting stage 2 versus stage 3 automation, respectively. So also is the distinction between some advanced synthetic vision systems (stage 2) and the highway-in-the-sky. The synthetic vision display depicts terrain and other environmental features, but does not in itself recommend an action. The highway in the sky display (stage 3) in contrast, recommends a flight path trajectory to be followed (Wickens & Alexander, 2009).

At each stage and level of automation, there are a variety of human-automation interaction issues that must be addressed in the actual implementation of the system. For example, in designing alerting devices and diagnostic automation (e.g., collision alerts, intelligent engine diagnostics; stages 1 and 2), a key element is setting the threshold whereby the automation "decides" how serious a situation must become before the pilot or controller is alerted. When the threshold is set too low, humans may not be warned of critical events or they are warned too late; but when set too high, false alarms proliferate. Possible ways to reduce the problem include: (1) providing the human user with access to the raw data in parallel with the automated alert, (2) assuring that the false alarm rate is not overwhelmingly high, and (3) introducing "likelihood alarms" (e.g., Sorkin, Kantowitz, & Kantowitz, 1988) that provide a more graded (rather than single state) estimation of the probability that a problematic or dangerous event will be encountered.

The following sections will discuss a wide range of issues related to human-automation interaction. These issues are, in part, a function of different types, stages, and levels of automation. They also involve two fundamental characteristics of automation design: observability and directability (Christoffersen & Woods, 2002). Observability refers to an operator's ability to see current and future targets and activities of the automation. Directability is concerned with enabling the user to (re)direct the automation when intervention is required. Aspects of the relationship between human and machine, such as trust and reliance, will be briefly reviewed also.

BREAKDOWNS IN PILOT-AUTOMATION INTERACTION: RESEARCH METHODS AND FINDINGS, OPERATIONAL PROBLEMS, AND PROPOSED SOLUTIONS

Since the early days of cockpit automation, a variety of reasons for breakdowns in the interaction between pilots and their automated systems have been identified, documented, and researched. Better insight into these issues has triggered research and development activities that try to correct current problems and identify and prevent possible future ones resulting from changes in technologies and procedures. The following section will discuss the forms of breakdowns that have received the most attention over the past 20 years. Common methods of investigation as well as reasons for, and proposed countermeasures to, these breakdowns will be reviewed.

Research Methods

A wide range of complementary methods has been employed to study pilot-automation interaction over the past 20 years. Databases of aviation incidents and accidents, such as NASA's Aviation Safety Reporting System (ASRS), and reports issued by agencies such as the National Transportation Safety Board (NTSB) have been scoured to identify factors that can contribute to breakdowns in joint system performance and to compare the contexts and events surrounding them (e.g., Eldredge, Mangold, & Dodd, 1992; Funk et al., 1999). Pilot surveys have revealed attitudes and preferences toward automation and suggested further contributing factors to breakdowns in pilot-automation coordination and communication (e.g., Hutchins, Holder, & Hayward, 1999; Sarter & Woods, 1992; 1997). Direct observations of flight crews from the jumpseat (e.g., Damos, John, & Lyall, 1999) or, more commonly, in flight simulators (e.g., Sarter & Woods, 1992; 1994; 2000) contributed to a better understanding of design-induced problems and training needs. Under more controlled conditions, laboratory experiments and simulator studies have provided a wealth of performance (e.g., detection rates and response time for various task-relevant alerts or events) and psychological (e.g., mental workload or situation/mode awareness) data. More recently, improvements in eyetracking technology have enabled researchers to monitor pilots' attention allocation strategies as they interact with automated systems (rather than having to infer their monitoring behavior from behavioral data) (e.g., Sarter, Mumaw, Wickens, 2007; Wickens, Goh, Helleberg, Horrey, & Talleur, 2003). By analyzing the cockpit as a sociotechnical distributed system with humans and automation providing complementary contributions, a deeper understanding has been achieved about the roles of each system component (e.g., Hutchins, 1995; Rasmussen, 1999).

In addition to this wide spectrum of empirical studies, various modeling techniques have been used to define engineering models and capture pilots' mental models of automated flight deck systems (in particular the FMS) in an effort to identify mismatches between these models that help explain and predict the likelihood of performance breakdowns. By modeling systems according to different types (e.g., information, decision, and action automation) and/or levels of automation (e.g., Billings, 1997; Parasuraman et al., 2000), researchers have attempted to identify appropriate configurations and roles for pilots and the automation. Formal and computational models have been used also to classify and analyze cognitive processes underlying pilot-automation interaction (e.g., Polson, Irving, & Irving, 1994), as well as predict pilots' mental workload and their focus of attention (e.g., Wickens et al., 2003).

Breakdowns in Pilot-Automation Interaction Related to System Observability

One major concern with cockpit automation has been its low observability (i.e., lack of adequate feedback about targets, actions, decision logic, or operational limits) in combination with its high degree of complexity and authority. Low observability has been shown to lead to a lack or loss of mode awareness, that is, a lack of knowledge and understanding of the current and future automation configuration and behavior (Billings, 1997; Funk et al., 1999; Sarter, 2008; Sarter & Woods, 1992, 1994; 1995; 1997). One manifestation of reduced mode awareness are automation surprises, in which pilots detect a discrepancy between actual and expected or assumed automation behavior (Sarter, Woods, & Billings, 1997), often resulting from uncommanded or indirect (i.e., without an explicit instruction by the pilot) mode transitions (Johnson & Pritchett, 1995; Sarter & Woods, 1992; Wiener, 1989). These uncommanded transitions may occur as a result of previously programmed instructions, exceedance of the design or flight envelope parameters for the current mode, or sensed environmental conditions (e.g., leveling off upon reaching a target altitude; Abbott et al., 1996; Sarter & Woods, 1997). Most annunciations associated with mode states and transitions tend to be cryptic alphanumeric indications that are too subtle to capture pilots' attention when the flight crew is busy performing other tasks or not expecting, and therefore not looking for, a change. Pilots tend not to monitor automation-related feedback to the extent expected of them, and they do not always process it with sufficient depth (Abbott et al., 1996; Sarter et al., 2007). Instead, they rely more heavily on the observation of an actual change in airplane behavior which puts them in a reactive state and may cause problems in case of slow smooth changes in heading or altitude (Sarter & Woods, 1995).

Reduced mode awareness can also increase the likelihood of mode errors. Mode errors of commission occur when a pilot executes an action that is appropriate for the assumed but not the actual current mode of the system. Mode errors of omission take place when the pilot fails to take an action that would be required given the currently active automation configuration and behavior (Abbott et al., 1996; Sarter, 2008; Sarter & Woods, 1992, 1994; 1997; 2000; Sarter et al., 1997, Wiener, 1989). Mode errors have been a major contributing factor in several aviation incidents and accidents (Billings, 1997; Funk et al., 1999). For example, in 1992, a highly automated aircraft crashed into the Vosges Mountains outside of Strasbourg, France, when the pilots changed the lateral navigation mode but did not realize that, due to system coupling, this resulted in a change of the vertical navigation mode also. When they entered the digits "33" for a 3.3 degree descent angle, the system instead interpreted their input as a rate of descent, and the aircraft started descending at a much faster than intended rate of 3300 feet/minute. This descent profile, in combination with other factors such as a very busy approach and pilots' lack of terrain awareness, resulted in the aircraft crashing into the mountainside (Investigation Commission of Ministry of Transport—France, 1993).

Several ideas have been proposed to better support mode awareness and reduce the likelihood of mode errors. To address feedback issues with indirect mode transitions, the design of flight mode annunciators (FMAs) has been examined (e.g., Abbott et al., 1996). One shortcoming of current FMAs is their exclusively visual nature, which implies that they compete with considerable visual attentional demands from other tasks and displays. Pilots' visual attention could be guided to these indications with 3D auditory cues, which have shown to improve visual detection of critical objects and information inside and outside of the cockpit (e.g., Begault & Pittman, 1996), and additionally may improve awareness and reduce workload by reducing some visual sampling requirements (Parker, Smith, Stephan, Martin, & McAnally, 2004). Multisensory notifications can also be used to supplement visual displays in order to more reliably communicate flight-relevant information to a pilot while reducing the demand on foveal visual resources (Sarter, 2000). For example, Nikolic and Sarter (2001) and Sklar and Sarter (1999) showed how peripheral visual and vibrotactile notifications, respectively, led to a substantial improvement in the detection of unexpected mode transitions, independent of concurrent task load. However, a note of caution is needed when designing nonvisual notification systems: because of the relatively lower bandwidth of nonvisual communication channels, there may be a tendency to underspecify the message, assuming pilots will consult more complete visual sources of information for details. This may leave the multisensory notification somewhat ambiguous: telling the crew that *something* happened/is wrong, but not clearly *what*, which could cost valuable time in identifying the problem (Woods, 1995). Underspecification of an auditory alert played a role, for example, in a 2005 crash north of Athens, Greece (Air Accident Investigation & Aviation Safety Board, 2006). Unknown to the flight crew, the automatic cabin pressurization system of the aircraft was disabled, which caused an altitude horn to activate shortly after takeoff. The ambiguous sound of the horn was misidentified as a false takeoff configuration warning and ignored. By the time the aircraft reached cruising altitude, all crew and passengers had been rendered unconscious due to hypoxia and none survived the subsequent crash. One way this particular auditory alert could have been more informative and/or less ambiguous would be to rely on intuitive mappings with the system being represented (e.g., Perry, Stevens, Wiggins, & Howell, 2007), for example, by replacing the horn sound with a synthesized sound of air rushing from a depressurized container.

More consideration needs to be given to display context also. While an abrupt onset or flashing visual cue can sometimes be sufficient to capture attention, properties of surrounding display elements, such as color similarity, movement of background elements, and display eccentricity from a pilot's gaze direction can significantly reduce detection of those indications (Nikolic, Orr, & Sarter, 2004). Pilots tend to be reluctant to visually sample disparate displays which they feel are irrelevant in the given context; this can be problematic since display relevancy is determined by mental models which are often inadequate (Sarter et al., 2007). Perceptual and attentional phenomena such as inattentional and change blindness can also pose a challenge for designing displays to effectively capture attention. Inattentional blindness describes how salient but unexpected visual events can go undetected even when within the field of view (e.g., Mack & Rock, 1998). Change blindness relates to the observation that people are surprisingly poor at noticing even large changes to their visual scene when the changes occur simultaneously with scene disruptions, which may include eye blinks or saccades between disparate displays (e.g., Durlach, 2004). These phenomena can lead to pilots missing not only visual warnings and automation-initiated display changes but also immediate threats such as obstructions on a runway during landing procedures (Haines, 1991; Ververs & Wickens, 1998). To combat the effects of change blindness, display designers may co-locate, integrate, or reduce the number of visual displays in order to reduce the number of saccades required in a visual sampling pattern. Durlach (2004) provides a list of training methods and design guidelines to facilitate detection of changes in critical sources of visual information, such as programmable change alerts and a dedicated change detection tool that logs changes in relevant systems.

Another approach to improving the observability of flight deck systems is to apply principles of ecological interface design (EID; Rasmussen, 1999). Some proposed designs that were created with these principles in mind have already shown success. For example, some researchers have looked at ways to integrate information regarding aircraft speed and altitude—which are interrelated but often subject to different constraints and involve separate displays and controls—into one "total energy" display (e.g., Amelink, Mulder, van Paassen, & Flach, 2005; Catton, Starr, Noyes, Fisher, & Price, 2007). Pilots operating while monitoring total energy displays can be more acutely aware of the energy state of the plane in relation to the coupled constraints for the speed and altitude measures, resulting in greater flight safety and efficiency. Hutchins (1996) demonstrated another way that information from cockpit automation displays could be integrated in the so-called IMMI (Integrated Mode Management Interface). The IMMI integrated the horizontal and vertical navigation displays, graphical information regarding the engaged and available modes for each, and the future flight paths for the current automation program.

The incremental addition of new technologies and display components to existing systems over time introduces another problem in providing adequate feedback to flight crews. Information that is critical for crews to make automation management decisions may be distributed across multiple cockpit locations and pages of the control display unit (CDU), the strategic interface for programming the automation (Sarter & Woods, 1992; 1997; Eldredge, Mangold, & Dodd, 1992). To adequately monitor autoflight performance, pilots may need to access and integrate information from the primary flight display (PFD), the mode control panel (MCP), the CDU, and the Engine Indicating and Crew Alerting System (EICAS; Johnson & Pritchett, 1995). This "fragmentation of feedback" can contribute to automation awareness issues.

In order to address system observability issues without exacerbating "fragmentation of feedback" problems, perhaps it is appropriate to explore more comprehensive, ground-up interface redesigns. One recent example of such a redesign of the FMS interface was undertaken by Boorman and Mumaw (2004), showing improved visibility of automated system behavior, in particular for vertical navigation functions, and providing more effective feedback about autoflight targets and future aircraft flight path changes. Simulator evaluations of this interface have shown initial success in that training can be completed in significantly fewer trials than with the traditional interface (Prada, Mumaw, Boehm-Davis, & Boorman, 2006).

Breakdowns in Pilot-Automation Interaction Related to Directability

Directability refers to the pilots' ability to (re)direct efficiently and safely the activities of the automation when his/her involvement becomes necessary. Directability requires that the interface is designed in such a way that it avoids what Norman (1986) has called the "gulf of execution," where an operator struggles with identifying and operating the proper controls and commands to translate an intended action into the machine's "language." One problem related to directability is a lack of standardization of the physical properties and arrangement of controls. Different aircraft employ automation controls with similar shape, feel, and/or location that activate different systems or require different manipulations. For example, there are different conventions for selecting or engaging modes from the MCP, such as pushing or pulling the corresponding button/knob (Abbott et al., 1996). Similarly, there are no standardized locations for some critical controls, such as the takeoff/go-around and autothrottle disconnect switches (Abbott et al., 1996). Combined, these inconsistencies can leave pilots who transition between aircraft or airlines highly vulnerable to errors, especially when under stress.

Spatial arrangements of controls are also an area of concern. One way real estate on the flight deck can be economized is through the use of multifunction knobs—which activate different functions depending on, for example, how far they are pushed in/pulled out. However, the design and configuration of these knobs are not always easily observable, which can lead to mode errors (Wiener, 1993). Without adequate spatial segregation between critical input devices, such as mode selection knobs, the controls can be confused or inadvertently activated. This problem is illustrated by a 1990 accident in Bangalore, India. One of the stated causes of the accident was that "The vertical speed and altitude selection knobs of the Flight Control Unit (FCU) are close to each other, and instead of operating the vertical speed knob, the pilot [...] had inadvertently operated the altitude selection knob" (Ministry of Civil Aviation—India, 1990). Similarly, one main contributor to a 1994 accident in Nagoya, Japan, was that the First Officer accidentally triggered the takeoff/go-around (TOGA) switch, located on the engine power lever, during the approach. The pilots, apparently assuming they were still in normal approach mode, fought the autopilot's actions to initiate a go-around, which eventually resulted in a controlled flight into terrain accident (Aircraft Accident Investigation Commission, 1996).

Another important challenge to directability is that, with the increasing flexibility and complexity of modern flight deck systems, pilots sometimes struggle with deciding which method or mode is best suited for accomplishing a desired task in the current context (Sarter & Woods, 1992). For example, on some flight decks, five different vertical navigation modes can be invoked, each with its own constraints and benefits. Making these choices is difficult because of limited training on some aspects of the automation, a lack of visualizations of the airplane trajectory resulting from mode activation, and because choices sometimes need to be made under time pressure. To address this problem, work has been done to reorganize the hierarchy of automation modes and procedures to make them clearer, grouped by similarity of function and accessed with common procedural steps (e.g., Boorman & Mumaw, 2004).

Control design not only affects individual pilot performance but also the coordination between crew members. Much like displays, the state and behavior of controls around a cockpit can serve as external memory aids and can support valuable forms of indirect communication between crew members (Hutchins, 1995). For example, a common complaint with the introduction of side-stick controls in fly-by-wire aircraft is that one pilot cannot easily see the position and manipulation of the other pilot's side-stick. Since the two side-sticks are not coupled, pilots do not receive any direct visual or tactile feedback of the other pilot's inputs (Abbott et al., 1996). Another side effect of the introduction of fly-by-wire controls can be an increased monitoring load. With older mechanical systems, pilots were able to notice even unexpected movements of the manual throttle levers based on peripheral visual cues, but thrust levers on fly-by-wire aircraft remain in detent positions even as thrust is being varied by the autothrust system. Thus, pilots need to actively search for information on electronic engine displays that are not very effective at capturing attention in case of unexpected changes (Sarter & Woods, 1997).

Unintended Performance Consequences of Cockpit Automation

Designers of automated systems strive to achieve a number of goals, including a reduction in pilot workload, relief from having to perform mundane tasks on a regular basis, and an extremely high level of system reliability. While these are well-intentioned driving forces behind the introduction of automation, they have resulted in some unexpected difficulties (e.g., Bainbridge, 1983) that will be discussed in the following sections.

Workload Imbalance

One of the primary goals of automating tasks in the cockpit was, and continues to be, a reduction of physical and cognitive workload. Early on, it appeared that this goal had been achieved to some extent (Wiener, 1988). However, as more experience with cockpit automation accumulated, it became clear that the overall amount of workload was not affected as much as it was re-distributed over crewmembers and over time. "Clumsy" automation, as Wiener (1989) called it, led to a further reduction in workload when it was already low (e.g., the cruise phase of flight), while during periods of high tempo and workload (e.g., the approach and departure phases), the need for instructing and monitoring the automation actually increased workload in some cases (Billings, 1997; Parasuraman & Riley, 1997; Wiener, 1989). Clumsy automation can therefore increase the risk of pilot error in two ways: through vigilance decrements during long periods of inactivity, and through inadequate monitoring or procedural errors when numerous tasks compete for attention during high tempo periods.

A tragic example of the consequences of increased workload and attentional demands while interacting with automation during critical phases of flight occurred in a 1989 airplane accident near Kegworth, England (Air Accident Investigation Branch, Department of Transport—England, 1990). While trying to accomplish an emergency landing because of a malfunctioning engine, both the captain and first officer repeatedly tried and failed to program the FMS to display landing patterns for a nearby airport. This activity consumed the first officer's attention for a full 2 minutes, and may have affected his ability to notice that the captain was about to shut down the wrong, healthy engine, which ultimately resulted in a catastrophic crash.

Deskilling

In addition to monitoring the automation, the other task left for pilots on highly automated flight decks is to take over from the automation in cases of failure or undesired system behavior

(Bainbridge, 1983). One problem with this task allocation is that, over time, continued and extensive use of automation can lead to overreliance on technological assistance and the loss of psychomotor and cognitive skills required for manual flight—a phenomenon referred to as deskilling. Thus, in those rare circumstances when pilots need to intervene and manually control the airplane, they may struggle, especially since they are now required to manually control a system that is not functioning properly (Damos et al., 1999; Hutchins et al., 1999; Sarter & Woods, 1997). Deskilling can lead to a "vicious cycle" of performance degradation (Parasuraman & Riley, 1997) when pilots' realization of their own skill loss leads to even heavier reliance on automation.

Deskilling may have played a role in a controlled-flight-into-terrain accident outside of Cali, Columbia in 1995 (Aeronautica Civil, 1996). In this case, the pilots, who were accustomed to relying heavily on FMS-generated assistance and displays for navigation, exhibited a diminished ability to recognize the proximity of terrain and to quickly determine that a waypoint they were attempting to locate was behind the aircraft—information which would have been more immediately realized with traditional methods of consulting flight charts.

Reliability, Reliance, and Trust Issues

Automated systems on modern aircraft are extremely reliable and will continue to improve as more sophisticated and precise sensor technologies are being developed, and as researchers attempt to more efficiently tune the automation, for example, by determining the most cost-effective signal/noise threshold for notifications/alerts by applying Signal Detection Theory (e.g., Parasuraman & Byrne, 2003). Even with these continuing improvements, however, malfunctions will continue to occur and affect pilots' trust in, and reliance on, their automated systems. If and when automation failures do occur, it can be especially problematic when the nature and extent of failures are not apparent to the human operator. For example, partial FMS failures can leave pilots unsure of which subsystems are still active and available, and unaware of how the failure may interact with the overall automation configuration (Sarter & Woods, 1992).

Independent of an automated system's actual reliability, the *perceived* reliability of a system has a strong influence on the amount of trust operators put in it, and consequently, the likelihood of its use (Lee & Moray, 1992; 1994). If the level of trust is inappropriate relative to the actual reliability of the system—in other words, if trust is "miscalibrated"—automation use may be inefficient and/or the negative effects of malfunctions may be exacerbated (Lee & See, 2004; Parasuraman & Riley, 1997). An excessive amount of trust in a flightdeck automation system may result in misuse of the system, in which crews may continue to rely on the automation after it malfunctions or has otherwise proven itself unreliable (Parasuraman & Riley, 1997). As a consequence, crews may be placed in a position of needing to intervene when they have not carefully monitored the state of the automated process, in a state sometimes referred to as "complacency". Conversely, an inappropriately low level of trust in a system can lead to disuse of an otherwise beneficial system. Disuse of automation can occur when systems have a high propensity for false alarms, often as a direct consequence of setting the alarm decision criteria fairly low because of the high cost of a missed warning (Parasuraman & Riley, 1997). A high false alarm rate for automated warnings in the cockpit can lead to pilots ignoring or disabling the alarms due to the "crying wolf" effect (Sorkin, 1988).

To help pilots appropriately calibrate trust with automated aircraft systems, researchers have taken a deeper look at the issues that affect trust in automation, and the contexts in which trust levels are most appropriate (e.g., Lee & See, 2004). For example trust degrades less, and "cry wolf" behavior is less, when errors of automation are smaller and more understandable (Lees & Lee, 2007; Wickens, Rice, Keller, Hutchins, Hughes & Clayton, 2009). One method that has been found to support trust calibration and increase the likelihood of correct compliance to automation-generated alerts is to display the underlying logic for the alerts. For example, when pilots are able to view displays depicting how the actions of surrounding aircraft triggered TCAS alerts, they show a higher rate of compliance and faster responses to the alerts (Pritchett & Hansman, 1997). Another promising approach is to move from binary alerts to a more continuous display of the automation's alerting logic, employing graded notification strategies (e.g., Lee, Hoffman, & Hayes, 2004).

This approach is exemplified in so-called likelihood-alarm displays, which can indicate the likelihood, rather than presence or absence, of a dangerous condition, ranked according to the automation's confidence in its own diagnostic capability (Sorkin et al., 1988). For example, Xu, Wickens, and Rantanen (2007) showed how a CDTI display that employs three levels of alerting for decreasing distance of approaching aircraft not only improved estimation of miss distance with the approaching aircraft, but led to a higher likelihood for pilots to review raw data, so that even when the automation was not completely reliable, task performance improved.

Another form of a likelihood-alarm display was successfully demonstrated by McGuirl and Sarter (2006), who asked pilots to fly a series of approaches in simulated icing conditions. A neural network-based automated decision aid assisted them in noticing the presence and location (wing or tailplane) of icing and, in one condition, recommended required responses to the icing condition. One group of pilots was provided with a continuously updated 5-minute trace of system confidence in its ability to diagnose the current icing condition. When compared to those who were told only the overall system reliability, the pilots who received the continuously-updated confidence information were faster and more accurate at calibrating their level of trust to the actual reliability of the decision aid, and also showed improved performance—experiencing significantly fewer icing-related stalls and showing a higher likelihood to correctly switch to alternative recovery actions when the decision aid's diagnosis of the location of icing was found to be incorrect.

LOOKING AHEAD: OPPORTUNITIES AND CHALLENGES ASSOCIATED WITH THE MOVE TOWARD THE NEXT GENERATION AIR TRANSPORTATION SYSTEM (NEXTGEN)

At the beginning of this chapter, we suggested that the human factors profession may soon again be struggling to catch up with the impact of technological advances, just like it was at the time Earl Wiener wrote his chapter on cockpit automation in 1988. This time, it is the envisioned Next Generation Air Traffic System (NEXTGEN; JPDO, 2007) that is likely to create new opportunities and demands for both airborne and ground-based operators and that will require the introduction of more advanced automation support. The challenge for the human factors community is that it faces an "envisioned world" problem (Hoffman & Woods, 2000). In other words, researchers need to figure out how to predict the interaction between technology and human operators in a domain that does not yet exist. They need to anticipate how technology will shape performance and how operators will adapt systems to their needs and preferences. NEXTGEN is still ill-defined, and the human factors community thus lacks a clear target to be sufficiently proactive and to anticipate and avoid likely challenges.

There is widespread agreement that NEXTGEN operations will involve a wide range of key capabilities, including (a) network-enabled information access to support distributed decision making, (b) real-time weather information, (c) broad-area precision navigation which allows pilots to define their own desired flight paths, (d) visual flight capabilities in Instrument Flight Rules (IFR) conditions, (e) knowledge of the relative location and movement of other aircraft in all visibility conditions, (f) reduced separation standards and maximized use of runways, and (g) trajectory- and performance-based operations. Trajectory-based operations are based on four-dimensional flight paths that, by taking into consideration time constraints, optimize both individual flights and the overall air traffic system. Performance-based navigation will utilize new navigation capabilities such as Area Navigation (RNAV) and Required Navigation Performance (RNP). RNAV supports point-to-point operations by allowing for flights along any desired flight path that is within coverage of ground- and space-based navigation aids. RNP adds onboard monitoring and alerting capabilities that track navigation performance and afford reduced obstacle clearance and closer route spacing.

Some potential difficulties with NEXTGEN operations can be predicted based on experiences with earlier automation technologies. For example, future aviation operations will depend heavily on the use of Automatic Dependent Surveillance Broadcast (ADS-B), which uses GPS satellite signals to provide air traffic controllers and pilots with highly accurate information to keep aircraft separated both in the sky and on the ground. A number of human factors challenges have been identified with respect to the future use of this particular technology (Funk, Mauro, & Birdseye, 2008). Pilots may become overconfident in, and overreliant on, ADS-B and ignore other available sources of information. Also, they may not adequately understand ADS-B capabilities and limitations. These kinds of problems have been encountered with earlier complex systems, such as the FMS.

A different, more challenging set of concerns is raised, however, by the significantly increased distributed and flexible nature of the envisioned overall air traffic system. These concerns are not related to any individual operator and technology but rather the coordination and collaboration between all players. More than ever before, a system-level approach to the design of technologies, tasks, and procedures will be needed. Among the critical questions that need to be addressed in the near future are: How can roles and responsibilities be shared dynamically between humans and automated systems and between airborne and ground-based personnel? And how can the considerable amount of uncertainty inherent in the future system be handled? For example, weather and wind create uncertainties for the prediction of flight paths which affects the appropriateness of separation decisions. Thus, new approaches will need to be developed for depicting dynamically the predicted and actual uncertainty of information and automation. One possible approach to this challenge—indications of system confidence in its own abilities—was described in the previous section (McGuirl & Sarter, 2006). Similar displays will need to be created to support pilots as they interact with the separation and navigation tools which are vital to NEXTGEN Operations.

SUMMARY AND CONCLUSIONS

In this chapter, we have provided an update and overview of the knowledge base on the nature of, reasons for, and countermeasures to breakdowns in pilot-automation interaction. Importantly, we highlighted that a more refined view of automation has emerged since the early days of advanced flight deck technologies. This more recent perspective considers the implications of introducing various stages, levels, and forms of automation that pilots need to dynamically configure and collaborate with, rather than adopt an all-or-none approach to the invocation of technology. The chapter also describes how, over the past 20 years, the use of a wide range of complementary research methods (such as mishap investigations, surveys, cognitive modeling, and experimental/simulation studies) has resulted in a better understanding of, and the design of promising countermeasures to, problems with system observability and directability that were observed in the early days of flight deck automation. For example, multimodal displays and EID-based interfaces have been shown to reduce the risk of breakdowns in mode and overall system awareness. Progress has been made also with respect to improving the directability of the automation. The need for increased standardization and reduced confusability of controls has been acknowledged. And recent attempts to simplify the mode structure of modern flight deck systems appear promising for assisting pilots in instructing the automation more efficiently. The chapter illustrates that the human factors community has "caught up" with many of the challenges posed by flight deck systems that have been in use for over 20 years. Most of the obstacles to reaping the benefits of the improved designs, training, and procedures resulting from this work are practical in nature. They include certification requirements, the limited number of new airplanes entering the market, and the costs associated with major changes to any existing flight decks (Sarter, 2008). In the last part of the chapter, we describe some of the new challenges and potential difficulties that appear on the horizon and may be encountered in the context of envisioned tasks, procedures, and technologies during future NEXTGEN operations. This section highlights the importance of being proactive and investing in research in advance of the planned transition to much more complex and demanding air transport operations rather than again struggling to catch up.

ACKNOWLEDGMENT

The preparation of this chapter was supported, in part, by the National Science Foundation [NSF] Graduate Research Fellowship Program (Recipient: Thomas Ferris; Coordinating Official: Erin

Cain), by a grant from the National Science Foundation (NSF grant # 0534281; Program manager Ephraim Glinert), and by a grant from NASA-Ames Research Center (NASA grant # NNX07AT79A; Technical Monitor: Dr. Michael Feary).

References

Abbott, K., Slotte, S., Stimson, D., Amalberti, R. R., Bollin, G., Fabre, F., Hecht, S., Imrich, T., Lalley, R., Lydanne, G., Newman, T., & Thiel, G. (1996). *The Interfaces Between Flightcrews and Modern Flight Deck Systems* (Report of the FAA Human Factors Team). Washington, DC: Federal Aviation Administration.

Aeronautica Civil—Republic of Columbia. (1996). Aircraft Accident Report. Controlled flight into terrain, American Airlines Flight 965, Boeing 757–223, Near Cali, Columbia, December 20, 1995. Bogota, Columbia: Aeronautica Civil.

Air Accident Investigation & Aviation Safety Board, Ministry of Transport & Communications—Hellenic Republic. (2006). *Aircraft Accident Report: Helios Airways Flight HCY522, Boeing 737–31S, at Grammatiko, Hellas, on 14 August 2005.* Available online at <http://www.moi.gov.cy/moi/pio/pio.nsf/All/F15FBD7320037284C2257204002B6243/$file/FINAL%20REPORT%205B-DBY.pdf>.

Air Accident Investigation Branch, Department of Transport—England. (1990). *Report on the accident to Boeing 737–400, G-OBME, near Kegworth, Leicestershire on 8 January 1989. Aircraft Accident Report No: 4/1990.* London: HMSO.

Aircraft Accident Investigation Commission, Ministry of Transport—Japan. (1996). *China Airlines Airbus Industrie A300B4–622R, B1816, Nagoya Airport, April 26, 1994.* Report 96–5. Ministry of Transport.

Amelink, M. H. J., Mulder, M., van Paassen, M. M., & Flach, J. (2005). Theoretical foundations for a total energy-based perspective flight-path display. *International Journal of Aviation Psychology, 15*(3), 205–231.

Bainbridge, L. (1983). Ironies of automation. *Automatica, 19,* 775–779.

Begault, D. R., & Pittman, M. T. (1996). Three-dimensional audio versus head-down Traffic Alert and Collision Avoidance System displays. *International Journal of Aviation Psychology, 6*(1), 79–93.

Billings, C. E. (1997). *Aviation Automation: The Search for a Human-Centered Approach.* Mahwah, NJ: Lawrence Erlbaum Associates.

Boorman, D. J., & Mumaw, R. J. (2004). A new autoflight/FMS interface: Guiding design principles. In A. Pritchett & A. Jackson (Eds.), *Proceedings of the International Conference on Human-Computer Interaction in Aeronautics [CD-Rom].* Toulouse, France: EURISCO International.

Catton, L., Starr, A., Noyes, J. M., Fisher, A., & Price, T. (2007). Designing energy display formats for civil aircraft: Reply to Amelink, Mulder, van Paassen, and Flach. *International Journal of Aviation Psychology, 17*(1), 31–40.

Christoffersen, K., & Woods, D. D. (2002). How to make automated systems team players. In E. Salas (Ed.), *Advances in Human Performance and Cognitive Engineering Research,* Volume 2: Automation (pp. 1–12). Oxford: England.

Damos, D. L., John, R. S., & Lyall, E. A. (1999). Changes in pilot activities with increasing automation. In R. S. Jensen, B. Cox, J. D. Callister, & R. Lavis

(Eds.), *Proceedings of the 10th International Symposium on Aviation Psychology* (pp. 810–815). Columbus: The Ohio State University.

Durlach, P. J. (2004). Change blindness and its implications for complex monitoring and control systems design and operator training. *Human-Computer Interaction, 19*(4), 423–451.

Eldredge, D., Mangold, S., & Dodd, R. S. (1992). *A review and discussion of Flight Management System incidents reported to the Aviation Safety Reporting System* Final report DOT/FAA/RD-92/2. Washington DC: U.S. Department of Transportation, Federal Aviation Administration.

Funk, K., Lyall, B., Wilson, J., Vint, R., Niemczyk, M., Suroteguh, C., & Owen, G. (1999). Flight Deck Automation Issues Available online at http://www.flight-deckautomation.com/fdai.aspx. *International Journal of Aviation Psychology, 9*(2), 109–123.

Funk, K., Mauro, R., & Birdseye, H. (2008). Identifying and addressing human factors issues of ADS-B and Cockpit Displays of Traffic Information. Presentation at Human Factors and NextGen: The Future of Aviation. University of Texas, Arlington, TX, May 28–29, 2008.

Haines, R. F. (1991). A breakdown in simultaneous information processing. In G. Obrecht & L. W. Stark (Eds.), *Presbyopia Research: From Molecular Biology to Visual Adaptation* (pp. 171–175). New York: Plenum.

Hicks, M., & DiBrito, G. (1998). Civil aircraft warning systems: Who's calling the shots?. In G. Boy, C. Graeber, & J. Robert (Eds.), *International Conference on Human-Computer Interaction in Aeronautics* (pp. 205–212). Montreal, Canada: Ecale Polytechnique de Montreal.

Hoffman, R., & Woods, D. (2000). Studying cognitive systems in context. *Human Factors, 42*(1), 1–7.

Hutchins, E. (1995). How a cockpit remembers its speeds. *Cognitive Science, 19*(3), 265–288.

Hutchins, E. (1996). *The Integrated Mode Management Interface* (NASA Contractor Report NCC 2–591). Moffett Field, CA: NASA Ames Research Center.

Hutchins, E., Holder, B., & Hayward, M. (1999). *Pilot Attitudes toward Automation.* Web-published at <http://hci.ucsd.edu/hutchins/aviation/attitudes/attitudes.pdf>.

Investigation Commission of Ministry of Transport—France. (1993). *Rapport de la Commission d'Enquete sur l'Accident survenu le 20 Janvier 1992 pres du Mont Saite Odile (Bas Rhin) a l/Airbus A.320 Immatricule F-GGED Exploite par lay Compagnie Air Inter.* Official English translation from the Ministere de l'Equipement, des Transports et du Tourisme, France. Ministere de l'Equipement, des Transports et du Tourisme.

Johnson, E. N., & Pritchett, A. R. (1995). *Experimental Study of Vertical Flight Path Mode Awareness.* Cambridge: Massachusetts Institute of Technology, Department of Aeronautics and Astronautics.

JPDO (Joint Planning and Development Office) (2008). Next Generation Air Transportation System Research and Development Plan FY 2009-FY 2013 (Draft 4).

Lee, J. D., Hoffman, J. D., & Hayes, E. (2004). Collision warning design to mitigate driver distraction. In *Proceedings of the SIGCHI Conference on Human Factors in Computing Systems, 6*(1), 65–72.

Lee, J. D., & Moray, N. (1992). Trust, control strategies and allocation of function in human-machine systems. *Ergonomics, 35*(10), 1243–1270.

Lee, J. D., & Moray, N. (1994). Trust, self-confidence, and operators' adaptation to automation. *International Journal of Human-Computer Studies, 40*(1), 153–184.

Lee, J. D., & See, K. A. (2004). Trust in automation: Designing for appropriate reliance. *Human Factors, 46*(1), 50–80.

Lees, N., & Lee, J. D. (2007). The influence of distraction and driving context on driver response to imperfect collision warning systems. *Ergonomics, 30*, 1264–1286.

Mack, A., & Rock, I. (1998). *Inattentional Blindness.* Cambridge, MA: MIT Press.

McGuirl, J. M., & Sarter, N. B. (2006). Supporting trust calibration and the effective use of decision aids by presenting dynamic system confidence information. *Human Factors, 48*(4), 656–665.

Ministry of Civil Aviation—India. (1990). *Report on accident to Indian Airlines Airbus A-320 aircraft VT-EPN at Bangalore, February 14, 1990.* Ministry of Civil Aviation, Government of India.

Mykityshyn, M. G., Kuchar, J. K., & Hansman, R. J. (1994). Experimental study of electronically based instrument approach plates. *International Journal of Aviation Psychology, 4*(2), 141–166.

Nikolic, M. I., Orr, J. M., & Sarter, N. B. (2004). Why pilots miss the green box: How display context undermines attention capture. *International Journal of Aviation Psychology, 14*(1), 39–52.

Nikolic, M. I., & Sarter, N. B. (2001). Peripheral visual feedback: A powerful means of supporting effective attention allocation in event-driven, data-rich environments. *Human Factors, 43*(1), 30–38.

Norman, D. A. (1986). Cognitive Engineering. In D. A. Norman & S. Draper (Eds.), *User Centered System Design: New Perspectives in Human-Computer Interaction.* Hillsdale, NJ: Lawrence Erlbaum Associates.

Parasuraman, R., & Byrne, E. A. (2003). Automation and human performance in aviation. In P. Tsang & M. Vidulich (Eds.), *Principles of Aviation Psychology* (pp. 311–356). Mahwah, NJ: Lawrence Erlbaum Associates.

Parasuraman, R., & Riley, V. (1997). Humans and automation: Use, misuse, disuse, abuse. *Human Factors, 39*(2), 230–253.

Parasuraman, R., Sheridan, T. B., & Wickens, C. D. (2000). A model for types and levels of human interaction with automation. *IEEE Transactions on Systems, Man, and Cybernetics, 30*, 286–297.

Parker, S. P. A., Smith, S. E., Stephan, K. L., Martin, R. L., & McAnally, K. I. (2004). Effects of supplementing head-down displays with 3-D audio during visual target acquisition. *International Journal of Aviation Psychology, 14*(3), 277–295.

Perry, N. C., Stevens, C. J., Wiggins, M. W., & Howell, C. E. (2007). Cough once for danger: Icons versus abstract warnings as informative alerts in civil aviation. *Human Factors, 49*(6), 1061–1071.

Polson, P., Irving, S., & Irving, J. E. (1994). Applications of formal models of human-computer interaction to training and use of the control and display unit. Boulder: University of Colorado *FAA Technical Report #94–08.*

Prada, L. R., Mumaw, R. J., Boehm-Davis, D. A., & Boorman, D. J. (2006). Testing Boeing's flight deck of the future: A comparison between current and prototype autoflight panels. In: *Proceedings of the Human Factors and Ergonomics Society 50th Annual Meeting* (pp. 55–58). Santa Monica, CA: HFES.

Pritchett, A. R., & Hansman, R. J. (1997). Pilot non-conformance to alerting system commands. In R. S. Jensen & L. Rakovan (Eds.), *Proceedings of the 9th International Symposium on Aviation Psychology* (pp. 274–279). Columbus: The Ohio State University.

Rasmussen, J. (1999). Ecological interface design for reliable human-machine systems. *International Journal of Aviation Psychology, 9*(3), 203–223.

Sarter, N. B. (2000). The need for multisensory interfaces in support of effective attention allocation in highly dynamic event-driven domains: The case of cockpit automation. *International Journal of Aviation Psychology, 10*(3), 231–245.

Sarter, N. (2008). Mode Errors on Automated Flight Decks: Illustrating the Problem-Driven, Cumulative, and Interdisciplinary Nature of Human Factors Research. *Human Factors (Special Golden Anniversary Issue), 50*(3), 506–510.

Sarter, N. B., Mumaw, R. J., & Wickens, C. D. (2007). Pilots' monitoring strategies and performance on automated flight decks: An empirical study combining behavioral and eye-tracking data. *Human Factors, 49*(3), 347–357.

Sarter, N. B., & Schroeder, B. (2001). Supporting decision making and action selection under time pressure and uncertainty: The case of in-flight icing. *Human Factors, 43*(4), 573–583.

Sarter, N. B., & Woods, D. (1992). Pilot interaction with cockpit automation: Operational experiences with the Flight Management System. *International Journal of Aviation Psychology, 2*(4), 303–321.

Sarter, N. B., & Woods, D. D. (1994). Pilot interaction with cockpit automation II: An experimental study of pilots' model and awareness of the Flight Management System. *International Journal of Aviation Psychology, 4*(1), 1–28.

Sarter, N. B., & Woods, D. D. (1995). How in the world did we ever get into that mode? Mode error and awareness in supervisory control. *Human Factors, 37*(1), 5–19.

Sarter, N. B., & Woods, D. D. (1997). Team play with a powerful and independent agent: Operational experiences and automation surprises on the Airbus A-320. *Human Factors, 39*(4), 553–569.

Sarter, N. B., & Woods, D. D. (2000). Team play with a powerful and independent agent: A full-mission simulation study. *Human Factors, 42*(3), 390–402.

Sarter, N. B., Woods, D. D., & Billings, C. E. (1997). Automation surprises. In G. Salvendy (Ed.), *Handbook of Human Factors & Ergonomics* (2nd ed.). New York: Wiley.

Sklar, A. E., & Sarter, N. B. (1999). Good vibrations: Tactile feedback in support of attention allocation and human-automation coordination in event-driven domains. *Human Factors, 41*(4), 543–552.

Sorkin, R. D. (1988). Why are people turning off our alarms? *Journal of the Acoustical Society of America, 84*, 1107–1108.

Sorkin, R. D., Kantowitz, B. H., & Kantowitz, S. C. (1988). Likelihood alarm displays. *Human Factors, 30*(4), 445–459.

Ververs, P. M., & Wickens, C. D. (1998). Head-up displays: Effect of clutter, display intensity, and display location on pilot performance. *International Journal of Aviation Psychology, 8*(4), 377–403.

Wickens, C. D. (2003a). Aviation displays. In P. S. Tsang & M. A. Vidulich (Eds.), *Principles and practice of Aviation Psychology* (pp. 147–200). Mahwah, NJ: Lawrence Erlbaum Associates.

Wickens, C. D. (2003b). Pilot actions and tasks: Selections, execution, and control. In P. S. Tsang & M. A. Vidulich (Eds.), *Principles and Practice of Aviation Psychology* (pp. 239–263). Mahwah, NJ: Lawrence Erlbaum Associates.

Wickens, C. D. & Alexander, A. L. (2009). Attentional tunneling and task management in synthetic vision displays. *International Journal of Aviation Psychology*, *19*(2), 182–199.

Wickens, C. D., Helleberg, J., & Xu, X. (2002). Pilot maneuver choice and workload in free flight. *Human Factors*, *44*(2), 171–188.

Wickens, C. D., Goh, J., Helleberg, J., Horrey, W. J., & Talleur, D. A. (2003). Attentional models of multitask pilot performance using advanced display technology. *Human Factors*, *45*(3), 360–380.

Wickens, C. D., Rice, S., Keller, D., Hutchins, S., Hughes. J. & Clayton, K. (2009, in press). False alerts in air traffic control Conflict Alerting System: Is there a "cry wolf" effect? *Human Factors*.

Wiener, E. L. (1988). Cockpit automation. In E. L. Wiener. & D. C. Nagel (Eds.), *Human Factors in Aviation* (pp. 433–461). San Diego: Academic.

Wiener, E. L. (1989). *Human Factors of Advanced Technology ("Glass Cockpit") Transport Aircraft* (NASA Contractor Report 177528). Moffett Field, CA: NASA Ames Research Center.

Wiener, E. L. (1993). *Intervention Strategies for the Management of Human Error* (NASA Contractor Report NCA2–441). Moffett Field, CA: NASA Ames Research Center Excerpts available online at http://www.flightdeckautomation.com/resourceevid.aspx?ID = 82.

Woods, D. D. (1995). The alarm problem and directed attention in dynamic fault management. *Ergonomics*, *38*(11), 2371–2393.

Xu, X., Wickens, C. D., & Rantanen, E. M. (2007). Effects of conflict alerting system reliability and task difficulty on pilots' conflict detection with cockpit display of traffic information. *Ergonomics*, *50*(1), 112–130.

16

Unmanned Aircraft Systems

Alan Hobbs Ph.D.

San Jose State University
Human Systems Integration Division
NASA Ames Research Center
Moffett Field CA 94035–1000

Unmanned aviation may appear to be a recent development, but its history stretches back to the beginnings of aviation. The first unmanned glider flew a century before the Wright Flyer (Pritchard, 1961), and unpiloted, powered aircraft made their appearance in the first decades of the 20th century. Over the last 100 years, a series of technological innovations have expanded the capabilities of unmanned aircraft to the point where they now fulfill an increasing range of civilian and military roles. The central message of this chapter is that the further development of unmanned aviation may be held back more by a lack of attention to human factors than by technological hurdles. This chapter begins with a brief overview of unmanned aircraft systems (UAS), from their historical beginnings to the present. Next, the accident record of unmanned aviation is reviewed. The emerging human factors of UAS operation are then examined under the broad headings of teleoperation, design of the ground control station (GCS), transfer of control, airspace issues, and maintenance. If unmanned systems are to reach their full potential, it will be necessary to address each of these issues.

The technologies needed for unmanned flight began to appear early in the history of powered aviation. During World War I, autopilots and stability control enabled Sperry's unpiloted aerial torpedo and the Kettering Bug to fly preset courses (Newcome, 2004) although in each case they were designed as flying bombs intended for one-way missions. In 1917, Archibald Low was the first to control an unmanned aircraft via radio signals (Bloom, 1958). The aircraft met an ignominious end during a demonstration for senior military staff, when, engine sputtering, it turned towards the viewing area. Generals scattered in all directions before the aircraft tore into the mud of Salisbury Plain. In subsequent years, radio controlled aircraft were widely used as target drones. By the 1960s developments in sensors enabled reusable unmanned aircraft to perform military photo-reconnaissance and electronic warfare roles (Barry & Thomas, 2008). The technology of commercial model aircraft was also advancing, with the mass-production of inexpensive radio control units, servos and engines. In recent years, the silicon revolution and the continuing impact of Moore's law have dramatically expanded the capabilities of even the smallest unmanned aircraft through powerful microprocessors, global positioning capabilities, miniaturized sensors, and improved batteries originally developed for cell phones and other consumer products.

Today, unmanned aircraft range from inexpensive, hand-launched micro air vehicles with endurance measured in minutes, to large, long endurance aircraft powered by jet turbine engines (see Figure 16-1). At one extreme are micro-electric helicopters that operate virtually silently, and are well suited for law enforcement, aerial photography, close area surveillance, and even indoor operations. At the other extreme are High Altitude Long Endurance (HALE) vehicles, exemplified by the Global Hawk, with a maximum take-off weight of 14,600 kg. In between is a diverse array of fixed wing, rotary wing and lighter-than-air vehicles, some powered by fuel-burning engines, others by electric motors, in some cases with fuel cells or solar cells to permit very long endurance flights. Despite their apparent differences, a characteristic common to all of these designs is that the aircraft is part of a wider system, including a GCS, communication links, and support equipment. Throughout this chapter, the term "Unmanned Aircraft" will be used to refer to the airborne element of the system, and the term "Unmanned Aircraft System" (UAS) will be used when the intention is to refer to the entire system, including ground-based components.

Most nonmilitary UAS are designed to gather information rather than transport people or cargo. The information may relate to the location and movements of people, traffic, fish or wildlife, the condition of power lines or other infrastructure, the location and intensity of wildfires, the quality of air, or the presence of natural resources.

FIGURE 16-1 The diversity of unmanned aircraft. The 14600 kg Global Hawk, the 20 kg RnR APV3, and the 900 gram microdrone helicopter.

FIGURE 16-1   (Continued)

Despite their potential, the commercial use of unmanned aircraft is still far from commonplace, partly due to regulatory restrictions that limit their access to airspace.

Military uses of unmanned aircraft include surveillance, weapons delivery, signals intelligence, and suppression of air defenses. There is a substantial history of unmanned operations by many of the world's militaries. However, senior political figures in the United States appear to have only recognized the full potential of unmanned aviation after conflicts in Kosovo in the 1990s, and 21st-century conflicts in Iraq and Afghanistan (Department of Defense, 2004). In 2007, 80% of U.S. military flights over Iraq were reportedly by unmanned aircraft (Wise, 2007). In the future, unmanned freighters and unmanned combat aerial vehicles (UCAV) may replace conventional military aircraft. Unmanned systems, as disruptive technologies, are not only changing the way wars are fought but are also changing the human experience of the fighter whose task may be to unleash deadly force on adversaries, while seated at a keyboard far from the conflict zone (Singer, 2009). While this chapter will draw on the military experience of unmanned flight, our principal concern is with those issues that apply to commercial or civilian operations. As will be seen, rather than eliminating the human factor, the removal of the pilot from the aircraft has sometimes amplified the impact of human fallibility on system performance, and given us a renewed appreciation of the contribution made by on-board pilots.

THE ACCIDENT RECORD

The accident rate for unmanned aircraft is significantly higher than for manned aircraft. In the period 1986–2002, three types of unmanned aircraft operated by the U.S. military—Predator, Hunter, and Pioneer—were lost in accidents at rates of 32, 55 and 334 per 100,000 hours respectively (Department of Defense, 2003). This compares unfavorably with the rate for general aviation of approximately one accident per 100,000 hours. The accident rate for Predator has reduced significantly since 2002, but still remains at about 10 times the general aviation rate (Nullmeyer & Montijo, 2009).

Tvaryanas, Thompson, and Constable (2006) analyzed 10 years' worth of unmanned aircraft mishaps in the U.S. military. Overall, just over 60% of mishaps were judged to involve human factors of one kind or another. The issues identified for Air Force UAS accidents included automation problems, inadequate instrumentation or feedback to the operator, and channelized attention. Within the Army, Navy, and Marines, UAS accidents were more likely to involve problems with procedures and guidance documents. Accident rates for civilian UAS operations are not yet available, however one manufacturer of small commercial unmanned aircraft has reported that it expects to lose one aircraft every 300 flight hours—approximately equivalent to a rate of 330 accidents per 100,000 hours.

A Landmark UAS Accident

A sentinel event for the UAS industry was the April 2006 crash of a General Atomics Predator B, operated by the U.S. Department of Homeland Security near the border with Mexico. When the facts of the accident began to emerge, media reports labeled it as a case of "pilot error." Although human actions undoubtedly contributed to the accident, the National Transportation Safety Board (NTSB) investigation uncovered deeper systemic issues that have relevance to the entire UAS sector (NTSB, 2006).

The Predator B is a turboprop powered aircraft with a maximum takeoff weight of approximately 4700 kg. The aircraft is operated from a ground control station (GCS) that has two side-by-side consoles. The aircraft controls on each console are identical, enabling control to be transferred between consoles if necessary. In normal operations, the pilot sits at the left console and a sensor operator seated at the right console controls a camera mounted on the aircraft. The condition lever shown in Figure 16-2, located to the left of the power lever on each console, operates in two modes. When the console is set to sensor mode, moving the condition lever forward opens the iris of the camera, moving it aft closes the iris, and placing it in the middle position locks the iris setting. When the console is set to pilot mode, however, the condition lever becomes a critical engine control. In the middle position the lever shuts off fuel to the engine, the aft position feathers the propeller.

NTSB records show that several hours into the April 2006 flight the pilot's console locked up, or "froze." The pilot decided to switch control to the right-hand console. The border patrol agent who had been seated at the right hand console stood back to enable the pilot to transfer control. Although there was a checklist to guide the transfer between consoles, the pilot reported that, due to time pressure, he did not refer to it. He also did not notice that the condition lever on the right-hand console was in the middle (fuel shut-off) position when control was transferred. A warning tone sounded when the engine stopped, however the ground station uses the same tone for all warnings. Personnel who heard the warning seem to have assumed that it indicated a noncritical situation, such as a temporary loss of satellite link.

The pilot noticed that the aircraft was losing altitude. Unaware that the engine had stopped, he turned off the data uplink to activate the aircraft's lost link procedure. This would have directed the aircraft to climb to an altitude of 15,000 feet and fly a predetermined course until control could be reestablished, but the aircraft continued to descend until it was below line-of-sight communications.

The aircraft was equipped with an engine auto-ignition system that relied on a secondary, satellite-based, communication link.

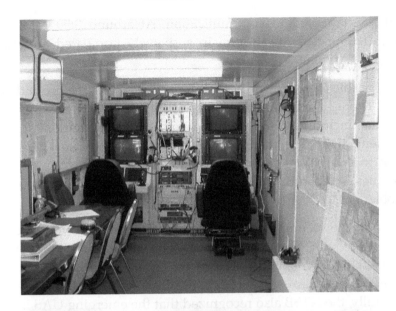

FIGURE 16-2   A general view of the Predator control station, and a close view of the throttle controls. The dual-mode condition lever is second from the left.

However, when the engine stopped, the aircraft automatically began shedding electrical load to conserve battery power. The satellite communication equipment was one of the systems that were shed. Consequently, the engine could not be restarted once the aircraft had lost line-of-sight communication. At around 3:50 A.M., the aircraft impacted the ground in a sparsely populated area near Nogales, Arizona.

The console lock-up was not the first to occur on the ground station. There had been 16 console lock-ups in the preceding four months, including two that occurred during preparations for the accident flight. The maintenance response to these lock-ups had been to swap the main processor circuit cards between the two consoles. It was not clear whether this was done to diagnose the problem or to clear the fault.

Some of the recommendations in the NTSB report were specific to the Predator B, and dealt with console lock-ups, inadvertent engine shutdowns, and the need for an engine restart capability even when line-of-sight communication is not available. More generally, the NTSB also recognized that the emerging UAS industry lacks a safety reporting system. A safety reporting system may have helped to avert the accident by drawing attention to a similar uncommanded fuel shutoff incident that had occurred to a Predator B two years previously, also when control was being switched between consoles (Williams, 2006).

The Nogales accident involved a large and sophisticated UAS. Despite the diversity of unmanned systems, many of the human factors of operations and maintenance apply across the range of UAS, from hand-launched micro air vehicles, to large high-altitude aircraft.

TELEOPERATION AND UNMANNED AIRCRAFT

Teleoperation is the control of machinery by a remotely located human operator. Teleoperation is used in a wide variety of applications including ocean or space exploration, mining, bomb disposal, and the handling of dangerous materials. Teleoperation can release people from dull, dirty, and dangerous jobs, but introduces a new set of challenges including time lags, limited perceptual cues, reduced situational awareness, and the potential for inconsistent mapping of control inputs to system behavior.

Teleoperation and Risk

Teleoperation decouples the safety of the operator from the fate of the vehicle. The operator of an unmanned aircraft, free from the constraints of shared fate, may choose to take risks with the survival of the aircraft, or even intentionally destroy it in extreme situations. Most UAS possess flight termination systems, typically an engine kill switch or a preprogrammed control surface configuration to rapidly bring down the aircraft. Although intended to manage risk, such systems also introduce the possibility of inadvertent activation. The first loss of a Global Hawk occurred when its flight termination system was activated by mistake (Department of Defense, 2003).

The absence of shared fate results in a significant shift of risk from aircraft operators to the general community. With the beneficiaries of flight safely out of harm's way, the risks associated with unmanned aviation are borne largely by non-involved individuals: occupants of conventional aircraft, people under the flight path of the aircraft, and owners of property that might be damaged in the event of an accident. The community tends to have a lower tolerance of risk when technologies are new, are not well understood, or where the targets of a hazard have little control over their level of exposure (Slovic, 2000). For these reasons, there may be less community tolerance of occurrences involving unmanned aircraft, such as near mid-air collisions or crashes in urban areas, than there would be for similar events involving aircraft with on-board pilots.

Teleoperation and Aircraft Control

The human controller of a teleoperated system faces a different set of demands, depending on the extent to which control is achieved directly or via automation. Sheridan (1997) identifies three levels of control in technological systems: manual, supervisory, and autonomous. Under manual control, the unmanned aircraft is controlled directly by a pilot, often via a handheld radio control box. Supervisory control occurs when the operator issues commands to an automated system, which then executes the commands. Autonomous control, as the name suggests, is when the aircraft is controlled entirely by automation. In practice, the level of control may change according to the phase of flight or other determinants, or there may be a blend, with some functions performed autonomously while others remain under manual control. The skills required to operate a UAS are closely related to the level of control involved. Not surprisingly, prior flight experience has been shown to be an advantage when operating an unmanned aircraft that requires stick and rudder inputs (Schreiber, Lyon, Martin, & Confer, 2002) but may be less relevant for systems that are largely controlled via a computer interface (Barnes, Knapp, Tillman, Walters, & Velicki, 2000).

A challenge common to all forms of teleoperation is the reduced set of perceptual cues available to the operator. The pilot of an unmanned aircraft is faced with reduced visual information, and the complete absence of the auditory, tactile, olfactory and vestibular cues familiar to pilots of conventional aircraft. The perceptual gulf between operator and aircraft is illustrated by reports from military UAS operators who have been unaware their aircraft was being targeted by ground fire until fuel was seen to splash on the camera lens. Future UAS may provide haptic feedback such as vibration to improve the operator's awareness of aircraft performance, turbulence, precipitation, or obstacles (Ruff, Draper, Lu, Poole, & Repperger, 2000; Lam, Mulder, & van Passen, 2007).

Manual Control

Direct manual control of unmanned aircraft can involve one of two scenarios. An external pilot may control the aircraft within visual range by observing the aircraft directly and making control inputs via a control box, or an internal pilot stationed at a GCS with no direct view of the aircraft may exercise direct control via stick and rudder inputs. As with any form of teleoperation, a time delay between control input and response can impede direct manual control. Lags of a second or more make manual control all but impossible, and even lags as brief as 50 msec can seriously degrade performance and lead to pilot induced oscillations (Welch, 2003). Time lags may be most problematic when control is via satellite link (Mouloua, Gilson, Daskarolis-Kring, Kring, & Hancock, 2001) or if the unmanned aircraft is being flown via a relay from another aircraft (Gawron, 1998).

Several UAS designs rely on an external pilot for takeoff and landing, with control being transferred to an internal pilot for the remainder of the flight. The direct visual control of an unmanned aircraft is a challenging task, heavily reliant on the skills of the individual. In the case of the Hunter and Pioneer, military unmanned aircraft that require an external pilot for takeoff and landing, 68% and 78%, respectively, of human factor accidents involve takeoffs or landings conducted by the external pilot (Williams, 2004). Autonomous takeoff and recovery systems have been found to significantly reduce UAS accident rates (Department of Defense, 2003).

One of the difficulties facing external pilots is that of control consequence incompatibility, particularly left-right reversals. The control movements required to turn the aircraft left or right when the aircraft is approaching the pilot, are the reverse of the movements required for the same outcome when the aircraft is flying away from the pilot. Unmanned helicopters are widely used in Japan to spray rice crops. An external pilot standing by the field controls the aircraft via a handheld radio control unit. Pilots are trained to fly a box pattern in which the helicopter is flown in a forwards direction when moving away from the pilot, is then flown sideways at the end of each spray run, before being brought back towards the pilot tail first. In this way, control-consequence compatibility is maintained, albeit at the expense of style (T. Suzuki, personal communication).

In some circumstances an internal pilot controls the aircraft manually. For example, the Predator GCS is fitted with standard stick and rudder controls to enable the internal pilot to perform takeoffs and landings with the aid of a nose-mounted camera. The so-called soda straw view from an on-board camera removes binocular cues and reduces monocular depth perception cues, including the flow of the visual field in peripheral vision. The elimination of peripheral vision can also limit an internal pilot's situational

awareness. The operators of unmanned helicopters face a particular problem because, although on-board cameras typically face
forwards, the helicopter can be maneuvered in any direction.

Supervisory Control

Virtually all modern UAS possess auto-flight systems that enable
supervisory control via a keyboard, or point and click inputs. UAS
with automated takeoff and landing systems have the potential to
be flown entirely under supervisory control. A risk associated with
supervisory control is that the human operator may adopt a more
passive role, and possess a lower level of situational awareness
than would be the case in a system under direct manual control
(Endsley, 1996). When glass cockpit aircraft were being introduced
in the 1980s, Wiener (1988) noted that " . . . the introduction of
automation tunes out small errors and creates the opportunities for large ones" (p. 453). Time-consuming and complex mission planning requirements for automated unmanned aircraft can
shift some of the human error risk from the in-flight phase to the
preparation stage The GCS can be a particularly unstimulating
environment on long duration missions, and UAS operating crews
have been found to experience greater levels of fatigue, burnout
and boredom than traditional aircrew (Tvaryanas, Lopez, Hickey,
DaLuz, Thompson, & Caldwell, 2006).

In a conventional aircraft, the on-board pilot is the last line of
defense against automation-related errors. Modern unmanned aircraft face the double challenge of high levels of automation, combined with a remotely located pilot who may be unable to detect
and respond to problems in sufficient time. On occasion, automated
unmanned aircraft have executed maneuvers that an on-board pilot
would have immediately discerned as inappropriate. In 1999 a
Global Hawk was severely damaged when, after landing, it taxied
at high speed and ran off the paved surface. Six months before the
flight, a touchdown elevation 614 feet higher than the actual runway elevation had been erroneously entered into the flight plan.
The mission planning software was designed to treat all altitude
changes as either climbs or descents, even when the aircraft was on
the ground. Because the touchdown elevation had a value higher

than the elevation of the first taxi waypoint, a descent speed of 155 knots was automatically entered into the mission plan for this taxi segment. The commanded speed for the remaining taxi segments then reverted to a normal six knots. The error in touchdown altitude was corrected some time before the flight, but the excessive taxi speed that had resulted from it remained in the plan (Department of Defense, 2000).

Control of Multiple Unmanned Aircraft

An irony of "unmanned" aviation is that even small UAS are typically supported by a team of pilots, sensor operators and support personnel. Many UAS users, particularly in the military, envision a future in which a single operator will be able to supervise multiple vehicles. Yet before this can occur, the operator-to-vehicle ratio must be brought down to 1:1.

Simulation studies suggest not only that performance degrades significantly when operators move from manually controlling one to two unmanned aircraft, but even with the addition of automation, controlling two aircraft is still significantly more challenging than controlling a single vehicle (Dixon, Wickens, & Chang, 2005; Ruff, Narayanan, & Draper, 2002). A solution may lie in automation that enables an operator to issue a single command to a swarm of unmanned aircraft, which then autonomously carries out a pre-programmed coordinated action sequence. This approach has been shown to reduce workload and improve performance when compared to the traditional approach in which the operator interacts with each aircraft separately (Fern & Shively, 2009). Ultimately, the feasibility of a single operator controlling multiple aircraft may be determined by a combination of the workload and the cost of failure. It may be perfectly feasible for a single operator to manage several autonomous HALE "pseudo-satellites" programmed to loiter on station, if the workload remains light and operations remain normal. It may also be an acceptable risk for an operator to supervise a swarm of small, expendable autonomous aircraft on noncritical missions. However, if the loss of the vehicle is deemed unacceptable, or if poor performance could result in loss of life or property on the ground, the control of multiple aircraft may be difficult to justify.

DESIGN OF THE GROUND CONTROL STATION

Ground control stations (GCSs) range from sophisticated purpose-built shelter trailers or control facilities, to commercial off-the-shelf laptop computers (see Figure 16-3). Teleoperation introduces a new set of critical parameters which must be displayed to the operator. In addition to information on aircraft location, flight path, and the status of on-board systems, the operator must have access to UAS-specific information including the strength of the communication link; the potential for interference from other users of the radio spectrum; the time lag between control and response; and the status of ground equipment, for example, the charge remaining on battery powered systems.

It appears that some military unmanned aircraft have been rushed into service, with GCS interfaces that violate established basic design principles. Problems include the use of difficult-to-read color combinations, the presentation of large amounts of data in text rather than graphical displays, and the placement of critical controls adjacent to non-critical controls (Pedersen, Cooke, Pringle, & Connor, 2006). Tvaryanas and Thompson (2008) identified a range of design deficiencies with military GCSs, including "non-intuitive automation, multifunctional controls and displays, hierarchical menu trees and nonintegrated data that provide . . . inadequate feedback to crewmembers on system settings and states, overload crew members with raw data and require sustained attention and complex instrument scans" (p 530). A survey found that Global Hawk operators were dissatisfied with the displays and controls in the GCS in use in 2001 (Hopcroft, Burchat & Vince, 2006). Problems included difficult-to-read fonts, unfortunate color schemes, and complicated retasking processes, including an inability to add waypoints once the flight was underway.

The ground stations for small commercial UAS tend to be standard laptop computers, to which the operator inputs commands via keystrokes or mouse clicks. If the laptop is used outdoors, a lightproof hood is commonly used to enable the screen to be read in daylight. In most cases, the control software runs on a standard commercial operating system. The user must be comfortable with the vagaries

FIGURE 16-3   Ground control stations (GCSs) range from commercial off-the-shelf laptops, to purpose-built facilities. Shown here is the GCS for the 7 kg MLB Bat (with car top launcher visible in background), and the GCS for the 14600 kg Global Hawk.

of mass-market computers, and prepared for the distractions of dialog or pop-up notification boxes that obscure the display until cleared. At the time of writing, no human factors analyses had been published on the diverse range of laptop UAS interfaces.

TRANSFER OF CONTROL

The discontinuity created by crew changes and shift handovers has been a source of difficulty in many industrial, medical and transport settings, including air traffic control (Parke & Kanki, 2008). The already reduced perceptual cues available to UAS operators mean that the communication of situational information during control transfers is critical. Williams (2006) describes four types of UAS control transfers: (1) between an external and an internal pilot, (2) between internal pilots in different GCSs, (3) between consoles in the same GCS, and (4) from one pilot to another without a console change, typically during a crew handover. In some cases, control will be transferred between personnel who are geographically separated. Although the in-flight handover of control between pilots is an everyday aspect of conventional aviation, UAS operators face several unique challenges in this area.

Off-Duty Crew Physically Absent from the Workplace

When watch changes or crew rotations occur on board ships and long-haul aircraft, the off-duty crewmembers remain on-board, and may be consulted or called upon if necessary. Unmanned aircraft may be unique as the only vehicle in which an entire control crew can complete a shift, and then leave the workplace, with the vehicle remaining in motion all the while.

Geographical Separation

For some military UAS operations, takeoff and landing is handled by a pilot near the conflict zone, who then hands the aircraft over to a GCS thousands of kilometers away. Geographical separation not only introduces challenges to situational awareness, but may also introduce disruptive time lags if communication is via satellite.

Mode Errors

The ability to switch between control stations increases the potential for mode errors, as illustrated by the Predator accident at Nogales. Checklists can help to manage this hazard, but only design solutions can address the root causes. For example, designers can limit the use of multifunction controls, minimize the number of available modes, and ensure that mode status is displayed clearly.

Flight Duration

As the endurance of unmanned aircraft increases from hours to months, transfer of control will become increasingly critical (Tvaryanas, 2006). Although long-duration missions do not necessarily increase the risk of error associated with each control transfer, they increase the number of control transfers per flight, thereby increasing *exposure* to risk. Future solar electric aircraft may offer virtually unlimited flight durations comparable to satellites, and will depend on effective long-term management by operators on the ground.

AIRSPACE ACCESS

The safe integration of unmanned aircraft into civilian airspace is one of the most difficult problems currently facing unmanned aviation. The FAA requires pilots to keep a lookout for other aircraft whenever weather conditions permit, even when flying under instrument flight rules (FAA, 2004) and has stated that unmanned aircraft must be able to demonstrate a level of safety equivalent to see-and-avoid (Kuchar, Andrews, Drumm, Hall, Heinz, Thompson, & Welch, 2004). The sense-and-avoid challenges for small and micro unmanned aircraft are potentially very different from those facing larger High Altitude Long Endurance (HALE) or Medium Altitude Long Endurance (MALE) operations.

Small and Micro Unmanned Aircraft

Small and micro unmanned aircraft are likely to spend virtually all their time at low level, outside controlled airspace, where the primary means of separation for conventional aircraft is the see-and-avoid principle, in some situations augmented by radio to broadcast position or arrange self-separation. If flown within visual range, an external pilot may be able to maintain a look out, however once beyond visual range, collision avoidance becomes more difficult.

It is not only the lack of an on-board pilot that complicates matters for small unmanned aircraft, but also the encounter geometry that results when two aircraft with significantly different speeds are on a collision course. The pilot of a slow moving aircraft has a more difficult see-and-avoid challenge than the pilot of a faster aircraft (Hobbs, 1991). This is because, in straight and level flight, the pilot of a conventional aircraft wishing to sight a slow-moving threat (such as a small unmanned aircraft) must pay attention to a narrow cone of airspace centered on the intended flight-path. However, the operator of a slow-moving aircraft must monitor all 360 degrees of the compass to detect the approach of a faster aircraft. It may be more realistic to consider slow-moving unmanned aircraft as passive obstructions that may be *collided with*, rather than active players capable of rapid avoidance action.

Two risk management approaches may be appropriate for small and micro unmanned aircraft outside controlled airspace. The first is to ensure that they remain clear of areas used by manned aircraft by staying at low level, using dedicated sites for takeoff and landing, and remaining clear of airports. Australia has adopted such an approach, by allowing the operation of small unmanned aircraft outside controlled airspace below 400 feet over unpopulated areas (Civil Aviation Safety Authority, 2003). A second approach for the smallest unmanned aircraft is to minimize the consequences of a collision by limiting their mass, and designing them to be "ingestible."

HALE and MALE Unmanned Aircraft

The operation of unmanned HALE and MALE aircraft introduces additional complications, but also potential risk mitigation approaches. Although such aircraft may need to climb and descend through noncontrolled airspace, most of their operations will be under Instrument Flight Rules in controlled airspace where see-and-avoid is not the primary means of collision avoidance.

Encounters between manned and unmanned aircraft may be complicated by the unconventional flight patterns of unmanned aircraft, characterized by steep climbs and descents, slow speeds and loitering, instead of point-to-point travel. Even when the flight plan of an unmanned aircraft calls for it to takeoff and land from dedicated airports, and cruise in controlled airspace at high altitude, an emergency may lead to unplanned interactions with other airspace users. For example, an engine failure is likely to result in an emergency descent, and a lost link may trigger the execution of a preprogrammed procedure, such as a return to the last waypoint at which communication was successful.

Equipping unmanned aircraft with Mode S transponders would give them a passive role in collision avoidance by identifying them to air traffic control radar and to the Traffic Collision Avoidance System (TCAS) of large conventional aircraft (Drumm, Andrews, Hall, Heinz, Kuchar, Thompson, & Welch, 2004). Whether TCAS is appropriate for unmanned aircraft is a matter of debate. Kuchar et al. (2004) note that TCAS was intended to *enhance* see and avoid, not replace it. TCAS traffic advisories aid the pilot in sighting potential threat aircraft, and would be of limited use when the pilot is located on the ground. Kuchar et al. also question how UAS operators could respond effectively to the more urgent resolution advisories (RA) issued by TCAS. There may be a discernable time lag in communication of an RA to the operator on the ground. There may then be an additional delay while an operator, who may have been controlling the automation in a supervisory mode, assesses the situation, executes a response, and transmits the instruction to the aircraft. If the response requires screen selections and mouse clicks, the UAS operator may take significantly longer to respond to a TCAS RA than would an airborne pilot in a similar situation. Given these limitations, the option of programming unmanned aircraft to respond autonomously to a TCAS RA must be considered (Drumm et al., 2004).

The reaction of other airspace users and air traffic controllers to unmanned aircraft remains to be seen. We do not yet know whether pilots of conventional aircraft will respond to a TCAS RA differently if they know that the threat aircraft is unmanned. It is also unclear how the task of air traffic controllers will change when communicating with a pilot who, like them, is situated in a control room on the ground. The reduced situational awareness of the UAS pilot is such that a midair collision may initially be indistinguishable from a loss of link. Particularly where satellite communication is used, verbal communication between the UAS operator and ATC could be disrupted by time lags of up to several seconds (Drumm, Andrews, Hall, Heinz, Kuchar, Thompson, & Welch, 2004; McCarley & Wickens, 2005). The challenges for controllers may become most noticeable for approach control and surface movements at airports, where there is a greater reliance on the ability of the pilot to visually sight other aircraft.

HUMAN FACTORS IN AIRWORTHINESS

The aviation industry was slow to recognize that human factors apply not only to pilots and controllers, but also to maintenance personnel (Reason & Hobbs, 2003). The maintenance of teleoperated systems introduces a new set of challenges (Hobbs, Herwitz, & Gallaway, 2009) not least because unmanned systems may be less able to withstand maintenance error than systems with an on-site operator who can respond rapidly to an anomaly. The absence of an on-board pilot can also make the detection and troubleshooting of problems more challenging. In-flight monitoring systems can provide a rich source of electronic data to maintenance personnel, but lack the qualitative information on handling, sounds, vibrations, and smells contained in a pilot report.

As Figure 16-4 illustrates, in conventional aviation, the responsibilities of the maintenance technician are limited to the airworthiness of an aircraft; however, the UAS maintenance technician is responsible for a complete system, comprising the aircraft, a diverse set of ground-based equipment, and the links between these elements. Large unmanned systems are generally maintained by specialist maintenance technicians. However, small commercial UAS are usually operated by generalist teams of multiskilled individuals who perform all ground tasks including assembly, flight preparation, in-flight operation, and maintenance. As a consequence, the distinction between pilot and maintainer that has existed since the beginning of aviation may not apply in the case of small unmanned aircraft.

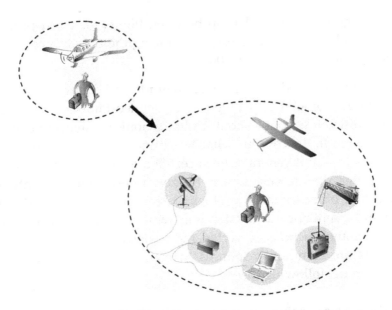

FIGURE 16-4   In conventional aviation, the responsibilities of the maintenance technician are limited to the airworthiness of an aircraft. The UAS maintenance technician is responsible for a system comprised of diverse elements, including the aircraft, communication equipment, modems, computers, radio control units, and launch/recovery devices.

Technical Skills

The UAS maintainer, whether a specialist or a generalist, requires an understanding of the technologies used to communicate with and control the aircraft. These include computer software and hardware, radio communication equipment, modems, and in some cases satellite phones. Ensuring the integrity of the data link between the aircraft and the ground takes on a level of criticality not present in conventional aviation because the loss of communication is more likely to result in the loss of the aircraft.

For most small UAS, the "cockpit on the ground" is a standard laptop or desktop computer exposed to the outdoor hazards of moisture, dust, and temperature extremes. Computer system administration now takes on flight safety importance, because system failures that would be minor irritations in an office environment (for example, screen lockups or software slowdowns) present significant hazards if they occur during a flight. Some operators use the ground control laptop between flights for word processing, e-mail, or Internet access, introducing the threat of computer viruses (Hobbs & Herwitz, 2008).

The software-based systems used to guide and control unmanned aircraft are sometimes plagued by ill-defined faults that are especially difficult to troubleshoot. Unlike avionics systems in conventional aircraft, the computer hardware and software in unmanned systems have not generally been certified for safety and reliability. Computer slowdowns, screen freezes, radio frequency interference, and other technical problems are sometimes resolved without the maintainer understanding why the fault occurred, and whether their actions corrected the underlying problem. System reboots are common responses to computer problems as illustrated in the following case.

> The desktop computer, which was serving as the ground control system, locked up while the unmanned aircraft was in flight. The PC-based computer was housed in the ground control station trailer. The only alternative was to re-boot the computer, and this took about two to three minutes before command-and-control was reestablished. The unmanned aircraft's flight path, however, was already uploaded so there was no effect on the flight sequence. Incident report from Hobbs and Herwitz (2008).

Repetitive Assembly and Handling

In contrast to conventional aircraft, most small unmanned aircraft are designed to be reassembled and disassembled before and after each flight. The frequent connection and disconnection of electrical, fuel, and data systems can increase the chances of damage and maintenance error, as illustrated by the following example.

> After departure the unmanned aircraft performed unusually slow rates of turn to the right and tight turns to the left and struggled to track as designated by the operator. Approximately seven minutes into the flight, the outboard section of the right wing separated from the centre wing section. The aircraft immediately entered a rapid clockwise spiral before impacting the ground. The most likely explanation for the crash was that the outboard section of the right wing was incorrectly attached during pre-flight assembly and from launch it flew with difficulty until the wing section eventually separated. Incident report from Hobbs and Herwitz (2008).

Maintenance of Ground Equipment While Missions are Underway

The cockpit of a conventional aircraft is beyond the reach of maintenance personnel once the aircraft has left the ground. In contrast, the ground-based elements of a UAS are always accessible to personnel who may be required to perform unscheduled maintenance while a flight is underway. For example, an in-flight problem may require troubleshooting of ground equipment, or a restart of the ground control computer. A maintenance technician interacting with a live system requires a clear understanding of the operational implications of the planned intervention, and must also consider the potential effects of errors. For example, even a brief interruption to a computer's power supply can have an extended impact if it leads to a slow reboot sequence.

CONCLUSION

After 100 years of gradual development, unmanned aviation is entering a period of rapid expansion. The current unmanned aircraft sector can be compared with the automobile industry of 100 years ago, characterized by a large range of manufacturers, little standardization of designs, a lack of supporting regulations, minimal infrastructure, and a high accident rate.

Despite being referred to as "unmanned," some of the major challenges confronting UAS relate to human factors. Indeed, the absence of an on-board pilot may make unmanned aircraft more fragile and less error-tolerant than conventional aircraft, particularly in the case of latent errors originating during flight planning, equipment assembly and maintenance. Ironically, the removal of the human operator from the aircraft may simultaneously increase the probability of accidents, while reducing their consequences, at least for the operator.

The development of unmanned aviation reflects two wider trends. The first is automation, a development that permeates virtually all sociotechnical systems. The second trend is teleoperation, as advances in communication allow equipment to be controlled by people located remotely. For the purposes of this chapter, human factors were considered under the broad headings of teleoperation, GCS design, transfer of control, airspace issues, and airworthiness. Ultimately, many of the challenges of UAS operations (such as reduced perceptual cues, time lags and interface design) also apply to other teleoperated systems, including unmanned ground vehicles, remotely operated mining equipment, and undersea exploration vehicles. However, unlike remotely operated equipment on the land or under the sea, when teleoperated systems take to the air, their mishaps have greater potential to affect community safety, and will inevitably attract greater public scrutiny. A high-profile accident involving an unmanned aircraft has the potential to set back commercial unmanned aviation, just as the 1979 nuclear accident at Three Mile Island led to a 30-year hiatus in the construction of nuclear power plants in the United States.

There is still much to learn about the human factors of unmanned systems, and progress will require the open sharing of information. Unfortunately, the civilian UAS sector is generally reluctant to reveal details of mishaps, whether for commercial reasons, or for fear of regulatory action. Furthermore, there is currently no incident and accident reporting system for UAS occurrences. Mishaps involving unmanned aircraft may not even be categorized as "accidents." The International Civil Aviation Organization's definition of an accident specifies that the occurrence must occur while people are on board the aircraft with the intention of flight, thereby excluding unmanned systems.

In one respect, autonomous aviation is not an entirely new development. Throughout history, carrier pigeons and falcons have served as living autonomous aerial systems for communication and hunting. Future aerial vehicles may possess similar levels of autonomy—able to fly a mission and return to their owner with minimal human intervention. Throughout this chapter, the terms "pilot" and "operator" have been used to refer to the controller of an unmanned aircraft, yet neither word is completely adequate. The word "pilot" has been borrowed from manned aviation, and had a maritime heritage before that. Continuing advances in unmanned aviation may call for a term that more accurately expresses the relationship between human and aircraft, perhaps something along the lines of "falconer." Just possibly, in years to come, autonomous unmanned aircraft will become the norm rather than the exception, and future generations will look back to a transitory period in history when members of a specialized profession rode aboard aircraft and tended them in flight.

References

Barnes, M. J., Knapp, B. G., Tillman, B. W., Walters, B. A., & Velicki, D. (2000). *Crew Systems Analysis of Unmanned Aerial Vehicle (UAV) Future Job and Tasking Environments* (Army Research Laboratory Report TR-2081). Aberdeen, MD: Army Research Laboratory.

Barry, J., & Thomas, E. (2008 June 9). The drone wars. *Newsweek*, 24–29.

Bloom, U. (1958). *He Lit the Lamp: A Biography of Professor A.M. Low.* London: Burke.

Civil Aviation Safety Authority. (2003). *Civil Aviation Safety Regulation, Part 101. Unmanned Aircraft and Rocket Operations.* Australian Capital Territory: Author.

Department of Defense. (2000). *United States Air Force Aircraft Accident Investigation Board Report. RQ-4A a Global Hawk Unmanned Aerial Vehicle Class a Mishap, 6 December 1999, Edwards AFB.* Washington, DC: Author.

Department of Defense. (2003). *Unmanned Aerial Vehicle Reliability Study.* Washington, DC: Author.

Department of Defense. (2004). *Unmanned Aerial Vehicles and Uninhabited Combat Aerial Vehicles* Defense Science Board. Washington, DC: Author.

Dixon, S. R., Wickens, C. D., & Chang, D. (2005). Mission control of multiple unmanned aerial vehicles: a workload analysis. *Human Factors, 47*, 479–487.

Drumm, A. C., Andrews, J. W., Hall, T. D., Heinz, V. M., Kuchar, J. K., Thompson, S. D., & Welch, J. D. (2004). *Remotely Piloted Vehicles in Civil Airspace: Requirements and Analysis Methods for the Traffic Alert and Collision Avoidance System (TCAS) and See-and-Avoid Systems.* Proceedings of the 23rd Digital Avionics Systems Conference, Salt Lake City, UT.

Endsley, M. (1996). Automation and situational awareness. In R. Parasuraman & M. Mouloua (Eds.), *Automation and Human Performance: Theory and Applications* (pp. 163–181). Mahwah, NJ: Erlbaum.

Federal Aviation Regulation, Part 91. General Operating and Flight Rules, Section 113.b (2004). Washington, DC: Federal Aviation Administration.

Fern, L., Shively, J. (2009), August). *A Comparison of Varying Levels of Automation on the Supervisory Control of Multiple UASs.* Paper presented at the Association for Unmanned Vehicle Systems International (AUVSI) Conference, Washington, DC.

Flight Safety Foundation. (2005 May). See what's sharing your airspace. *Flight Safety Digest, 24*(5), 1–26.

Gawron, V. (1998). *Human Factors in the Development, Evaluation, and Operation of Uninhabited Aerial Vehicles.* Proceedings of the Association for Unmanned Vehicle Systems International, 431–438.

Hobbs, A. (1991). *Limitations of the see-and-avoid principle*. Canberra, Australian Capital Territory: Bureau of Air Safety Investigation.

Hobbs, A., Herwitz, S. (2008). *Maintenance Challenges of Small Unmanned Aircraft Systems.—A Human Factors Perspective*. Final report to Federal Aviation Administration under inter-agency agreement DTFA01–01-X-02045. Moffett Field, CA: NASA Ames Research Center.

Hobbs, A., Herwitz, S., Gallaway, G. (2009). *Human Factors in the Ground-Support of Small Unmanned Aircraft Systems*. Proceedings of the 15th International Symposium on Aviation Psychology, Dayton, OH.

Hopcroft, R., Burchat, E., & Vince, J. (2006). *Unmanned Aerial Vehicles for Maritime Patrol: Human Factors Issues*. (DSTO publication GD-0463). Fishermans Bend, Victoria, Australia: Defence Science and Technology Organisation.

Kuchar, J., Andrews, J., Drumm, A., Hall, T., Heinz, V., Thompson, S., & Welch, J. (2004). *A safety analysis process for the Traffic Alert and Collision Avoidance System (TCAS) and see-and-avoid systems on remotely piloted vehicles*. Proceedings of the American Institute of Aeronautics and Astronautics, 3rd Unmanned Unlimited Technical Conference, Chicago, IL.

Lam, T. M., Mulder, M., & van Passen, M. M. (2007). Haptic interface for UAV collision avoidance. *International Journal of Aviation Psychology, 17*, 167–195.

McCarley, J. S., & Wickens, C. D. (2005). Human Factors Implications of UAVs in the National Airspace. (Technical Report *AHFD-05–05/FAA-05–01*). Atlantic City, NJ: Federal Aviation Administration.

Mouloua, M., Gilson, R., Daskarolis-Kring, E., Kring, J., & Hancock, P. (2001). *Ergonomics of UAV/UCAV Mission Success: Considerations for Data Link, Control, and Display Issues*. Proceedings of the Human Factors and Ergonomics Society 45th Annual Meeting, pp 144–148.

National Transportation Safety Board. (2006). Accident to Predator B, Nogales Arizona (NTSB report CHI06MA121). Washington, DC: Author.

Newcome, L. R. (2004). *Unmanned Aviation: A Brief History of Unmanned Aerial Vehicles*. Reston, VA: American Institute of Aeronautics and Astronautics.

Nullmeyer, R., Montijo, G. (2009). *Training Interventions to Reduce Air Force Predator Mishaps*. Proceedings of the 15th International Symposium on Aviation Psychology, Dayton, OH

Parke, B., & Kanki, B. (2008). Best practices in shift turnovers: Implications for reducing aviation maintenance turnover errors as revealed in ASRS Reports. *International Journal of Aviation Psychology, 18*, 72–85.

Pedersen, H. K., Cooke, N. J., Pringle, H. L, & Connor, O. (2006). UAV human factors: Operator perspectives. In N. J. Cooke, H. L. Pringle, H. K. Pedersen, & O. Connor (Eds.), *Human Factors of Remotely Operated Vehicles* (pp. 21–33). Oxford: Elsevier.

Pritchard, J. L. (1961). *Sir George Cayley: The Inventor of the Aeroplane*. New York: Horizon Press.

Reason, J., & Hobbs, A. (2003). *Managing Maintenance Error*. Aldershot, UK: Ashgate.

Ruff, H. A., Draper, M. H., Lu, L. G., Poole, M. R., Repperger, D. W. (2000). Haptic feedback as a supplemental method of alerting UAV operators to the onset of turbulence. *Proceedings of the Human Factors and Ergonomics Society 44th Annual Meeting*, San Diego, CA.

Ruff, H. A., Narayanan, S., & Draper, M. H. (2002). Human interaction with levels of automation and decision-aid fidelity in the supervisory control of multiple simulated unmanned air vehicles. *Presence, 11*, 335–351.

Schreiber, B. T., Lyon, D. R., Martin, E. L., & Confer, H. A. (2002). *Impact of Prior Flight Experience on Learning Predator UAV Operator Skills* (Report No. AFRL-HE-AZ-TR-2002–0026). Mesa, AZ: Air Force Research Laboratory.

Sheridan, T. B. (1997). Supervisory control. In G. Salvendy (Ed.), *Handbook of Human Factors and Ergonomics* (pp. 1295–1327). New York: Wiley.

Singer, P. W. (2009). *Wired for War: The Robotics Revolution and Conflict in the Twenty-first Century.* New York: Penguin.

Slovic, P. (2000). *The Perception of Risk.* London: Earthscan.

Tvaryanas, A,P. (2006). *Human Factors Considerations in Migration of Unmanned Aircraft System (UAS) Operator Control* (USAF Performance Enhancement Research Division Report No. HSW-PE-BR-TE-2006–0002). Brooks City, TX: United States Air Force.

Tvaryanas, A. P., Lopez, N., Hickey, P., DaLuz, C., Thompson, W. T., & Caldwell, J. L. (2006). *Effects of Shiftwork and Sustained Operations: Operator Performance in Remotely Piloted Aircraft* USAF Performance Enhancement Research Division Report No. HSW-PE-BR-TR-2006–0001). Brooks City, TX: United States Air Force.

Tvaryanas, A. P., & Thompson, B,T. (2008). Recurrent error pathways in HFACS data: Analysis of 95 mishaps with remotely piloted aircraft. *Aviation Space and Environmental Medicine, 79*, 525–532.

Tvaryanas, A. P., Thompson, B. T., & Constable, S,H. (2006). Human factors in remotely piloted aircraft operations: HFACS analysis of 221 mishaps over 10 years. *Aviation Space and Environmental Medicine, 77*, 724–732.

Walker, L. A. (1997). *Flight Testing the X-36—The Test Pilot's Perspective.* Moffett Field, CA: NASA Ames Research Center (NASA Report No. CR-198058).

Welch, R. B. (2003). Adapting to telesystems. In L. J. Hettinger & M. Haas (Eds.), *Virtual and Adaptive Environments* (pp. 129–167). New York: Erlbaum.

Wiener, E. L. (1988). Cockpit Automation. In E. L. Wiener & D. C. Nagel (Eds.), *Human Factors in Aviation* (pp. 463–494). San Diego: Academic Press.

Williams, K. W. (2004). *A Summary of Unmanned Aircraft Accident/Incident Data: Human Factors Implications* (Technical Report No. DOT/FAA/AM–04/24). Washington, DC: Federal Aviation Administration.

Williams, K. W. (2006). Human factors implications of unmanned aircraft accidents: Flight control problems. In N. J Cooke, H. L. Pringle, H. K. Pedersen, & O. Connor (Eds.), *Human Factors of Remotely Operated Vehicles* (pp. 105–116). San Diego: Elsevier.

Wise, J. (2007 April). Civilian UAVs: No pilot, no problem. *Popular Mechanics*, 64–69.

Crew Station Design and Integration (for Human Factors in Aviation Chapter 18)

Alan R Jacobsen,
The Boeing Company P.O. Box 3707, MC 02–59 Seattle, WA

David A. Graeber and
The Boeing Company P.O. Box 3707, MC 42–53 Seattle, WA:

John Wiedemann
The Boeing Company P.O. Box 3707, MC 02–59 Seattle, WA

INTRODUCTION

There seems to be universal agreement that integration is necessary when designing complex systems such as crew stations, where humans are an integral part of the overall system (Proctor & Van Zandt, 1994). There's also general recognition that an effective, efficient, and safe crew station results when design integration is done well (Sanders & McCormick, 1993). Conversely, when not done well, the results are costly where the price is measured in terms of rework, increased training requirements, expensive hardware or software fixes, and unsafe systems. Yet there doesn't seem to be general consensus on the how and what of actually performing crew station design integration.

This chapter serves as a starting point that both highlights significant issues as well as identifies practical aspects of crew station design and integration. Clearly, crew station design and integration is not a standalone effort, nor is it a final check on human performance after the rest of the system is designed. It is also not simplistic or linear. The first step is the recognition that the crew is an integral part of the larger system. When done well, it considers the crew in all aspects and phases of system design through an iterative and cyclical process. It involves a series of trade-offs and decisions achieved through analysis, test, evaluation and demonstration. In the end, it is about good business and good engineering. Three themes emerge from this chapter. The first is a historical context of the last 30 years of crew station evolution representing first commercial and then military crew stations. A second theme revolves around the issues that significantly impact crew station design. It focuses specifically on competing constraints and on the growth of information management and integration complexity. Finally, the third theme is a discussion of tools and methods for design and integration. No one will become an expert after reading this chapter, but rather the reader will gain an appreciation for the larger picture of what crew station design and integration means.

EVOLUTION OF CREW STATION DESIGN

Commercial Transport Evolution

Over the last 30 years, the aviation industry has witnessed an evolution of information integration, increased automation and new forms of information management, resulting in an evolving role of the pilot. Early on, cockpit designs consisted of multiple non-integrated instruments, direct cable controls and low level control of systems (Wiener, 1989). The average transport aircraft in the 1960/1970s had more than 100 cockpit instruments and controls. Figures 17-1 and 17-2 depict examples of these: a vintage Boeing Stratocruiser and a somewhat more modern 747-200 flight deck. These instruments ranged from basic flight instruments such as attitude direction indicators, altimeters, and airspeed indicators to navigation instruments such as radio distance magnetic indicators, horizontal situation indicators, and autopilots to systems instruments such as engine instruments, fuel gauges, and a myriad of annunciators. To monitor and decipher all this information, the

FIGURE 17-1   Flight deck of the Boeing Stratocruiser. Image courtesy Boeing.

FIGURE 17-2   Flight deck of the Boeing 747-00. Image courtesy Boeing.

flight crew, especially for long range flights, consisted of a Captain, Co-pilot, Navigator, and Flight Engineer.

The crew members were required to perform all of the mission tasks as well as most, if not all, of the integration of information and control inputs. Today, that same mission is performed by two crew members and usually under low workload conditions. This crew complement evolution occurred through a number of different mechanisms, the first being systems simplification. The radio operator and navigator were eliminated as navigation and radio equipment became easier to use, and airplane systems simplification eliminated the need for a flight engineer. Also introduced during this time frame were new integrated flight management and navigation displays as well as centralized engine indication and crew alerting displays. These first generation glass cockpits allowed for a significant step forward in the integrated presentation of information to the flight crew and the glass cockpit has become standard equipment in commercial transports, military aircraft, regional jets and general aviation (Figure 17-3).

FIGURE 17-3   Flight deck of the Boeing 757. Image courtesy Boeing.

Over the last 30 years a profound change has occurred in the level of sophistication of flight deck instruments and the systems that support the pilot in accomplishing flight tasks. The safety and efficiency benefits of these advanced systems are well documented (Bureau of Air Safety Investigation, 1998); a moving navigation map can be readily seen and interpreted by the crew as opposed to having to create a mental model from a variety of discrete instruments. At the same time, the evolutionary approach taken often resulted in these new electronic instruments merely duplicating older electromechanical instruments. For example, the Electronic Attitude and Direction Indicator (EADI) in its earliest stages wasn't much more than an electronic copy of the electromechanical version. In some instances, significant steps forward were made, for example, in navigation, where sophisticated flight management computers and map displays were introduced. Flight crews could now input a flight plan and performance data and the computer would calculate and display an optimal lateral and vertical plan that could be coupled to the autoflight system. As systems became more automated a new design philosophy emerged for flight decks, particularly the quiet dark cockpit, where crew alerting and messaging systems were developed to augment the flight crew in monitoring systems. All these technological improvements were the foundation to establishing greater safety, reduced crew workload and two-crew operations in the first-generation glass cockpits.

The second-generation glass cockpit started during the late 1980s with aircraft such as the A irbus A320, Boeing 747-400, Beachcraft, and Starship (Figures 17-4 and 17-5). The highlight of these flight decks was the introduction of large CRTs and fly-by-wire systems that brought about further information integration, enhanced automation, and new ways for the pilot to perform old tasks. Examples of these include fly-by-wire flight controls, integrated primary flight displays, more automated airplane systems, such as environmental control system control, datalink displays and control, as well as the introduction of multifunction displays. The new systems allowed designers to push more integrated information to the flight crew allowing for greater safety, efficiency and reduced costs. Basic flight instruments now evolved to the point where integration of data is the norm. Primary flight displays, navigation displays, multifunction displays and systems displays are designed to maximize situational awareness for the flight crew.

FIGURE 17-4   Flight deck of the Boeing 747–400. Image courtesy Boeing.

FIGURE 17-5   Flight deck of the Airbus A-320. Image copyright Airbus.

A third generation glass cockpit as seen in the Boeing 777, Boeing 787, and Airbus A380 aircraft has brought about the increased use of multifunction displays and cursor control devices for display interaction, more software based controls, and data fusion, for example, weather, terrain, and traffic information on the moving map (Figures 17-6, 17-7, 17-8). These advances have created a more efficient and effective crew interface enabling the crew to "see at a glance" the state of not only the external world but airplane systems as well. It has also brought the crew closer and more in touch with actual information needed to effectively perform their mission, both in terms of display and control. This third generation glass cockpit continues to evolve and brings about even more display interactivity and multifunction windowing. Multiple functions appear simultaneously on a single screen and the flight crew has more flexibility in bringing specific information to the forefront at any given time. Large-scale systems integration is now the norm as designers look for new ways to consolidate individual and discrete electronic boxes thereby reducing weight and power consumption. New pressures face flight deck designers as we move to the next level. Airlines face enormous business pressures. Designers must now take this new reality into account when designing flight deck equipment. Good human factors must be employed in order to balance optimal human system interaction with the cost limitations facing the airline industry. One example of this is the increased use of soft controls and multifunction displays that can greatly decrease the cost and weight of separate hardware controls and display. Sufficient task analysis and user interface evaluations are necessary to ensure that the move to soft controls and multifunction displays aid the crew rather than make their jobs more difficult. For example, it is crucial that quick access information always be available to the crew without having to call up multiple formats or menus.

At the same time as the industry has been evolving in terms of system automation and integration, we find substantial growth of new airlines in developing countries. The flight crews trained to fly these airplanes are challenged by smaller training footprints, English as a second language, very sophisticated equipment, and operating in

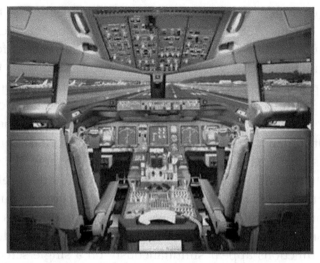

FIGURE 17-6   Flight deck of the Boeing 777. Image courtesy Boeing.

FIGURE 17-7   Flight deck of the Boeing 787. Image courtesy Boeing.

an evolving air traffic environment. As these challenges become known, human factors has taken a prominent position in seeking solutions that benefit all parties. These new generation flight decks are achieving an unseen level of efficiency and safety.

© AIRBUS S.A.S. 2007 _ photo by e*m company / H. GOUSSÉ

FIGURE 17-8   Flight deck of the Airbus A-380. Image copyright Airbus.

Military Crew Stations

Similar to changes in the flight deck of commercial aircraft, military, nonflight deck, crew stations have also evolved over the last 20 years with respect to information presentation, management, and controls. At the root level, a substantial portion of crew station design evolution is analogous to the aforementioned flight deck enhancements due to a leveraging of common technological advancements. However, military crew station design has capitalized on these advancements for different purposes commensurate with their mission objectives and crew roles. The primary areas of change have been seen in the maturation of digital displays, benefits of data fusion, increases in the amount of information presented and communication media, and shift toward flexible multirole workstations. As digital displays have evolved, legacy analog (knobs, switches, dials) and monochrome user interfaces have transitioned to primarily software based, context sensitive controls that capitalize on high resolution color displays and vector graphics. Advancements in data fusion algorithms have reduced operator workload with respect to managing the battlespace by leveraging automation to present a single, valid, trustworthy data point that previously required the operator to cognitively fuse inputs from multiple sensors and cluttered the user interface. However, the spare cognitive capacity resulting from data fusion has been taxed by the increasing volumes of information and diversification of communication media enabled by advancements in secure networks and communication bandwidth. Legacy crew stations primarily dealt with information loaded prior to the mission, radio communications, and small data packets, but, today, this has expanded to include e-mail, instant messaging, internet, chat, imagery, video, and shared displays. Finally, the network advancements have also resulted in a migration over the last 20 years from platform centric crew stations designed to perform specific tasks by specially trained crew to multirole crew stations staffed by cross-trained crew capable of meeting the dynamic missions today's platforms perform in a network-centric warfare environment.

Focusing on the significant changes in military command and control philosophy, as well as crew station capabilities over the last twenty years, a subset of key new military crew station design challenges are discussed below and how they are driven by system-of-systems complexity, expanding information management needs, growth in software user interface capabilities, and manpower reduction goals. Today's military operations are increasingly leveraging a system-of-systems environment where diverse, multirole platforms contribute to achieving a shared intent within a dynamic, fast paced, networked battlespace (Bolia et al., 2008). The impact on crew station design has been substantial due to the concept of operations paradigm shift in how the warfighter contributes to mission effectiveness in a network centric operations environment. Crew station design for a system-of-systems environment has to account for providing the operator commander's intent, rather than traditional specified directives. This nuance is further complicated with the recent integration of unmanned assets into the system-of-systems where maintaining a shared awareness of intent between warfighter and autonomy is an area requiring further maturation. The system-of-systems environment also requires crew station design to account for enabling the operator's ability to maintain shared situation awareness of an expansive battlespace that rapidly changes due to an ever-quickening operations tempo. Providing the warfighter a crew station design that affords understanding the collateral impact of decision making throughout the system-of-systems and making an effective decision in a timely manner is a constant challenge of utmost importance.

With the growth of network centric warfare capabilities, information management presents continually evolving challenges for crew station design. The quantity of information shared in today's operations and the diverse communication media utilized has resulted in design issues driven by both display real estate limitations and the need for distributed teams to efficiently and effectively perform in time sensitive situations. In essence, a classic design problem in a modern context where the goal is to help crew members benefit from a wealth of information while avoiding cognitive processing and decision making bottlenecks due to information overload, key information occluded in the user interface, or the lack of a shared view of the operational environment. Information integration (e.g., data fusion) approaches have been effective in generating a pared-down representation of various data sources, but to achieve maximum benefit it is critical that the output is presented in intuitive and actionable formats that engender trust.

The growth in software user interface capabilities has been both a tremendous enabler of improved crew station designs, but also created new challenges. The ubiquitous Windows© operating system has matured crew station user interface designs from analog to digital solutions, however, it has strengthened the importance of accounting for operator workflow and intent when developing underlying software navigation architectures. The pervasiveness of Windows© has also resulted in saturation of the visual channel and the risk of poor user interface designs burying an operator in cascading windows. Conversely, tapping additional human sensory channels to reduce cognitive workload has not been widely utilized in production systems despite the known benefits (e.g., spatial audio).

Finally, today's military is trying to minimize manpower requirements for its systems and flexible, efficient, and effective crew station designs are a critical enabler. As a result, modern crew station design challenges include embedding training, effective parsing of tasks to automation while maintaining crewmember situation awareness, and supporting diverse users performing a variety of roles with varying levels of expertise and cognitive capacity. This challenge is realized in manned platforms via reduced crew targets, swapping crew workstations for additional sensor payload, or increasing the breadth of workstation capabilities (e.g., adding control of unmanned systems). It's also realized in future autonomous systems where the goal is to allow a single operator to efficiently and effectively control multiple, dissimilar unmanned systems in heterogeneous environments. For autonomous system crew station design, this creates an interesting design challenge with respect to conveying and sharing intent in a natural manner similar to teams of warfighters.

WHAT IMPACTS TODAY'S CREW STATION DESIGN AND INTEGRATION?

Competing Constraints

Crew station design and integration in practical terms focuses on finding an optimized solution to a variety of competing objectives and requirements. Unfortunately, the optimal solution cannot be determined merely by identifying the right set of mathematical expressions and parameters, at least not today. However, determining an optimal solution does necessitate identification of the appropriate set of parameters to consider in the trade space. Figure 17-9 depicts a small subset of generic competing requirements one needs to consider when integrating humans into the flight deck.

In addition to these generic competing requirements and objectives, every crew station will have its own unique set as well. This list normally includes requirements and objectives related to:

- The specific mission and intended functions
- Personnel training and experience
- Customer expectations
- Infrastructure and physical space constraints

Flight deck engineering compromises

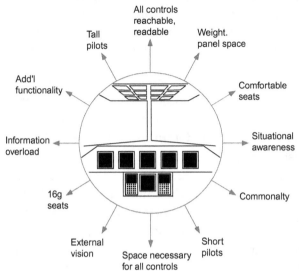

FIGURE 17-9 Various conflicting and completing flight crew requirements that must be accommodated by flight deck design.

- Human characteristics (strengths and weaknesses of human to fulfill task)
- Other real world constraints (such as time, money, and regulations)

A modern flight deck/crew station design brings together many stakeholders. From an airplane level perspective, a balance must be attained to achieve a design which satisfies all requirements and the business case. From weights to cost to human factors, each stakeholder must first define a set of validated requirements. Systems designers then trade various aspects of the design, resulting in the end product. As seen in Figure 17-9 and the list here, there are many factors to consider when designing an integrated flight deck.

Omitting any of these requirements can result in a less than optimal or even unacceptable design. For example, one design limitation in today's flight deck is to account for pilot's heights in a range of 5′2″ to 6″4′. This has an impact on the placement of the seats, forward instrument panel, control size and location, and so on. Human factors personnel work as team members to then ensure that the final design is safe and certifiable by conducting many types of analyses. These analyses can range from full scale physical mock ups to computer man-model simulations.

As the aerospace industry becomes more competitive a further emphasis on cost and schedule reductions has emerged, thereby necessitating analysis of development and life-cycle costs. The systems engineering model examines cost versus value, both for the customer as well as the developer. It is imperative for the design team members to consider all of these sometimes conflicting aspects of the design in determining the "best" solution for the entire industry. It is also important to understand the goals and perspectives of production program management as they strive to meet their cost and schedule targets that ensure profitability for the manufacturer. To a program chief engineer, human system integration (HSI) engineering is sometimes viewed as a driver of program risk, rather than risk mitigation. Often, particularly in military programs, program contractual requirements do not adequately capture HSI requirements with respect to being comprehensive or in a manner that affords objective verification and validation. The impact is that it allows the chief engineer to downplay the need to completely address areas of concern that HSI surfaces, or creates an atmosphere of hesitancy when the only means for verification and validation of requirements is subjective in nature. This has been a classic barrier hindering the HSI skill from adding maximum value to system design, but must be accounted for in the politics and organizational structure of a production environment. Its importance is highlighted by that realization that actions during production have a lasting impact on the customer's life-cycle cost.

How one balances among all of these overlapping and often conflicting requirements is what happens during iterative design and integration through analysis, prototyping, test, and evaluation. The first step, however, is always to identify the entire solution space and boundaries, that is, requirements and objectives, and then move to iterative design. The trade space is wide and includes both "human factors" as well as mission and customer factors.

While "real-world constraints" is identified as the last key parameter to impact crew station design and integration, these are often the starting point for the design effort. There are never infinite financial or time resources to complete the design and integration effort. In addition, a new design will often entail additional and costly items, such as training, to the operator. Or the new design may have new and unique infrastructure requirements, such as a new way to load data. On the other hand, there are some significant real world values that appropriate crew station design and integration can yield. These include cost of airline operations through interfaces that require less time and resources to train, reducing the cost and weight of expensive parts through integration of functions in a single piece of hardware, and allowing the flight crew to reduce the time it takes to completes tasks where these actions represent a bottleneck. An example of this kind of value-added time-savings is reducing the time it takes the flight crew to complete the turn around process at the gate. Other opportunity areas for achieving value in crew station design include finding ways to reduce the cost of parts that go obsolete or need to be upgraded with emerging functionality and capabilities. Achieving the "optimal" solution is not just about fitting the human into aircraft system but also allowing the customer to "close the business case" for the design.

For a new concept to make its way into a production crew station these various perspectives need to be accounted for and the collateral impact of integrating a new technology or enhancing a crew station's capability understood. Traditional items that need to be vetted include the impact on crew training and experience, airline and airspace infrastructure, and certification and regulatory aspects. Crew training for both military and commercial market segments is an important item from a cost avoidance perspective for both the crew station manufacturer and their customers. For the manufacturer it requires an initial cost to create and certify the requisite training regimens and associated materials, and a recurring cost to keep the materials current. However, the greater costs is realized by the customer who has to put its crews through initial training and the recurring training costs to keep the crews current per regulatory requirements.

Crew system design changes can also have a less tangible, but just as important cost where valuable knowledge accumulated over years of experience are negated by design modifications. The heuristics and lessons learned that contribute to a crews' expertise can be negated by a substantial design change or even worse, result in increased error rates due to applying previous knowledge and experience to a new design that doesn't adhere to the prior mental model.

In considering the cost of integrating a new crew station design into the airline or military infrastructure, in addition to the impact on training, the designer needs to consider how the crew station design will impact communications, between crew members as well as between air traffic control and the airline operations center. How data and information gets transmitted among all of these stakeholders as well as the protocols used can have a distinct impact on crew station design and if not addressed appropriately yield unnecessary costs. Another infrastructure constraint is database management. Many of the functions, such as navigation and charts, have databases that must be kept up to date. As well, there may be multiple databases for similar information. Adding another database or requiring another separate update cycle entails costs to the airlines. Lastly, the amount of downtime for an airline to install new hardware and/or software should never be underestimated. All of these factors have clear and significant impact on the design and cost of the crew station design.

A significant impact on crew station design and integration, especially from the commercial transport perspective, are all of the certification and regulatory agency rules and guidance pertaining to commercial aircraft. These regulations and guidance are meant to ensure that all designs will yield safe operation of the aircraft. While the basic regulations have been in effect for many years, just as the crew station design has evolved, so have the regulatory agency rules and guidance. Regulators have recognized the importance of such advancements as automation and integration of function. This has yielded new guidance and rules on alerting and display of information as well as on methods for evaluating the design from the users' perspective. As the amount of information available to the crew has increased and information is more easily modified via electronic displays, there has also been increased pressure on defining standard symbology. With increased integration of function and large scale integration of systems, came the requirement to analyze, evaluate and assess the impact of the ability of the crew to handle single failures that can have either cascading or multiple effects. And as with many aspects of engineering and design, the regulators are giving a more critical review of design methods and processes instead of just relying on final test and evaluation. This is true especially in the area of human error where analysis of what kinds of errors might be made as well as the error tolerance of the design is just as or more important as the finite number of tests conducted during airplane certification and validation.

The Growth of Information Management and Integration

In both commercial and military applications, the aviation industry has witnessed significant increases in system complexity and the need to manage information. As previously noted, the advent of computers and increased automation has resulted in higher levels of crew control. We find that with that higher level of crew control, there is often ensuing questions on the part of the crew wondering about the systems and asking, "What's it doing now?" Allocating appropriate tasks to the crew versus the airplane and mission systems is a partial solution to this problem. Giving the crew adequate knowledge of the status and predictability of future system actions is also required.

Examples of the integration of information was the introduction of the Engine Indication and Crew Alerting System (EICAS) display as well as the electronic check list Electronic Check List (ECL). In the former, key engine and crew alerting information is presented to the crew such that system status and health can be quickly known. The EICAS display replaced the myriad of separate engine gauges and indicators, an example of which is shown in Figure 17-10. As additional awareness of system status, system synoptics are often used to facilitate the crews understanding of health and operation of various airplane systems such as fuel, hydraulics, electrical, and flight control systems to name just a few. The ECL helps guide the crew through ensuring that all appropriate steps have been performed in both normal and non-normal systems operations. By linking the call-up of appropriate information from these three systems, that is, EICAS, ECL, and system synoptics, the airplane computer system can aid in the crews understanding of airplane state as well as an understanding of what to do next (Figure 17-11).

In more recent times, we have seen new channels of information being delivered to the crew. The traditional channels, radio for auditory communication and displays for visual communication, are being used for more types and sources of information. Weather information doesn't just come from on-board weather radar but now weather information from far down the flight plan can be uplinked and displayed to the crew. In addition, there is more sharing of information and decisions among the various stakeholders that include the flight crew, air traffic management personnel, and airline operations personnel. This cooperative decision making promises only to increase in the future. Deciding on how and what information these shareholders need to have and how they are to communicate is an important issue for crew station design and integrations.

In all of this evolution, the question that remains on the forefront is: how does the crew integrate into the overall system design? Understanding the system requirements as well as how the crew

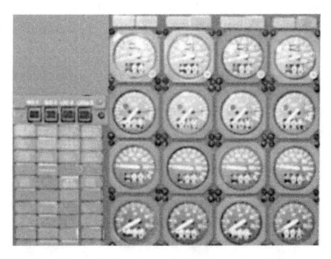

FIGURE 17-10   Typical classic style engine instrument and alerting panel versus modern integrated EICAS display. Image courtesy Boeing.

should interact with the system, in terms of both knowledge of and control of, is key to integrating the crew into the design. As computing technology progressed to the point of being applicable to commercial flight decks, an emphasis emerged to reduce part counts and provide the flight crew with a more integrated picture of his primary flight parameters. The first attempts to replace electromechanical instruments with displays resulted in a duplication of the basic Attitude and Direction Indicator (ADI). In the 1980s studies were conducted at various locations to produce a Primary Flight Display (PFD), which would include all the basic flying instruments in one display surface (Konicke, 1988). The functions integrated into the PFD included the ADI, altimeter, airspeed, heading, vertical speed, flight modes, radio altitude, and a variety of new data.

To integrate these related but different flight instruments into a single format, numerous evaluations were conducted. Previous instruments depicted in a round dial format had to be converted to a vertical tape presentation. Classic display range vs. resolution studies for various flight scenarios were carried out. Philosophies/guidelines

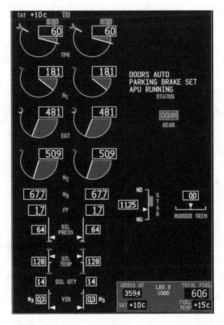

FIGURE 17-11   Flight control synoptic display. Image courtesy Boeing.

were developed to ensure that colors, fonts, highlights, etc were used consistently across the PFD and all other display formats. New display techniques for grouping categories of information were defined. Not only is color and location critical for grouping, but it also was discovered that the use of shading could facilitate quick scanning of the data (Figure 17-12) (Konicke, 1988).

Finally, as described in the introduction section of this chapter, the history of crew station design has been characterized by evolution rather than revolution. This is often seen as being driven by two commonly held notions: (1) evolutionary change builds on the existing experience and expertise of the crew, and (2) evolutionary change tends to be safer than larger revolutionary changes. The fundamental issue, however, is that in order to maintain appropriate safety margins, the bigger the change, the larger the amount of analysis, test, and evaluation that will normally be required. Even training may be positively impacted with large or revolutionary changes. The impact of change on training has more to do with how the crews experience and expertise is brought into the new design as well as how intuitive the new design appears to the crew. However, ensuring the adequacy of the design from the perspective of training also means additional analysis, test, and evaluations. All of these activities have associated costs and hence must be

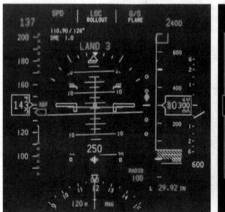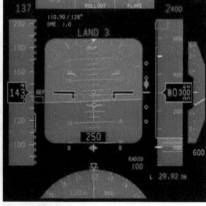

FIGURE 17-12 Example of how shading facilitates integration of information on the Primary Flight Display. Image courtesy Boeing.

off-set with the value that the new design brings to the fleet. Larger changes also typically entail larger infrastructure changes, again with the associated costs of implementation. Hence, what we've witnessed in commercial aviation tends to be evolutionary rather than revolutionary as safety is of utmost concern, along with the economics of introducing new concepts and capabilities.

The design of crew stations has historically entailed the integration of the flight crew into the overall design of flight decks and cockpits. Little attention was paid to how the flight crew and crew station were integrated into a larger system. This has changed over the last 20 or so years with the increased pressure to create more efficiency and greater capacity within the operational airspace, both in the air and on the ground. For the flight deck, this has meant a greater emphasis on design solutions that can, for example, reduce turn-around times at the airport gate. It also means that flight coordination is not just about crew resource management, that is, how the flight crew members divide and cover all of the tasks necessary to fly, navigate, communicate, and mange airplane systems, but now also entails a greater emphasis on coordination among other stakeholders in efficient and higher capacity operations. These include air traffic control (ATC) and airline or company operations centers, both of whom take significant responsibility for safe and efficient flight operations. This has resulted in a greater emphasis on designs that can facilitate better collaboration and communication among the flight crew, ATC, and operations centers. Examples of this include designs that support datalink operations as well as the uplinking of efficient flight plans such as exhibited in the emerging concept of Tailored Arrivals (Mead et al., 2008).

TOOLS AND METHODS FOR DESIGN AND INTEGRATION

Over the years, various conventions and concepts have gained popularity in describing the tools and methods for the integration of the human into the overall design. These have included the notions of human factors engineering, user and/or human centered design (UCD or HCD), and human system integration (HSI). Each carries its own nuances but all are related to the tools and methods for ensuring the human is adequately addressed during system design. The emphasis on Systems Engineering as a methodology for successfully accomplishing large-scale systems design and integration has seen a corollary focus on the concept of HSI. This can be seen as a formalization of the UCD process as it contains all of the key attributes of ensuring that the user is considered throughout the design and integration process.

HSI represents a systematic approach for incorporating the crew into the overall systems design and involves the integration across multiple domains, such as training, staffing, safety, traditional human factors, and so on, as well as the integration of these within the rest of the program disciplines and functions. The domains included are analogous to the traditional Department of Defense (DoD) organizational groups that are stakeholders in integrating the human into a design. These include: Manpower (staffing), Personnel (skills and abilities), Training, Human Factors (e.g., cognitive, physical, sensory characteristics), Safety and Occupational Health, Personnel Survivability, and Habitability. By ensuring integration across all of these domains as well as integration of the HSI domains within the overall systems engineering effort, the crew's impact on and by the system should be adequately addressed.

The more traditional approach to integrating the crew into the design is often referred to as UCD or HCD. This approach is not in conflict with HSI but just views the crew integration from a different perspective. UCD and HCD are both processes that relate to how the user is included in the design process but is typically more specific in focus as it doesn't contain the aspect of integration across the multiple HSI domains. One view of the UCD process includes the following steps: definition of design objectives, definition of operational requirements and concept of operations, definition of crew functional and information requirements as well as user tasks (Task Analysis), definition of display and control requirements, development of conceptual design and various design options, conduct analysis of design concept and options, conduct prototype, test and evaluation where analysis is inadequate, and resolve an open issues. While this appears to be a very linear process, it is actually iterative and spiraling with feedback given to previous steps as one iterates through the design process (Figure 17-13).

For UCD, one of the significant events required to ensure a successful program is determination of the level of effort required to yield an acceptable design. Key decision criteria are the user interface's novelty and complexity. To determine novelty of a design, the following set of questions is useful: (1) does it represent a new technology?, (2) does it involve new or unusual procedures?, (3) does it introduce new ways for the crew to interact with the system?, or 4) does it involve new uses of existing systems that change crew tasks or responsibilities? As novelty and/or complexity of the design increase, the need to increase the focus, rigor, and resources applied to UCD increases.

The design tools used within HSI and UCD have the objectives of reduced cycle time and cost to develop the design, reduced risk of poor and/or unacceptable user designs, enhanced product usability and functionality, reduced risk of potential sources of crew error, increased crew efficiency and effectiveness. Tools vary over phase of program and the whole process tends to be iterative.

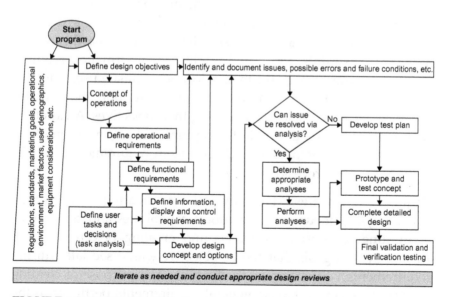

FIGURE 17-13   Process flow diagram of a User Centered Design method.

Design tools and methods can be broken into several distinct categories: (1) those used during design to ensure a user-centered design approach, (2) those tools and methods that are specifically related to integration, and (3) the concept of human-system integration as a process for ensuring the appropriate integration of the crew into all aspects of system design.

Crew station design generally begins with using analysis tools and methods, assuming that initial requirements have already been captured. Traditional analysis tools include task analysis and mission analysis to begin the concept development phase. Much has already been said about this in other chapters but the main point here is that it is a critical stage for crew station design and integration. As rapid prototyping tools have increased in capability and ease of use, there has been a tendency to jump to these before adequate analysis is accomplished. This is usually to the detriment of system and concept design. The emphasis on systems engineering helps to ensure that good analysis is completed first, that is, understanding mission, system and user needs, capabilities, and requirements before any design is actually begun.

Mission analysis is critical as a key first step in the UCD process in establishing baseline system requirements based on an integrated understanding of mission objectives, the system's context of use and work environment, end user characteristics, hardware and software constraints, associated workflows, information required by the end user to perform their role, and requisite user interface components and navigation architecture to support the operator in effectively and efficiently achieving their objectives. A variety of tools exist today for aiding in the capture and organization of the information that helps define UCD system requirements, but the focus here is on process steps.

Requirements analysis typically involves creating user profiles, task analyses, documenting platform capabilities and constraints, accounting for design principals and creating objective usability and system goals that can be objectively assessed later during requirements verification and validation. The purpose of creating user profiles is to establish general requirements pertinent to the system's use for categories of users that can be applied to overall interface look and feel. Generally, a user profile covers the following characteristics: psychological and physiological (fatigue, attitude, motivation, etc.), knowledge (areas of expertise, skill sets, degree of experience, etc.), task characteristics (frequency of use, basic task structure, etc.), and unique physical characteristics.

Task analysis is performed to capture a user-centered model of how work is performed today with respect to how the user thinks about, talks about, and performs their role in their work environment. This provides a basis for identifying areas of process improvement, potential impact of new user interface concepts, and making sure unwritten procedures and jargon are accounted for during design. Task analysis also helps in establishing a baseline for dealing with the system design trade space. Several trade space aspects have been discussed previously, but it is worth noting this could also include appropriate use of automation and revising processes to meet business objectives.

Documenting platform capabilities and constraints and applying solid design principles is not discussed here due to its straightforward nature. However, it is worth noting that system constraints typically include real-world hardware limitations (e.g., screen size and resolution, weight and power limits, physical volume, etc.) and software constraints (e.g., design flexibility, underlying database architectures, communication bandwidth, etc.). The capturing and analyzing of the various requirements drivers is often culminated in not only a set of requirements, but also usability goals that are applied in the design decision trade space. These requirements and usability goals must be concise, objective, testable, and most importantly agreed upon by more than just the UCD team. Defining usability goals serves two purposes: (1) helps focus design decisions by giving concrete requirements that can also be used to assess design prototypes, thereby streamlining design activities as well; and (2) establishes acceptance criteria and clear means for verification and validation that the system has met its UCD requirements.

Modeling of both the human and system is often pivotal in understanding basic operation and characteristics of the human and the interaction with the system. Today, models, whether they are electronic or just theoretical exist for many human physical characteristics. These include anthropometry as well as perceptual aspects such as vision, auditory, and tactile. However, fewer and lower fidelity modeling and analytical tools available for the cognitive aspects of human behavior and interactions. This has been recognized and the industry has been working on developing these tools (John & Suzuki, 2009).

Prototyping is an invaluable tool for iterative design process and it allows many forms of crew interface evaluation. The use of prototypes runs the gamut from part-task, desktop types prototypes to foam-core and cardboard mock-ups to part-mission pilot in-the-loop simulations to full-task mission simulation. In addition, both objective and subjective measures are employed in using prototypes to evaluate designs. A critical part of the use of prototype evaluations is scenario development and definition. The scenarios must be generated around interface and concept issues and risks. In addition, an effective scenario ultimately looks to test for corner conditions as well as normal conditions to test and evaluate.

Another useful user centered design method is the use of guidelines and philosophies. These are effective in facilitating decisions, maintaining consistency of design, and enabling efficient development. These methods allow both lessons learned and tribal knowledge to be employed effectively during the crew station design process. Examples of these are industry guidelines, top level philosophies, design principles, and style guides (Aerospace Recommended Practice 5056, 2006).

Integration is an important part of the system design process because it accounts for impacts related to subsystems interacting at the system level as well as the user experience when interacting with the system as a whole (e.g., subsystems may be designed well, but when merged at the system level create an overall poor design). Integration testing and evaluation with respect to UCD is often done at multiple levels of fidelity and system maturity. For example, at early stages of design, integration evaluations may leverage low-fidelity part-task simulation facilities and progress through full-mission simulations and ultimately flight tests as the system matures and subsystems are integrated. At each phase of integration testing as the system matures, it is important to scrutinize the user experience with respect to overall system design and back-drive consistency across subsystems where needed while understanding the impact on the user's workflow. The goal is to create an effective and efficient system that ensures mission success.

Another important UCD impact to integration during system design is that it provides an opportunity to evaluate overall system failure scenarios that cannot be analyzed at the subsystem level. This type of analysis has to be done at the system level to understand cascading impacts on not only subsystems but also how the user may react to various failure scenarios that subsequently may impact design, operational procedures, and training. To perform this type of analysis, the UCD process makes use of "what if . . ." scenarios and scripted failure/multifailure scenarios ranging from more common to extreme instances. Testing the system and user jointly in these diverse situations it not only critical for creating a usable and effective system but, more importantly, to ensure safety.

CONCLUSIONS

All of these design tools and methods are targeted at generating crew station design and integration that will ensure the user needs are met, crew error is mitigated, and system design can be accomplished as cost-effectively as possible. Crew station design and integration is best done as part of the overall system design and in fact is most useful when user and system requirements are completed up front and then considered throughout the whole design process. Above all, crew station design and integration is about considering the crew in all aspects and phases of design. As crew stations continue to gain in complexity the importance of using the right tools and methods at the right time through-out the design process to achieve the desired goal of an efficient, effective and safe crew station design also increases. It is less expensive in the long run to perform design and integration well. However, doing this well doesn't mean using all of the tools and methods all the time but, rather, judiciously applying them throughout and ensuring that a process exists from day one to include crew considerations in the overall system design.

References

Proctor, R. W., & Van Zandt, T. (1994). *Human Factors in Simple and Complex Systems*. Needham Heights, MA: Allyn and Bacon.

Sanders, M. S., & McCormick, E. J. (1993). *Human Factors in Engineering and Design*. United States: McGraw-Hill.

Wiener, Earl L. (1989). *Human Factors of Advanced Technology ('Glass Cockpit') Transport Aircraft*. Moffett Field: NASA-Ames Research Center.

Bureau of Air Safety Investigation. (1998). *Advanced Technology Aircraft Safety Survey Report*. Department of Transport and Regional Development, Bureau of Air Safety Investigation.

Bolia, R., Vidulich, M., Nelson, T., & Cook, M. (2008). A history lesson on the use of technology to support military decision making and command and control. In M. Cook, J. Noyes, & Y. Masakowski (Eds.), *Decision Making in Complex Environments* (pp. 191–199). Burlington, VT: Ashgate Publishing Company.

Konicke, M. L. (1988). *747-400 Flight Displays Development* (AIAA Paper 88-4439), AIAA/AHA/ASEE Aircraft Design and Operations Meeting, Atlanta, Georgia. American Institute of Aeronautics and Astronautics 370 L'Enfant Promenade, S.W., Washington, D.C. 20024.

Mead, R., Coppenbarger, R., & Sweet, D. (2008). *Field Evaluation of the Tailored Arrivals Concept for Datalink-Enabled Continuous Descent Approach* (AIAA Paper 2007-7778), AIAA Air Transportation Systems and Operations Meeting, Anchorage, Alaska. American Institute of Aeronautics and Astronautics 370 L'Enfant Promenade, S.W., Washington, D.C. 20024.

John, B. E., & Suzuki, S. (2009). Toward cognitive modeling for predicting usability. In J. A. Jacko (Ed.), *Human-Computer Interaction, Part I, HCII2009, LNCS 5610* (pp. 267–276). Berlin: Springer-Verlag.

Aerospace Recommended Practice 5056. (2006). *Flight Crew Interface Considerations in the Flight Deck Design Process for Part 25 Aircraft*. SAE International, SAE World Headquarters 400 Commonwealth Drive, Warrendale, PA 15096-0001 USA.

PART V

VEHICLES AND SYSTEMS

18

The History in the Basics and the Challenges for the Future

Captain William Hamman

William Beaumont Hospital Center of Excellence for
Simulation Education and Research

Introduction by Captain William
L. Rutherford MD

Vice president Flight Standards and Training (retired)
United Airlines

The past 20 years have seen a tremendous advancement in technology, training, assessment, and monitoring of the performance of airline flight crews. Three-person crews have been replaced by two-person flightdecks, analog aircraft have been replaced with glass fightdecks monitored by hundreds of computers, and the availability of information to the flight crew has increased exponentially. This chapter will look at the impact of the integration of human factors and flightdeck complexity into the world of the airline pilot. The chapter will move from the very basics of the simple two-challenge rule for pilot response to the complex, team, system analysis, and risk management of the modern flightdeck. The chapter will identify the modern-day challenges to crews and if the current science of human factors is mitigating these challenges becoming the next airliner accident in the view from the flight deck. This chapter is an operational look at the output of human factor research.

INTRODUCTION

This is a story that is best understood from the top down. Critical components are lost by narrow focus on the specifics of the procedures or the mechanics of individual evolving solutions—it is the forest that needs to be understood, not the individual trees. Like real ones, metaphorical trees change with time and circumstances. It is the functioning of the culture that is effective. Details of "the rest of the story" that sometimes contribute to a broader understanding can be instructive, but for the most part the story of crew resource management (CRM) is the "big picture."

Motivation to the panoply of tasks aviation accomplishes often arises in the romantic fantasies and longings of young restless minds. We will be neither successful nor forgiven should we fail in our reverence for these seminal phenomena. Economics and industrial culture militate powerfully against them! Understanding the importance of maintaining this perspective and a sense of humor will enable a better understanding of CRM fundamentals.

Errors and accidents are inevitable in human activities. When the stakes are high, the consequences can be disastrous. Beyond sharing tasks and physical workload, cognitive efforts within coordinated groups of people can create teams, in some contexts called "crews." But these patterns of effective interaction do not spontaneously arise with reliable regularity. This failure of situational team formation is demonstrated in the story behind the culture of aviation safety and the culture of CRM.

Erroneous "facts," incorrect assumptions, miscommunications, misunderstandings, increasing disconnection between individually generated "situational awareness" and "truth"—real-world reality—are insidious enemies. Hierarchical social, workplace, and management cultures—and ego—are their powerful allies. Probably most important to understanding the source of CRM's power is to recognize that it is based on fundamentals honesty, humility, respect for others—"others" of all ranks, because the person at the bottom of the local "totem pole" may have the key fact or idea that prevents

YOU from making the fatal mistake. Effectively mobilizing and enabling the intellectual resources of all participants in a complex task is the goal of CRM, it requires education, training, enculturation, and derivative equipment and procedure adjustments. Eventually, with time, it assumes the appearance of a subculture, the self-sustaining and reinforcing power of the herd.

CRM emerged and ripened in an aviation community that in its early years was focused on the practical problems of flight—airframe design, aerodynamics, propulsion systems, materials, meteorology—that challenged the existing limits of their disciplines. Not that these elements have been mastered but, over time, solutions were developed that allowed a degree of aviation system adaptation and tolerance of them. As the other components of the system became more predictable, the accident history came more and more to identify "pilot error" as a final determination of the official cause of crashes. The official investigations (Department of Commerce, Civil Aeronautics Board, National Transportation Safety Board) tended to identify pilot error as an acceptable final determination. Surviving pilots and their organizations, such as the Air Line Pilots Association (ALPA), were uncomfortable with what they viewed as incomplete explanations. On their own, they looked further, beyond "what the pilot did" to understand "why the pilot did it." Underlying this search was the concern that since respected colleagues could have stumbled into what often looks like very bad decisions are there environmental or system traps that caught them and may await others?

The arrival of the Boeing 707 and Douglas DC-8, in 1958 and 1959, respectively, changed the face of commercial aviation. They were not only bigger and faster and more comfortable, they incorporated reliable technology that failed much less frequently than their predecessors. Turbine power plants particularly were vastly more reliable than the previous generation of reciprocating engines they were replacing that had been stretched to their limits. The greater size, complexity, and cost of introducing this equipment into an aviation system that had evolved to meet the requirements of the "recips," caused the Federal Aviation Administration (FAA) to require three pilots in each jet airliner's cockpit and to impose the notorious "Age 60 Rule" requiring airline pilots to retire earlier than practitioners of other trades or professions. But the extra member of the crew did not reduce the proportion of crashes that resulted from the notorious pilot error.

In the late 1960s concern developed around the vulnerability of airline crews to pilot incapacitation as the pilot population aged. A diagnosis of coronary artery disease brought mandatory denial of medical certification to fly, but the condition is nearly ubiquitous, often silent, and medical diagnostic and therapeutic technology was primitive. Was there a way to mitigate the risk of pilot incapacitation and capitalize on the safety opportunity created by the third crewmember? United Airlines, for one, addressed the dual problems of older pilots' vulnerability to a very common condition that can suddenly incapacitate and the authoritarian cockpit culture that prevailed. If the captain gets incapacitated, how will it be recognized and handled?

Looking back, the "Two Communication Rule" that United developed in its simulators can be seen as the first systematic CRM thinking. The rule required all cockpit crewmembers to be aware of the position and condition of flight at all times. They would know the expected flow of the activities and recognize when the flight was not proceeding as planned or expected. Serious deviations from standard profiles (those that could potentially affect safety) required the noncaptains to call them to the captain's attention. If the captain did not respond with either a satisfactory explanation or correction of the condition, crewmembers were required to repeat the alert. If no satisfactory response was then forthcoming, the co-pilot was *required* to take physical control of the airplane from the captain. This prescribed step of the procedure constitutes the first chink in the armor of absolute authority vested by law, tradition and culture in "the Captain." The procedure was trained and exercised in annual recurrent training during the 1970s with no aircraft accidents resulting from the few occasions it was performed in real life. But its reach was limited by the definition of the precipitating circumstances—the pilot is, or appears to be, "incapacitated." "Captain's Authority" would not yield so gracefully to subsequent migration of the paradigm to less morbid flight situations.

As the aviation community was struggling to understand how human errors occurred it became apparent that though accidents were rare there must have been many more similar events that for some reason did not ripen into tragedy—discipline, procedure, equipment, fate, or just plain luck? What was going on out there in the airspace? The FAA has responsibility for regulatory enforcement so they were interested in the answers, but hampered by their "traffic cop" image in the community. Who is going to admit to errors/potential violations to an enforcement officer?

On August 15, 1975, the FAA and the National Aviation and Space Administration (NASA) signed the first of several memoranda of understanding that created a unique tool for answering questions about the daily realities of life "on the line"—the "NASA Program," formally the Air Space Reporting System (ASRS) (FAA, 1975). That understanding has been modified only in September 1983, August 1987, and January 1994, while receiving over 700,000 voluntary individual reports of the human vulnerability and hazards in the airspace system. Under the memorandum of understanding, people reporting hazards and their own errors are shielded from any punitive action by the FAA unless the transgression was a willful violation of law. Aviation managers and regulators gained tools to modify procedures and facilities based on facts from system users. System understanding was growing, but incidents and accidents continued.

A series of joint industry/government safety meetings in the late 1980s produced a number of insights into the problem. Prominent among them was a recognition that the FAA-approved and required recurrent proficiency checks were forcing airlines intent on efficiency in the training systems to "dummy down" their training/checking to allow the completion of the FAA-mandated maneuvers in the allotted simulator time (a precious commodity). The net result was that while CRM/team skills were drilled in recurrent training, on the final day of the multi-day event the *check ride* was conducted as a solo demonstration of rote flying skills. In effect the system was saying, "CRM is key to operational safety, but a pilot's individual flying skills are really more important since that is what passing the check requires." That had to change. Since the FAA recognized its requirements were inducing the defective training model, the FAA approved the exploration of a new one—one that would integrate technical/piloting performance with CRM behaviors in crew training/checking on the condition that an "equal or greater" level of safety could be assured. The Advanced Qualification Program (AQP) was conceived. AQP gestation was protracted as the lists of required tasks and skills for each aircraft type and crew position were created.

Specific markers of individual behaviors that could be recognized, taught and evaluated—and are conducive to effective team performance—had to be developed. These markers also had to be credible to skeptical crewmembers who were now to be evaluated on their integration of these behaviors, not just in practice but in their licensing checks, in addition to their required demonstration of technical mastery of their airplane. This work was done in simulation by recreating events that had occurred in the airspace system as reported to the NASA ASRS. The scenario/event set was devised to extract understandings of these behaviors by engaging flight crews in credible, realistic replications of reported events. Details of this work can be found in publications by Hamman and others (Hamman, 1994).

"CRM crew" behavior has been shaped by practical CRM research, adaptive crew education, and modified aircrew training/checking over at least three decades. Indeed, it is the evolution of an aviation system safety culture, centered on the cockpit culture of CRM, that is most remarkable. The notion is crudely relevant that "herd immunity" resulting from the "vaccination" of a sufficient subset of members is protective across the subject population, the more correct inference would be that everybody in the system knows the importance of safety, the disputes are over priorities and what is "safest."

It is hoped that by understanding this bigger picture, practitioners of other professions would appreciate the time horizon over which this "culture" emerged and the multiplicity of factors determining the shapes of its many iterations. Intentional culture change is a tedious process of relentless commitment to a shared core set of values and objectives and is never complete.

EVOLUTION OF EVIDENCE-BASED SAFE PRACTICES IN THE FLIGHTDECK

Understanding Pilot Error

The world of airline travel has seen a tremendous improvement in flight safety over the past 20 years. The reasons are multiple from more reliable technology, training crews as teams, feedback of performance information on pilot performance, and near miss reporting. These systems have been recognized for their effectiveness and attempts to emulate them in other industries such as healthcare are being attempted. As can be seen from the incident of accident reports (Figure 18-1) (Boeing, 2007), there has been a drastic decrease in hull loss in airline operations. In Figure 18-2, the causation factors for accidents can be seen. Leading these causes is the important issue of crews and the relationships related to human factors and fragility of human beings. Loss of control in flight (LOC I) and Controlled Flight into Terrain (CFIT) lead the list. Over 70% of the accidents are caused by failed communication by flight crews. This has led to multiple new tools and processes to understand these errors and to mitigate errors of flightcrews. The NASA ASRS reporting system has been in place the longest and has several thousand reports on human error. Although a very powerful tool for analysis the impact of errors and system issues, the generic nature of the NASA reports made them hard to interpret for an individual airline. In 1994 an industry task force began the important but difficult task of moving from anonymous reporting to NASA to direct identified reporting to the pilot's airline identifying errors that have occurred. This program became

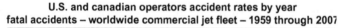

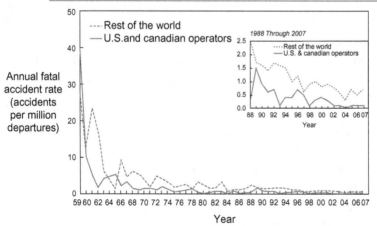

U.S. and canadian operators accident rates by year
fatal accidents – worldwide commercial jet fleet – 1959 through 2007

FIGURE 18-1   Boeing aircraft.

TABLE 18-1   Key Issues from an Airline Pilot's Perspective.

Issue	Subtopic Skills
Back to basics	Basic skills of flying aircraft or managing
Automation	Technology placed in between the pilot and the aircraft operation
Experience	The knowledge and skill in flying the aircraft over time being in it or exposed to it over time
Fatigue	Sleepiness, as governed by the circadian pacemaker and the homeostatic drive for sleep, determine the level of alertness and performance associated with rest-activity patterns
Risk Management	Proactive management of information to mitigate risks in flight operations becoming a loss of an aircraft

a relationship between the FAA, Airline Management, and Pilot Representation.

The program is known as the Aviation Safety Awareness Program (ASAP) and provides a wealth of information on threats, errors, and risk occurring on the line in the operational environment (FAA, 2002). The program has a detailed process for the event review committee to decide the actions if any will be taken against the pilot. In most of the reports the action is an e-mail for verification of completion of the report and a thank you. This occurs in over 99% of the reports and only if there is an intentional violation of Standard Operating Procedures (SOP), alcohol, drugs, or repetitive occurrences will action be taken against a pilot. For the pilot, it is still the vehicle for protection against FAA action if an error has occurred.

After the report is processed by the airline it is deidentified and sent to the Aviation Safety Reporting System at NASA. For the airline, ASAP provides a wealth of information for their operation. If done correctly, the warehousing of the data can track trends of data for improvements caused by training interventions or changes in operating procedures. The risks are drowning in the data, and not having

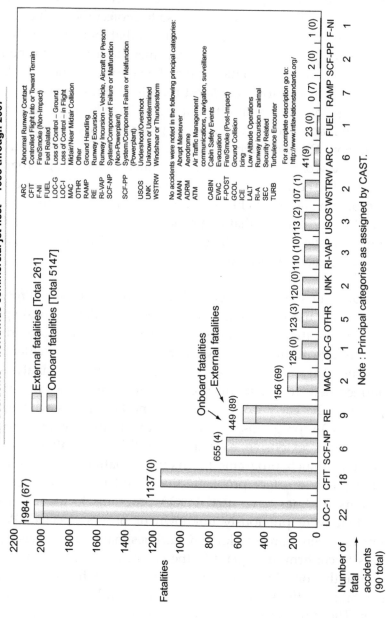

FIGURE 18-2 Boeing aircraft.

the communication or trust of the pilots to effectively provide the information in a meaningful manner to the line pilot. If airlines are not careful the airline pilots do not see any meaningful change from providing a report and will only do so to protect themselves rather than to send reports on all operational risk that they incur while flying the line. For example, the management at a major airline and the pilot union representation have allowed the airline's trailblazing ASAP to end during the latest rounds of contract negotiations. "For years, the airlines and its employees have been leaders in the field of aviation safety with its ASAP program," noted Flight Safety Foundation FSF President and CEO William R. Voss. "Airlines around the world modeled their own internal reporting programs after ASAP. The Flight Safety Foundation has publicly supported this program and others like it as an important tool to prevent accidents. We are alarmed that either side would allow this incredibly important safety program to fall victim to distrust between labor and management" (FSF, 2008).

The object of a Flight Operational Quality Assurance Program (FOQA) is to use flight data to detect technical flaws, unsafe practices, or conditions outside of the desired operating procedures early enough to allow timely intervention to avert accidents or incidents. This is in addition to already using the analysis of data recorded during flight in determining the causes of aircraft accidents. The early experience of domestic airlines with established FOQA programs, as well as the testimony of non-U.S. airlines with extensive experience in this area, attests to the potential of such programs to enhance aviation safety by identifying possible safety problems that could lead to incidents or accidents. Airlines have used FOQA programs to identify problems that were previously unknown or only suspected. Where problems were already known, airlines have used these programs to confirm and quantify the extent of the problems. More importantly, on the basis of analyses of flight data, airlines have taken actions to correct problems and enhance aviation safety. The data provided by FOQA is the most accurate available in how the airline is being flown. However, in this accuracy, the story of why the airline is being flown in a particular manner is lost. The way FOQA works the data from the flight data recorders is downloaded on a regular basis. Software programs have been developed to decide what parameters are to be monitored, what defines an exceedance, and levels of exceedance. The output reports define the information as data points but without any story behind the points. For example, the airline may have criteria for stabilized approaches at 1000 feet above the ground. The airline has a policy that if the aircraft is not stabilized by 1000 feet a go-around must be executed. The FOQA data may show that 10% of the approaches are not stable based on airspeed, configuration criteria etc but only 1% of the aircraft have executed a missed approach. On the surface this seems critical. However, several questions must be asked in order to determine if there is a real issue:

- Is exceedance criteria data too sensitive?
- What were the operational conditions?
- How was the approach and landing actually flown?

For example, in one case of exceedances of airspeed for stabilized approaches, it was the design engineering of the Airbus aircraft to fly a managed speed rather than a set speed that resulted in so many hits. Although this is a perfectly safe approach FOQA did not differentiate. If an airline reacts too quickly, false positives may occur. The other issues missing from FOQA are the crews flying the aircraft. There is absolutely no answer to why the parameters were exceeded provided by FOQA. Only in the most severe exceedence do some airlines have the capability to contact the crews for the human element of the exceedance. This makes it critical to link FOQA with other safety programs such as ASAP.

For the pilots, the balance between safety and invasions of privacy remains the issue along with how does their airline process the data to meaningful improvement. In October 1997, FAA Administrator Jane Garvey said that the agency would issue a Notice for Proposed Rule Making to codify the limitations on the FAA's use of FOQA information; federal officials said in August 1998 that work was continuing on a solution to the regulatory issues of FOQA. The FAA's delay in promulgating an enforcement regulation has hampered the ability to reach agreement with some pilot unions and threatens the continuance of agreements already reached. One of the issues facing the FAA is how broad the enforcement protection should be. FAA attorneys have concluded that it is beyond the scope of the FAA's authority and in violation of its statutory duties to issue a regulation that precludes the agency from taking action if FOQA data reveal that an airplane was not in a condition for safe flight or that a pilot lacked qualifications (FSF, 1998). On May 9, 2002, part 14 CFR was amended and now states "Enforcement. Except for criminal or deliberate acts, the Administrator will not use an operator's FOQA data or aggregate FOQA data in an enforcement action against that operator or its employees when such FOQA data or aggregate FOQA data is obtained from a FOQA program that is approved by the Administrator" (CFR, 2002). Although the airlines pilots have been tolerant of these programs any false step for enforcement will probably end the programs.

EVOLUTION OF FLIGHTCREW SYSTEMS-BASED TRAINING AND ASSESSMENT

The Change in Airline Pilot Training

For years the focus, training, and assessment of airline pilots was on individual technical skills. These skill proficiencies (engine failures on takeoff, engine out hand-flown instrument approaches, etc.) were the gold standard for evaluation and the vehicle for type certification in the United States. In the late 1980s, it was identified that crew human factors and system issues were the main cause of airline accidents. The data demonstrated that over 70% of airline hull losses were because of crews and dynamics such as communication, workload management, and situational awareness. However, the airline training centers were still focusing on individual pilot technical performance. Even though most airlines had implemented a Crew Resource Training Model (CRM) and integrated Line Operational Flight Training (LOFT), there was no significant improvement in crew caused accidents. The FAA, Air Transport

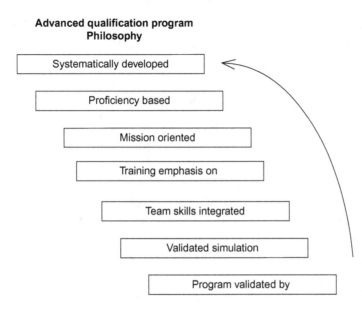

Advanced qualification program Philosophy

Systematically developed

Proficiency based

Mission oriented

Training emphasis on

Team skills integrated

Validated simulation

Program validated by

FIGURE 18-3   AQP boeing aircraft.

Association (ATA), and the flightcrew unions came together to address the problem. The Advanced Qualification Program (AQP) was developed as a vehicle to change training and evaluation of airline crews. This state-of-the art airline crew proficiency is conducted on the AQP template. Three consistent and intertwining trends can be identified that accompany this development of characteristics that, in sum, constitute the "safety culture": (1) reporting systems collect real-world system performance data in finer and finer detail, and these data are shared across the industry; (2) standard operating procedures—best practices—are developed based on the knowledge extracted from these "lessons learned"; and (3) increasing attention is given to understanding and integrating team performance into daily operations and overall professional requirements. The overview of a typical AQP program is shown in Figure 18-3.

These team skills were identified by modeling using simulation scenarios employing the event set design process. Specific skills (technical and team) are identified by criticality measurement and task analysis methodology. The skills are defined as a "tool box" of processes, strategies and tactics empirically identified and characterized in studies of crew behaviors conducted in simulated flight. Once the analysis of these skills is complete, the skills become the focus for evaluating proficiency and measuring system reliability. In turn, training and reinforcement/remediation using simulation is distributed throughout the team member's career.

High-fidelity training simulations do not automatically yield effective training. Simulations, at their best, reproduce realistic tasks and afford trainees practice that enhances learning (Cannon, 1997). In a team setting, simulators allow teams the opportunity to practice both team- and task-related skills. Context-specific cues imbedded within the simulation provide trainees with signals that activate trained behaviors. In addition, simulators provide opportunities for team members to receive feedback on the actions, activities or strategies performed or not performed. A supplementary benefit of simulation training is that it allows training instructors to identify performance decrements and particular situations that require further training. Under AQP the primary unit of both simulator scenario model design and assessment is the scenario event set, a group of related events that are part of the scenario that are inserted into a simulator session for specific objectives (FAA, Line Operational Simulations, 2004). The event set is made up of one or more events including an event trigger, distracting events, and supporting events. The event trigger is the condition (or a condition), which fully activates the sequence. The distracters are conditions inserted within the event set time frame that are designed to divert the team's attention from actual or incipient flightdeck events. Finally, supporting events are other events taking place within the event set designed to further the training objectives (Hamman, 1994). This process of design of science-based team skills and safety science centered on simulation tools creates improved team recognition of "threats" and errors permitting early interdiction of "accident-chain" sequences. When "teams" are trained in this manner artificial barriers to information-sharing are reduced as the "authority gradient" is flattened.

The program was met with both enthusiasm and concern during the early 1990s. The first AQP was performed in December of 1993 and was favorably received. All the concern of the late 1980s and early 1990s abated. The issues for the airline pilot was the process of AQP and the change from one final checkride to a series of validation points followed by the final evaluation known as a Line Operational Evaluation (LOE) (FAA, AQP, 2004; ATA, 1994). Was there only one checkride or was the pilot now faced with up to six evaluations during the program? Most education to the line pilots was initiated to shift the paradigm to this new understanding. The validation point should not have any job or career implications if the pilot fails and, rather, is a test of the training as much as it is an assessment of the pilot performance. One of the core elements of AQP is a program validated by data. The validation points provided this data on pilot performance. If success was low or a trend downward in performance was indicated, than the footprint of the training curriculum is too short and should be lengthened. In contrast, if the validation success was excellence, this was an indication that the curriculum could be shortened. This provided a strong financial incentive for airlines to participate for the potential of reduced training and assessment intervals that would save millions of dollars. Thus, if a pilot fails a validation point such as aircraft knowledge, or maneuver performance more training is provided to the pilot before they move onto the next phase of training. There was overall acceptance of this by the line pilots and a study performed by the FAA in the early 2000s validated these results. The results suggested that most pilots regardless of whether they were trained under AQP or traditional training were satisfied with their CRM training and found it useful. However, AQP pilots did rate their CRM training as more useful than traditional trained pilots. Finally, the results suggest that training programs, which integrate CRM principles throughout the entire curriculum, are perceived more favorably than stand-alone CRM training courses. In comparison to traditional trained pilots AQP training programs are more likely to include:

- the integration of CRM concepts throughout the training curriculum;
- the requirement for pilots to demonstrate proficiency on technical and CRM skills in LOE scenarios prior to certification;
- the requirement that CRM evaluation focus on specific, observable behaviors that were derived from the task analysis; and

- the requirement that check airmen receive special training in evaluation pilot's CRM skills (Beaubien, 2001) (FAA, CRM, 2004).

In the early years of AQP, a significant amount of training was provided to the instructors and evaluators for the airlines. It may be because this training has been reduced or the true philosophy of AQP has been lost over the years, but there has been a shift in the perceptions of validation points and the instructors now perceive them as evaluation for their crews. This has resulted in skewing of the data of the validation points in a program because the instructors do everything possible inside or outside of the curriculum to make sure their crews are successful. The result is 100% success at the validation points that indicates to the airline they can reduce the program. Because of this training programs are drastically reduced but with no change in validation performance because of the "teaching the test" commitment of the instructors. In many cases the complaints are significant for a program because of their length. However, the airlines do not respond because of the strong validations and meeting data collection requirements of the FAA. It has become a vicious cycle and has reduced the quality of the AQP training programs significantly.

EVOLUTION IN THE FLIGHTDECK DESIGN AND OPERATION

The Tremendous Increase in Flightdeck Complexity and Information Overload

In the past twenty years there has been a tremendous change in the technology, and the information management of flightdecks. Global Positioning Systems (GPS), Satellite Communication Systems, electronic checklist, electronic flight bags, reduced vertical separation minimums (RVSM), and internal navigational systems have changed the way aircraft are operated. Additionally, major differences in aircraft manufacture design philosophies for the interface of the crews with the aircraft computer systems have created significant changes and challenges to the job task of the airline pilot. There has been an explosion in new technology, information available during flight, and new critical skills to fly advanced highly auto-mated aircraft. Additionally, technology has led to new procedures to maximize airspace efficiency and aircraft handling capabilities. In these changes there are inherent risks and there are significant dif-ferences in the two major producers of aircraft (Boeing and Airbus) in how the automation is managed. In the Boeing system, the pilot has final authority over automation and can override the computer inputs. The flow is traditional pilot in command of the aircraft with aid from the autoflight systems. In the Airbus philosophy, the final authority is the computers managed by the pilot. The crew has been put one step further back in the control of the aircraft. The final authority is given to the autoflight systems to control the aircraft. Some airlines have a mix of aircraft types and the pilots have flown both. There are pros and cons to either system, but the critical issue for the airlines is the human factors, critical task for each type, and implications for training. The stress of economic times has caused some airlines to cut training programs that are putting pilots out on the line feeling very uncomfortable for the first year on the aircraft before they understand all of the quirks of the aircraft.

The propagation of GPS is changing the skies we fly in, how we fly approaches, and how we manage the aircraft. Throughout the United States the FAA has begun programs to decommission ground navigational aids such as Automatic Directional Finders (ADF) and in their place the approach is generated internal to the aircraft and position reference is performed by GPS. Included in these internal approach capabilities is the ability to generate glide paths for a constant vertical descent path to the runway. Once again, although a tremendous new tool and safety device (moving away from the old descend and level off "dive and drive" nonpreci-sion approaches), the need for strong procedures, situational aware-ness, and crew discipline is critical for safe operation. Adding to the safety of the operation are new hardware such as heads up night vision devices, better runway markings and lighting, and a host of devices too large to be discussed in this chapter. Figure 18-4

Pilots have never had more devices at their side to aid in flying a safe airplane in most any conditions. However, in these choices there are human factor dangers that still need to be addressed. Pilots' initial training in autoflight technology with these devices

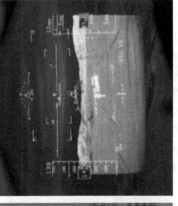

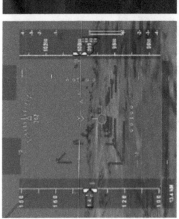

Primary flight display   Head up display

Reduced visibility affects the safety and efficiency of nearly all flight operations. As a result, researchers have looked for ways to improve and/or provide a vision capability to pilots that is independent of actual visibility or weather conditions. In recent years, research has focused on two technologies - Synthetic and Enhanced Vision Systems (SVS/EVS). SVS technology provides pilots with a virtual visual depiction of features in the external environment superimposed with relevant aircraft state, guidance, and navigation information. In an SVS, the geographic location and dimensions of many of these features are stored in on-board databases or models.

In contrast, EVS technology uses imaging sensors that attempt to "see through" obscurations such as those produced by darkness, cloud, or fog. Raw, or processed, video images derived from EVS sensors are typically presented to pilots on Head-Up Displays. A third construct - integrated SVS/EVS - is also emerging that takes advantage of the complementary characteristics of each. For example, SVS is independent of weather effects and provides unlimited field-of-regard; while EVS is independent of navigation system and geo-spatial model failures or errors. The idea is that although the methods, architectures, and operational issues associated with SVS and EVS are significantly different, integrating the two approaches can provide a more robust capability.

FIGURE 18-4   Advanced flightdeck.

was an add-on to just flying the airplane to make your job easier and safer. However, over time this mind-set need to shift to a new way of looking at how pilots fly airplanes. There is a different set of skills for a pilot to be managing over 300 computers on the flightdeck, in a complex air traffic control environment. These skills are critical in RSVM or programming a computer generated vertical descent to the ground. Although, the safety of these devices are great the margin for error has been drastically reduced! Rules and procedures for autoflight must be developed and followed (Young, 2008).

For example, you do not solve a problem with programming the autoflight by going to a more complex level of autoflight engagement. These and other standard operational procedures (SOPs) have slowly come to light over the years. The main feeling of the line pilot is disengagement with the design process. New devices show up designed by an engineer with very little understanding of the line operational environment. This has led to the affectionate saying about automation: Automation is an amplifier because it amplifies the low workload times in the flightdeck and it amplifies the high workload times.

EVOLUTION OF THE AIRLINE ENVIRONMENT

The Challenge of the World We Work In

There have been tremendous changes incurred by the airline industry in the number and type of airlines operated in today's world. These include airlines going out of business, consolidation, the continued proliferation of low cost airlines, and the complex issue of security since September 11, 2001. The impact of 9/11 will never be fully understood, but it has completely changed the world of the airline pilot. Words cannot express the loss of our brothers and sisters, the loss to their families, and the vulnerability it has brought to the world. It can best be surmised by a call to an airlines employee assistance hotline shortly after 9/11. A captain called the hotline and said that their "five-year-old daughter had just crawled in bed with them crying and did not want mommy to go fly because she was afraid mommy would never come back." The captain did not know what to do. A part of all of us died on September 11, 2001. Security changed drastically and for the first time pilots could carry weapons in the flightdeck. The government also reactivated armed federal air marshals. This has all led to a flurry of new procedures and policies for the aircraft. Included in this were new policies for pilots to ride in the flightdeck when they are going to work. (jump seating) Jump seating is the term for sitting in the flightdeck extra seat. It is used by pilots to go to and from work. When it was made more restrictive it had a significant impact on pilots ability to commute to work.

Although necessary, it has caused an increase in workload for the crews, when these issues are added to the pressure for on-time departures for the airline. There have been several reports of illegal dispatches as pilots try to manage all the elements of complexity caused by a post-9/11 world.

The airline world has been a flurry of economic activity post-9/11. Buyouts, mergers, foreclosures, fuel cost, and economics of the world have changed the world in which we work. The market pressure for cheap air travel has put tremendous pressure on the safety systems instituted during the 1990s and early 2000s. One of the greatest changes has been the growth of the regional jets (RJ) providing transportation. The regional carrier route structure and numbers of aircraft have increased tremendously in the past 10 years.

The RJs were never huge profit-makers in the best of times. Rather, they were promoted as economical substitutes for 110- to 140-seat jetliners on thin-demand routes because their ownership and operating costs were lower per mile flown. They were also marketed as more consumer-friendly than smaller, noisier, slower, and less comfortable turboprop planes. Airlines invested heavily in them starting in the late 1990s, primarily to feed more travelers from small markets into their hubs and onto their more profitable mainline flights. As a result, one in four commercial takeoffs today is made by a 50-seat or smaller RJ. This has led to crowded skies, increased delays, and changing the demographics in pilot hiring. In the early years of the regional carriers, they were perceived as a step in the advancement of the airline pilot carrier. Because of this, the work environment was poor, benefits were very few, and attrition was large. However, as the regional carriers increased their market share

	Minimum		Competitive	Preferred	
Total time	500		2000		
Last 6 months:	0		100		
	PIC	SIC	PIC	SIC	
Multi time:	100	0	350	0	
Turboprop time:	0	0	0	0	
Jet time:	0	0	0	0	
Certificate	Commercial		ATP		ATP
CFIIME			Yes		Yes
ATP written	No		Yes		Yes
Regional jet standards certification	No		No		Yes
Education	No college required		2-Year degree		4-Year degree
Age:	21				

FIGURE 18-5   Qualifications.

there was stagnation at the major carriers. This led to a change, and the regional airlines were not a step in an airline pilot's carrier but became the final career move for the pilot. Because of this, pilots organized at the regional airlines, and working conditions and benefits improved. However, during 2006 and 2007, hiring movement, retirements, and early buyouts at the major airlines once again led to a tremendous need for hiring of pilots by the regional carriers. To fill the need, several regional airlines dropped their standards to some of the lowest on record. Pilots with 500 hours' total time and no multiengine time were being recruited by the regional airlines. In some cases, regional airlines were approaching students still in college to persuade them to quit college and come to work for the airline. They would provide the training, multiengine rating, and qualifications to move into the right seat of a regional jet. The days of building time, gaining experience, and watching the old salts is over. The minimum requirements for one regional airline are listed in Figure 18-5 (ATP, 2008).

There is no doubt they have passed the checkride and can fly the aircraft. However, where does experience come in the areas of decision making, risk assessment, and error management? To the seasoned pilots, it appears to be a game of chance and probabilities. A low-time inexperienced crew on a good day, no problem, but add bad weather and some unexpected mechanical issue, and this could be a roll of the dice for which nobody has planned.

Time will tell if these are true issues or just not letting go of the "old way of doing things." However, there must be human factors studies to proactively look at these issues and with the cutbacks in funding these do not seem to be occurring. These same low-experienced pilots will also be reaching the majors in larger proportions in the future. Training will have to be modified to address these issues. There will be some mitigation of these issues based on the recent change of retirement age requirements of pilots.

EVOLUTION AND CHANGE IN THE ENVIRONMENT WE FLY IN

The Health of the Airline Pilot in Today's World of Airliners with 17-hour Cruise Range

As technology has advanced in health care, there has been tremendous improvements in the prevention and treatment of disease processes, which in the past would have been disqualifying for airline pilots. There are hundreds of pilots flying post-myocardial infarction, coronary bypass surgery, and seizures. The recent movement of the mandatory retirement age to 65 epitomizes these advancements in health care. However, within this arena of advancements, tremendous challenges still exist in the interface of the human being and the challenges of air travel.

There are 24 time zones covering 360° of the globe and, although there is a new time zone every 15° of longitude, these divisions do not cover the Earth's surface in a straight line, pole to pole, or at 90° to every degree of latitude. Indeed, the time zones are man-made and vary in width, thereby representing arbitrary, albeit convenient methods of relating activities in different parts of the globe.

Time zones themselves therefore possess no physical or physiological properties. However, feelings of jet lag after a transmeridian flight are very real and may be due to a number of factors in addition to the crossing of time zones. These include pretrip activities, in-trip activities, noise, vibration, low humidity, and the new light/dark relationship. The effect of all these is such that any performance change will be consequent upon the fatigue of the journey together with the resultant phase asymmetry of the circadian rhythms.

The term "jet lag" has been given to a variety of symptoms experienced when making a transmeridian flight, but there are two main components to the syndrome:

a) Stress effects that extend from the physical and psychological aspects of the flight itself (Carruthers, Arguelles, & Mosovich, 1976), and

b) Effects that are due to the disruption of the internal biological clock (Winget, DeRoshia, Markley, & Holley, 1984).

Both are key to the maintenance of alertness, and, although man remains central to safe aviation operations, both need to be considered.

The effects that may be expected due to the flight itself include tiredness, malaise, nausea, headaches, and aching joints. They are a function of the duration of the flight rather than the number of time zones crossed (Monk, 1983) and seldom last more than few hours after the end of the journey. However, their importance during the flight may be crucial.

Other effects stem from the need to reset the biological clock to the new time zone. They are much longer lasting (Klein, Wegmann, & Hunt, 1972), with the most intrusive problem being sleep disturbance with periods of wakefulness occurring during the normal sleep pattern (Weitzman, Kripke, Goldmacher, McGregor, & Nogeire, 1970). This may result in partial sleep deprivation and a consequent decrease in daytime alertness, an increase in irritability, and impairment of performance efficiency. All of these factors can impair performance, reduce alertness, and ultimately compromise safety. However, there is a large volume of scientific literature that can be used to assist in assessing the requirements for sleep, the impact of performance and alertness, together with the strategies that may be used to understand and monitor the state of awareness.

Human error at work is the most frequently identified cause of accidents and contributes to between 30% and 90% of all serious incidents across industries (Reason, 1990; (Senders & Moray, 1991). In aviation, according to Nagel (1988), 68% of all accidents are attributable to performance errors by flight crew. Unfortunately, there is no currently agreed reliable estimate of the extent to which fatigue and states of awareness contributes to the incidence of human error related accidents. There is, however, ample scientific evidence that sleepiness, as governed by the circadian pacemaker and the homeostatic drive for sleep, determine the level of alertness and performance associated with rest-activity patterns. Consequently, sleepiness and the diminished performance capability that accompanies it, are programmed to occur in the brain of every human regardless of training, occupation, education, motivation, skill level, intelligence, or the commitment of that person to maintaining high levels of waking alertness (Dinges, 1995).

Tasks requiring sustained attention, such as those frequently found on the modern flightdeck, are particularly vulnerable to performance impairment as a result of sleepiness. Whereas the contribution of boredom or distraction to the causation of operator inattention is difficult to qualify, there is ample scientific evidence regarding the impact of sleepiness and the resultant increase in performance errors.

The operation of an intercontinental airline is by definition a 24-hour activity. Indeed, the increasing demand for long-haul, short-haul, and domestic operations continues to increase the demand and hence a requirement for shift work, night work, irregular duty rosters, and the crossing of time zones will remain inevitable for the majority of flight and cabin crew. Flight operations are therefore not always conducive to a regular sleep/wake schedule and can affect sleep and circadian physiology. This may be as a result of duty periods occurring at unusual or changing times in the day/night sleep cycle or because of the effect of time zone crossings leading to conflict between the environmental time (in the case of unusual or changing work schedules) and/or local time (in the case of changing time zones) and body time with resultant circadian disruption. The requirement for a prolonged period of continuous wakefulness may itself also lead to sleep loss and it is therefore clear that a protracted duty period can create fatigue by extending wakefulness and decreasing sleep and, could also involve circadian disruption. Finally, in many flight operations, the time available for sleep is constrained by a number of factors and, if the physiological timing for sleep does not coincide with the scheduled sleep opportunity, then inadequate sleep may be taken with the resultant penalty of an accrued sleep debt.

Any or all of these factors individually or in combination may result in jet lag, sleepiness, and fatigue, and if they impair performance could pose a risk to safety. Alertness management strategies are therefore aimed at minimizing the adverse effects of sleep loss and circadian disruption and to promote optimal alertness and performance during the work period. However, sleep and circadian physiologies are complex, there is significant individual variation from person to person and operating schedules, even within a single airline, are extremely varied. As a result, no single strategy will fully eliminate fatigue and the aim must therefore be to promote and optimize alertness.

The application of "strategic countermeasures" in alertness management involves three components (Rosekind, Smith, Miller, Co, Gregory, Webbon, Gander, & Lebacqz, 1995):

a) Understanding the physiological principles related to sleep and circadian rhythms.
b) Determining the specific alertness and performance requirements of a given operation.
c) Taking deliberate actions to apply the physiological principles to meet the operational requirements.

The resulting strategies can then be categorized into preventative and operational (Rosekind, Gander, & Dinges, 1991). Preventative strategies, which may be used before a duty period, are aimed at addressing the underlying physiological mechanisms associated with sleep and circadian factors. Operational strategies, on the other hand, may be used during a duty period and are intended to maintain performance alertness.

In response to a 1980 U.S. congressional request, the NASA Ames Research Center in California initiated a fatigue/jet lag program to examine fatigue, sleep loss, and circadian disruption in aviation. The program later evolved into the Fatigue Countermeasures Program to emphasize its efforts on the development and evaluation of strategies to maintain alertness and performance in operational settings. From the inception of the program, one of the principal goals was to feed back the information learned to the operational community and the objectives of the Education and Training Module (Rosekind, Gander, Connell, & Co, 1994) was to explain what had been learned about the physiological mechanisms underlying fatigue, demonstrate the application of this information in flight operations and to offer some specific fatigue countermeasure recommendations.

Although, there has been a tremendous amount of research and measurement of crew states of awareness there has not been a concise process to link this physiology to actual performance. Thus, what are the main predictors of crew alertness? How should the new ultralong haul aircraft (cruise range over 17 hours) be staffed? On one side of the debate are the manufactures with a very different view of the risk as compared to the employee groups who will staff and fly these aircraft day in and day out. There have been no real human factor discussions in this debate on the human under extreme conditions induced by the work schedule imposed by the airline and contract. The concerns are real whether looking at long-haul flying on the backside of the clock or multiple flights being flown by a regional airline crew in one day. What is the safe limit? How can we better predict the capabilities of human beings under difficult conditions? Although a tremendous amount of work has been done on how fatigue reduces human function and ways to mitigate its effect, there has been very little connect to the flight-decks. Naps are not allowed, crew breaks on long-haul flights are taken by tradition and not science, and the question of the ability to have restorative sleep on aircraft are still unanswered. Will there ever come the day where a system represented in Figure 18-6 will provide better feedback of the alertness of the crew while respecting individual privacy of the crews (Michigan Aerospace, 2008)?

EVOLUTION INTO THE FUTURE OF AIR TRAVEL

The current technologies of airline aircraft are at the cusp of an explosion in new design, missions, and capabilities. Next generations of air/space travel will have new composite materials, free

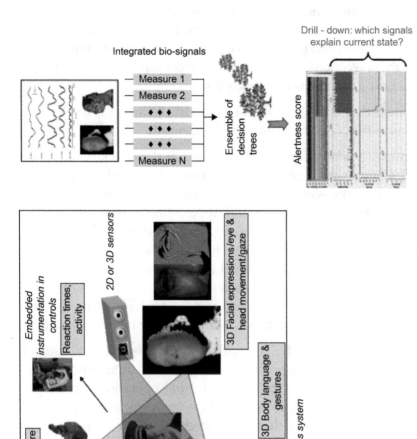

FIGURE 18-6 Fatigue monitoring.

flight capabilities, and the ability to travel in a sub-orbit escaping the friction of air. Additionally, there will be advancements in technology, regulations, and procedures to control aircraft from the ground. All of these technologies will provide opportunities and challenges for the next generations of airline pilots.

The Integrated Intelligent Flight Deck (IIFD) Project is one of four projects in the NASA Aviation Safety Program.

IIFD research is based on a vision for future flightdeck systems that includes systematic incorporation of integrated displays and interactions, decision-aiding (decision-support) functions, information management and abstraction, and appropriate human/automation function allocations. The future flightdeck system is aware of the vehicle, operator, and airspace system state, and responds appropriately. The system senses internal and external hazards, evaluates them, and provides key information to facilitate timely and appropriate responses. The system is robust and is adaptable to the addition of new functions and information sources as they become available.

To achieve this vision, IIFD comprises a multidisciplinary research effort to develop mitigations for operator-, automation-, and environment-induced hazards for future operational concepts. IIFD leverages, and depends on, others to develop mitigations of other hazard categories that may affect the flight deck—such as degraded vehicle health states.

IIFD addresses, from the flight deck safety perspective, the integration of IIFD-developed capabilities with future communications, navigation, Air Traffic Management (ATM), and vehicle technologies being investigated by others either within NASA, other government agencies, or industry (NASA, 2008). Will the pilot become extinct in the future? The FAA and NASA to understand controlling aircraft from the ground are performing significant work. For example is it better to control from a lap top computer or should the ground module look like a flightdeck? Although currently the work is for military drones and border patrols, it is only a matter of time before control from the ground will have implication for commercial air travel.

Recently there was an ad in the travel magazine *Virtuoso Life* to book your reservation for suborbital space travel before gliding safely back to earth. The ad goes on to say that: "By participating, you are paving the way for safer, less expensive, and more environmentally friendly space tourism." And long-term plans include the development for hypersonic point-to-point flights that would enable people to travel across the world in just a few hours with minimal environmental impact. Physiology testing in altitudes chambers capable of altitudes of over 100,000 feet is being conducted to see the effects on humans and survivability in this environment. Although years away from the line pilots, the work needs to be performed now to understand these future challenges.

SUMMARY

There has been an explosion in technology, capabilities, and opportunities for airline pilots in the past 20 years. There also have been frustrations, dramatic changes to the profession, and continual threats to our existence. The study and implementation of the science of human factors has done a tremendous job in mitigating the challenges faced by the airline crews flying millions of passengers millions of miles every year. However, we need to work closer with manufactures, designers, airline management, pilot unions, scientific community, and our own workforce to assure we continue the culture of safety and continuous improvement necessary to reduce the chance of the next accident occurring. From an airline pilot's view:

1. Human factors design engineers need to improve the involvement of the "human pilots" in the next generation of aircraft. Greater automation is not always the answer.
2. The balance and the interplay between economic pressures and flight safety are not fully understood. The safeguards in place by the regulatory agencies are not sufficient to understand these dynamics.
3. The stresses and challenges to airline crews have changed drastically over the past 20 years. However, the way that we address these issues has been retrospective rather than prospective in nature. For example, the way that we train and assess new pilots entering the field of aviation is the same as we did 60 years ago.

Flight safety is as much a human performance topic as it is a design topic. Success will come from interdisciplinary efforts using methods to understand and enhance human expertise in the context of diverse and dynamic air carrier settings.

References

Airline Pilot Training & Pilot Career Development, PilotJobs.com http://www.pilotjobs.com/

ATA, AQP Subcommittee (1994). Line operational simulation: LOFT scenario design and validation. Washington, DC: Author. WR Hamman.

Beaubien, B. (2001). Airline Pilots' Perceptions of and Experiences in Crew Resource Mangement (CRM) Training 2002–01–2963 grant 99-G-048 from the office of the Chief Scientific and Technical Advior for Human Factors (AAR-100).

Boeing Commercial Airplane Group. (2007). *Statistical summary of commercial jet airplane accidents* Worldwide operations 1959–2007. Seattle, WA: Author.

Cannon-Bowers, J. A., & Salas, E. (1997). Teamwork competencies: The interaction of team member knowledge, skills, and attitudes. In H. F. O'Neil (Ed.), Jr., *Workforce readiness: Competencies and assessment* (pp. 151–174). Mahwah, NJ: Erlbaum.

Carruthers, M., Arguelles, A. E., & Mosovich, A. (1976). Man in transit: biochemical and physiological changes during intercontinental flights. *Lancet*, 1, 977–981.

Dinges, D. F. (1995). An overview of sleepiness and accidents. *Journal of Sleep Research*, 4, 4–14.

FAA AC 120–54A Advanced Qualification Program (2006) http://rgl.faa.gov/Regulatory_and_Guidance_Library/rgAdvisoryCircular.nsf/MainFrame?OpenFrameSet

FAA. Aviation safety reporting program (ASRP). Advisory Circular 00–46 Issued May 9, 1975.

FAA AC 120–51E Crew Resource Management Training (2004) http://rgl.faa.gov/Regulatory_and_Guidance_Library/rgAdvisoryCircular.nsf/MainFrame?OpenFrameSet

FAA AC 120–35C Line Operational Simulations (2004) http://rgl.faa.gov/Regulatory_and_Guidance_Library/rgAdvisoryCircular.nsf/MainFrame?OpenFrameSet

FAA. Order 8000.82, *Designation of Aviation Safety Action Program (ASAP) Information As Protected From Public Disclosure Under 14 CFR Part 193*. Sept. 3, 2003.

Hamman, W. R. (1994). Crew training and assessment in airline training. Proceedings of the 21st Conference of the European Association for Aviation Psychology (EAAP) March 1994, Dublin, Ireland, Trinity College.

Hamman W, Seamster T, Smith K, Lofaro R. The LOE Worksheet developed to provide clear structure to the assessment of both CRM and technical crew performance. ATA Conference 1993.

Klein, K. E., Wegmann, H. M., & Hunt, B. (1972). Desynchronization of body temperature and performance circadian rhythm as a result of outgoing and home-going transmeridian flights. *Aerospace Medicine, 43*(2), 119–132.

Michigan Aerospace. (2008). *Crew Alertness Monitoring System and Evaluation Framework*. NASA SBIR topic A1.05 "Crew Systems Technologies for Improved Aviation Safety" National Aeronautics and Space Administration Small Business Innovation Research & Technology Transfer 2008 Program Solicitations http://sbir.gsfc.nasa.gov/SBIR/sbirsttr2008/solicitation/SBIR/TOPIC_A1.html#A1.05

Monk, T. H., Weitzman, E. D., Fookson, J. E., Moline, M. L., Kronauer, R. E., & Gander, P. H. (1983). Task variables determine which biological clock controls circadian rhythms in human performance. *Nature, 304*, 543–545.

Nagel, D. C. (1988). Human error in aviation operations. New York: Academic Press.

National Aeronautics and Space Administration Aviation Safety Program. (2008) Integrated Intelligent Flight Deck Technologies

Part 14 CFR Ch. 1 (1–1-03 Edition) 13.401 http://electromagnetics.larc.nasa.gov/agency_programs/iifd.htm

Reason, J. (1990). *Human Error*. Cambridge University Press, The Code of Federal Regulations (CFR) http://www.gpoaccess.gov/cfr/about.html.

Rosekind, M. R., Gander, P. H., Connell, L. J., & Co, E. (1994). Crew factors in flight operations: X. Alertness management published in flight operations. *NASA Technical Memorandum*. Moffett Field, CA: NASA Ames Research Center.

Rosekind, M. R., Gander, P. H., & Dinges, D. F. (1991). Alertness management in flight operations: Strategic napping. SAE Technical Peper Series. Warrendale, PA: Society of Automotive Engineers, #, 912138.

Rosekind, M. R., Smith, R. M., Miller, D. L., Co, E., Gregory, K. B., Webbon, L. L., Gander, P. H., & Lebacqz, J. V. (1995). Alertness management: strategic naps in operationa; settings. *Journal of Sleep Research, 4*(Suppl 2), 62–66.

Senders, J., & Moray, N. (1991). Human Error: Cause, Prediction, and Reduction. Hillsdale, NJ: Lawrence Erlbaum Associates.

Flight Safety Foundation, Flight Safety Digest Vol. 17 No. 7–9 July–September 1998.

The Flight Safety Foundation, Safety Bulletin October 20, 2008, Alexandria, VA

Weitzman, E. D., Kripke, D. F., Goldmacher, D., McGregor, P., & Nogeire, C. (1970). Acute reversal of the sleep-waking cycle in man: Effect on sleep stage patterns. *Archives of Neurology, 22*, 483–489.

Wilkinson, R. T., Archives of Neurology Edwards, R. S., & Haines, E. (1966). Performance following a night of reduced sleep. *Psychonomic Science, 5*(12), 471–472.

Winget, C. M., DeRoshia, C. W., Markley, C. L., et al. (1984). A review of human physiology and performance changes associated with desynchronisis of biological rhythms. *Aviation, Space, and Environmental Medicine 1984, 55*, 1085–1096.

Young, S., & Bailey, R. Aviation, Space, and Environmental Medicine (2008). Technical Excellence: Technical Seminar Series, National Aeronautics and Space Administration. Aspects of Synthetic Vision Display Systems and the Best Practices of the NASA's SVS Project Bailey, Randall E.; Kramer, Lynda J.; Jones, Denise R.; Young, Steven D.; Arthur, Jarvis J.; Prinzel, Lawrence J.; Glaab, Louis J.; Harrah, Steven D.; Parrish, Russell V. Langley Research Center, May 2008 20080018605, Report number L-19422; NASA/TP-2008-215130

General Aviation

Stephen M. Casner
National Aeronautics and Space Administration

It is often said that there are three classes of aviation: military, airlines, and everybody else. Everybody else is better known as general aviation, and on any given flying day, it might include flight training, pleasure flying, police and sheriff air patrol, aerial firefighting, aerial burial, crop dusting, gliding, aerobatics, air charter, air racing, air ambulance, ballooning, insect control, photography, surveying, sightseeing, logging, parachuting, search and rescue, traffic watch, utility line maintenance, and the opportunity to join the "mile high club" (Yancy, 2006). It should come as no surprise that general aviation is not only the largest but also the most diverse sector in aviation.

General aviation accounts for roughly 87% of the almost 600,000 U.S. certificated pilots, and makes use of more than 13,000 public and private airports. General aviation pilots log roughly 27 million hours per year. There are about 221,000 general aviation aircraft based in the U.S. and they carry more than 166 million passengers each year, making the GA fleet the largest air carrier in the world (FAA, 2007; GAMA, 2007; AOPA, 2007).

General aviation pilots vary widely in their experience: from highly experienced corporate pilots to new student pilots who excitedly log their first minutes of flight time. General aviation aircraft can range from the most primitive amateur-built craft, to the highest-tech glass cockpit jets, to the futuristic folding-wing airplane that can be driven home after landing (Mone, 2008).

General aviation accidents occur at a rate of roughly 6 accidents per 100,000 flight hours, while accidents involving fatalities occur at slightly more than 1 per 100,000 hours. Although these numbers show a continuing overall trend toward improvement, general aviation remains the riskiest of the three classes of aviation. It comes as no surprise that safety remains a top concern for human factors professionals who work in the aviation field.

In this chapter, we will look at the human factors issues that impact safety in the three largest sectors within general aviation: (1) personal flying, (2) flight training, and (3) business flying.

Personal flying makes up the largest sector within general aviation and includes the hundreds of thousands of pilots who fly mainly for pleasure. Since fun fliers are often the least experienced pilots, we will closely examine the relationship between flight experience and overall safety. Since recent changes in demographics have increased the average age of the U.S. pilot, we look at studies that have examined the relationship between age and pilot skill and safety. Next, we will see how the introduction of technology in the general aviation cockpit has changed how pilots fly and has presented new safety challenges for flights conducted by a single pilot. Lastly, we look at how researchers are turning their attention to the study of pilot judgment and decision-making processes in search of clues about how general aviation pilots arrive to accident situations.

Flight training describes the activities of more than 60,000 flight instructors and roughly 60,000 pilots who earn new certificates and ratings each year in the United States alone, along with the countless other pilots who train to maintain or improve their skills. These numbers alone make the flight training sector the vanguard of general aviation safety. In our discussion of flight training, we will look at how pilots are trained to make good decisions, and how the arrival of automation to the general aviation fleet has affected the skill set that pilots must now master. We will look at how advances in simulator technology have led to an increased use of simulators, computer-based training tools, and communications tools in the training environment. We will take a closer look at social learning processes, how student pilots informally learn from one another, and how technology might be affecting the way student pilots interact.

Business flying includes those who operate aircraft for work but who are not part of a scheduled air carrier (airline) operation. We will see how the less predictable scheduling of business flight operations can introduce fatigue in the cockpit, and look at the effectiveness of fatigue countermeasures that have been proposed by human factors experts. We also look at how the complexities of equipment, weather, and flight planning complicate the decision-making process for business fliers.

Our discussion of safety-related human factors is designed to be of interest to researchers, managers, policymakers, as well as present and future general aviation pilots. It is only through a common understanding of these human factors issues, shared throughout the general aviation community, that we can hope to maintain and continue to improve upon the safety record held by the general aviation industry today.

PERSONAL FLYING

Personal flying, the largest sector within general aviation, accounted for roughly 48% of all GA operations in 2006, but nearly 72% of pilot-related accidents (AOPA, 2007). Clearly, this sector of general aviation accounts for more than its fair share of accidents. Here we look at some of the human factors that seem to have the biggest impact on safety for this half of general aviation.

Flight Experience

When we read a general aviation accident report, a first question that naturally comes to mind is that of how much experience the pilot had. In aviation, there are a number of ways of measuring pilot experience.

Total Flight Time

Total flight time is perhaps the most coveted and respected measure of piloting experience among pilots, instructors, and managers alike. From what we hear and what we read, we are tempted to believe that more flight time is better, and that pilots with more flight time are less likely to be involved in an accident. Is this belief accurate? A quick look at the accident data reveals that the relationship between total flight time and accident likelihood is not so simple. Figure 19-1 shows the proportion of accidents attributed to pilots in command at each level of total flight experience. Indeed, the graph in Figure 19-1 shows that almost 34% of all accidents involve a pilot-in-command with fewer than 500 hours of total flight time. However, roughly 34% of all certificated pilots have accumulated less than 500 hours of total flight time (AOPA, 2007). Hence, the proportion of accidents occurring in this segment of the pilot population is no greater than that occurring among pilots at all higher experience levels considered together.

This question was further investigated by O'Hare and Chalmers (1999) who closely examined logbook data provided by 8500

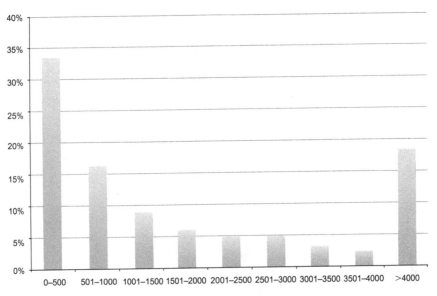

FIGURE 19-1   Proportion of accidents by total flight experience.

general aviation pilots. These researchers were able to compare the flight experience of pilots who were involved in accidents with the proportion of pilots in the overall population at each level of flight experience. They found that although roughly 21% of all accidents involved pilots in the 100 to 300 hours of flight experience range, a figure that roughly agrees with the numbers we see in Figure 19-1. However, the proportion of pilots in the population with this level of experience was roughly 26%. Their findings again reject the claim that there is a period of special vulnerability for pilots at this experience level, and that the data trended toward the idea that these lower-experience pilots were perhaps *less* likely to be involved in an accident than more experienced pilots.

These studies hardly demonstrate that experience is worthless, but rather that the effects of experience are nuanced. Li et al. (2001) looked at the link between total flight time and specific types of accidents. They found that general aviation pilots at the highest experience levels were less likely to be involved in accidents under visual meteorological conditions (VMC), but just as likely to be involved in accidents under instrument meteorological conditions. In their study of pilot weather decision making, Stokes et al. (1992) found that experienced pilots used more sources of information and generated more alternative courses of action. Goh and Wiegmann (2002b) found that more experienced pilots feel they are better at recognizing problems and implementing solutions, but do not feel more able to diagnose underlying causes. Stokes et al. (1990) found that greater flight experience buffers pilots against negative effects of stress during in-flight planning.

Does having more total flight time make pilots better learners? Casner (2004) found no correlation between total flight experience and time to learn to use GPS for IFR flight. In that study, it appeared that learning to use an advanced cockpit system was a separate skill that did not seem to be leveraged by previous experience. Casner, Heraldez, and Jones (2006) found no effect of total flight experience on long-term retention of aeronautical knowledge.

Overall, the research we have so far suggests that while accruing flight time appears to season pilots in interesting ways, it in itself is not a guarantee of knowledge and skill nor a safeguard against accidents.

Relevance of Experience

The tentative link between total flight time and safety invites us to consider whether or not there is a protective effect of flight experience that is directly relevant to specific flight situations. For example, it is reasonable to ask if hours of IFR cross-country experience will contribute to safety during IFR cross-country operations more so than do smaller amounts of local-area VFR flying experience. In a simulated weather decision-making task, Wiggins and O'Hare (1995) found that performance on selected cross-country flying decision tasks varied more as a function of cross-country flight experience than it did with total flight experience. In their analysis of VFR to IMC accidents, Goh and Wiegmann (2002a) found that low amounts of instrument flight time and the presence of passengers were common among this type of crash.

Experience beneficial in some situations might be less beneficial in others. Goh and Wiegmann (2002b) found a negative correlation between instrument time and vigilance in monitoring for traffic. Pilots with more instrument time felt they were less vigilant in monitoring out-the-window traffic.

These findings suggest that what is needed next is a study that compares accident characteristics with more detailed data about the experiences that pilots collect during their previous flights.

Practice Time

Another way to consider the relevance of flight experience is to ask each pilot how much of their flight time was devoted explicitly to the task of improving their skills, rather than simply making use of their existing skills. Kershner (2002) makes the distinction of a pilot with 10,000 hours and a pilot with 1 hour, 10,000 times. Pilots who passively exercise their skills may simply be getting better at exercising their bad habits. A number of studies have looked at the effects of **deliberate practice** (Ericsson et al., 1993). In these studies, differences in ability between individuals are closely related to amounts of deliberate practice. Future aviation human factors studies might investigate links between accidents and incidents and the amount of time that pilots devote to honing their skills.

Time in Type

Another twist on the idea of relevance of experience is to look at aircraft-specific experience. AOPA (2007) provides a breakdown of time in aircraft type and proportion of accidents, shown in Figure 19-2.

The graph in Figure 19-2 clearly shows that the greatest proportion of accidents happens among pilots who have less than 100 hours in type. The open question, of course, is what proportion of pilots has less than 100 hours in the aircraft type they are operating on any given day. As O'Hare and Chalmers (1999) point out, what is needed is a more detailed record of pilots' experience that can only be gleaned from pilots who choose to share their logbooks in the interest of human factors research.

Recent Flight Experience

Yet another way to look at flight experience is to consider how much flying a pilot has done *lately*. Again, our intuition quickly suggests that an abundance of recent flight experience means sharp

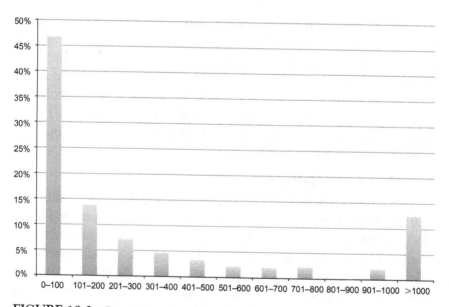

FIGURE 19-2   Proportion of accidents by time in aircraft type.

skills and safe flight. A few studies have looked at the relationship between recent flight experience, pilot proficiency, and accident likelihood, and once again, these studies suggest that the relationship is a subtle and nuanced one.

O'Hare et al. (2001) found that pilots who had more recent flight experience were more likely to be involved in VFR to IMC accidents that the researchers characterized as "controlled exposure to risk" accidents. That is, these more current pilots were in some cases more willing to take on risk, and ultimately paid a price for it. The researchers found that reports of this type of accident were significantly more likely to cite "overconfidence in personal ability" as a contributing factor than did reports of other types of accidents.

An early study by Mengelkoch et al. (1971) asked the simple question of how long it takes for the skills of the average pilot to deteriorate, given no further practice. These researchers found that manual control skills were well retained even after a 4-month hiatus, demonstrating a "just like riding a bike" effect for manual control skills. Procedural sequences, however, were found to fade more quickly. Studies such as these helped establish the recent flight experience rules that pilots operate under today.

Turning to the retention of aeronautical knowledge, Casner et al. (2006) found that recent flight experience was somewhat associated with better retention of aeronautical knowledge, but that the particulars of each pilots' everyday flight environment seem to play an important part in determining what is remembered and forgotten. As is the case for total flight experience, recent flight experience may not be enough to guarantee that pilots will remember what they have learned.

Certificates and Ratings

A few studies have looked at differences in accident rates, attitudes, and abilities among pilots at different levels of certification. Figure 19-3 shows the proportion of U.S. pilots at each certificate level, along with the proportion of accidents involving pilots at each certificate level. Surprisingly, commercial pilots made up roughly 21% of the pilot population yet accounted for a disproportionately high 35% of all accidents (AOPA, 2007). This may happen because many

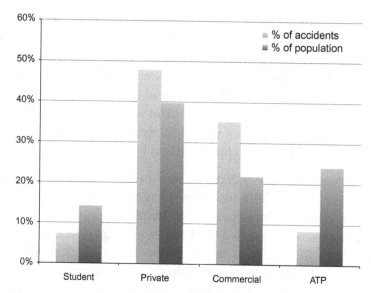

FIGURE 19-3   Proportion of accidents by pilot certificate held.

commercial pilots operate in challenging conditions outside of the highly supervised world of scheduled airline service.

Other studies have looked at characteristics of pilots at different levels of certification. Hunter (2006) found that holders of commercial and airline transport pilots reported lower levels of perceived risk. Looking at pilot ability, Taylor et al. (2005) found that pilots who hold higher grades of certificates were better able to follow ATC instructions (controlling for age). As is the case with total flight experience, it seems that as pilots progress through the certificate ranks, they acquire characteristics that can help and hinder in subtle ways.

Age

The past 50 years has seen an increase in the median age for most worldwide populations (UNDP, 2005), and a corresponding increase in the average of pilots (AOPA, 2007). How does getting older affect us as pilots? Hardy et al. (2007) tested 220 medically qualified pilots from age 28 to 62 and found a gradual and linear decline in psychomotor and information processing speed, attentional ability, verbal and visual learning, and memory.

Although changes associated with aging seem to be inevitable, there remains the question of to what extent they impact pilot performance. Taylor et al. (2005) noted a decline in general aviation pilots' ability to follow ATC instructions, and linked this performance decrement to age-related decline in working memory span. Tsang and Voss (1996) found that time-sharing ability for particular cockpit tasks decreases with age. Rebok et al. (2005) noted an increase in violation rates for the 40s to late 50s age groups.

Despite these demonstrated declines in pilot performance, other researchers such as Salthouse (1987) have proposed that age effects might be defrayed by expertise, or by changing the task to minimize age effects, when clear links between age effect and task are understood. Indeed, in a recent study, Taylor et al. (2007) found that knowledge of aviation procedures and principles may indeed compensate for decline in general abilities and speed of processing. Checklists and other external memory devices are known to make human performance improvements for tasks that require memory and recall. An interesting topic for future research is to further explore ways in which the strengths of age and experience might be leveraged against noted declines by making changes in procedures or the cockpit environment.

Advanced Cockpit Systems

Perhaps the most striking change to general aviation aircraft used for personal flying has been the arrival of advanced cockpit systems such as GPS navigation computers, autopilots, electronic flight instruments, traffic collision avoidance systems, terrain and weather displays. Through these innovations, general aviation aircraft now represent the highest tech fleet in civil aviation.

Advanced cockpit systems have gradually automated many cockpit tasks that were traditionally performed by general aviation pilots, such as flight planning calculations (wind, fuel, time, courses, etc.), aircraft control, en route progress calculations, position fixing, and many others. A wealth of research has been directed at learning more about how the use of these advanced systems affects pilots and the way they do their job.

Awareness

Advanced cockpit systems such as colorful moving map displays now present aircraft position, surrounding terrain, nearby aircraft, and weather to the pilot in one integrated and convenient picture. In the margins of these displays, pilots can often find power settings, engine health parameters, ground speed, wind direction and velocity, along with the time, distance, and fuel required to arrive at the next waypoint in the planned route. Couched in such an information-rich environment, it is tempting to think that any pilot now flies, as one manufacturer puts it, "at the pinnacle of situational awareness."

The effect of advanced cockpit systems on pilot awareness has been investigated in a number of studies and the results have been mixed. Casner (2005a) asked two groups of general aviation pilots to navigate over unfamiliar terrain. One group used a paper chart and traditional navigation technique (pilotage), while the other group used a GPS navigation computer, a color moving map display, and the same paper chart. At the conclusion of the cross-country flight, both groups of pilots were asked (to their surprise) to once again navigate over the same route, only this time without the use of any navigation equipment or charts. Pilots who used charts and traditional navigation performed well. Pilots who used GPS performed significantly worse, making large navigational errors, and in some cases were unable to find their way back to the point of origin. It was hypothesized that the reason for the drop in awareness among GPS users was the reduced involvement associated with passively following one's progress while a computer performs most of the work required to find one's way from point to point. A similar hypothesis was made in an earlier study by Endsley and Kaber (1999). As a direct test of this hypothesis, Casner (2006) asked a third group of pilots to use GPS but to also engage in a simple "tour guide" task. As pilots made their way from waypoint to waypoint along the route, they were asked to point out three landmarks of interest on the ground. When later surprised with the task of navigating without the GPS, these pilots navigated with significantly more accuracy, restoring their performance to the level of that of pilots who used the traditional navigation method. It seems that even this simple task prompted pilots to stay sufficiently engaged in the navigation task.

Sarter and Woods (1995, 1997) have argued that the complexity of "mode-rich" advanced cockpit systems not only lowers pilots' awareness of where they are, but also of how the advanced systems are configured and what they intend to do next. The source of this lowered awareness seems to stem from incomplete understanding of how the automation works, and poor feedback given to pilots by the automation (Norman, 1990). Earl Wiener was first to point out the three most common questions asked in the advanced cockpit: (1) Why did it do that? (2) What is it doing now? and (3) What is it going to do next? (Wiener, 1985).

The effect of advanced cockpit systems on pilot awareness is far from straightforward. It appears that pilot awareness is a fragile phenomenon, and that just because information is available in the cockpit, there is no guarantee that it is actively circulating in the head of the pilot.

Workload

Advanced cockpit systems are widely touted to reduce pilot workload. Studies of pilots using advanced systems have generally failed to support this idea. Casner (2009) directly tested the effects of advanced cockpit systems found in a general aviation glass cockpit airplane on pilot workload. Pilots flew two four-leg instrument flights: one flight in a conventional cockpit airplane, and one flight in a glass-cockpit airplane. Pilots were required to fly each aircraft using particular combinations of conventional and advanced systems in an attempt to isolate the effects of each system on pilot workload. The results showed that using a GPS navigation computer reduced workload during some phases of flight but raised it during others. Using the autopilot reduced workload in the glass cockpit airplane but did not appear to do so in the conventional cockpit. Despite this modest showing for advanced systems, pilots still stated an overwhelming preference for using them during all phases of flight.

Some pilots have suggested that the increased monitoring requirements incurred when advanced systems are used can often raise workload levels to those experienced when the advanced systems are not used. In two attitudinal surveys, pilots were split in their responses to the probe: "I sometimes spend more time setting up and monitoring the autopilot than I would just hand-flying the aircraft" (Casner, 2008; Wiener, 1985).

Others have argued that workload is in some cases spiked by "automation surprises" that arise from the complexity of the advanced systems (Sarter & Woods, 1997). Pilots are surprised by an unexpected response from an advanced system and are suddenly cast into a high workload situation.

Error

Casner (2009) examined error rates in advanced cockpit general aviation airplanes and found that even pilots who were experienced with advanced systems still made more errors than they did in conventional cockpits. Casner (2004) studied pilots who were first learning to use advanced systems such as GPS navigation computers. Pilots who were new to these systems committed very few navigational errors, but made large numbers of errors that arose from the many operational modes of the navigation computer—errors that are unique to the advanced system. Wickens (2000) argues that advanced cockpit systems do not eliminate pilot error, but rather "relocates the sources of human error to a new level."

Wiener (1980) pointed out that advanced cockpit systems sometimes hide the consequences of errors made by pilots and inflate simple errors such as mistyped waypoint names into large blunders. A number of techniques have been proposed to better manage or mitigate error in the advanced cockpit. As one study of altitude deviations suggests: to err is human, to be error-tolerant is divine (Palmer et al., 1993).

Alerts and Alarms

General aviation cockpits are making increasing use of visual and audible alerts and alarms designed to capture the attention of the pilot. At least three human factors issues associated with the use of alerts and alarms in the cockpit have been identified.

Primary/Secondary Task Inversion Wiener (1985) was first to give a name to the problem of overreliance on alerting systems such as the altitude alerter provided by most every autopilot system. Wiener pointed out that pilots could excuse themselves from the task of monitoring their altitude in exchange for the task of simply monitoring the altitude alerter. When the alert goes off, the pilot could then redirect their attention to altitude, knowing that a target altitude was soon coming up. Wiener named this phenomenon **primary/secondary task inversion**, and pointed out a number of problems associated with it. For example, suppose the target altitude was mis-set, or there is some type of malfunction with the system? Palmer et al. (1993) describe instances of altitude deviations during airline flights that occurred for reasons such as these.

Startle Several researchers have made a distinction between bringing information to the attention of pilots, and startling them. There is a wide literature on human startle response: the deleterious effects of startle (Sorkin, 1988; Peryer et al., 2005), and the physiological indications of startle (Wilkins et al., 1986). This knowledge readily allows particular examples of cockpit alerts and alarms to be evaluated. Alerts can be measured for intensity, rise time, and duration, while pilots' reactions to the alerts can be measured to determine if they physiologically indicate startle. Despite the potential for the startling effects of alert and alarm intensity, audibility continues to be a primary concern in the design of alerting and alarm systems (Beringer et al., 1998).

Mistrust Several authors have examined the effects of alerts and alarms in response to situations that pilots that pilots do not think are significant. It is well known that false alarms can quickly erode pilots' trust and confidence, resulting in what is often called the **cry wolf phenomenon** (Breznitz, 1984). When the reliability of alerts and alarms are perceived to be low, pilots respond more slowly and less frequently when alerts and alarms occur (Bliss, 1997). In the presence of nuisance alarms, pilots sometimes adopt the practice of disabling the alarm system, or poising their fingers over an alarm cancel button when an alarm is expected (Sorkin, 1988).

Single Pilot Operations

Single pilot operations have long been a part of personal flying in general aviation. What have changed are the demands that are now placed on general aviation pilots: enhanced aircraft performance, increasingly busy and complex airspace, exposure to unfamiliar weather patterns, automation, and so on. Are general aviation pilots able to meet these demands and fly these aircraft safely?

A first question to consider is whether or not a single pilot can simultaneously perform all of the tasks required by a complex and fast-moving aircraft. Multitasking in humans has been well studied and found to be mostly impressive. Given sufficient practice, skills can be performed seemingly at the same time with minimal interference with one another (Hoover & Russ-Eft, 2005; Spelke, Hirst, & Neisser, 1976; Schneider & Shiffrin, 1977).

A well-argued benefit of having two pilots in the cockpit is that one pilot can catch any errors made by the other, or help remember things that the other pilot has forgotten to do. Dismukes and Nowinski (2006) point out that this second type of memory task becomes particularly challenging when pilots are interrupted, either by external events, or as part of their own multitasking behavior when they must suspend performance of one task to complete another. In a study of pilot error in automated cockpits, Skitka et al. (2000) found that single pilots were no worse than crews with respect to errors made in an advanced cockpit. They found that crews made fewer errors of commission (taking a wrong or inappropriate action), but about the same number of errors of omission (forgetting or neglecting to take a needed action) as single pilots. This is an interesting result that shows that having two heads in the cockpit is no guarantee that things will be remembered.

Having two pilots present can sometimes lead to improved decision making performance. The so-called two heads are better than one effect has been demonstrated in several studies, when the decision makers are collaborating (Klein & Zsambok, 1997) and even when they are not (Surowiecki, 2005). How to best support single pilots in their decision-making processes is a topic deserving of further research.

Judgment, Risk, and Decision Making

When we read an account of an accident or incident, it is common to ask the question: What was he or she thinking? A number of studies have looked at the decisions that general aviation pilots make and at the decision-making process.

The most popular way of studying pilot decision making has been to place pilots in a simulator, set them out on a cross-country flight, introduce a precarious situation such as deteriorating weather, and see what they decide to do. This paradigm has produced a variety of interesting results that have helped us to characterize good and bad decisions and the pilots who make them. For example, O'Hare (1990) found that pilots who elected to continue into deteriorating conditions during a simulated decision-making task were younger, had more total flight time, greater willingness to take risk, and higher scores on tests of invulnerability.

Aside from finding examples of poor decisions and identifying the characteristics of pilots who might be prone to making them, other researchers have aimed to understand the processes by which pilots arrive at these decisions. Orasanu et al. (2001) offer one explanation for pilots' decisions to continue into adverse weather—what they term **plan continuation errors**. Pilots commit this type of error when they form an initial plan and are biased toward continuing with that plan, even in the face of new information that might suggest that the plan is no longer the best option.

O'Hare and Smitheram (1995) looked at general aviation pilot decision processes in terms of **prospect theory** (Kahneman & Tversky, 1979). They found that pilots tend to be risk-averse when they see their flight situation in terms of gains, and risk takers when they view their situation in terms of losses. Of course, any number of details about any given flight could potentially lead pilots to view a situation in terms of gains or losses. A student on a cross-country flight might see a flight as a gain that could easily be had on another day in the near future. The same student who has a check ride scheduled in three days and is squeezing in a last flight to meet the minimum flight experience requirements might view the same flight as a loss.

Other studies suggest that details about a decision-making situation might prompt pilots to lean in one direction or another. Beringer et al. (2004) studied the effect of weather data (NEXRAD) display resolution on pilot decision making. They found that colorful weather displays that offered higher resolutions invited pilots to continue into challenging weather when displays with lower resolutions did not.

Orasanu and Connolly (1993) point out that one obstacle to better understanding pilot decision making lies in the way we presently characterize the decision making process. Under classical views of the decision making process, pilots have clear-cut alternatives set before them, and proceed to ponder the pluses and minuses of each alternative, eventually choosing one. Orasanu and Connolly point out that real decision-making situations are usually ill-defined, and uncertain to the point that only some things (if any) are known. Pilots' goals are constantly shifting as they operate under time pressure, and in the presence of competing immediate and high-level goals. Means et al. (1993) argue that sometimes people don't even perceive these situations as a "choice." The key to better understanding the way pilots decide what to do may be to altogether change the way that we think about the decision making process.

Amateur-Built Aircraft

General aviation pilots are in some cases permitted to build and maintain their own aircraft. These pilots are typically not certified maintenance technicians but rather pilots who take the same pride and interest in working on aircraft that they do flying them. How safe is this practice? Nelson and Goldman (2003) report that although roughly 3% of all general aviation hours are flown in amateur-built aircraft, these operations account for an average of 14% of all maintenance-related accidents. These numbers clearly demonstrate that work done by trained and certified mechanics yields better results than work done by amateurs. While the human factors issues that arise among certified mechanics have been studied (see Chapter 21), little attention has been given to the underlying causes of these amateur-built aircraft accidents, or to developing mitigation strategies that might help amateur mechanics fly more safely.

FLIGHT TRAINING

Before 1990, it was estimated that roughly 80% of all airline pilots were military trained. By 2001, this number had dropped to under 40%. General aviation has rapidly become the primary training ground for most commercial airline pilots, and now the training platform for most all U.S. certificated pilots (GAMA, 2007). General aviation pilot training has undergone some remarkable changes, and here we look at some of the most important human factors issues that have arisen from them.

Teaching and Learning Decision-Making Skills

We have already identified an important link between pilot decision making and general aviation safety. This raises the question of how to teach decision-making practices and good judgment to pilot trainees, and the question of whether or not such an endeavor is even possible. Many approaches to improving decision making in pilots have been proposed and studied.

Advice: A number of researchers, reviewed in Bonaccio and Dalal (2006), have investigated the usefulness of offering advice. An example of advice might be to tell pilots to beware of overreliance on a GPS navigation computer—that they may tend to follow their progress less closely when their position is keenly displayed on a large colorful moving map display. There is good evidence that people naturally gather advice to improve decision outcomes, and also to share accountability when things go wrong. Nevertheless, a robust "egocentric advice discounting" has been found by many researchers. That is, people tend to discount the advice given by others in favor of their own ways of thinking. There are a number of explanations for this phenomenon. One theory points out that advice-takers seldom have access to advice-givers' reasoning and therefore tend to favor their own reasoning (Yaniv & Kleinberger, 2000). Another theory postulates that people begin with their own reasoning as an anchor that must then be adjusted by advice (Tversky & Kahneman, 1974). Gino (2008) found an interesting effect: people are more likely to accept advice that they pay for than they are advice that they get for free.

Formal models: General aviation pilots are offered a variety of formal methods of arriving at a decision. The FAA (2003) proposes the D-E-C-I-D-E method as a general model for arriving at decisions. Similarly, the I-M-S-A-F-E method intends to help pilots decide whether or not they are fit to fly. More recently, the P-A-V-E model uses a series of numerical estimations and computations to arrive at measures of risk.

Studies of decision makers in real-world settings have found that formal decision models such as those taught during training seldom get used in real-world situations (numerous cited in Means et al., 1983). Research has shown that people tend to reject formal models as too complex or time-consuming (Means, 1983). Furthermore, there is little evidence to suggest that these formal methods will be useful across a variety of domain-specific tasks (Means et al., 1993).

Bias reduction training: Another approach to teaching pilot judgment is to alert pilots to specific pitfalls in human judgment and reasoning such as hazardous attitudes, confirmation bias, the overgeneralization of specific available cases, and so on. Again, the available literature cited in Means et al. (1983) "does not inspire optimism about this kind of training."

Cases: O'Hare and Wiggins (2004) looked at the role of prior cases in guiding the decision making of individual pilots. By prior cases, these authors mean situations drawn from the pilot's own experiences rather than situations or examples concocted by others. O'Hare and Wiggins found that roughly half of all pilots they surveyed were able to recall using a prior case drawn from their experience when they later needed to respond to a critical flight even such as an adverse weather encounter or equipment failure. Furthermore, the researchers found that the use of prior cases was increasingly common in pilots who were older and had more flight experience. The results suggest that training techniques that incorporate the use of cases could be potentially useful.

Overall, the available research suggests that improving decision making skills among pilot trainees might be the greatest challenge facing aviation educators, and one deserving of ongoing investigation.

Learning to Fly in the Age of Automation

Among the most fast-breaking changes happening in general aviation training is the deployment of advanced cockpit systems in general aviation aircraft. Learning to fly in the presence of advanced cockpit systems has focused on three topics.

How to work it: At a minimum, pilots must master the sequence of knob twists and button pushes required to activate the many features offered by advanced cockpit systems—a topic that several authors have labeled *knobology* (Overgard et al., 2007). A few researchers have studied how well pilots are able to master the steps required to use advanced avionics, and have attempted to link successes and failures to specific features of the user interface. Fennell et al. (2006) and Casner (2005b) found that pilots' ability to master an avionics device was highly linked to the presence or absence of external cues about how to accomplish the goals that the pilot has in mind. For example, the presence of a button marked Flight Plan allows new pilots to make immediate progress when presented with a flight-planning task. System functions that require pilots to memorize sequences of button pushes for which there is little apparent guidance in the interface are known to cause ongoing problems. Another challenge facing pilots is the increasing number of features being added to advanced cockpit systems that must ultimately be mastered.

How it works: There is good evidence to suggest that understanding how a device works allows users to more easily recall procedures, reason their way through difficulties they encounter, and learn similar systems (Kieras & Bovair, 1984; Casner, 2005b). Sarter and Woods (1995) trace the occurrence of "automation surprises" to gaps in pilot understandings of how a system works. For pilots learning to fly in the advanced cockpit, time invested in learning how advanced cockpit systems work is likely to pay future dividends.

Automation pitfalls: Another way to understand advanced cockpit systems is to be aware of the deleterious effects it can sometimes have on pilot performance, such as lowered awareness, sudden spikes in pilot workload, confusion, complacency, etc. How concerned are pilots about these issues? In a survey of general aviation pilots, Casner (2008) found that pilots were generally aware of these issues but were more likely to attribute "automation pitfalls" to other pilots than they were to themselves. This raises the question of whether or not it is beneficial to teach pilots directly about potential automation pitfalls. There has been little evidence presented thus far to support idea that explicitly teaching automation pitfalls will lessen the chances of them occurring. Skitka et al. (2000) found that explicit teaching about automation complacency reduced the number of errors of commission, but did not reduce the number of errors of omission.

Another important question related to training in the age of automation is what to do with those piloting skills that are being assumed by technology. As an extreme example, modern navigation computers automatically determine the heading that must be flown to achieve a desired course, in the presence of wind, variation, and compass deviation. Should pilots continue to learn the formula for determining compass heading?

> True Course +/− Wind Correction Angle
> = True Heading +/− Magnetic Variation
> = Magnetic Heading +/− Deviation
> = Compass Heading

Even if a skill like this is mastered during training, there is good evidence that pilots' ability to recall the skill after a matter of months will be greatly deteriorated (Casner et al., 2006). Continuing to teach these skills might only be worthwhile if we also create an environment in which pilots are required to routinely practice these skills and keep them fresh.

What do pilots think about this issue? Seventy-three percent of survey respondents in Casner (2008) agreed or strongly agreed with the statement: "New pilots that learn to fly only in advanced cockpit aircraft are going to be lacking some important piloting skills."

Role of Simulators

Simulators continue to grow in sophistication and play an increasing role in the training of general aviation pilots. General aviation instructors and students now routinely use fixed-based simulators that are equipped with the latest advanced cockpit systems and feature realistic out-the-window visual systems. An important research question is that of how much of a flight training curriculum can be accomplished in these simulators.

Simulators based on personal computers, **personal-computer aircraft training devices** (PCATD), have also grown in sophistication and it is natural to consider what role they might play in the training process. Talleur et al. (2003) tested the efficacy of PCATDs for maintaining instrument currency and found that these simulators were as effective as larger simulators for this purpose. Johnson and Stewart (2005) found that PCATDs also supported instrument training for helicopters, especially in training navigation techniques.

The expanding capabilities of all types of simulators are a resource that can be exploited to: (1) bring more of the everyday flying experience to pilots, outside of operating an aircraft; and (2) to create learning experiences that are difficult to create in an aircraft.

Electronic Learning

In parallel to the deployment of computers in the cockpit and the use of high-tech simulators, general aviation has also seen a rush of computer-based learning products. Texts and pictures that were once only presented in books can now be found on CDs, DVDs, and Internet sites along with videos and other interactive media. The relevant human factors question is, What learning benefits do these new media offer?

Research about the effectiveness of new media as learning technologies dates back as far as the invention of the overhead and film strip projectors (Mialaret, 1966). For the most part, the idea of simply transporting existing learning materials into different media fares poorly in research studies. Howell et al. (2003) found no inherent benefits of rendering text on a computer screen rather than a page, or with replacing a human speaker with a video or audio clip. The first generation of multimedia learning materials has been generally limited to rendering on a computer screen what we can already access in books and from human instructors, and it is unclear what benefits they offer.

Norman (1993) argues that computers are beneficial to learning when they can make available stores of information or interactive mechanisms that support exploration and hypothesis formation and testing. Ultimately, as Norman points out, whatever environment is used must support the activities that are known to lead to learning: intense, devoted concentration that leads to reflection and tuning of the students' knowledge and skill.

There are ongoing efforts to build intelligent tutoring systems that more closely manage the student's progress as they acquire new knowledge and skills. In some cases, these systems attempt to ascertain what the student currently knows and diagnose misunderstandings by constructing customized teachings or problems that deliver what the student needs next in their learning trajectory (Polson & Richardson, 1988).

Social Learning (Hangar-Flying)

Much of what is taught to student pilots is seldom found in books. Students and instructors routinely pass along accounts of personal experiences, along with their understandings of, attitudes about, and approaches to virtually everything they do, as students congregate on and off of the airport to exchange information. In aviation, this type of information exchange is known as "hangar flying" and it is widely regarded as an important part of any pilot's educational experience (Flying Magazine, 2007).

Of particular interest is how the arrival of computer-based instructional materials, the Internet, and social media is affecting the way that information passes between students. One concern about computer-based instruction is that it may cut down on face-to-face time spent with other students. On the other hand, blogging and social networking sites allow rapid sharing of information between pilots, from any locale.

Interestingly, few researchers have yet directed their attention to understanding and making the most of social learning resources in the aviation training sector.

Abstract Principles Versus Concrete Examples

There has been much discussion about the use of scenarios in flight training. While good instructors have always made ample use of scenarios, some have argued that a focus on basic flight maneuvers might even be discarded in favor of an entirely scenario-based training syllabus in which all flying skills are taught in context. This discussion parallels a long-standing argument that has occurred throughout all of education. Carroll (1990) points out that years go by in which theorists argue the futility of teaching abstract knowledge and skills founded on the hope that students might later be able to generalize them to particular circumstances. These theorists argue that only practical skills that are learned and practiced in particular contexts will lead to usable training once the student has left the training environment. A period such as this came to a screeching halt in the 1980s when researchers found that students were lacking in basic skills and unable to recite facts such as the significance of the July 4th holiday in the United States. This discovery in turn launched a "back-to-basics" movement in which students were once again directed to learn basic skills and facts that contributed to cultural literacy. Now, again riding the crest of a skills-in-context training wave, researchers have recently found that students who learn mathematics principles in the abstract performed better than students who learned them by working particular problems (Kaminski et al., 2008). As Carroll (1990) points out, education is "chronically subject to trends" and that the treatment that the hapless student can expect to receive might largely be a function of the decade in which he or she was born.

Looking at the research, it seems that a good case can be made for the benefits of learning abstract principles and skills as well as concrete ones that are couched in specific circumstances. Countless studies have shown that students benefit from abstract principles, guidance in applying the principles to particular situations, and from ample practice in real and abstracted situations.

Flight Instructing

For many pilots, flight instructing is a stepping-stone to a corporate or air carrier job. Thus one threat to the quality of flight instruction is the limited time that many instructors remain in the teaching role. A short tenure as a flight instructor often prevents pilots from attaining a high degree of teaching competence. How then do we achieve and maintain a corps of proficient flight instructors? As human factors practitioners, we have little ability to change the market conditions that determine how long any pilot will likely remain in a teaching role (e.g., salaries, hiring trends, etc.). At best we can attempt to get the most from any instructor during their tenure by making improvements in the way they are trained and used.

Bertrand (2005) attempted to discover differences between practices of high and low-time flight instructors. Using informal interview techniques, Bertrand found that high-time flight instructors focused less on FAA practical tests, were more likely to include students in their other flying activities, felt strongly about the need for students to learn to control the aircraft to the edge of the flight envelope, focused on weather and traditional navigation skills, and emphasized emergency procedures and preparedness.

In a survey of flight instructor training and evaluation, Henley (1991) found that the training of instructors largely relies on traditional methods of rote learning and mimicking. Perhaps training methods or teaching environments can be used to help instructors become more effective more quickly.

Culture

Another unique characteristic of general aviation training is its culture. Different from air carrier pilots who are selected and carefully managed throughout training, general aviation pilots decide for themselves to become pilots, while instructors are free to craft training programs that are loosely based on the regulations. Recently, general aviation training has seen a trend toward an airline type of training culture. More closely scrutinized training programs, such as those authorized under Part 141 in the United

States, have grown in popularity. Manufacturers have begun to create aircraft or product-specific training programs. Some manufacturers have developed standard operation procedures (SOP) that are similar to those used by air carriers. Although numerous studies have looked at the role of training culture among air carriers (von Thaden et al., 2003; Chute & Wiener, 1995), there has been very little research to look at the impact of these cultural changes on general aviation safety.

BUSINESS FLYING

General aviation aircraft are used for a wide range of business purposes: from air charter operations who offer on-demand air taxi services, to air freight operations, to individuals who purchase their own aircraft and hire pilots to fly them. Business flying removes some of the flexibility of personal flying, as pilots begin to fall under the direction of others who operate under their own constraints and follow their own agenda. We look at some of the human factors issues that affect safety in this sector of general aviation.

Fatigue

Pilots who fly for business must deal with the same fatigue issues faced by air carrier crews. Corporate pilots often deal with long duty hours, early shifts, and abrupt changes in work and sleep schedules. Rosekind et al. (1997) surveyed corporate pilots and found that 61% felt that fatigue was a common problem. Eighty-five percent agreed that fatigue is a moderate or serious safety issue.

Fortunately, corporate pilots have the same fatigue countermeasures that are available to airline pilots. Regulatory agencies typically impose duty limits (e.g., rest periods, total flight hours per month and year) for certain types of air charter operations. Private business flights often have no such restrictions. In the case of private business flights, duty shifts are entirely determined by the aircraft owner—who is frequently a non-pilot.

21% of the corporate pilots surveyed by Rosekind et al. (1997) reported that their department offered some kind of training about fatigue. For the rest, Mallis et al. (2003) describes an on-line training module designed specifically to educate corporate pilots about fatigue and fatigue countermeasures.

Caffeinated beverages are a first line of defense against the deleterious effects of fatigue and their effectiveness has been well investigated (Caska & Molesworth, 2007; Deixelberger-Fritz et al., 2003). These researchers point out that, to achieve maximum benefits, caffeine must be used strategically. Simply drinking caffeine all day in hopes of achieving an ongoing state of alertness has been shown to be a poor strategy. Rather, caffeine should be used periodically, to maximize alertness during critical periods. For example, a pilot might minimize the use of caffeine during a long, low-workload cruise phase and then drink caffeine prior to descent and arrival into busy terminal airspace.

Decision Making

The business flying environment presents pilots with several unique decision making challenges. For example, larger and more complex aircraft typically make use of minimum equipment lists (MEL) that spell out complex rules about which systems can be inoperative during which types of flight operations. In the presence of any inoperative system, the pilot in command must decide whether or not that type of flight operation is authorized and safe. Decisions about inoperative equipment are difficult in general aviation because individual pilots seldom have access to the kinds of resources available at an airline company. General aviation business pilots typically dispatch themselves, have no central maintenance operation that they consult with questions (either prior to departure or when airborne), and have no layers of management above them that scrutinize the entire operation.

The decision process is further complicated by the fact that the aircraft owner, often a non-pilot, may exercise a strong influence on making decisions. The NTSB cited pressure from a charter customer as a contributing factor for the crash of a Gulfstream III in Aspen, Colorado, in March 2001 (NTSB, 2001—DCA01MA034)

Training

Training of corporate pilots can be vastly different from the way airline pilots are trained. Airline carriers carefully prepare SOPs that detail most every anticipated normal, abnormal, and emergency scenario. SOPs are designed to train pilots to be, as one airline pilot puts it, "virtual clones of one another." This type of standardization is typically absent in the business flying environment. The cockpit of a corporate aircraft will sometimes combine two pilots who work for two entirely different companies. To further complicate matters, many corporate pilots operate more than one type of aircraft. This burdens pilots with having to remember the systems, performance, and other characteristics of more than one aircraft at a time. A systematic study of the effect of training type on business flying safety is needed.

AFTERWORD

Although general aviation is relatively safe and shows an overall trend toward increasing safety, our review of the research suggests that further improvements could be made through the systematic study of human factors that affect the outcome of general aviation flights. Further attention should be given to understanding the kinds of aviation experience that lead to improved safety, how an aging population affects safety, and how advanced cockpit systems are changing they way we fly. Improvements in pilot training might follow from continued study of how technology can best be used in the training process, how pilots' own experiences can be used as training resources, how pilots exchange information with one another, and how to best train flight instructors. The shift toward an airline-type training culture for technically advanced aircraft deserves closer study for its effect on overall safety. More studies are needed of the decision-making processes that govern the actions pilots take, how best to support the single pilot of an increasingly complex aircraft and airspace, and ways to help pilots deal with job pressures such as fatigue.

Improvements in safety will be in part made possible through a continued focus on operationally oriented research conducted by government agencies, university labs, and other aviation organizations working together. But of equal importance are the thousands of general aviation pilots, maintenance technicians, air traffic controllers, and others who take the time to participate in research studies, share their thoughts or experiences in an open and cooperative environment, or to educate themselves about the ways we have discovered to be more safe. This sense of community in general aviation is perhaps our most valuable asset, and one that must be nurtured and protected at all costs as exciting new changes come along and we move forward into the future.

ACKNOWLEDGMENTS

I thank Colleen Geven, Immanuel Barshi, Daniel Heraldez, Michael Feary, Bob Wright, and Jeffrey McCandless for valuable comments.

References

AOPA. (2007). *2007 Nall report: Accident trends and factors for 2006*. Frederick, MD: Aircraft Owners and Pilots Association: Air Safety Foundation.

Beringer, D. B., & Ball, J. D. (2004). *The effects of NEXRAD graphical data resolution and direct weather viewing on pilots' judgment of weather severity and their willingness to continue a flight*. Washington, DC: U.S. Department of Transportation, Federal Aviation Administration.

Beringer, D. B., Harris., Howard, C., Jr., & Joseph, K. M. (1998). *Hearing thresholds among pilots and non-pilots: Implications for auditory warning design*. Proceedings of 42nd Annual Meeting of the Human Factors and Ergonomics Society, Santa Monica, CA.

Bertrand, J. E. (2005). Practices of high-time instructors in Part 61 environments. *International Journal of Applied Aviation Studies, 5*(1), 41–52.

Bliss, J. (1997). Alarm reaction patterns by pilots as a function of reaction modality. *International Journal of Aviation Psychology, 7*(1), 1–14.

Bonaccio, S., & Dalal, R. S. (2006). Advice taking and decision making: An integrative literature review, and implications for the organizational sciences. *Organizational Behavior and Human Decision Processes, 101*(2), 127–151.

Breznitz, S. (1984). *Cry wolf: The psychology of false alarms*. Hillsdale, NJ: LEA.

Caska, T. J., & Molesworth, B. R. C. (2007). The effects of low dose caffeine on pilot performance. *International Journal of Applied Aviation Studies, 7*(2), 244–255.

Casner, S. M. (2004). Flying IFR with GPS: How much practice is needed? *International Journal of Applied Aviation Studies, 4*(2), 81–97.

Casner, S. M. (2005a). The effect of GPS and moving map displays on navigational awareness while flying under VFR. *International Journal of Applied Aviation Studies, 5*, 153–165.

Casner, S. M. (2005b). Transfer of learning between a small technically advanced aircraft and a commercial jet transport simulator. *International Journal of Applied Aviation Studies, 5*(2), 307–319.

Casner, S. M. (2006). Mitigating the loss of navigational awareness while flying with GPS and moving map displays under VFR. *International Journal of Applied Aviation Studies, 6*(1), 121–129.

Casner, S. M., Heraldez, D., & Jones, K. M. (2006). Retention of aeronautical knowledge. *International Journal of Applied Aviation Studies, 6*(1), 71–97.

Casner, S. M. (2008). General aviation pilots' attitudes toward advanced cockpit systems. *International Journal of Applied Aviation Studies, 8*(1), 88–112.

Casner, S. M. (2009). The effect of advanced cockpit systems on pilot workload and error. *Applied Ergonomics, 40*(3), 448–456.

Chute, R., & Wiener, E. L. (1995). Cockpit-cabin communication: I. A tale of two cultures. *International Journal of Aviation Psychology, 5*(3), 257–276.

Deixelberger-Fritz, D., Tischler, M. A., & Kallus, K. W. (2003). Changes in performance, mood state, and workload due to energy drinks in pilots. *International Journal of Applied Aviation Studies, 3*(2), 195–206.

Dismukes, R. K., & Nowinski, J. (2006). Prospective memory, concurrent task management, and pilot error. In D. W. A. Kramer & A. Kirlik (Eds.), *Attention: From Theory to Practice*. New York: Oxford University Press.

Endsley, M. R., & Kaber, D. B. (1999). Level of automation effects on performance, situation awareness and workload in a dynamic control task. *Ergonomics, 42*(3), 462–492.

Ericsson, K. A., Krampe, R. T., & Clemens, T. R. (1993). The role of deliberate practice in the acquisition of expert performance. *Psychological Review, 100*(3), 363–406.

FAA. (2003). *Pilot's handbook of aeronautical knowledge*. Washington, DC: Author.

FAA. (2007). *Estimated active airman certificates held*. Washington, DC: Author.

Fennell, K., Sherry, L., Roberts, R. R., & Feary, M. (2006). Difficult access: The impact of recall steps on flight management system errors. *International Journal of Aviation Psychology, 16*(2), 175–196.

GAMA. (2007). *2007 general aviation statistical databook and industry outlook*. Washington, DC: Author.

Gino, F. (2008). Do we listen to advice just because we paid for it? The impact of advice cost on its use. *Organizational Behavior and Human Decision Process, 107*(2), November 2008, 234–245.

Goh, J., & Wiegmann, D. A. (2002a). Human factors analysis of accidents involving visual flight rules flight into adverse weather. *Aviation, Space, and Environmental Medicine, 73,* 817–822.

Goh, J., & Wiegmann, D. A. (2002b). Relating flight experience and pilots' perceptions of decision-making skill. The 46th Annual Meeting of the Human Factors and Ergonomics Society, 81–85.

Hardy, D. J., Satz, P., D'Elia, L. F., & Uchiyama, C. L. (2007). Age-related group and individual differences in aircraft pilot cognition. *International Journal of Aviation Psychology, 17*(1), 77–90.

Henley, I. (1991). The development and evaluation of flight instructors: A descriptive survey. *International Journal of Aviation Psychology, 1*(4), 319–333.

Hoover, A. L., & Russ-Eft, D. F. (2005). Effect of concurrent task management training on single pilot task prioritization performance. *International Journal of Applied Aviation Studies, 5*(2), 234–252.

Howell, C. D., Denning, T. V., & Fitzpatrick, W. B. (2003). Traditional versus electronic information delivery: The effect on student achievement. *International Journal of Applied Aviation Studies, 3*(2), 207–214.

Hunter, D. (2006). Risk perception among general aviation pilots. *International Journal of Aviation Psychology, 16*(2), 135–144.

Johnson, D. M., & Stewart, J. E., II (2005). Utility of a personal computer-based aviation training device for helicopter flight training. *International Journal of Applied Aviation Studies, 5*(2), 288–306.

Kahneman, D., & Tversky, A. (1979). Prospect Theory: An Analysis of Decision under Risk. *Econometrica, XLVII,* 263–291.

Kaminski, J. A., Sloutsky, V. M., & Heckler, A. F. (2008). The advantage of abstract examples in learning math. *Science, 320,* 454–455.

Kershner, W. K. (2002). *Logging flight time: And other aviation truths, near-truths, and more than a few rumors that could never be traced to their sources.* Newcastle, WA: Aviation Supplies & Academics, Inc.

Kieras, D. E., & Bovair, S. (1984). The role of a mental model in learning to operate a device. *Cognitive Science, 8,* 255–273.

Klein, G., & Zsambok, C. E. (1997). Naturalistic decision making. Mahwah, NJ: Lawrence Erlbaum Associates.

Li, G., Baker, S. P., Grabowski, J. G., & Rebok, G. W. (2001). Factors associated with pilot error in aviation accidents. *Aviation, Space, and Environmental Medicine, 72,* 52–58.

Mallis, M., Co, E., Rosekind, M. R., Neri, D., Oyung, R., Brandt, S., Colletti, L., & Reduta, D. (2003). *Evaluation of a web-based fatigue education and training module in the general aviation (GA) population.* Moffett Field, CA: NASA Ames Research Center.

Means, B. (1983). *How to choose the very best: What people know about decision-making strategies.* Montreal, Canada: Annual Meeting of the American Educational Research Association.

Means, B., Crandall, B., Salas, E., & Jacobs, T. O. (1993). Training decision makers for the real world. In J. O. G. A. Klein, R. Calderwood, & C. E. Zsambok (Eds.), *Decison making in action: Models and methods* (pp. 306–326). Norwood, NJ: Ablex Publishing.

Mengelkoch, R. F., Adams, J. A., & Gainer, C. A. (1971). The forgetting of instrument flying skills. *Human Factors, 13*(5), 397–405.

Mialaret, G. (1966). *The psychology of the use of audio-visual aids in primary education.* Paris: United Nations Education, Scientific and Cultural Organization.

Mone, G. (October, 2008). The flying car (driving airplane) gets real. *Popular Science.*

Nelson, N. L., & Goldman, S. M. (2003). *Maintenance-related accidents: A comparison of amateur-built aircraft to all other general aviation.* Denver, CO: Human Factors and Ergonomics Society Annual Proceedings 191–193.

Norman, D. A. (1993). *Things that make us smart: Defending human attributes in the age of the machine.* Reading, MA: Addison-Wesley Publishing Company.

Norman, D. A. (1990). *The problem of automation: Inappropriate feedback and inter-action, not over-automation.* Philosophical Transactions of the Royal Society of London.

NTSB. (2001). *Accident report, NTSB Identification DCA01MA034.* Washington, D.C.: National Transportation Safety Board.

O'Hare, D. (1990). Pilots' perception of risks and hazards in general aviation. *Aviation, Space, and Environmental Medicine, 61,* 599–603.

O'Hare, D., & Smitheram, T. (1995). Pressing on into deteriorating conditions: An application of behavioral decision theory to pilot decision making. *International Journal of Aviation Psychology, 5,* 351–370.

O'Hare, D., & Chalmers, D. (1999). The Incidence of Incidents: A Nationwide Study of Flight Experience and Exposure to Accidents and Incidents. *International Journal of Aviation Psychology, 9*(1), 1–18.

O'Hare, D., and Owen, D., & Wiegmann, D. (2001). The 'where' and 'why' of cross-country VFR crashes: Database and simulation analyses. 45th Annual Meeting of the Human Factors and Ergonomics Society.

O'Hare, D., & Wiggins, M. W. (2004). Remembrance of cases past: Who remembers what, when confronting critical flight events. *Human Factors, 46,* 377–387.

Orasanu, J., Martin, L., & Davison, J. (2001). Cognitive and contextual factors in aviation accidents. In E. Salas & G. Klein (Eds.), *Linking expertise and naturalistic decision making* (pp. 209–226). Mahwah, NJ: Erlbaum.

Orasanu, J., & Connolly, T. (1993). The reinvention of decision making. In J. O. G. A. Klein, R. Calderwood, & C. E. Zsambok (Eds.), *Decison making in action: Models and methods* (pp. 3–20). Norwood, NJ: Ablex Publishing.

Overgard, K. I., Fostervold, K. I., Bjelland, H. V., & Hoff, T. (2007). Knobology in use: an experimental evaluation of ergonomics recommendations. *Ergonomics, 50*(5), 694–705.

Palmer, E. A., Hutchins, E. L., Ritter, R. D., & van Cleemput, I. M. (1993). *Altitude deviations: Breakdowns of an error-tolerant system. NASA Technical Memorandum 108788.* Moffett Field, CA: National Aeronautics and Space Administration.

Polson, M. C., & Richardson, J. J. (1988). *Foundations of intellligent tutoring systems.* Mahwah, NJ: Erlbaum.

Peryer, G., Noyes, J., Pleydell-Pearce, K., & Lieven, N. (2005). Auditory alert characteristics: A survey of pilot views. *International Journal of Aviation Psychology, 15*(3), 233–250.

Rebok, G. W., Qiang, Y., Baker, S. P., McCarthy, M. L., & Guohua, L. (2005). Age, flight experience, and violation risk in mature commuter and air taxi pilots. *International Journal of Aviation Psychology, 15*(4), 363–374.

Rosekind, M. R., Gregory, K. B., Miller, D. L., & Neri, D. F. (1997). A survey of fatigue factors in corporate/executive aviation operations. *Sleep Research, 26*(213).

Salthouse, T. A. (1987). Age, experience, and compensation. In C. S. K. W. Schaie (Ed.), *Cognitive functioning and social structure over the life course* (pp. 142–157). Cambridge, MA: Ablex.

Sarter, N. B., & Woods, D. D. (1995). How in the world did we ever get into that mode? Mode error and awareness in supervisory control. *Human Factors, 37*(1), 5–19.

Sarter, N. B., & Woods, D. D. (1997). Teamplay with a powerful and independent agent: A corpus of operational experiences and automation surprises on the Airbus A-320. *Human Factors, 39,* 553–569.

Schneider, W., & Shiffrin, R. M. (1977). Controlled and automatic human information processing: I. Detection, search, and attention. *Psychological Review, 84*(1), 1–66.

Skitka, L. J., Mosier, K. L., Burdick, M., & Rosenblatt, B. (2000). Automation bias and errors: Are crews better than individuals? *International Journal of Aviation Psychology, 10*(1), 85–97.

Sorkin, R. D. (1988). Why are people turning off our alarms? *Journal of the Acoustical Society of America, 84*(3), 1107–1108.

Spelke, E., Hirst, W., & Neisser, U. (1976). Skills of divided attention. *Cognition, 4,* 215–230.

Stokes, A. F., Belger, A., & Zhang, K. (1990). *Investigation of factors comprising a model of pilot decision making: Part II. Anxiety and cognitive strategies in expert and novice aviators.* Urbana-Champaign, IL: Institute of Aviation, University of Illinois, Urbana-Champaign.

Stokes, A. F., Kemper, K. L., & Marsh, R. (1992). *Time-stressed flight decision making: A study of expert and novice aviators.* Arlington, VA: Office of Naval Research.

Surowiecki, J. (2005). *The wisdom of crowds,* Anchor.

Talleur, D. A., Taylor, H. L., Emanuel, T. W., Jr, Rantenen, E., & Bradshaw, G. (2003). Personal Computer Aviation Training Devices: Their Effectiveness for Maintaining Instrument Currency. *International Journal of Aviation Psychology, 13*(4), 387–399.

Taylor, J. T., O'Hara, R., Mumenthaler, M. S., Rosen, A. C., & Yesavage, J. A. (2005). Cognitive ability, expertise, and age differences in following air-traffic control instructions. *Psychology and Aging, 20*(1), 117–133.

Taylor, J. T., Kennedy, Q., Noda, A., & Yesavage, J. A. (2007). Pilot age and expertise predict flight simulator performance: A 3-year longitudinal study. *Neurology, 68,* 648–654.

Tsang, P. S., & Voss, D. T. (1996). Boundaries of cognitive performance as a function of age and flight experience. *International Journal of Aviation Psychology, 6*(4), 359–377.

Tversky, A., & Kahneman, D. (1974). Judgment under uncertainty: Heuristics and biases. *Science, 185,* 1124–1131.

UNDP (2005). *Human development report.* United Nations Development Program.

von Thaden, T. L., Wiegmann, D. A., Mitchell, A. A., Sharma, G., Zhang, H. (2003). *Safety culture in a regional airline: Results from a commercial aviation safety survey.* 12th International Symposium on Aviation Psychology, Dayton, OH.

Wickens, C. D. (2000). *Engineering psychology and human performance.* Upper Saddle River, NJ: Prentice-Hall.

Wiener, E. L. (1989). *Human factors of advanced technology glass cockpit transport aircraft. NASA Contractor Report 177528.* Moffett Field, CA: National Aeronautics and Space Administration.

Wiener, E. L. (1985). Human factors of cockpit automation: A field study of flight crew transition, National Aeronautics and Space Administration: 118.

Wiener, E. L. (1980). Flight deck automation: Problems and promises. *Ergonomics, 23,* 995–1011.

Wiggins, M., & O'Hare, D. (1995). Expertise in aeronautical weather-related decision making: A cross-sectional analysis of general aviation pilots. *Journal of Experimental Psychology: Applied, 1,* 305–320.

Wilkins, D. E., Hallett, M., & Wess, M. M. (1986). Audiogenic startle reflex of man and its relationship to startle. *Brain, 109,* 561–573.

Yancy, K. B. (September 7, 2006). *A flight that goes all the way.* USA Today.

Yaniv, I., & Kleinberger, E. (2000). Advice taking in decision making: egocentric discounting and reputation formation. *Organizational Behavior and Human Decision Processes, 83,* 260–281.

Air Traffic Management

Ann-Elise Lindeis
NAV CANADA 77 Metcalfe Street Ottawa,
Ontario Canada

INTRODUCTION

Aviation continues to evolve in every sphere imaginable. The remarkable advances made in areas such as aircraft design, manufacture, performance, maintenance, and on-board communication and navigation technology have opened the doors to more aircraft flying to more destinations with new means of getting there. Achieving the full potential of these advances relies on equally remarkable achievements in providing pilots with safe and efficient air navigation services.

Air navigation service providers are in the process of making significant changes in technology. These systemwide changes present a unique opportunity to shape the future design, maintenance, and operation of air navigation services. This chapter provides the reader with the link between air traffic management and flying. More importantly, the reader will be presented with the critical issues in air traffic management that Human Factors researchers and practitioners are tackling to meet the demands of air travel.

The chapter is divided into three main sections. The first section provides an overview of the technology and procedures used to provide air navigation services. This section also highlights significant changes in aviation technology that have created the opportunity to reshape air traffic management. The second section highlights areas in current operations where safety or efficiency is compromised due to less than optimal system design. The third section describes the progress that researchers and practitioners have made in incorporating human factors knowledge, tools, and techniques into air traffic management. There have been many excellent book chapters, research papers, articles, and comprehensive books written on human factors and air traffic control. This chapter provides only a sampling of the work being conducted.[1]

ELEMENTS OF THE AIR NAVIGATION SYSTEM

In order to meet the goal of providing safe, effective, and efficient air navigation services, each air navigation service provider[2] must ensure that a number of elements are integrated: (a) a network of communication, navigation, and surveillance (CNS) systems; (b) methods and systems for air traffic management (ATM); and, most importantly, (c) humans to develop, manage, deliver and maintain the products and services, for example, airspace design specialists, standard and procedures specialists, air traffic controllers, flight service specialists, air traffic flow management specialists, installation and maintenance technologists, engineers, and safety specialists, to name but a few. International harmonization of these elements is conducted through the auspices of the International Civil Aviation Organization (ICAO) and most countries have regulations that are consistent with ICAO standards and recommended practices.

[1] Noticeably absent from this chapter is any discussion on training, selection, and licensing of ATS personnel, as well as human factors considerations for maintenance and installation of systems.

[2] Some air navigation service providers are owned or operated by government (e.g., FAA) or military organizations (e.g., Brazil), and other are completely privatized (e.g., NAV CANADA) or a combination of private and government (e.g., NATS).

An overview of each element is provided to set the framework for the subsequent discussion on human factors issues. These elements provide the context that shapes the daily work of pilots and controllers.

Communication

The safety and efficiency of air navigation services depends on reliable and high-quality voice and data communication capabilities. A number of voice communication systems are used by controllers. Telephones landlines are used when controllers are talking to controllers within the unit or at another unit, while VHF or HF radio[3] is used for voice communication between controllers and pilots.

Another way that controllers and pilots communicate is through text messages. The controller can choose a predefined message or enter free text and then send it to the pilot, where it is displayed on a computer screen in the cockpit. This capability is referred to as controller-pilot data link communication (CPDLC). CPDLC reduces the amount of controller-pilot voice communication and is being used in an increasing number of areas around the world. In addition to replacing voice communication, CPDLC can also transmit information about the aircraft, such as position and speed, derived from on-board systems. CPDLC is a critical component in the implementation of new forms of surveillance, in particular Automatic Dependent Surveillance-Broadcast (ADS-B). ADS-B is described in the section on Surveillance.

Pre-Departure Clearance (PDC) is another form of data-link communication. This system allows pilots to obtain IFR clearances prior to taxi-out, thus eliminating the need for the controller to verbally communicate the departure clearance information. A PDC also provides the flight crew with more time to prepare for departure.

[3] Very High Frequency (VHF) radio is the primary method for voice communication, except in areas where VHF communications are not effective, such as over oceans and other remote areas. In these areas, High Frequency (HF) or satellite phone is used for voice communication.

Navigation

For pilots, navigation is about knowing how to get to final destination from their current position. The pilot's flight plan indicates to air traffic services the planned route to reach the destination point. There are several different methods for pilots to navigate from departure to destination. The decision of what method to use is constrained by the availability of navigation aids and on-board equipment. The on-board equipment interprets signals from the navigation aids and allows pilots to know where they are and what direction they need to go.

Traditional navigation requires the pilot to follow a network of designated routes, called airways. Airways are laid out between ground-based navigation aids (e.g. VHF Omni-directional Range [VOR] and Non-Directional Beacon [NDB]). In addition to traditional navigation methods along airways, many aircraft are now able to navigate a route between any two points. This point-to-point navigation is referred to as Area Navigation (RNAV). RNAV is possible because of on-board avionics that automatically determine the aircraft's position from one or more sources, such as VORs, DMEs, and satellites. The on-board systems compute distances and estimates to selected points along a route, and provide the pilot with guidance on how to stay on the selected route. The affordability and availability of satellite based navigation equipment has opened the doors to all air operators and general aviation. Satellite is rapidly replacing traditional ground-based navigation aids as the preferred system for navigation.

Some airports have published routes for departures and arrivals, called Standard Instrument Departures (SIDs) and Standard Terminal Arrival Routes (STARs). A SID route is a published departure procedure that takes an aircraft from the departure runway to the enroute phase. Similarly, a STAR takes the aircraft from top of descent enroute flight to the final approach of a runway. SIDs and STARs eliminate a considerable amount of verbal communication between controllers and pilots because the path is predefined. RNAV SIDs and STARs are similar to conventional SIDs and STARs except the pilot allows the on-board avionics to follow a route based on preprogrammed waypoints, altitudes, and speeds.

Over the North Atlantic Ocean, the published routes are called "tracks." Because there are no ground-based navigational aids available over the ocean, the location of the tracks are based on waypoints defined by latitude and longitude. The track structure changes from day-to-day depending on the trade winds. The concept of flexible tracks is also being applied to certain areas over land, such as northern Canada.

Surveillance

Surveillance refers to methods of keeping track of an aircraft's position. In situations where there are no surveillance technologies available, the controller keeps track of where the aircraft is through pilot position reports and estimates on where the aircraft is expected to be in the near future. The controller uses a flight progress strip or a computerized display of the reported and estimated positions to track the aircraft.

Traditionally, radar systems have been the primary tool used in areas of the world where there is sufficient air traffic to justify the cost of the system. In many locations, radar is an essential tool for managing the safe and expeditious flow of traffic. Radar data is processed and displayed to the controller on a computer screen. A radar display typically represents the aircraft with a symbol, and each aircraft symbol has a tag attached to it that contains data about the flight, such as aircraft identification, altitude, and speed. Most of the radars are for the terminal and enroute phases of flight, but there are also some radar systems that support surveillance at airports, such as the Airport Surface Display Equipment (ASDE) radar.

A relatively new method of surveillance is Automatic Dependent Surveillance-Broadcast (ADS-B). ADS-B is based on an aircraft's on-board equipment broadcasting information about the flight such as position, speed, and altitude. The broadcast information is received and processed on the ground and then displayed to the controller in a manner very similar to a radar display. ADS-B is currently being implemented in areas such as Hudson's Bay in northern Canada and the north-Atlantic ocean. The accuracy of ADS-B allows for much smaller separation standards between aircraft compared to the standard required with no radar coverage.

ATM

Air traffic management (ATM) refers to all aspects of managing the traffic, including airspace management, Air Traffic Services (ATS), and Air Traffic Flow Management. The following paragraphs provide an overview of each of these areas.

Airspace Management

Airspace around the world is divided in to Flight Information Regions (FIRs). Within each FIR there are areas designated as "controlled" and "uncontrolled" airspace. Within controlled airspace, the controller provides some form of separation between aircraft and other aircraft, obstacles, and terrain. Controllers also provide clearances, which are permissions to aircraft to proceed under certain conditions. The amount of ATS control for a flight depends on the class of airspace[4] and the flight rules[5] under which the pilot is flying. Some controlled airspace have special control areas and control zones designed to help the coordination of traffic into and around certain aerodromes. Pilots and aircraft must meet specific requirements to operate in these control areas and zones.

Air Traffic Services

The ability to deliver air navigation services and products relies on individuals with expertise in a wide variety of disciplines. This section will focus exclusively on the controller, who remains an essential part of delivering air traffic control services.

ATC Standards and Procedures The primary goal of ATS is to provide a safe and expeditious flow of traffic. Standards have been designed to separate aircraft from other aircraft, protected airspace, vehicles operating on and around runways, and from the ground. In an IFR environment, the standards provide each aircraft with a

---

[4] ICAO Class A-G.

[5] For example, Instrument Flight Rules (IFR), Visual Flight Rules (VFR), Special VFR (SVFR), and Controlled VFR (CVFR).

protected area of airspace (longitudinal,[6] vertical, or lateral). The controller's main objective is to ensure the required standards of separation are established and maintained. In order to accomplish this objective, the controller must know the aircraft's position, the controller and pilot must be able to communicate with each other, and the pilot must follow controller clearances and instructions throughout the route of flight.

Generally speaking, as the reliability and accuracy of communication, navigation and surveillance increase, the standards permit aircraft to fly closer together. For example, two IFR enroute aircraft under radar coverage in some airspace require 5 miles of longitudinal distance and 2000 feet of vertical distance unless the aircraft meet certain navigation performance requirements. If the aircraft has been certified for the more precise navigation performance, then the controller may reduce the vertical separation to 1000 feet. In contrast, if the controller is relying on pilot position reports, the separation standard will result in much greater distances between aircraft (e.g., 80 miles of longitudinal distance). There are also separation standards applied to departing and arriving aircraft to ensure that the following aircraft is not affected by wake vortex turbulence generated by the preceding aircraft.

The controller plans, executes, and monitors the standards through a complex set of procedures and techniques. Throughout each phase of a flight, controllers issue clearances and instructions to the pilot. These clearances contain the conditions under which the pilot should proceed, such as the altitude, speed and direction or route. The ability of pilots to comply with the clearances and instructions are critical to the overall management of the air traffic.

The Controller's Workstation An aircraft departing a medium-sized airport is first handled by the ground controller, who is responsible for movement on the maneuvering area. The ground controller then

---

[6]Longitudinal separation may be distance- or time-based. For example, a distance based longitudinal separation standard might read "Aircraft on the same track using DME shall be separated by 20 miles," while a time based longitudinal separation standard might read "Aircraft on the same track shall be separated by 15 minutes".

passes control for the aircraft over to the airport controller, who is responsible for takeoffs and landings and other aircraft operating within their control zone. The transition from the airport to enroute is handled by the terminal (departure/arrival) controller, and finally, a series of enroute controllers control aircraft as they transition through low-level and high-level airspace. Aircraft traveling across the oceans are handled by another specialty, called oceanic controllers.

Each controller has a different workstation configured for the position's tasks and information requirements. All controllers have access to a system for voice communication that includes telephone landlines and a number of VHF or HF channels. The airport controller has a view of the approach, departure and maneuvering areas of the airport, as well as access to a number of computer systems. The computer systems provide information such as weather, status of navigation aids, and approach information. Some airport controllers have access to electronic surveillance information, where ground or airborne aircraft are displayed as targets on a computer display. Airport controllers receive advance information about each flight in the form of electronic or paper flight data strips. The flight strip contains key information about the aircraft and route which assists the controller in planning activities related to the flight. The controller also uses the flight strip to track new information, such as cleared altitudes, and to coordinate the flight with other controllers. Traditional flight strips are printed strips of paper that the controller writes on. Electronic strips are viewed on a computer display and the controller inputs information and manipulates the flight strip through touch screen, keyboard and mouse.

The terminal and enroute controller workstation consists of a number of displays (see Figure 20.1). One computer display shows the location of the aircraft as seen by radar (or other surveillance). Each target on the screen has a data tag that contains the most important information about the flight, such as the aircraft call sign, altitude and speed. The complete flight plan information is available to the controller, usually on a separate display system. The controller also has access to a visual display of weather and restricted airspace. Terminal and enroute controllers use either electronic or paper flight strips for displaying flight information. Controllers often

FIGURE 20-1   Workstation for Enroute controller. © NAV CANADA 2008.

organize the flight strips of the aircraft under their responsibility, grouping the strips by point of departure or expected time of arrival at a waypoint. New and updated information related to the flight, such as altitude or holding instructions can be entered on the flight strip. Electronic flight strips allow the automatic transfer of updated flight information to downstream controllers.

Automation in ATM Systems Technological advances in ATM systems have focused on integrating flight data with surveillance data. These integrated systems have allowed for a number of automated tools to assist the controller in planning, executing and monitoring separation. Traditionally, a controller's radar display identified an aircraft's actual position, and the controller was required to compute flight estimates to predict where the aircraft would be in the near future. Integrated systems can automatically compute flight estimates and display the flight planned routes thereby assisting the controller to plan separation.

A controller must also transfer control of a flight to the adjacent controller as the aircraft transits from one sector to the next. In the past, the coordination of the flight was conducted with a verbal "hand-off" between two controllers. Advanced ATM systems have automated the hand-off, thereby reducing the time necessary to coordinate the flight between controllers.

Monitoring a flight's progress along the planned route can also be assisted through automatically generated alerts, and a number of alerting tools are now available, such as Conflict Alert and Minimum Safe Altitude Warning. Conflict Alert provides the controller with an alert (auditory and/or visual) a predetermined time before a predicted loss of separation between two aircraft. Similarly, Minimum Safe Altitude Warning provides an alert to the controller if the software predicts that the aircraft will descend below the minimum safe altitude.

The preceding paragraphs outline the enormous impact of CNS/ATM technology on how controllers provide air traffic control services. The following section outlines an area of growing importance in ATM called flow management.

Flow Management

Every airport has a finite capacity for the number of aircraft that can land or depart within a given time period. Capacity depends on factors such as the availability of runways, airport layout, type of traffic, and weather conditions. Flow management refers to the methods used to optimize air traffic flows and reduce delays for aircraft arriving at their destination airport.

Recent developments in sharing flight data between systems have allowed for remarkable advances in managing the flow of traffic. Systems are now available for identifying the flights scheduled to depart or land in airports from various regions around the world. These systems also provide real-time information on each aircraft's location (see Figure 20.2).

The flight data is used by central and local flow management units to assist decisions on where and when to impose restrictions on the arrival flow. There are various types of flow control methods, including: (a) a ground delay, where aircraft destined for congested airports are held on the ground at the airport of departure, rather than encountering delays in the air; (b) a reservation system which requires all pilots to obtain, prior to arriving or departing

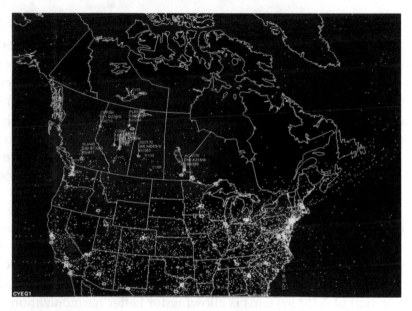

FIGURE 20-2 Enhanced Traffic Management System's view of North American traffic flow. © NAV CANADA 2008.

from an airport, a time slot for their flight during certain times of the day; and (c) a metering system that allows the unit to balance the arrival rate into an airport. It has been argued that airport capacity is becoming the limiting factor to overall system performance and the concept of gate-to-gate air traffic management requires the integration of airports, air navigation service providers, and airlines (Eurocontrol).

HUMAN FACTORS AND OPERATIONAL ISSUES

The previous sections described the basic elements of the air navigation system, including some of the recent developments that are revolutionizing the provision of air navigation services. This section is intended to highlight some of the Human Factors issues that currently exist in ATM. Many issues are relatively new, and are the result of how the various elements of the air navigation system have evolved. Other issues are long-standing and require more focused efforts to adequately address.

ATC and Flightdeck Procedures

Predictability in flight operations has always been important. From the pilot's perspective, flying becomes more manageable when expectations are met in regards to weather, aircraft performance, on-board automation, and air traffic control clearances. The controller equally relies on predictability, in terms of the pilot complying with air traffic control clearances and instructions. The dynamic nature of flying makes it difficult to attain complete certainty on all variables. However, well-integrated ATC and flightdeck procedures provide controllers and pilots with fewer surprises.

The concern over incompatibilities between ATC procedures and cockpit procedures was highlighted in a comprehensive human factors report in 1996 (Federal Aviation Administration, 1996). These problems persist in today's environment, although the modernization of ATM systems is allowing for better harmonization of procedures.

Highly automated cockpits have the potential to reduce flight crew workload by pre-programming on-board flight management systems (FMS) with departure and approach paths, altitude constraints, and enroute navigation. However, workload increases in situations where the crew must rapidly reprogram information based on revised air traffic control clearances and instructions. The FAA report classified the incompatibilities between ATC and cockpit procedures in three general areas. The first area concerned clearances that are difficult for most aircraft, and particularly challenging for highly automated aircraft. This category includes ATC clearances that take the aircraft toward the limits of its performance capabilities, such as approaches requiring a high rate of descent, or maintaining an approach speed higher than permitted for landing configuration. Another challenging clearance is a last-minute runway change for takeoff or landing. These events represent extremely busy periods for the crew, and a last minute change in runway further condenses the time to complete all the required tasks. For example, a change in runway and SID requires that the flight crew calculate the new takeoff performance, conduct crew briefings and reprogram the new SID in the FMS. Incident analyses have shown that it is not uncommon for the pilot to forget to enter the new SID. For ATC, these last-minute changes are used primarily as a means to enhance the number of arrivals and departures at the airport. However, there are other situations that may require issuing last-minute changes, such as unplanned vehicles or objects on the runway, or an aircraft taking an unusually long time to depart or climb.

The second category identified in the FAA report is described as clearances that work well for aircraft with no or low levels of automation, but that are challenging for aircraft with advanced automation. These include tracking outbound on a VOR radial, back course approaches, and issuing a complex go-around that differs in altitude and heading from the published missed approach. Go-arounds are a particularly important area for compatibility in pilot/controller procedures because of the critical time and phase of flight at which they occur. A published missed approach is important in circumstances where communication is lost between the pilot and the controller. However, controllers require the flexibility to modify the missed approach when the surrounding traffic situation makes the published missed approach a less desirable solution.

The third category of incompatible procedures identified by the FAA report was for clearances that do not take advantage of the aircraft's FMS. Examples include headings to intercept radials at an arbitrary point or flight over radio navigational aids versus direct routings to a waypoint. These clearances are sometimes the result of preferred routings used by ATS units to manage traffic in and out of busy airports or around restricted airspace. A number of clearances result in inefficient climb and descent profiles in comparison to the trajectories computed by the aircraft's FMS.

The effect on safety of these incompatibilities in procedures is that flight crews find it more difficult to comply with ATC clearances, and controllers must focus more attention on monitoring aircraft. Various sources of data[7] indicate that incompatibility between current ATC

[7] Sources include pilot incident reports, investigations, and LOSA (Line Operations Safety Audit) reports.

procedures and highly automated flightdecks is often a contributing factor in altitude deviations[8] and course deviations.[9]

Pilot-Controller Communication

Although the vast majority of pilot-controller communications are accurate and successful there are a number of critical pieces of information which if misunderstood by the controller or pilot can increase workload or jeopardize safety (Morrow & Rodvold, 1998). Given the critical nature of communication, aviation has established specific radiotelephony (R/T) phraseology and procedures. These phraseologies are found in the Aeronautical Information Manual (AIM) of each country. In most countries, accepted phraseology is based on ICAO standard aviation terms and phrases. Exceptions do exist across regions, and there is continued pressure to harmonize phraseology internationally in an effort to reduce confusion for pilots flying into different parts of the world.[10] Many of the features described for proper R/T include specific words as well as the pronunciation, timing of communications and requirements for readbacks from the pilot so that the controller can "hearback" (i.e., listen for the correct readback). The challenge in many terminal areas and airports is that the volume of ATS-Pilot communications results in blocked, stepped-on, or garbled transmissions. It is also not uncommon that in high traffic environments controllers and pilots compensate for the airway congestion by increasing speech rate, dropping call signs from readbacks, or in some cases not requiring readbacks. While this expedites traffic, it also increases the risk of miscommunication. The reliance on one-at-a-time radio transmissions has been described as one of the weakest areas in surface operations (for a review of runway safety, see Cardosi, 2005).

---

[8] An altitude deviation (also called a level bust) is when an aircraft deviates 300 feet or more from the assigned level.

[9] A course deviation is when an aircraft deviates from the assigned or planned track or heading.

[10] For example, in 2008 Canada replaced the term "taxi to position and wait" with the ICAO recommended phraseology of "line up and wait."

Another long-standing area of concern in many areas of the world is call sign confusion (CAA, 2000). The same controller is often handling aircraft identifications that either look similar or sound similar, for example, ABC3313 and ABC3133. This creates situations in which a controller may inadvertently issue a clearance to the wrong aircraft, or where a pilot may accept a clearance intended for another aircraft. The potential for these situations is increased in hub operations, where an airline has many different flights arriving and departing from the same airport. These situations require the controller to focus more attention on the pilot's readback and on monitoring the aircraft for compliance.

A more recent area of concern is the similar sounding names of SIDs/STARs and waypoints for RNAV procedures around airports (Eurocontrol, 2008a). The similarity in names has resulted in some pilots inadvertently using the waypoint name as the SID, and consequently not flying the correct SID. In some cases, the problem is not a misunderstanding between ATS and the flight crew, rather the problem is the pilot unintentionally selecting the incorrect but similar sounding or looking SID from the on-board FMS database.

It is generally acknowledged that miscommunication between pilots and controllers will continue to be an area of concern, even with the advent of data-link technology. Pilots and controllers using data-link systems (CPDLC and Pre-Departure Clearances) have reported misreading conditional altitude clearances and misunderstanding free text, and pilots have logged aircraft onto the data link system with the wrong aircraft identification (Eurocontrol, 2008b).

Controller Automated Decision Support Tools

A number of automated tools have been developed to assist controllers in planning, executing and monitoring separation. The extent to which these tools deliver the expected support to controllers depends on their usability, suitability for the operation, and general workforce acceptance (International Civil Aviation Organization, 2000). A large volume of research exists on human factors and automation, and specifically on automation in ATM. ICAO identified a set of general objectives for ATC automation: transparency of underlying operations, error-tolerance and recoverability, consistency with controllers' expectations, compatibility with human capabilities and limitations, ease of revision to lower levels of automation and of returning to higher levels of automation, and ease of use and learning (International Civil Aviation Organization, 2000). The following paragraphs illustrate how automation objectives can be compromised.

Responding to Automated Alerts

Conflict Alert and Minimum Safe Altitude Warning are just two examples of ATM automated tools that assist the controller's monitoring tasks through the provision of alerts or alarms. The success of these and other automated tools is directly related to the trust the user has in the tool (Wickens et al., 1998a). Trust in alerting tools is largely predicated on differences between how the system is intended to work and how the controller expects it to work (Benel & Benel, 1998). Nuisance alerts refer to alerts that are valid according to the alert algorithm, but which do not provide any useful information to the controller. The history of airborne systems such as the ground proximity warning system (GPWS) and the traffic alert and collision avoidance system (TCAS) illustrated how the initial implementation with an excessive nuisance alert rate led pilots to mistrust the system (Wickens et al., 1998b). One study found that over 80% of alerts for conflict alert and minimum safe altitude warning were nuisance alerts or unnecessary alerts (Allendoerfer et al., 2007). These alerts create interruptions and can desensitize the controller to responding to a real alert.

Automated alerts also require careful integration of ATC procedures with pilot procedures. Minimum Safe Altitude Warning is a tool that provides an alert to controllers if it appears that the aircraft will impact elevated terrain or obstacles. Procedures must address the question of what action the controller takes when an alert is generated, that is, should the controller issue a traffic advisory to the pilot or a control instruction? Similarly, what procedures should the pilot follow if the controller's instructions differ from cockpit advisories? Conflicting controller/pilot procedures were a contributing factor in a mid-air collision in 2001, where the pilot of one aircraft was unsure whether to follow the TCAS Resolution Advisory or the instructions of the controller (German Federal Bureau of Aircraft Accidents Investigation, 2004).

Responding when Automated Tasks Fail

Passing estimates from tower to terminal, computing flight estimates and displaying the authorized route are tasks that have
been automated in modernized ATM systems. The automation of
these manual processes provides the opportunity for the controller to focus on other tasks. However, on those occasions when the
automated task is not performed, it is essential that the controller
receives adequate feedback (Hopkins, 1999). This type of monitoring for low frequency events is not well suited to humans, and
when anomalies are detected the controller's workload can dramatically increase during the attempt to troubleshoot and remedy
the problem.

TABLE 20.1   Key Human Factors Issues in Air Traffic Management

Communication

- What information should be communicated by technology (on-board
  aircraft, ATM systems) versus humans (controllers, pilots)?
- What form of communication (voice, text, data) is optimal, and under
  which operational circumstances?

Navigation

- How can pilot and ATC procedures be better integrated to make it easier
  for pilots to comply with ATC clearances and instructions?

Surveillance

- What tasks in ATM should be automated to better support the controller
  plan, execute and monitor aircraft separation?
- How should anomalies in automated functions be identified and managed
  so that automation "surprises" are avoided and spikes in controller
  workload are minimized?
- How can controller automated decision-support tools be transparent,
  error-tolerant, easily recoverable, and consistent with controller
  expectations?

Measurement

- How can deviations from expected system performance be detected?
- What risk management techniques can be used to test proposed changes
  to processes, operating standards, procedures, technology, and local
  workplace factors to prevent the inadvertent drift into organizational
  accidents?

A number of full and partial system disturbances have been reported (Eurocontrol, 2004). Full system disturbances that affect automated functions include loss of radar, workstation failures and flight plan processing failures. In most cases there are obvious indicators to the controller when full functionality has failed. Of greater concern is the effect of a partial loss of data or function. In these cases, the quality and integrity of the data driving the automation is in question. For example, a processing error that results in the on-screen display of one aircraft target's data tag being linked with another aircraft target can take considerable time for the controller to identify and interpret, and draws attention away from monitoring other areas of the display.

This section has outlined a number of challenges in current operations (see Table 20.1). The next section describes human factors methods and techniques used to develop ATM systems, as well as tools to identify and manage operational safety issues.

HUMAN FACTORS METHODS AND TECHNIQUES

Integrating Human Factors into the Management of Operational Changes

There are a number of reasons why changes are made to how air traffic is managed. In some cases, the changes are driven by the need for better safety nets (e.g., conflict alert), cost-effectiveness (e.g., ADS-B surveillance to support more flexible, fuel-saving routes), or efficiency (e.g., automation of manual processes to support the handling of more traffic). Many of the changes have benefits in more than one area. In recent years, the two main drives for change have been the need to handle increasing amounts of air traffic and the need to update legacy systems which can no longer be supported. The purpose of this section is to identify how human factors approaches, tools and techniques help shape the development and implementation of these changes.

Applying a Human-Centered Approach

A human-centered approach to automated ATM systems requires identifying how to support the controller, rather than automating functions simply because the technology is capable of performing the task. The application of a human-centered approach to ATM system design is based on the underlying assumption that the controller remains responsible for traffic separation and the safe flow of traffic, and therefore must remain in command of air traffic (International Civil Aviation Organization, 2000).

As the complexity and interconnectivity of ATM systems increase, there has been growing recognition of the need for a systems approach with greater planning efforts devoted to the integration of the human component with equipment, procedures, and training. Some air navigation service providers have introduced human factors policies requiring human factors to be integrated into all aspects of system development, acquisition and operation[11]. This type of policy helps to ensure both a systems and a human-centered approach to design. For a comprehensive review of case studies in which human factors has been applied to operational ATM systems, see Kirwan et al. (2005).

The impact of proceeding without due consideration to all these elements is that the expected benefits of the change is not achieved. For example, new functionality may be developed without clear procedural guidance to the controller on how and when to use the feature or the amount of training required may be prohibitive. These factors can result in a decision not to deploy the new software/functionality despite the considerable effort, time and costs in developing the product (Kjaer-Hansen, 1999). From a safety perspective, managing change without a human-centered systems approach increases the likelihood that latent unsafe conditions are introduced into the operation.

It has also been argued that a critical component of developing human-centered design is the explicit specification of the design rationale and the assumptions underlying the design choices

---

[11] For example, FAA Human Factors Policy 9550.8 establishes policy, procedures, and responsibilities for incorporating and coordinating human factors considerations in Federal Aviation Administration programs and activities to enhance aviation safety, capability, efficiency, and productivity (FAA, 1993).

(Daouk & Leveson, 2001). Documenting the decision-making rationale may help reduce the likelihood that future design modifications set the stage for an accident. The recognition that complex systems may inadvertently drift into failure has led safety experts to call for a new approach to managing safety, often referred to as resilience engineering (Dekker, 2005a).

Integrating Human Factors into System Requirements

The life cycle of a system can be described as having five phases: concept, development, deployment, operation, and decommissioning. During the concept and the development phases, one of the key processes is defining requirements. Requirements describe how a system is to function (inputs, behavior, outputs) and identify constraints on the design (performance or reliability). Writing effective requirements (i.e., unambiguous, complete, traceable and testable) is a challenging task, but this is how software programmers and engineers turn concepts into products or functionality. Consequently, usability requirements that specify "the graphical user interface must be user friendly" are not particularly effective unless accompanied by testable criteria.

The body of knowledge generated through Human Factors research has resulted in various standards, some of which have specifically been designed for ATM systems, such as the *Human Factors Design Standard* (Ahlstrom & Longo, 2003). Standards play an important role in designing systems consistent with human factors principles. The development of ATM requirements also relies on a close partnership between those who operate the systems and those who design them. There are a number of extremely valuable human factors tools and techniques for gathering input from subject matter experts in a manner that can be translated into requirements. For example, defining the high-level business requirement in the concept stage can be assisted by techniques for conducting focus groups with users, conducting interviews, and defining success metrics. In the development stage, tools include task analysis, user profiling, usability walk-throughs, expert evaluations, and assessing against standards. During the deployment stage and postcommissioning life cycle, tools focus on conducting field assessments and obtaining user-feedback.

There are a wide variety of styles for writing human factors requirements, and careful consideration must be given to which style is appropriate for a given project (Allendoerfer, 2005). Comprehensive compilations of various human factors standards, tools and techniques are available on Eurocontrol and FAA Web sites. In addition to usability requirements, strategies have been developed for writing requirements that reflect total system performance (human, equipment, procedures, and training), thereby ensuring a comprehensive systems approach to system development (Burrett et al., 2005a).

Human Factors Methods for Identifying Operational Issues

Managing safety in ATM has always been a fundamental focus. However, the complexity and inter-connectivity in today's aviation world has led to a more integrated approach to managing safety. The term Safety Management System (SMS) is used to describe the approach many organizations, including air navigation service providers, have implemented to make a safe system even safer. This section describes some of the traditional as well as newer tools in the SMS toolbox.

Reporting Systems

One of the most common safety management tools is a process for people on the front lines to report events or conditions that impact safety. There are a variety of reporting systems which differ in terms of what and how reports are made. Many of the government-operated voluntary confidential programs are open to all aviation personnel, such as the Transportation Safety Board of Canada's SECURITAS (Transportation Safety Board of Canada, 2008), the United Kingdom's Confidential Human factors Incident Reporting Program (CHIRP) and the Aviation Safety Reporting System (ASRS) in the United States. The reporting of human factors issues is encouraged in these systems, and the reporting forms for CHIRP and ASRS provide prompts to the submitter to consider issues such as communication, perception, decision making, task allocation, and other factors affecting the quality of human performance. Some air navigation service providers also run their own voluntary confidential reporting systems.

Many air navigation service providers have mandatory reporting systems that require particular types of events to be reported internally and/or to the Regulator. The specific reporting requirements differ from one air navigation service provider to the next, but at the very least they require reporting the breach of minimum separation standards. Some air navigation service providers rely entirely on self-reporting, while others have software detection programs that identify loss of separation events. These mandatory reporting systems usually provide very limited information about events and are typically used to trigger further investigations or analyses. Increasingly, airlines and air navigation service providers are sharing pertinent information from the events or subsequent internal investigations, for example, an air navigation service provider and an airline may discuss a particular altitude deviation to reach a more comprehensive understanding of the contributing factors and to identify effective risk control measures.

As a stand-alone tool, incident reporting systems have a number of limitations with respect to capturing the full spectrum of contexts that shape human performance (International Civil Aviation Organization, 2002). One of the limitations relates to the actual reporting form. Most incident reports capture basic tombstone variables, such as experience, time on duty, phase of flight, and weather conditions, followed by a free-text area to explain what happened and why. This free-text is unstructured, with very little guidance provided to the individual reporting the event. The result is that very often the report is limited to *what* happened, as opposed to *why* an incident happened, although some reporting tools have been developed to improve the quality of incident reports (Weigman & von Thaden, 2001).

The importance of front-line reporting is highlighted by recent efforts in ATM to promote a Just Culture (GAIN, 2004). Just Culture describes an environment where people are confident they can participate in safety initiatives without fear of reprisal and where they know they will be treated fairly in the event that they are involved in an incident or accident. The concept has been described as "creating an atmosphere of trust in which people are encouraged for providing essential safety-related information, but in which they are also clear about where the line must be drawn between acceptable and unacceptable behaviour" (GAIN, 2004). The ultimate benefit of operating in a Just Culture is that potential hazards and risks do not remain hidden. The formalization of a policy and procedures describing how individuals will be treated when they are involved in an incident helps establish a common understanding of how "human error" is viewed and how accountability is handled in an organization. Establishing procedures will also result in more consistent and transparent handling of individual treatment.

ATS Investigations

A number of air navigation service providers have moved away from the old view of human error being the "cause" of incidents, to the new view of human error as a "symptom" of something deeper in the system (Dekker, 2005b). This approach is characterized by the integration of human factors frameworks into ATM incident and accident investigations. A wide variety of frameworks are in use that analyze human error and the organizational and local workplace factors that shape error. Published frameworks for human error management in ATM include HERA-JANUS (Isaac et al., 2003), Threat and Error Management (International Civil Aviation Organization, 2008a), TRACer (Burrett et al., 2005b), and the Operations Safety Investigation process (Lindeis, 2003). Investigations conducted with a Human Factors framework help ensure that the right questions are asked to understand the contributing factors (context) that led to the error. Applying a human factors framework also allows analysts to code and database the contributing factors in a manner that permits subsequent analyses to identify trends in terms of interface issues (i.e., incompatibilities between the controller and the tools, tasks, operational, and organizational environment). This approach has proven more effective than categories in which the contributing factors are unhelpful labels, such as "inattention" or "failed to monitor."

Monitoring Normal Operations

Many of the methods used to identify potential safety issues are triggered by an event (i.e. an incident or accident). However, it has been argued that ATM requires more methods for monitoring normal operations. What takes place during routine operation provides an extremely valuable window into the ways controllers adapt to less than perfect operating environments. The reasons for the drift from ideal to real system performance include "technology that does not operate as predicted; procedures that cannot be executed under dynamic operational conditions; regulations that do not reflect contextual limitations; introduction of subtle changes to the system after its design; addition of new components to the system without an appropriate safety assessment; or the interaction with other systems" (International Civil Aviation Organization, 2008b).

One methodology developed for monitoring normal operations is the Normal Operations Safety Survey (NOSS). The NOSS methodology involves training controllers to apply the Threat and Error Management framework to perform unobtrusive over-the-shoulder observations in the operational environment. The observations are analysed and a report is produced that describes the most prevalent threats, errors and undesired states as well as how these situations were managed (International Civil Aviation Organization, 2008b). This "snapshot" of the operation is extremely rich in contextual information.

Converging lines of evidence from normal operations monitoring programs, incident reports and incident or accident investigations provide organizations with data-driven information that can feed into design consideration for airspace, technology, and procedures. Information on potential areas of risk can also be provided to controllers though safety or procedures committee meetings, recurrent training, bulletins, and articles. The common use of the Threat and Error Management framework in NOSS and the airline equivalent, Line Operations Safety Audit (LOSA), allows for a very productive exchange of safety issues between an air navigation service provider and an airline.

Approaches to Managing Fatigue

Air traffic control services are required 24 hours a day and 7 days a week. Consequently, many ATS units require staffing on shifts to cover continuous operation. The primary concern of shiftwork in a safety-sensitive position is that it can result in fatigue, and fatigue can have a number of negative effects on performance. The accumulated research over the past 40 years has provided an understanding of the broader effects of fatigue, including internal processes in the body, family life and human performance (Della Rocco & Nesthus, 2005).

More recent efforts have focused on applying the scientific knowledge toward the management of fatigue. It is generally acknowledged that a fatigue management program must be comprehensive and integrated, and consist of activities in education, alertness strategies, and scheduling practises (Transport Canada, 2001).

Scheduling in many ATS units is currently a complex balancing of operational demands, individual preferences, contractual agreements and labour laws governing hours of work and economic considerations. Therefore, education about fatigue and its effects provides a critical foundation for the various groups and individuals making decisions that impact scheduling. Individuals who work the shifts require a basic understanding of the physiological basis of fatigue (i.e., sleep and circadian disruption), the effects on performance and alertness, and practical tips on how to prevent fatigue or apply alertness solutions in cases where fatigue is present. The education foundation forms the basis of the application of alertness strategies and scheduling activities, so that the basic scientific principles behind the strategies and approaches are clearly understood.

Alertness strategies include things that individuals can do at home, such as protecting opportunities for sleep, sleep hygiene, exercise, and nutrition. Other alertness strategies are designed for application at work, such as strategic use of caffeinated beverages, strategic napping, and light exercise.

Until fairly recently, the practical application of scientific knowledge on fatigue to an operation was challenging. The accumulated science made it relatively easy to identify where schedules strayed from the ideal schedule in terms of shift length, shift timing, direction of shift rotation and time for sleep. However, there was less guidance available for assessing how much risk a particular schedule introduced, and no practical tools for the operational floor that could assist individuals in determining shift solutions for a particular operational environment. Recently a number of computerized tools have been developed for assessing shift schedules. Current assessment tools must be applied judiciously, as the applicability of the underlying assumptions and algorithms may not be appropriate for the operation. The future holds promise for refining these tools to provide guidance to shift schedulers as well as ATS personnel seeking to reduce shift-induced fatigue.

The diverse operational environments and traffic patterns in ATS, coupled with individual differences in human physiology and lifestyle preferences has led many researchers to conclude that a wide variety of approaches is necessary to optimally manage fatigue. Education, alertness strategies, and shift scheduling are interdependent activities that must be coordinated to effectively manage fatigue. The interdependence of these activities underscores the need for organizations and individuals to accept their respective responsibilities.

SUMMARY

Technological advances in the past decade have dramatically transformed methods for providing air traffic services. Automated tools, systemwide information management, collaborative decision making, and integrated procedures are being implemented to allow for better and earlier flight planning, management and monitoring. A human-centered approach to the design of these new tools and processes is essential to support the human operators that are critical components of the overall system. The application of human factors principles, tools and techniques in all phases of the life cycle is necessary to ensure that the expected performance benefits are achieved.

The industry must continue to develop new methods for detecting and resolving situations where system performance is drifting from expected performance. The definition and measurement of "drift" in system performance are supported by various sources of information, including normal operations monitoring, pilot and controller reports, accident/incident investigations, and technical system performance monitoring. The concept of drift is particularly important with respect to organizational decisions to modify operational standards, procedures, and technology. Drift into failure has been identified by some as the greatest residual risk in the aviation system (Dekker, 2005a). The growing field of resilience engineering offers promise in delivering solutions to the industry's evolving understanding of accident causation.

The technological changes being undertaken to modernize ATM systems also provide an opportunity to effectively resolve operational issues that have arisen due to past incompatibilities between air navigation service providers, air operators, aerodrome operators, and manufacturers. A systems approach, with input from all aviation stakeholders, is necessary to develop ATM systems that meet the future demands of air travel.

References

Ahlstrom, V., & Longo, K. (2003). *Human Factors Design Standard* (HF-STD-001). Federal Aviation Administration William J. Hughes Technical Center, Atlantic City International Airport.

Allendoerfer, K. L. (2005). An analysis of different methods for writing human factors requirements. In: *Proceedings of the mini-conference on human factors in complex sociotechnical systems.* Atlantic City, NJ: Federal Aviation Administration (pp. 14-1–14-5).

Allendoerfer, K. L., Friedman, F., & Pai, S. (2007). *Human factors analysis of safety alerts in air traffic control* DOT/FAA/TC-07/22. Atlantic City, NJ: Federal Aviation Administration.

Benel, R. A., & Benel, D. R. (1998). A systems view of air traffic control. In M. W. Smolensky & E. S. Stein (Eds.), *Human factors in air traffic control*, p. 25. San Diego, CA: Academic Press.

Burrett, G., Weston, J., & Foley, S. (2005a). Integrating human factors into company policy and working practice. In B. Kirwan, M. Rodgers, & D. Schafer (Eds.), *Human factors impacts and air traffic management*, p. 510. Burlington, VT: Ashgate.

Burrett, G., Weston, J., & Foley, S. (2005b). Integrating human factors into company policy and working practice. In B. Kirwan, M. Rodgers, & D. Schafer (Eds.), *Human factors impacts and air traffic management* (pp. 515–516). Burlington, VT: Ashgate.

Cardosi, K. (2005). Runway safety. In B. Kirwan, M. Rodgers, & D. Schafer (Eds.), *Human factors impacts and air traffic management* (pp. 43–70). Burlington, VT: Ashgate.

Civil Aviation Authority. (2000). *ACCESS—Aircraft call sign confusion evaluation safety study* CAP 704. Norwich, UK: The Stationary Office.

Daouk, M., & Leveson, N. G. (2001). *An approach to human-centered design.* Presented at the Workshop on Human Error and System Development, Linkoping, Sweden, June 2001.

Dekker, S. W. A. (2005a). Why we need new accident models. In D. Harris & H. C. Muir (Eds.), *Contemporary issues in human factors and aviation safety* (pp. 181–198). Burlington, VT: Ashgate.

Dekker, S. W. A. (2005b). *Ten questions about human error: A new view of human factors and system safety.* Mahwah, NJ: Lawrence Erlbaum Associates.

Della Rocco, P. S., & Nesthus, T. E. (2005). Shiftwork and air traffic control: transitioning research results to the workforce. In B. Kirwan, M. Rodgers, & D. Schafer (Eds.), *Human factors impacts and air traffic management* (pp. 243–278). Burlington, VT: Ashgate.

Eurocontrol. *The Last Barrier? Eurocontrol.* http://www.eurocontrol.int/corporate/public/standard_page/focus_airports.html. Accessed November 15, 2008.

Eurocontrol. (2004). *Managing system disturbances in ATM: Background and contextual framework* HRS/HSP-005-REP-06. Brussels: Eurocontrol.

Eurocontrol. (2008a). *SID Confusion Safety Alert.* Eurocontrol http://www.skybrary.aero/index.php/SID_Confusion_Safety_Alert. Accessed October 20, 2008.

Eurocontrol. (2008b). *CPDLC incorrect callsign on log in Safety Alert.* Eurocontrol http://www.skybrary.aero/index.php/CPDLC_Incorrect_Call_Sign_on_Log-on_Safety_Alert. Accessed November 27, 2008.

Federal Aviation Administration. (1996). *The interfaces between flight crews and modern flightdeck systems.* Washington, DC: Author.

Federal Aviation Administration. (1993). *Human factors policy order 9550.8.* Washington, DC: Author.

German Federal Bureau of Aircraft Accidents Investigation. (2004). *Investigation Report AX001–0-2/02.* Braunschweig: Author.

Global Aviation Information Network. (2004). *A roadmap to a just culture: Enhancing the safety environment.* McLean, Virginia: GAIN.

Hopkins, D. (1999). Air traffic control automation. In D. J. Garland, J. A. Wise, & V. D. Hopkin (Eds.), *Handbook of aviation human factors,* p. 512. Mahwah, NJ: Lawrence Earlbaum Associates.

International Civil Aviation Organization. (2000). *Human factors guidelines for Air Traffic Management (ATM) systems* Doc 9758. Montreal: Author.

International Civil Aviation Organization. (2002). *Line Operations Safety Audit (LOSA)* Doc 9803. Montreal: Author.

International Civil Aviation Organization. (2008a). *Threat and Error Management (TEM) in air traffic control* Cir 314. Montreal: Author.

International Civil Aviation Organization. (2008b). *Normal Operations Safety Survey (NOSS)* Doc 9910. Montreal: Author.

Isaac, A., Shorrock, S. T., Kennedy, R., Kirwan, B., Andersen, H., & Bove, T. (2003). *The Human Error in ATM Technique (HERA-JANUS)* HRS/HSP-002-REP-03. Brussels: Eurocontrol.

Kirwan, B., Rodgers, M., & Schafer, D. (2005). *Human factors impacts and air traffic management.* Burlington, VT: Ashgate.

Kjaer-Hansen, J. (1999). *A business case for human factors investment.* Brussels: Eurocontrol.

Lindeis, A. (2003). Human error management within NAV CANADA's safety management system. In: *Proceedings of the 21st system safety conference.* Unionville, Virginia: System Safety Society (pp. 236–243).

Morrow, D., & Rodvold, M. (1998). Communication issues in air traffic control. In M. W. Smolensky & E. S. Stein (Eds.), *Human factors in air traffic control* (pp. 421–456). San Diego, CA: Academic Press.

Transportation Safety Board of Canada. (2008). *Securitas—Confidential reporting.* Transportation Safety Board http://www.tsb.gc.ca/eng/securitas/index.asp. Accessed November 14, 2008.

Transport Canada. (2001). *Report to the tripartite steering committee on ATC fatigue* TP 13742E. Ottawa, Ontario: Author.

Weigman, D. A., & von Thaden, T. L. (2001). *The Critical Event Reporting Tool (CERT).* Technical Report ARL-01-7/FAA-01-2. Atlantic City, NJ: Federal Aviation Administration.

Wickens, C., Mavor, A., Parasuraman, R., & McGee, B. (1998a). *The future of traffic control.* Washington, DC: National Academy of Sciences.

Wickens, C., Mavor, A., Parasuraman, R., & McGee, B. (1998b). *The Future of Traffic Control.* Washington, DC: National Academy of Sciences.

Maintenance Human Factors: A Brief History

Barbara G. Kanki

NASA Ames Research Center—MS 262–4, Moffett Field, CA

INTRODUCTION

On April 28, 1988, at 13:46, a Boeing 737–200 ... operated by Aloha Airlines ... experienced an explosive decompression and structural failure at 24,000 feet ... Approximately 18 feet from the cabin skin and structure aft of the cabin entrance door and above the passenger floor line separated from the airplane during flight. ... The National Transportation Safety Board determines that the probable cause of this accident was the failure of the Aloha Airlines maintenance program to detect the presence of significant disbonding and fatigue damage which ultimately led to failure of the lap joint a S-10L and the separation of the fuselage upper lobe. (NTSB, 1989, Executive Summary)

The structural failure and separation of a large piece of cabin structure was a shocking event to the public and industry alike, and the Federal Aviation Administration (FAA) immediately established the Aging Aircraft Program of research to look more closely at fatigue damage. However, the probable cause raised by the National Transportation Safety Board (NTSB) took issue with the operator's maintenance and inspection processes including the training and qualifications of mechanics and inspectors. In the words of the then FAA Associate Administrator for Regulation and Certification,

> ... the airplane was inspected and an airworthiness action performed just a few months before it had the tragic inflight episode. Then, just recently we found another airplane, with another airline, which had about 50,000 to 55,000 cycles and had developed a major crack and a number of smaller ones. This airplane also had been inspected earlier, with its cracks discovered only as it was going in for repainting. So here we have two airplanes ... for which somehow the system did not work. We have professionals involved in engineering and professionals involved in maintenance and yet cracks developed undetected. ... We must develop an improved approach to the inspection process and, more important, it must be an organized approach. We need to take a technological approach, break the process into its components, and then examine each component to see if we can build a body of knowledge that will apply. (Broderick, 1990, p. 4)

The Aloha accident led to some major realizations. Accident causes distinguished "aircraft" from "maintenance and inspection" factors, but compared to the knowledge about the reliability of aircraft systems, knowledge about the human reliability of maintenance and inspection processes was limited. We were learning the hard way that maintenance and inspection failures could result in injuries, fatalities and aircraft damage not only on the ground, but in flight; yet we did not fully understand the human factors that could lead to such failures. This was a wakeup call to fully acknowledge the critical role that maintenance and inspection played in ensuring flight safety. It was obvious that poorly performed maintenance and inspection could result in rework, aircraft damage, and injury to maintainers. But the chain of events leading to in-flight accidents and incidents was not clear. In many cases, inspection and maintenance errors resulted in latent conditions that did not cause immediate damage; failure might not occur for days, weeks, or ever. The complicated relationship between the latent nature of maintenance errors and adverse outcomes required systematic investigation and the establishment of principles and data that could direct the development of effective corrective and preventive strategies for managing maintenance risks.

This chapter will chronicle the 20-plus year journey of the community of operators, regulators, and researchers who embarked upon the task of identifying critical human factors of maintenance and inspection, establishing a database and research tools, and developing practical strategies for reducing the risks of maintenance and inspection errors. We will take a historical perspective, developing a timeline that is largely driven by key events such as accidents and regulatory/government actions and initiatives. Part 1 will cover 1988 through 1999 and Part 2 will cover 2000 through 2008.

PART 1: BUILDING THE FOUNDATION OF MAINTENANCE HUMAN FACTORS, 1988–1999

Following the landmark Aloha accident, the FAA called a meeting of aviation industry representatives to discuss problems associated with aging aircraft. Much of this meeting addressed issues of hardware, metal fatigue, and corrosion, but an additional discussion of human factors in maintenance reflected the growing interest in the contribution of human factors to aviation safety. Thus, in October, 1988, the FAA called an industry meeting that marked the beginning of a systematic review of Human Factors in Aircraft Maintenance and Inspection. From 1988 to 1999, the FAA sponsored 13 annual industry meetings that covered a wide range of human factors topics (see Table 21.1).

In addition to these meetings, the FAA began a program of research to investigate issues and recommendations generated by the community at large. The community was diverse including regulators, operators, union organizations, manufacturers, accident investigators, and researchers from both civil and military institutions. Soon, participants were coming from countries outside the United States eventually leading to a joint working agreement among the FAA, CAA-UK, and Transport Canada for the annual meetings. This expanded community greatly contributed to the research and regulatory progress as well.

The timeline in Table 21.1 shows key events (e.g., accidents, incidents, and government actions) in the left-hand column that serves as a context for the items in the right-hand column (e.g., industry meetings, research and guidance documents). To illustrate, the Aloha accident and FAA initiation of Industry meetings and research program (in the left-hand column), resulted in the series

TABLE 21-1   Maintenance Human Factors Timeline: 1988–1999

Accidents/Incidents (*) and Government Actions	Year	Industry Meetings, Research and Guidance Documents
*Aloha Airlines Flight 243 4/1/1988	1988	
FAA/Industry Meetings begin 10/1988		1 FAA/Industry Meeting: *Human Factors Issues in Aviation Maintenance and Inspection*
*United Airlines Flight 232 7/19/1989	1989	2 FAA/Industry Meeting: *Information Exchange and Communication*
	1990	3 FAA/Industry Meeting: *Training Issues*
		4 FAA/Industry Meeting: *The Aviation Maintenance Technician*
*Continental Express Flight 2574 9/11/1991	1991	5 FAA/Industry Meeting: *The Work Environment in Aviation Maintenance*
	1992	6 FAA/Industry Meeting: *Maintenance 2000*
		7 FAA/Industry Meeting: *Science, Technology and Management*
	1993	8 FAA/Industry Meeting: *Trends and Advances in Aviation Maintenance Operations*
	1994	9 FAA/Industry Meeting: *Human Factors Guide for Aviation Maintenance*
	1995	
*ValuJet Airlines Flight 592 5/11/1996	1996	10 FAA/Industry Meeting: *Maintenance Performance Enhancement and Technician Resource Management*
		Boeing's Maintenance Error Decision Aid (MEDA)
		NASA Aviation Safety Reporting System (ASRS) introduces Maintenance Form
White House Commission on Aviation Safety 2/12/1997	1997	11 FAA/Industry Meeting: *Human Error in Maintenance*
FAA Strategic Program Plan: Research 1988-1997		NASA Kennedy Space Center Human Factors Workshop I: *Human Error Analysis* Workshop II: *Human Factors Training*

TABLE 21-1   (*Continued*)

Accidents/Incidents (*) and Government Actions	Year	Industry Meetings, Research and Guidance Documents
	1998	US Navy's Human Factors Accident Classification System-Maintenance Extension (HFACS-ME)
		NASA Kennedy Space Center Human Factors Workshop III: *Procedures and Work Instructions*
		12 FAA/CAA/Transport Canada Industry Meeting: *Human Factors in Maintenance* (expanded sponsorship)
*NASA Shuttle STS-93 In Flight Anomaly 7/23/1999	1999	13 FAA/CAA/Transport Canada Industry Meeting: *Providing Hands-on Experience in Human Factors Topics*
		ATA Specification 113

of meetings that identified Human Factors issues in Maintenance and Inspection (listed in the right-hand column). In many cases, accident findings led to a specific emphasis such as a focus on training or the consideration of organizational factors and safety culture. The same community of researchers, regulators, and operators worked together to conduct the relevant research to develop and implement human factors solutions.

By the mid-1990s, a strong interest in collecting and analyzing incident data prompted the development of event databases. With the use of investigation and error analysis tools, human factors issues could be prioritized on the basis of trends in actual data. By the end of the 1990s, the FAA had reviewed many topics, developed many interventions, and renewed its plan to continue the research program and industry meetings. As the program and knowledge matured, NASA, motivated by its own maintenance issues, began extending the human factors principles, research and training to space operations (shuttle processing). It was also at this time that the industry began to capture its knowledge and lessons learned in guidance materials (e.g., ATA Specification 113).

Identifying Human Factors in the Maintenance and Inspection Work Domain

The series of FAA industry meetings beginning in 1988, revealed many perspectives and generated many useful recommendations. While some of the changes were directed at the industry and regulator level (e.g., better communication of information and development of an industry-wide database), an equally important objective was to build knowledge at the ground level. Where exactly were maintenance and inspection tasks failing? What factors contributed to maintenance and inspection errors or non-compliance? Researchers of "pilot error" were trying to get away from a "blame the pilot" mentality by investigating root causes of errors. Similarly, the maintenance community began systematically considering the numerous factors that could potentially affect maintenance and inspection performance. The first five FAA/industry meetings met this challenge by having experts in the field talk about the physical work environment, ergonomic factors, documents, procedures and training, in addition to individual factors such as fatigue, worker qualifications and skills.

Scarcely had the program gotten started when on July 18, 1989, the United Flight 232 crash landing at Sioux City, Iowa, captured the spotlight. At flight level 370, there was a catastrophic failure of the #2 tail-mounted engine, which led to loss of the three hydraulic systems that powered the flight controls. The flight crew experienced severe difficulties controlling the aircraft, which subsequently crashed during an emergency landing at Sioux City Gateway Airport.

> The National Transportation Safety Board determines the probable cause of this accident was the inadequate consideration given to human factors limitations in the inspection and quality control procedures used by United Airlines engine overhaul facility which resulted in the failure to detect a fatigue crack originating from a previously undetected metallurgical defect located in a critical area of the stage 1 fan disk that was manufactured by General Electric Aircraft Engines. (NTSB, 1990, Executive Summary)

The probable cause traced back to human factors limitations in inspection and quality control procedures, but the accident became famous for the skilled teamwork of the flight crew in executing the emergency landing and the extraordinary prepared-ness of ground personnel in helping survivors. As a classic exam-ple of Crew Resource Management (CRM), the captain and crew made effective use of all resources to make the best landing possi-ble. Although there were 111 fatalities, 175 passengers and 10 crew members survived.

Many lessons about flight crew CRM and survivability were learned from the Sioux City accident, but for the maintenance community, it was a painful reminder that human factors associ-ated with inspection processes, could significantly downgrade inspector performance. Again, the Safety Board raised the issue of human factors and inspector reliability, just as it had with the Aloha accident the year before. Acknowledging that the FAA had begun its series of industry meetings, the urgency of conducting these activities was reinforced. In addition, the Safety Board reiter-ated the need for an FAA research program that would not only build a solid understanding of human factors but could focus on emerging technologies (e.g., Non-Destructive Inspection tools) that could simplify, automate, or enhance the inspection process.

A Focus on Training

The Aloha and Sioux City investigations led the NTSB to issue two training recommendations to the FAA, namely:

A-89–56: Require formal certification and recurrent training of aviation maintenance inspectors performing non destructive inspection func-tions. Formal training should include apprenticeship and periodic skill demonstration.

A-89–57: Require operators to provide specific training programs for main-tenance and inspection personnel about the conditions under which visual inspections must be conducted. Require operators to periodically test per-sonnel on their ability to detect the defined defects. (NTSB, 1990, p. 88)

In addition to ensuring better technical training, the emphasis on human factors promoted the training of team skills similar, in theory, to CRM training that flight crews had come to accept. While flight crew CRM was a concept that had been developed at the end of the 1970s, it was implemented as classroom awareness training and integrated into line-oriented flight training in high-fidelity simulation by the end of the 1980s. But the work of maintainers and inspectors differed in significant ways from pilots' work including their organizational structure, certification process, documents and procedures. Thus, the maintenance community took great care to adapt team training to fit their own tasks, organizations, and workplace.

> The maintenance industry responded to the need for systemic solutions by first focusing on developing awareness-level training programs. The flight crew was already immersed in a proactive error management program—mostly focused on interpersonal communication and teamwork—called Cockpit [later called Crew] Resource Management (CRM). The maintenance community adapted the relevant portions of this CRM program to meet the perceived needs of their aircraft maintenance professionals and called such programs Maintenance Resource Management (MRM). (Patankar & Taylor, 2008, p. 62)

Maintenance Resource Management (MRM) training had immediate appeal because it was largely directed toward "protecting" the maintenance technician and inspector from human error. For many training practitioners, MRM was synonymous with Human Factors in these early years. Topics within the MRM training curriculum often included what is known as the "Dirty Dozen " which was developed in Canada and popularized world-wide. These included: (1) Lack of Communication, (2) Complacency, (3) Lack of Knowledge, (4) Distraction, (5) Lack of Teamwork, (6) Fatigue, (7) Lack of Resources, (8) Pressure, (9) Lack of Assertiveness, (10) Stress, (11) Lack of Awareness, and (12) Norms (Dupont, 1997).

At this early phase of MRM development and implementation, the topics were somewhat generic and quite similar to the topics discussed in CRM. However, MRM practitioners established credibility and relevance by interpreting these concepts within the context of maintenance and inspection operations. For example, pilot communication often focused on Captain and First Officer, occasionally extending to the cabin crew, passengers and Air Traffic Control. In maintenance, relevant roles in the workplace could involve mechanics, lead mechanics, company inspectors, FAA inspectors, supervisors, maintenance control, stores, and engineering. Often communications were needed to transfer information from one shift to the next. These could be face-to-face, or written on a task card or shift turnover form, thus accomplishing two functions, documenting and communicating. In short, maintenance tasks and work environments could present some unique human factors challenges as described here:

> In the early days of flight, pilots endured noise, wind and extreme temperatures as an accepted part of aviation. Maintenance technicians must still contend with the elements in ways that few airline pilots are required to do. A maintenance worker may be required to perch high above the ground, perhaps in rain and darkness, communicating by hand signals through deafening noise. ... Maintenance is different in other ways as well. An air traffic controller can unplug from the console at the end of the day, knowing that the day's work is finished. When the flight crew leaves the aircraft at the end of a flight, the chances are that any mistakes they made affected that flight only (unless they damaged the aircraft). But when maintenance personnel head home at the end of their shift, they know that the work they performed will be relied on by crew and passengers for days, weeks or even years into the future. (Hobbs, 1995, p. 4)

In the U.S. Navy, MRM training was never a standalone program; rather it was integrated into a more comprehensive safety program, a forerunner of sorts of safety management systems that have been growing in use since the mid-2000s. Components of the integrated MRM training were: (1) discussion of safety data from their own organizations, (2) best practices benchmarking with emphasis on operational risk management, and (3) safety climate assessment using the Maintenance Climate Assessment Survey (Schmidt & Figlock, 2001). This integrated approach showed that MRM training could be a highly effective mechanism for providing safety feedback to the workforce and developing preventive strategies for the future. It also could include built-in metrics for monitoring attitudes, risk and safety improvements.

Maintenance Accident Investigation

In spite of the surge of industry interest in maintenance human factors, there was little data to indicate trends and high risk issues. The primary data source was accident data. For example, Boeing statistics that covered the period between 1959 and 1989, indicated that of the 109 accidents with known causes (excluding sabotage and military action), 69.7% were attributed to cockpit crew as

Hull loss accidents -U.S. commercial jet fleet

Primary factor	Number of accidents		Percentage of total accidents with known causes
	1959–1989	1990–1999	10 20 30 40 50 60 70
Cockpit crew	76	13	69.7% / 54.2%
Airplane	11	3	10.1% / 12.5%
Maintenance & Inspection	4	2	3.7% / 8.3%
Weather	8	1	7.3% / 4.2%
Airport/ATC	5	1	4.6% / 4.2%
Miscellaneous/other	5	4	4.6% / 16.7%
Total with known causes	109	24	
Unknown or awaiting reports	3	8	
Total	112	32	

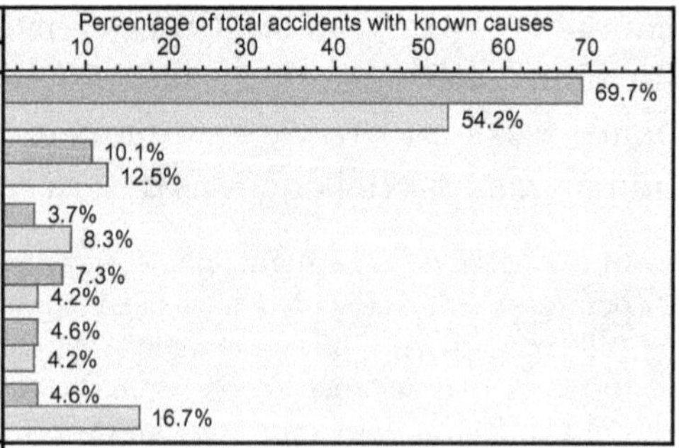

Excludes:
• Sabotage
• Military action

Legend:
■ 1959 through 1989
□ 1990 through 1999

FIGURE 21-1   Primary factors causing hull loss accidents in the U.S. commercial jet fleet (Boeing, 2003).

primary cause. In contrast, maintenance and inspection during this timeframe was a very small 3.7%, which was even lower than airport/ATC at 4.6% and weather at 7.3% (Boeing, 2003) (Figure 21.1).

While these were relatively small percentages, accident analyses pointed to the possibility of severe consequences tied to maintenance and inspection. Within only two more years, September 11, 1991, another accident involving maintenance occurred— Continental Express Flight 2574. As it would turn out, this accident became a classic for two reasons. First, it highlighted the difference between "procedures as written" and "procedures as performed;" that many factors (such as communication across shifts, or the clear delineation of technician versus inspector roles and responsibilities), introduced variability and complications in the way procedures were followed. Second, it introduced the concept of organizational culture; that individual actions were performed in the context of organizational norms and practices.

Organizational Factors and Safety Culture

On the day of the accident, Continental Express Flight 2574, an Embraer 120 operating under Part 135, experienced a sudden in-flight loss of a partially secured left horizontal stabilizer leading edge, leading to immediate severe nose-down pitchover, breakup of the airplane, and subsequent crash near Eagle Lake, Texas.

> ... probable cause of this accident was determined to be the failure of Continental Express maintenance and inspection personnel to adhere to proper maintenance and quality assurance procedures for the airplane's horizontal stabilizer deice boots that led to the sudden in-flight loss of the partially secured left horizontal stabilizer leading edge and the immediate severe nose-down pitchover and breakup of the airplane. Contributing to the cause of the accident was the failure of Continental Express management to ensure compliance with the approved maintenance procedures, and the failure of FAA surveillance to detect and verify compliance with approved procedures. (NTSB, 1992, Executive Summary)

Thus, probable cause was attributed to maintenance and inspection personnel; contributing factors involved management and regulators. The relevant company procedures, namely, the procedures for shift turnovers (including the documentation of incomplete work on work cards), were not found to be inadequate. Rather, the company was faulted for not considering the removal and replacement of the horizontal stabilizer leading-edge deice boot as a Required Inspection Item.

Of particular note, one NTSB member provided a dissenting opinion that stated probable cause should be: (1) the failure of the company to establish a corporate culture, which encouraged and enforced adherence to maintenance and quality assurance procedures, and (2) the consequent string of failures of the personnel to adhere to the approved procedures for replacement of the horizontal stabilizer deice boots. This opinion was an indicator of the growing appreciation for what was to be called the "safety culture" of an organization. This concept was often explained according to James Reason's "Swiss Cheese" model which illustrated how failed organizational barriers fit into an event occurrence with adverse consequences (Reason, 1990). The aviation maintenance community found Reason's model compelling because it depicted safety as a system and interpreted individual actions (e.g., maintainers and inspectors) in the context of local workplace conditions and within an organization that reflected management decisions, organizational processes, corporate culture, and so on. The complex network of organizational roles and responsibilities, regulations, and procedures was typical in maintenance and inspection processes in comparison with the seeming autonomy of pilots' actions. Furthermore, the ambiguity of roles and oversight responsibilities pertaining to the use of contract maintenance would become an issue as outsourcing became more prevalent. But even during those less complicated times, another accident pointed to organizational and oversight failures that contributed to a catastrophic outcome.

On May 11, 1996, ValuJet Flight 592 (DC-9–32) crashed into the Everglades about 10 minutes after takeoff from Miami International Airport. This outcome resulted from a fire in the airplane's class D cargo compartment that was initiated by the actuation of improperly carried oxygen generators. Probable causes were attributed to three failures: (1) the failure of SabreTech to properly prepare, package, and identify unexpended chemical oxygen generators; (2) the failure of ValuJet to properly oversee its contract maintenance program to ensure compliance with maintenance, maintenance training, and hazardous materials requirements and practices; and (3) the failure of the FAA to require smoke detection and fire suppression systems in class D cargo compartments.

> Contributing to the accident was the failure of the FAA to adequately monitor ValuJet's heavy maintenance programs and responsibilities, including ValuJet's oversight of its contractors, and SabreTech's repair station certificate; the failure of the FAA to adequately respond to prior chemical oxygen generator fires with programs to address the potential hazards; and ValuJet's failure to ensure that both ValuJet and contract maintenance facility employees were aware of the carrier's 'no-carry' hazardous materials policy and had received appropriate hazardous materials training. (NTSB, 1997, Executive Summary)

Maintenance Error Analysis

As accident investigations were providing a window into the causes and contributing factors of human error events, the maintenance community could see the value of evaluating their own smaller events and close calls. Such a system would help to structure event databases from which trends could be determined and corrective actions could be developed. Whether implemented on a company basis or industry-wide, a standard system for collecting maintenance error information would be needed for identifying current key issues and for developing and tracking strategic and systemic interventions. By the mid-1990s several serious efforts toward developing maintenance error investigation and analysis tools were developed. These efforts greatly enhanced the maintenance community's understanding of human error in maintenance and inspection. Often incorporated was Reason's model that differentiated latent factors, failed barriers, and active errors.

One of the most complete systems, Boeing's Maintenance Error Decision Aid (MEDA), was developed as an investigation tool for operators; providing a guide for collecting information about an event and the factors that contributed to its occurrence (Rankin & Allen, 1996). In addition, it provided a taxonomy and standard definitions of the elements of the event model and philosophy. The four-page data collection Results Form provided a structured approach to characterizing the event and for developing error prevention strategies. Boeing introduced this tool in 1996 and it became well known in the industry. Even if it was not completely implemented by operators, it enhanced the understanding of maintenance human error and provided a vocabulary for the maintenance community, as well as a model and tool for researchers. For example, by using the MEDA tool (with the collaboration of many of their airline customers), Boeing was able to determine the number of in-flight shut downs that were due to maintenance error and whether some airlines had more problems with human error than others (Rankin, 1999, Rankin & Sogg, 2003). In Europe, similar tools such as the Aircraft Dispatch and Maintenance Safety (Russell, Bacchi, Perassi & Cromie, 1998, McDonald, 1998) and company-developed tools such as British Airways' MEI (Maintenance Error Investigation) tool also followed a structured approach to investigation and analysis of error events.

At roughly the same time, the Human Factors Accident Classification System (HFACS) was developed by the U.S. Naval Safety Center to analyze pilot errors contributing to Naval Aviation mishaps. The original framework, subsequently adapted for maintenance events, was called HFACS-Maintenance Extension (ME). It was field tested by the Navy to ensure that the four error categories: Supervisory Conditions, Maintainer Conditions, Working Conditions, and

Maintainer Acts would be relevant and appropriate for use in maintenance operations (Schmidt, Schmorrow & Hardee, 1998). In the Navy implementation of HFACS-ME (similar to the use of MEDA), the tool was used to aid the investigation process and develop corrective actions. In addition, the safety results fed into Human Factors training (Navy version of MRM), providing data that pointed out key safety issues and content for the training course itself.

Maintenance and the National Database

The NASA Aviation Safety Reporting System (ASRS) was designed to provide a nation-wide repository of event data, thus giving aviation personnel, a vehicle for reporting unsafe occurrences and hazardous situations. Between its inception in 1976 and the end of 2008, this voluntary, confidential system received nearly 800,000 incident reports and currently grows by more than 4,200 reports per month. In 1996, ASRS introduced a specialized maintenance reporting form and began to actively encourage the reporting of maintenance incidents. Since that time, there has been a steady intake of maintenance reports, providing several advantages over other maintenance error databases. First, because it is a voluntary system it could capture close calls as well as events that would otherwise not be reported. Therefore the volume of ASRS reporting could greatly exceed the less frequently occurring incidents and accidents that are required to be investigated. Second, because it collects reports nationwide, it can detect trends or patterns that may not be apparent within a single organization (e.g., Aviation Safety Reporting System, 2002). Finally, the event reported is not influenced by an investigator's point of view; rather it is the reporter's own personal account of the event. In the second part of this chapter, we will return to the ASRS database as a source for establishing industry baselines and for reflecting new issues.

Government Reviews: Maintenance Human Factors and Aviation Safety

White House Commission on Aviation Safety and Security. Although there was, by now, increased awareness about maintenance human error and aviation safety, the 1997 White House Commission on

Aviation Safety and Security reinforced this linkage. In the released report, a National Goal was established to reduce the fatal accident rate by 80% in 10 years (White House Commission on Aviation Safety and Security, 1997). According to Boeing projections, the accident rate would be unacceptably high if increases in travel doubled in twenty years. In response to these concerns, NASA developed its Aviation Safety Program which included a focus on Maintenance Human Factors. The first phase of this program which took place from 2000 to 2005 will be described in Part 2.

FAA Strategic Program Plan When the White House Commission Report came out, the FAA could brief their 10-year progress and establish its new Strategic Program Plan (FAA, 1998). The plan described the objectives of the applied research of maintenance and inspection safety issues and the development of practical solutions to enhance training, job aiding, and information systems for aviation maintenance personnel. The progress of the program from 1989 to 1997 was summarized as follows:

> Since 1989, the research program has conducted annual conferences attended by thousands of participants, and has generated more than 400 technical reports, journal articles, and presentations. The research program has the international reputation of representing the "real world" of aviation maintenance and addressing maintenance human factors issues accordingly. It has raised the awareness of the importance of human factors to the aviation industry, and a number of organizations are implementing programs specifically designed to reduce maintenance errors." (FAA, 1997, Executive Summary)

Products and planned activities covering seven primary areas are listed here:

1. *Maintenance Resource Management (MRM)*—guidelines, training and reference materials for MRM through extensive cooperation with the airline industry (Federal Aviation Administration, 2000),
2. *Maintenance Error Reduction*—proactive reduction of maintenance errors by developing maintenance error reporting systems and self-disclosure programs,
3. *Job Task Analysis in Aviation Maintenance Environments*—an objective basis for structuring the maintenance curriculum and supporting of regulatory changes,

4. *Maintenance and Inspection Training*—improvements for maintenance training curriculum and new technologies for training delivery systems,
5. *Job Aids for Maintenance and Inspection*—electronic performance systems for inspectors, job aids to help design ergonomically efficient procedures, electronic checklists for auditing suppliers, design aids for document writers to ensure that procedures follow human factors conventions, software tools for tracking the repair and return of jet engine parts back to service.
6. *Information Dissemination*—research products disseminated through conferences and Web sites (Federal Aviation Administration, HFAMI website),
7. *Communication and Harmonization*—coordination with international organizations, regulators, and airlines.

Maintenance Human Factors in Space Operations

During the mid- to late 1990s the FAA and ever-growing maintenance human factors community were meeting regularly, conducting research, and steadily cultivating maintenance human factors knowledge, resources and solutions. The surge in awareness and knowledge reached the NASA space program where the space shuttle ground processing workforce found common ground with the issues being raised. Shuttle processing technicians, inspectors and engineers from Kennedy Space Center joined the aviation maintenance community and presented their human factors needs and initiatives at the 12th FAA/CAA/Transport Canada Symposium held in London (Kanki, Blankmann-Alexander & Barth, 1998). In addition, NASA coordinated several workshops that brought the aviation maintenance and shuttle processing communities together to share information and discuss human factors issues. Three workshops during 1997–1998 focused on: (1) Analysis of Errors, (2) Human Factors Training, and (3) Procedures and Work Instructions.

NASA Shuttle In-Flight Anomaly In 1999, the NASA Space Shuttle ground operations experienced their own human error event that resulted in a serious in-flight anomaly.

During the launch of STS-93 in July, 1999, two serious in-flight anomalies occurred. The first occurred five seconds after lift-off when a primary and back-up main engine controller on separate engines dropped offline due to a power fluctuation. Post-flight inspection revealed a single 14 ga. polyimide wire had arced to a burred screw head. The second anomaly was a liquid oxygen low-level cutoff 0.15 seconds before the planned Main Engine Cut Off (MECO). Post flight inspection of the affected engine indicated that a liquid oxygen post pin had been ejected and had penetrated three nozzle coolant tubes, causing a fuel leak and premature engine shut-off. (National Aeronautics and Space Administration, 2000, p. 8)

While the second anomaly could be considered a "design" issue resulting in "internal FOD (foreign object debris," the first anomaly (on which we will focus), describes a wire chafing event that could only have resulted from collateral damage incurred during ground processing. A NASA team reviewed the Space Shuttle systems and maintenance practices with the charter to investigate NASA practices, shuttle anomalies, and other civilian and military experiences. Detailed inspection of the shuttle payload bay cable tray revealed that a single 14-gauge polyimide wire had arced to a burred screw head causing a short circuit in two separate main engine controllers. An orbiter has more than 300 miles of wires such as those shown in the cable tray inside Columbia's payload bay (see Figure 21.2). The damaged wire was located in the aft left-hand mid-body lower wire tray which is normally covered. Records indicated that the last time covers were removed was four years earlier during a maintenance down period. Since there was no evidence of generic chafing, the root cause was determined to be work-induced collateral damage. Contributing to the maintenance error was the inconsistent specification of wire protection application. This event again highlights how maintenance errors may be latent (four years) and hidden from detection (unopened cable trays).

The shuttle fleet was grounded until inspections of all four shuttles were conducted, analysis of the root cause and contributing factors was completed, and corrective actions were taken. Immediate corrective actions included:

- Extensive wiring inspection, repair, and additional wire protection to critical wiring,

FIGURE 21-2   Space Shuttle STS-93 in-flight anomaly caused by wire damage.

- Redefinition and standardization of wire inspection criteria,
- Maximum feasible separation of redundant systems (redundancy of circuits had been compromised by placement in the same wire bundle).

In addition to immediate actions, a wide range of other factors, similar to those in aviation accidents, included organizational issues (reduced NASA oversight relationship to the contractor, reduction in workforce), procedural problems (compliance-based procedures that encouraged workarounds), metrics that discouraged the reporting of maintenance errors, and lack of standardization and communication across maintenance organizations. Additionally, major issues in risk management philosophy and practices lead to an overreliance on past successes and lack of data on smaller events. As in aviation maintenance, risk management tended to be understood as "design risk," which is addressed during the certification phase. Unfortunately, this perspective tends to overlook operational risks that may occur on a daily basis and have the potential to compromise flight safety even though they are hidden from obvious view and may lie dormant for years.

Key Industry Guidance

It was now just over ten years and the aviation maintenance community had documented many lessons learned. An industry working group (ATA Maintenance Human Factors Subcommittee) developed an Air Transport Association Specification (ATA Spec 113) that would pool their collective knowledge and experiences to date. In many respects this document provides a good summary of the first ten years of maintenance human factors programs from an operational perspective. The Subcommittee was made up of Human Factors representatives from manufacturers, regulators, operators, repair stations, unions and research organizations. The original release of the document was in January 1999 though it was subsequently revised in 2002 (Air Transport Association, 2002).

ATA Specification 113: Maintenance Human Factors Program Guidelines ATA Spec 113 provided definitions of pertinent maintenance human factors terms (e.g., human factors, ergonomics, error management systems) followed by a review of the scope and placement of currently active maintenance human factors programs. It was noted that Human Factors practitioners carry out their work by interfacing with many departments within the organization including line and base maintenance, quality assurance, safety and training departments. At the same time, a Boeing survey of customer data found that the Human Factors function could reside within a number of organizations: 58% were in Quality Assurance/Quality Control, 30% were in Maintenance Control, and 12% were in Other Departments. Companies were scoping and placing their programs to fit their organizational structure and needs, and not according to one set way.

The next four sections of ATA Spec 113 discussed the current state of the art of maintenance human factors programs. While acknowledging human factors training (e.g., MRM) was one important element, additional elements such as maintenance error management and ergonomics were called out, as well as the need for program element interaction. For instance, the data generated from error management systems could suggest developing ergonomic improvements or process and procedural enhancements. Additionally, maintenance error data could be the basis for developing specific training content, similar to the process described earlier in the U.S. Navy's program. Detailed guidance on developing training, error management, and ergonomic programs were provided based on the best practices of the day. For example, training program guidance included how to conduct a needs assessment, what should be accomplished in the design phase, components of a basic curriculum, developing and validating a prototype program, and steps toward adopting, implementing, and evaluating the program. Guidance for developing an error management and ergonomics programs was similarly described in a detailed, step-by-step manner.

After the initial release of ATA Spec 113, it became evident that would-be implementers needed guidance on how to calculate Return on Investment (ROI). While they were eager to develop human factors programs, they needed to develop financial justifications for management. Thus, the revised ATA Spec 113 added a chapter called Return on Investment (ROI): Design, Measurement, and Use of ROI Tools in Maintenance Human Factors Programs, thus giving developers some tools and ideas for presenting a business case to the company.

PART 2: DEVELOPING METHODS AND TOOLS TO MEET NEW CHALLENGES (2000–2008)

After 10 years, the state-of-the art of Maintenance Human Factors appeared to have achieved momentum in the United States and internationally, but more needed to be done. The White House Commission on Aviation Safety and Security set a goal of decreasing fatal aviation accidents and reinforced the need to include a focus on maintenance human factors. The maintenance community continued to develop and implement human factors programs, sharing their experiences at annual conferences, and producing research and guidance materials. But the next decade started with an aircraft accident that played out yet another variation on the maintenance error theme.

On January 31, 2000, just north of Anacapa Island, California, Alaska Airlines, Flight 261 crashed into the Pacific Ocean. The crew and passengers were killed and the airplane was destroyed by impact forces.

> The National Transportation Safety Board determines that the probable cause of this accident was a loss of airplane pitch control resulting from the in-flight failure of the horizontal stabilizer trim system jackscrew assembly's acme nut threads. The thread failure was caused by excessive wear resulting from Alaska Airlines' insufficient lubrication of the jackscrew assembly.

> Contributing to the accident were Alaska Airlines' extended lubrication interval and the Federal Aviation Administration's (FAA) approval of that extension ... and Alaska Airline' extended end play check interval and the FAA's approval of that extension, which allowed the excessive wear of the acme nut threads to progress to failure without the opportunity for detection. Also contributing to the accident was the absence on the McDonnell Douglas MD-80 of a fail-safe mechanism to prevent the catastrophic effects of total acme nut thread loss. (NTSB, 2002, Executive Summary)

Since the probable cause was failure of lubrication and inspection of the jackscrew, it could not be tied to a specific error event; rather, a nonaction built up over time set up the failure conditions for the accident. Numerous contributing factors were cited, including the jackscrew assembly overhaul procedures, design, and certification of the MD-80 horizontal stabilizer trim control system, Alaska Airlines' maintenance program, and the FAA's oversight of Alaska Airlines, thus reinforcing the need to research human factors issues related to procedures, workplace, organizational factors and oversight.

NASA Aviation Safety Program: Maintenance Human Factors Task

During the same year, planning for the NASA Aviation Safety Program[1] started (see Table 21.2). The program was justified by priorities identified by a group composed of government, industry, and academic members of the Aviation Safety Investment

---

[1]It would become the Aviation Safety and Security Program (AvSSP) after the terrorist events of 9/11 in the United States.

TABLE 21.2   Maintenance Human Factors Timeline: 2000–2008

Accidents, Incidents (*) and Government Actions	Year	Industry Meetings, Research and Guidance Documents
*Alaska Airlines Flight 261 1/31/2000	2000	14 FAA/CAA/Transport Canada Industry Meeting: Safety Management Theory to Practice
NASA AvSSP Maintenance Human Factors Program begins 2000-2005		FAA AC 120-72: Maintenance Resource Management Training 9/28/2000
9/11 Terrorist Events in U.S.	2001	15 FAA/CAA/Transport Canada Industry Meeting: Practical Solutions for a Complex World
	2002	16 FAA/CAA/Transport Canada Industry Meeting: Enhancing Human Performance
		CAA-UK CAP 716: Aviation Maintenance Human Factors (EASA / JAR 145 Approved Organisations)
		FAA AC 120-66B Aviation Safety Action Program 12/1/2002
*Air Midwest Flight 5481 1/8/2003	2003	ICAO Human Factors Guidelines for Aircraft Maintenance Manual
		17 FAA/CAA/Transport Canada Industry Meeting: Safety Management in Aviation Maintenance
		FAA AC 120-79A: Continuing Analysis and Surveillance System (CASS) 4/21/2003
	2004	
	2005	FAA Operators Manual: Human Factors in Aviation Maintenance (last update 10/6/05)
	2006	18 ATA/FAA Maintenance and Ramp Safety (change of meeting sponsorship)
		FAA AC 120-92: Introduction. to Safety Management Systems 6/22/2006
	2007	19 ATA/FAA Maintenance and Ramp Safety
	2008	20 ATA/FAA Maintenance and Ramp Safety

Strategy Team (ASIST) who identified more than forty high priority safety areas, many of which were relevant to maintenance human factors. Some were generic human factors topics such as: Human/Task Metrics and Models for Evaluation, Task Selection and Training, Skill Proficiency, Organizational Culture for Safety, Procedures Design Methods and Design to Support Teamwork. In addition, maintenance-specific areas included:

- Maintenance Teamwork Procedures and Roles/Responsibilities
- Maintenance Training
- Maintenance Task Procedures

The goals of the NASA AvSSP Maintenance Human Factors (MHF) task were: to provide guidelines, recommendations and tools directly to maintenance personnel and managers, through a better understanding of human error and human reliability, and to develop interventions that would enhance safety and operational effectiveness. Of critical importance to the success of the research was to follow an approach that involved operational partnership through all phases of the research:

1) Identification or validation of safety needs
2) Development of methods and tools
3) Development of human factors interventions
4) Validation of product implementation in operations

The FAA program of research had provided a 10-year foundation of research and industry involvement, so the NASA program easily complemented existing work by focusing on remaining industry-identified issues and expanding the research focus to address longer term goals.

For example, emphasis was placed on enhancing the use of maintenance error databases, developing generic methods and tools to solve a variety of safety needs, and developing interventions that included forward-looking new technologies. The MHF task was to be accomplished during the 2000 to 2005 timeframe according to the model in Figure 21.3, encouraging operational partnerships that would result in the validation of product implementation.

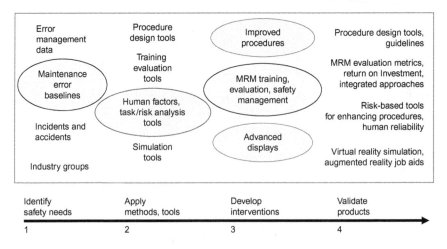

FIGURE 21-3   NASA research approach for Maintenance Human Factors.

Identification of Safety Needs

A fundamental challenge to working in the field of maintenance human factors was the general lack of human factors data. Clearly, the high visibility accidents over the last 10 years left no doubt that maintenance errors could lead to serious consequences and many contributing factors were identified. Reason's concepts and various systematic methods to classify, analyze, and summarize maintenance error events provided a common vocabulary and model for discussing and understanding maintenance human error. But the challenge of actually building databases that could be easily summarized and acted upon would take time. These efforts had to transform themselves from being purely reactive systems for large events (e.g., accidents) to becoming more proactive systems for collecting and analyzing lower-level events and close-calls that could identify potential precursors of accidents. This depended on cultivating voluntary reporting systems, similar to the NASA ASRS program but ones that could be maintained on a company basis. In 2002, the FAA published *Advisory Circular 120–66B: Aviation Safety Action Program,* (Federal Aviation Administration, 2002) guidance that could help maintenance organizations develop their own in-house voluntary reporting. While the FAA's Aviation Safety Action Program (ASAP) had already gained some success with the pilot community, but there was no guarantee that maintenance personnel would trust the system enough to disclose events that would have had them punished in the past.

During this timeframe, significant International progress was being made. For instance, European human factors guidance was developed to be consistent with the European Aviation Safety Agency (EASA) Part-145 Human Factors and Error Management Requirements. A comprehensive guidance document was completed in 2003 by the CAA-UK, the *Civil Aviation Publication 716, Aviation Maintenance Human Factors (EASA/JAR145 Approved Organisations).* (Civil Aviation Authority, 2003) During the same year, the International Civil Aviation Organisation published *Human Factors Guidelines for Aircraft Maintenance Manual* (International Civil Aviation Organisation, 2003).

In spite of newly available, comprehensive guidance materials, a continuing challenge was the nature of maintenance errors themselves. As accidents and incidents showed, maintenance errors did not always result in immediate outcomes, but could lie undetected for an indefinite period of time. In other cases, errors were discovered in the course of performing other maintenance tasks, and as a matter of course, technicians fixed them without documenting the error. Thus, it could be difficult to determine the root causes of errors and difficult to know how to prevent future risks if they were never documented.

A Focus on Procedural Error One area that seemed to defy solution, was procedural error as evidenced in the Air Midwest Flight 5481 accident that occurred on January 8, 2003. The Beechcraft 1900D with 19 passengers and 2 crew, lost pitch control during takeoff and crashed killing all on board. Probable cause was determined to be the incorrect rigging of the elevator control system compounded by the airplane's center of gravity, which was substantially aft of the certified aft limit.

> Contributing to the cause of the accident were (1) Air Midwest's lack of oversight of the work being performed at the … maintenance station; (2) Air Midwest's maintenance procedures and documentation; (3) Air

Midwest's weight and balance program at the time of the accident; (4) the Raytheon Aerospace's quality assurance inspector's failure to detect the incorrect rigging of the elevator control system; (5) the Federal Aviation Administration's (FAA) average weight assumptions in its weight and balance program guidance at the time of the accident; and (6) the FAA's lack of oversight of Air Midwest's maintenance program and its weight and balance program. (NTSB, 2004, Executive Summary)

While probable cause was traced to individual actions, the contributing factors assigned responsibility to numerous organizations: the operator, maintenance contractors, manufacturer and regulator. Specific procedure-related recommendations were based on an examination of current task documents. In Figure 21.4, the document on the left is from the operator's detailed inspection work card; the document on the right is from the manufacturer's Aircraft Maintenance Manual. In each case, it was felt that document inadequacies contributed to the failure of the mechanic, quality assurance inspector, and foreman on site, to detect the maintenance errors (i.e., incorrect rigging of the elevator control system).

The deficiencies led to the following requirements for manufacturers and operators of Part 121 aircraft:

- Manufacturers required to identify appropriate procedures for a complete functional check of each critical flight system; determine which maintenance procedures should be followed by such functional checks; and modify their existing maintenance manuals, so that they contain procedures at the end of maintenance for a complete functional check of each critical flight system.
- Part 121 air carriers also required to modify their existing maintenance manuals, so that they contain procedures at the end of maintenance for a complete functional check of each critical flight system.
- Part 121 air carriers required to implement a program in which air carriers and aircraft manufacturers review all work card and maintenance manual instructions for critical flight systems and ensure the accuracy and usability of these instructions so that they are appropriate to the level of training of the mechanics performing the work.

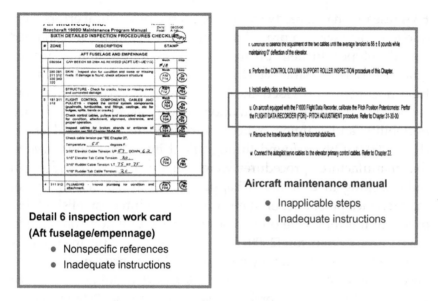

FIGURE 21-4   Deficient documents that contributed to the Air Midwest Flight 5481 accident.

In addition to these requirements, the NTSB report noted that many of the air carrier deficiencies should have been identified through their Continuing Analysis and Surveillance System (CASS) program. This was also the case in the Alaska Airlines Flight 261 accident earlier, and the FAA was working on a revision of the original CASS Advisory Circular to include human factors. In April, 2003, the enhanced Advisory Circular, *AC 120–79: Developing and Implementing Continuing Analysis Surveillance System* (Federal Aviation Administration, 2003) was published.

Maintenance Error Revisited As accidents and incidents continued to point to maintenance errors that jeopardized safety, Phase 1 research in the NASA MHF task tried to establish error descriptions based on incident data, asking for instance, what are the error types, the contributing factors, the contexts in which they occur, and their consequences? Systematic studies took advantage of recently developed error analysis tools, such as MEDA and HFACS-ME as well as the ASRS maintenance database that had been steadily growing.

FAA research had already shown that some procedural errors were due to poorly written procedures and the Document Decision Aid was developed to help document writers follow human factors guidance (Drury, Sarac, & Driscoll, 1997). Analyses of manufacturer documents provided another angle on procedural error. Hall reported that outdated information, as well as access, readability, portability and training issues contributed to procedural errors (Hall, 2002). Others found through field interviews and surveys, that manufacturer procedures were usually seen as accurate, but sadly lacking in usability (Chapparo & Groff, 2002). Surveys of maintenance personnel on their use of procedures established that procedural errors were often cases of procedural non-compliance. Hobbs and Williamson (2000) reported that 80% of the maintainers surveyed, reported that they had deviated from procedures at least once in the past year and nearly 10% reported doing so often or very often. McDonald, Corrigan, Daly, and Cromie (2000) reported that 34% of routine maintenance tasks were performed contrary to procedures.

The yet untapped NASA ASRS maintenance database quickly became a valuable source of additional insights on procedural error. In addition to being a testbed for developing error analysis tools (Hobbs & Kanki, 2003), substantive studies were also conducted. In the area of procedural error, studies confirmed that the causes of procedural error came from a variety sources; sometimes the procedure content (correctness, completeness, ambiguity, or conflicting information), and sometimes due to the usability or the norms and safety culture governing its use (Patankar, Lattanzio, Munro & Kanki, 2003; Kanki, 2005). Other problem areas were researched in the ASRS database such as the use of the Minimum Equipment List (Munro & Kanki, 2003), and the performance of shift turnovers (Parke, Patankar & Kanki, 2003).

Development of Methods and Tools

Maintenance Human Factors was still a relatively new research domain that lacked methods and tools, particularly in the area of process, task and risk analysis, so these topics were researched in the NASA MHF program. Such generic tools would not only aid the research community but could be adapted for use by human factors practitioners in assessing how specific processes and tasks can lead to errors and increased risk.

Risk and Process Modeling Tools Even though visual inspection for aircraft damage is determined by the inspection intervals of maintenance program requirements, a quality assurance department can optimize their resources and training if they track their own reliability data. Applying a risk-based model for estimating probability of visual detection could allow a company to track deviations from baselines and to allocate resources to inspection where it is most needed. Knowing one's own probability of detection can be an effective risk management tool for monitoring the safety margin. (Ostrom & Wilhelmsen, 2008).

Process modeling tools are extremely useful for systems engineers and researchers, but may also prove useful to human factors practitioners who are faced with error-prone processes or the need to build in error capture controls (Eiff & Suckow, 2008). In addition to revealing process traps and inefficiencies, process maps also clarify the interactions and dependencies among organizations. Taking process modeling one step further by the addition of a risk assessment element allows the likelihood of particular error outcomes to be quantified. Where traditional Probabilistic Risk Assessment (PRA) generally models a technical system with some input of human errors, sociotechnical PRA methods model human processes in technical systems. They can help both researcher and practitioner identify the error potential of certain processes and to develop interventions that can be shown to reduce the likelihood of error outcomes (Marx & Westphal, 2008).

Human Factors Interventions

The third element of the NASA MHF approach focused on developing human factors interventions that addressed safety needs identified. But interventions also needed to be validated with metrics that could demonstrate its effectiveness in operations.

MRM Training Revisited Among the interventions addressed by the MHF program was Maintenance Resource Management (MRM) Training as it had been implemented by U.S. operators since the late 1980s. Ten years of development and implementation had shown there to be considerable industry-wide variation. Thus, the research in this program focused on training effectiveness (both short and long term) by means of metrics that could be collected from programs that were just starting as well as those that had been running for many years. The researchers dealt with organizations on an individual basis giving each personalized assessments of their programs. Eventually a large database was established and both longitudinal changes over time and variations across programs could be analyzed. In fact, four generations of MRM training marked the evolution of training that started as an adaptation of CRM (Cockpit Resource Management), to other variations that better served the unique needs and culture of the maintenance community (Taylor & Patankar, 2001). These four variations included:

1. CRM-based Training in Communication Skills and Awareness.
2. Training that Directly Addressed Communicating and Understanding Maintenance Errors.
3. Maintenance Training for Individual Awareness and Readiness.
4. Integrated, behavior-based MRM programs.

Other Process Interventions Other interventions were researched by the analysis of incident and accident data from the ASRS database, followed by experimental studies. For example, research on shift turnovers (Parke & Kanki, 2008), resulted in recommendations based on an experiment that systematically compared multiple communication modes. Another project focused on improving the communication between airline mechanics and pilots via the Maintenance Log (Munro, Kanki & Jordan, 2004). This survey-based research focused on a well-known problem area that could potentially result in serious miscommunications about aircraft discrepancies. Because the Maintenance Log is often the primary means of communicating critical information between pilots and mechanics, the recommendations provided human factors strategies for reducing this type of risk.

Advanced Displays/Technologies The development of new technologies was intended to meet longer term objectives since implementation in operations was thought to be farther off in the future. However, technology advances so quickly that what was a long-term product in 2005 has already changed drastically. For example, a form of video-mediated technology to promote collaboration (e.g., technicians or inspectors with engineering support) was developed, tested, evaluated, and demonstrated in an operational setting (Majoros, 2008). While successfully completed during the MHF program timeframe, today's technologies have already evolved into a new set of information technologies, such as computers linked with real-time audio and video communications, or portable Internet access for shared resources and applications. Nevertheless, the demonstration of the collaboration-supported interactions between engineers, technicians, and inspectors was valuable because the organization could see firsthand how it could enhance their processes. In the context of geographically distributed locations, video-mediated (and other information technology-supported) collaborations could be seen as a way to help resolve a variety of communication, procedural, and technical issues in a timely and effective way.

Virtual reality (VR) technologies to supplement training demonstrated its usefulness for providing accelerated training to aircraft inspectors in what has been traditionally obtained through long years of on-the-job experience (Bowling, Khasawneh, Kaewkuekool, Jiang & Gramopadhye, 2008). These studies examined the impact of using VR to perform general and detailed inspection under time-constrained and non-time-constrained conditions. Analyzing a variety of measures (performance and process measures, subjective ratings, and measures of presence) showed that the VR training provided most benefit for detailed inspections under non-time-constrained conditions. If the appropriate development of a training regime was developed to fit the visual inspection needs of a company, VR training could be a useful addition. Further development is needed for specific uses, but one promising direction examined was the use of VR technology with expert inspectors and novices working together. In the VR environment, enhanced performance (correct detections) and process (effective tracking/scanning techniques) could be imparted to the novice through an expert's demonstration and interactive feedback.

Validation and Implementation of Products

The fourth element of the research approach was to validate and implement products in operations with metrics that are relevant to a company's business case. Since products could be methods or tools, human factors interventions such as training or process improvements, or technologies such as video-mediated devices or virtual reality training, validation of products would be achieved through a variety of means. For example, performance metrics were developed to test for significant benefits of improved procedures, the use of communication aids for shift turnover and VR training. In the case of MRM training, a family of metrics was developed. At the individual level, short-term and long-term metrics—both attitudinal and behavioral—were developed and systematically collected for 15 companies over many years. At the organizational level, return-on-investment metrics were developed and safety data was tracked. In some cases, prototypes were developed and evaluated (e.g., video-mediated collaboration tool) or analysis tools (e.g., risk, process analysis) were demonstrated in actual operations.

While demonstration of products was easily achieved, implementation on a continuing basis was more difficult. Post-9/11, the air transport industry received demanding, security requirements that left many operators under extreme economic strain. Cutbacks, layoffs, and other drastic measures were taken in order to stay in business. New projects were put on hold and, due to security policies, access to operations was more difficult for researchers. Even the annual FAA industry meetings were suspended in 2003 for a couple of years. Since MRM had been a stand-alone program, it was vulnerable to management changes and cuts in resources. However, other human factors products, particularly those that were tools and methods to support improvements (e.g., procedures, risk assessment, process mapping) could be maintained as best practices. Fortunately, many of the human factors principles and lessons learned had already been incorporated into guidance documents and integrated into ongoing programs such as the Continuing Analysis and Safety Surveillance and Safety Management Systems (Federal Aviation Administration, 2006). Implementation of human factors in maintenance was severely tested but the basic knowledge and tools were documented and available.

CHALLENGES FOR THE FUTURE

In spite of challenging times for the aviation industry in general, the state of the art and state of the practice of maintenance human factors has made impressive progress in the last 20 years. The maintenance community enjoyed years of active industry participation and over time, identified key human factors issues in maintenance and inspection, researched many of these issues and delivered many operational solutions. Researchers were welcomed into the operational environment in order to understand the nature of maintenance and inspection processes, procedures and norms. International collaboration and military participation also contributed to the momentum and to the knowledge base. Even though the maintenance community may itself go through active and latent phases, two key concepts must be sustained: maintenance error management and the relationship of maintenance error to safety of flight.

Maintenance Error Management

While specific types of maintenance error change over time, (1) there is a shared consensus on many key problem areas, (2) tools for characterizing maintenance error events are available, (3) guidance for developing human factors programs including error management have been developed, (Patankar, 2004) and (4) lessons learned have emphasized organizational pre-requisites for implementing human factors programs and instilling a safety culture. Accident investigations may have initiated the discussion of maintenance factors, but the expanding focus from individuals to workplace, to organizations, and finally to the system of operators, third party vendors, manufacturers and regulators has resulted in a more complete understanding of root causes, contributing factors, and safety culture issues. Error analysis methods that identify systemic problems as well as local issues can result in more appropriate and effective corrective actions. But in order to be proactive, an organization needs to be able to track its own safety data. The advancement of maintenance human factors has put great emphasis on the collecting of safety information whether it is through a national voluntary reporting system or through company-specific reporting programs.

Maintenance Error and Safety of Flight

Continued vigilance to guard against maintenance and inspection failures has twofold significance. First, maintenance and inspection errors have consequences within technical operations (e.g., repair and rework, component damage, personal injury). In addition, accidents have provided dramatic and devastating proof that maintenance and inspection errors may also negatively impact flight operations. In the IATA Safety Report of 2003, the contribution of maintenance and technical failures was found to be 26% of 92 accidents worldwide. But the "accident scenarios ... are ... often a combination of the precipitant technical failure and the handling of the technical failure by the flight crew" (International Air Transport Association, 2003). As human error in maintenance compromises aircraft reliability, the hazard potential for flight crews increases. Conversely, a reduction in maintenance error safeguards airworthiness and flight operations. While flight crews must be prepared for emergency situations, effective maintenance, and inspection can ensure that many of those situations will never happen.

Concluding Remarks

The foundation of Maintenance Human Factors knowledge and programs was established in a short timeframe. The development of longer term products and highly adaptable generic methods and tools was also begun. But the most critical lessons were learned when organizational changes and economic uncertainty threatened the continued support of maintenance human factors in operations and research. The availability of a knowledge base and tools cannot overcome the lack of an organizational will to value a safety culture. In short, the prerequisites for managing maintenance error risks in ground and flight safety are the following.

1. In spite of economic strain and organizational changes (including major changes in outsourcing), organizations need to uphold a safety culture in order to ensure that safeguards to maintenance and inspection errors are in place, and personnel have an avenue for reporting safety concerns without fear of reprisal.
2. A reporting culture makes it possible to collect and analyze data that forms a basis for developing effective corrective actions and proactive interventions. Although types of errors are likely to change over time, an error management system can help an organization remain current with their safety needs and to have a less reactive, more proactive safety management system.
3. Establishing an error management database is essential for tracking the effects of future changes:
    *   in aircraft structure (e.g., the inspection and repair of new composite material),
    *   in tools (e.g., new nondestructive inspection capability and health monitoring systems),
    *   in organizational responsibilities (e.g., outsourcing and complex organizational oversight relationships).

Much like the field of human factors in general, maintenance human factors issues will persist but will shift form as the workforce, workplace, regulations and resources change. Therefore it is crucial to keep the knowledge base active and growing, and to be sure that the maintenance community has a voice in the aviation safety enterprise. We should not need another accident to remind us that promoting the effective management of maintenance error enhances safety of flight.

References

Air Transport Association. (2002). *Air Transportation Association Specification 113*, http://hfskyway.faa.gov/2007/ATA%20Spec%20113_HF%20Guidelines.pdf Accessed November 2008.

Aviation Safety Reporting System. (2002). *An analysis of ASRS maintenance incidents*. Mountain View, CA: Author.

Boeing. (2003). *Statistical summary of commercial jet aircraft accidents*. Seattle WA.

Bowling, S. R., Khasawneh, M., T., Kaewkuekool, S., Jiang, X., & Gramopadhye, A. K. (2008). Evaluating the Effects of Virtual Training in an Aircraft Maintenance Task. In *Special Issue on Aircraft Maintenance Human Factors, International Journal of Aviation Psychology*, 18(1), 104–116.

Broderick, A. J. (1990). Meeting Welcome. In: *Meeting Proceedings First Federal Aviation Administration Meeting on Human Factors Issues in Aircraft Maintenance and Inspection*. Washington, DC: Federal Aviation Administration/Office of Aviation Medicine (pp. 3–5).

Chapparo, A., & Groff, L. S. (2002). User Perceptions of Aviation Maintenance Technical Manuals. In: *Proceedings of the 16th Symposium on Human Factors in Aviation and Maintenance*. Washington D.C.: FAA/Office Aviation Medicine.

Civil Aviation Authority. (2003). *CAP 716: Aviation Maintenance Human Factors* (EASA/ JAR 145 Approved Organisations) Guidance Material on the UK CAA Interpretation of Part-145, Human Factors and Error Management Requirements. http://www.caa.co.uk/docs/33/CAP716.PDF Accessed November, 2008.

Drury, C. G., Sarac, A., & Driscoll, D. M., (1997). Documentation design aid development, In *Human Factors in Aviation Maintenance, Phase 8: Progress Report*, DOT/FAA/AM-97XX, Springfield, VA: National Technical Information Service.

DuPont, G. (1997). The dirty dozen errors in maintenance. In: *Proceedings of the 11th FAA Meeting on Human Factors Issues oin Aviation Maintenance and Inspection: The work environment in aviation maintenance*. Washington D.C. FAA: Office of Aviation Medicine (pp. 173–184).

Eiff, G., & Suckow, M., (2008). Reducing Accidents and Incidents Through Control of Process. In Special Issue on Aircraft Maintenance Human Factors, International Journal of Aviation Psychology, 18(1), 43–50.

Federal Aviation Administration HFAMI Website http://hfskyway.faa.gov/hfskyway/Library.aspx Accessed November, 2008.

Federal Aviation Administration. (1997). Strategic Program Plan. Internal Document.

Federal Aviation Administration. (2000). *Advisory Circular 120-72, Maintenance Resource Management Training*. Washington DC: The Federal Aviation Administration.

Federal Aviation Administration. (2002). *Advisory Circular 120-66: Aviation Safety Action Program*. Washington DC: The Federal Aviation Administration.

Federal Aviation Administration. (2003). *Advisory Circular 120-79: Developing and Implementing Continuing Analysis and Surveillance System*. Washington DC: The Federal Aviation Administration.

Federal Aviation Administration. (2006). *Advisory Circular 120-92: Introduction to Safety Management Systems*. Washington DC: The Federal Aviation Administration.

Hall, D. A. (2002). Aviation Maintenance Documents and Data in the United Kingdom. In: *Proceedings of 16th Symposium on Human Factors in Aviation and Maintenance*. San Francisco: CA.

Hobbs, A. (1995). Human Factors in Airline Maintenance. *Asia Pacific Air Safety, pp. 2–6, March, 1995.*

Hobbs, A., & Kanki, B. G. (2003). A correspondence analysis of ASRS maintenance incident reports. In: *Proceedings of 12th International Symposium on Aviation Psychology*. Dayton: OH: Wright State University.

Hobbs, A., & Williamson, A. (2000). *Aircraft Maintenance Safety Survey: Results.* Canberra: Australian Transport Safety Bureau.

International Air Transport Association (2003) Annual Safety Report.

International Civil Aviation Organisation. (2003). *Human Factors Guidelines for Aircraft Maintenance Manual* 9824, AN/450. Montreal, IATA Annual Safety Report.

Kanki, B. G. (2005). Managing Procedural Error in Maintenance. In: *Proceedings of the 58th Annual International Air Safety Seminar*. Alexandria: VA: Flight Safety Foundation (pp. 233–244).

Kanki, B. G., Blankmann-Alexander, D., & Barth, T. S. (1998). Human Factors in Aerospace Maintenance: Perspectives from NASA Research and Operations. In: *Proceedings of the FAA/CAA-UK/Transport Canada, 12th Symposium on Human Factors in Aviation Maintenance and Inspection.*, Gatwick Airport, UK.

Majoros, A. (2008). Video-Mediated Collaborative Engineering Support In Special Issue on Aircraft Maintenance Human Factors. *International Journal of Aviation Psychology, 18*(1), 117–134.

Marx, D., & Westphal, J., (2008). Socio-technical Risk Assessment in Maintenance. In *Special Issue on Aircraft Maintenance Human Factors, International Journal of Aviation Psychology*, 18(1), 51–60.

McDonald, N. (1998). Human factors and aircraft dispatch and maintenance safety. Paper presented at the Nouvelle Revue D_aeronautique et d_astronautique. 3 Aero Days Post-Conference Proceedings.

McDonald, N., Corrigan, S., Daly, C., & Cromie, S. (2000). Safety management systems and safety culture in aircraft maintenance organizations. *Safety Science, 34*, 151–176.

Munro, P. A., & Kanki, B. G. (2003). An Analysis of ASRS Maintenance Reports on the Use of Minimum Equipment Lists. In: *Proceedings of the 12th International Symposium on Aviation Psychology*. Dayton: OH: Wright State University.

Munro, P. A., Kanki, B. G., & Jordan, K. (2004). Reporting discrepancies: An assessment of the informational needs of airline pilots and mechanics. In: *Proceedings of the Safety Across High-Consequence Industries Conference*. St Louis, MO: St Louis University.

National Aeronautics and Space Administration. (2000). *Shuttle Independent Assessment Team Report*, Office of Space Flight. http://www.hq.nasa.gov/osf/shuttle_assess.html

National Transportation Safety Board. (1989). *Aircraft Accident Report: Aloha Airlines, Inc., Flight 243, Boeing 737-200, N73711, near Maui, Hawaii, April 28, 1988.* (NTSB/AAR-89/03, PB89-910404), Washington DC: Author.

National Transportation Safety Board. (1990). *Aircraft Accident Report: United Airlines Flight 232, McDonnell Douglas DC-10-10, Sioux Gateway Airport, Sioux*

City, Iowa, July 19, 1989. (NTSB/AAR-90/06, PB90-910406), Washington DC: Author.

National Transportation Safety Board. (1992). *Aircraft Accident Report: Britt Airways, Inc., d/b/a Continental Express Flight 2574, In-flight Structural Breakup, EMB-120RT, N33701, Eagle Lake, Texas, September 11, 1991.* (NTSB/AAR-92/04, PB92-910405), Washington DC: Author.

National Transportation Safety Board. (1997). *Aircraft Accident Report: In-Flight Fire and Impact with Terrain, ValuJet Airlines Flight 592, DC-0-32, N904VJ, Everglades, Near Miami, Florida, May 11, 1996.* (NTSB/AAR-97/06, PB97-910406), Washington DC: Author.

National Transportation Safety Board. (2002). *Aircraft Accident Report: Loss of Control and Impact with Pacific Ocean, Alaska Airlines Flight 261, McDonnell Douglas MD-83, N963AS, About 2.7 Miles North of Anacapa Island, California, January 31, 2000.* (NTSB/AAR-02/01, PB2002-910402), Washington DC: Author.

National Transportation Safety Board. (2004). *Aircraft Accident Report: Loss of Pitch Control During Takeoff, Air Midwest, Flight 5481, Raytheon (Beechcraft) 1900D, N233YV, Charlotte, North Carolina, January 8, 2003,* (NTSB/AAR—04/01, PB2004-910401), Washington DC: Author.

Ostrom, L. & Wilhelmsen, C., (2008). Developing Risk Models for Aviation Maintenance and Inspection. In Special Issue on Aircraft Maintenance Human Factors, *International Journal of Aviation Psychology,* 18(1), 30–42.

Parke, B., Patankar, K., & Kanki, B. G. (2003). Shift Turnover Related Errors in ASRS Reports. In: *Proceedings of the 12th International Symposium on Aviation Psychology.* Dayton, OH: Wright State University.

Patankar, M. (2004). *Development of guidelines and tools for effective implementation of an Aviation Safety Action Program (ASAP) for aircraft maintenance organizations.* FAA Grant Number 2003-G-013.

Patankar, K., Lattanzio, D., Munro, P. A., & Kanki, B. G. (2003). Identifying Procedural Errors in ASRS Maintenance Reports using MEDA and Perilog. In: *Proceedings of the 12th International Symposium on Aviation Psychology.* Dayton, OH: Wright State University.

Patankar, M., & Tayor, J. C. (2008). MRM Training, Evaluation and Safety Management In *Special Issue on Aircraft Maintenance Human Factors, International Journal of Aviation Psychology,* 18(1), 61–71.

Rankin, W. L. (1999). *Investigating Maintenance Error Using the Maintenance Error Decision Aid (MEDA) Process.* Pre-conference workshop at the International Aviation Forum entitled Human Factors in Aircraft Maintenance—Integrating Safety Management Systems to Minimise Risk, pp. 26–28, April 1999, The Grand Hotel, Amsterdam, The Netherlands.

Rankin, W., & Allen, J. (1996). Boeing introduces MEDA, Maintenance Error Decision Aid. *Airliner, April–June,* 20–27.

Rankin, W. L., & Sogg, S. L. (2003). *Update on the Maintenance Error Decision Aid (MEDA) Process.* Paper presented at the MEDA/MEMS Workshop and Seminar. May 21–23, 2003, Aviation House, Gatwick, UK, 2003.

Reason, J. (1990). *Human error.* New York: Cambridge University Press 199–209.

Russell, S., Bacchi, M., Perassi, A., & Cromie, S. (1998). *Aircraft Dispatch And Maintenance Safety (ADAMS) reporting form and end-user manual* (European

Community, Brite-EURAM III report. BRPR-CT95-0038, BE95-1732). Dublin, Ireland: Trinity College.

Schmidt, J., & Figlock, R. (2001). *Development of MCAS: A web based maintenance climate assessment survey*. Paper Presented at 11th International Symposium on Aviation Psychology. Columbus, Ohio.

Schmidt, J. K., Schmorrow, D., & Hardee, M. (1998). *A preliminary human factors analysis of naval aviation maintenance related mishaps* (SAE Technical Paper 983111). Warrendale, PA: Society of Automotive Engineers.

Taylor, J. C., & Patankar, M. S. (2001). Four Generations of Maintenance Resource Management Programs in the United States: An Analysis of the Past, Present, and Future. *Journal of Air Transportation World Wide, 6*(2), 3–32.

White House Commission on Aviation Safety and Security. (1997). Final Report to the President, February 12. Washington, D.C. Available on the internet: http://www.fas.org/irp/threat/212fin~1.html

PART VI

NEXTGEN
COMMENTARY

Commentary on NextGen and Aviation Human Factors

Paul Krois, Dino Piccione and Tom McCloy
Federal Aviation Administration

The Federal Aviation Administration (FAA) Next Generation Air Transportation System (NextGen) poses unprecedented changes in the roles and responsibilities traditionally held by pilots and air traffic controllers (FAA, 2009b) dating back to the late 1920s when Archie League served as one of the first flagmen at the St. Louis, Missouri, airfield (Kraus, 2009). These changes are being enabled by implementation of new technologies, concepts, and procedures. The goals driving these changes are to improve capacity of the National Airspace System (NAS) and reduce flight delays while maintaining the strong safety record of civil aviation. Human factors has a great opportunity to contribute to and fashion these changes to ensure that pilots and controllers can perform their tasks within the envelope of effective and efficient human performance in a manner conducive to accruing these benefits intended with NextGen. One of several important human factors challenges is that as a system becomes more tightly coupled to ensure efficiencies, the less time there is available to respond to non-normal situations.

NextGen will provide operational improvements changing the jobs of pilots and air traffic controllers. Concepts for the mid- and far-term will introduce changes to the roles and responsibilities between pilots and controllers such as in relation to delegation of separation responsibility as pilots use satellite-based surveillance information shown on a cockpit display of traffic information (CDTI). New technologies will enable changes in the allocation of functions between humans and automation so that controllers can shift toward more strategic handling additional aircraft while pilots are delegated certain tactical responsibilities.

Changes in roles and in allocations of function necessitate new procedures that integrate air and ground responsibilities. Procedures are important to ensure operations proceed as expected under normal conditions and a range of non-normal conditions such as associated with equipment degradation or system failure, no matter how reliable the system is designed to operate.

Key to human performance is ensuring that design mitigates the potential for human error recognizing that new automation and procedures may introduce new kinds of errors, and that new technologies and kinds of airspace may introduce new types of threats to normal operations, and if not effectively managed these errors and threats may lead to new types of undesired states. Pilots and controllers will need to maintain situation awareness under new or different operational conditions while adjusting to changes in workload, crew size, and flight time.

Pilots and controllers will see changes in different phases of flight. For example, flight planning will be different with use of digital information exchange conveying airspace limitations associated with current or forecast weather or with traffic congestion. During push-back, taxi, and departure, pilots will use moving map displays with own ship position to ensure situation awareness and runway safety. Pilots will use multiple different precise departure paths from each runway end to ensure safe separation between aircraft and from wake vortices or preceding aircraft. Pilots will use accurate satellite-based position information in conjunction with reduced separation standards for expeditious climbs as air traffic controllers delegate spacing responsibility to the flight as it climbs past other flights to merge into the overhead stream at cruising altitude. During descent and approach pilots will, with properly equipped aircraft, fly conflict-free precise vertical and horizontal paths called optimized descent profiles that reduce flight time, fuel and emissions. In landing, taxi and arrival phases the ground system will recommend the best runway and taxi path for controllers use in issuing clearances. Pilots and controllers will monitor aircraft movement using specialized displays to assure runway safety.

Human factors are being addressed in NextGen using a human-system integration (HSI) approach. The HSI roadmap is one of several roadmaps comprising the enterprise architecture for the National Airspace System (NAS), which is accessible on the Web (FAA, 2009a). The roadmap couples human factors throughout NextGen development by identifying operational improvements aligned with other roadmaps (e.g., automation, avionics, air-ground integration) for which human performance is important. Through this early planning, we know the time frames for when our research and development outputs need to be provided.

For controllers, NextGen poses important opportunities to improve efficiency. Human factors considerations need to be addressed in the convergence of automated tools in en route, terminal, and tower controller workstations. Transitions of these tools and associated new procedures should consider maintaining situation awareness and workload and mitigating the potential for human error. In a broader vein, criteria for selecting applicants for controller jobs will be updated corresponding with changes in aptitudes needed for using new automation and for different roles and responsibilities. New training techniques will be needed to ensure proficiency with changes in tasks and tools.

For pilots, NextGen will bring critical challenges with regard to ensuring safety and the integration of air and ground systems. This includes defining, understanding, and developing regulatory guidance for successful implementation of changes in roles and responsibilities between pilots and controllers and in changes in how functions are allocated between humans and automation that are required for NextGen capabilities and applications. Research needs to define human and system performance requirements and guidance for the design and operation of aircraft and air traffic management systems, including identifying information needs, human capabilities, interface design, and system integration issues. For airport movements, research will evaluate and recommend minimum display standards for use of enhanced and synthetic vision systems, as well as airport markings and signage to conduct movements across a range of visibility conditions. Research will also provide guidance for the design of cockpit displays and alerts to support delegated separation responsibility and evaluate the impact and potential risks associated with use of Traffic Alert and Collision Avoidance System (TCAS) in NextGen procedures. These efforts will be undertaken in concert with FAA regulatory offices and industry groups to ensure the scientific and technical foundation of human factors and human performance information can support development of industry standards and aerospace recommended practices.

HUMAN FACTORS AND NextGen FLIGHTDECKS

NextGen will bring many changes to the flightdeck for properly equipped aircraft and trained aircrews, and the NAS will adjust to accommodate other aircraft not fully equipped. There is a body of human factors knowledge about pilot performance and equipment design for legacy systems. Advances introduced by NextGen will create gaps between existing knowledge and what is needed to understand crew performance using new technology, automation, and procedures including in the context of human error and pilot performance during both normal and non-normal operations.

One of the biggest changes in roles and responsibilities for the flight crew involves reduced aircraft separation and a shift toward flightdeck separation assurance responsibilities. Human factors considerations involve understanding changes in roles, responsibilities, displays, and alerts for collision avoidance and separation assurance. This includes the impact of human behavior relative to trust, use, and compliance with collision avoidance advisories and alerts including automated maneuvers. Pilot performance using closely spaced parallel runway operations should consider potential for missed approach procedures and human error, which poses a particular air-ground integration challenge in relation to potential differences in timing of conflict alerts associated with blunders. Similar pilot performance considerations may be associated with oceanic in trail procedures as well as merging and spacing operations.

The classic allocation of function between pilots and automation in today's flightdecks changes depending upon the environmental conditions and desired aircraft performance or state. Implementation of NextGen capabilities with new automation necessitates understanding the substantial changes in which some new systems may be nondeterministic in nature and automate pilot cognitive functions. Current flight management systems and other flightdeck displays have performance capabilities that could be leveraged by more complex NextGen operational procedures although the user interfaces are not necessarily intended for use in more complex procedures. With NextGen automation and procedures enabling pilots to interact with pilots of other aircraft and with controllers in new ways, there may be potential for pilots to dynamically make unintended use of new systems in ways not foreseen during design and implementation of NextGen technology and procedures and having potentially undesired consequences.

It is generally recognized that human performance and human error are the largest contributors to variability in system performance. Major changes in NextGen flightdecks provide new opportunities for error as well as a change in the nature and frequency of existing error patterns. Understanding system risk and developing effective error mitigation strategies necessitates modeling and analysis addressing error detection, identification, recovery, and response times. This includes interactions between pilots and controllers in use of automation and procedures in terms of data and voice communications and how data are used. Harmonization of applications and operations to mitigate potential international differences can be done by identifying potential differences, assessing the impact of these differences, and developing potential mitigation strategies.

Transitions of digital data communications technology will enable flightdeck automation and avionics to support flight crew decision making, receipt of clearances from controllers, and changes to four dimension trajectories. Voice communications will continue to be used for unusual circumstances such as when time is an important factor. Human factors considerations involve the adequacy of the data communication message set, displays to ensure shared situation awareness, negotiation, and other procedures, and circumstances in which data communication use should be avoided.

During ground surface movement and runway occupancy especially in low visibility conditions, Enhanced Flight Vision Systems (EFVS) and Synthetic Vision Systems (SVS) will enhance situation awareness and information sharing. Human factors considerations with EFVS and SVS include transitioning between surface and airborne modes, their integration with CDTI, and procedures involving degraded data quality. Other aspects of ground operations include airfield signage, lighting, and ground collision avoidance.

Human factors considerations are important to accruing benefits of trajectory based operations (TBO). TBO requires pilot negotiation, selection, implementation, and monitoring of aircraft state relative to position and time. To accommodate pilot tasks and information needs, navigation displays will integrate time information with position information and navigation targets as well as vertical situation displays. This information will be interoperable with time-based display information used by air traffic controllers. Air and ground procedures will be compatible. Also, implementation of TBO will require the flight crew to detect, plan, and execute avoidance maneuvers in weather events where minimum separation may be lost such as due to aircraft nonconformance or changes in turbulence, convection, and other weather conditions.

Specific instrument procedures for NextGen operations will be developed using human factors guidance in relation to special aircraft equipage, pilot training, operational experience, and proficiency. Human factors considerations pertain to the design of instrument procedures and associated charts to be usable and flyable by appropriately qualified pilots in a manner mitigating potential error. NextGen procedures will introduce new charting capabilities such as with terminal and en route phases of flight becoming more closely integrated as exemplified with the optimized descent profile being one continuous procedure from en route through terminal arrival and approach. In addition, electronic flight bags and other displays may have a limiting effect on the amount of detail and layout of procedures that can be displayed at one time. New formats may be needed for effective crew use.

The changes in roles, responsibilities and tasks translate into a need for the flight crew to develop and maintain appropriate skills to use new automation in conducting NextGen operations. These changes pose a potential for skill loss as new NextGen automation decreases use of manual skills. Training mitigations will need to be identified to prevent loss of safety-critical flight crew skills. At the same time, crews will be trained on new systems and procedures in relation to aircraft-specific equipage and performance requirements, and types of airspace to be accessed such as NextGen high-density terminal airspace. The effectiveness of flight crew training techniques for mitigating human error in pilot performance will be extended to crew certification for NextGen equipment and operations.

Many NextGen applications involve a shift away in some manner from direct aircraft control and toward cognitive activities and decision making. At one level, the timing, content, and procedures for use of shared information by the flight crew in effective decision making will be supported through training and effective design. At another level, collaborative decision making between the flight crew, dispatchers at airline operations centers, and traffic management specialists can use distributed information on current and forecast weather, traffic congestion, and airspace for efficient planning and rerouting.

Lastly, single pilot operations will operate in high-density terminal areas and other airspace in which NextGen procedures will be in effect with some aircraft using NextGen automation. Flightdeck management, operations, and training challenges for the crew of single-pilot aircraft, such as Very Light Jets (VLJs), are significant. Human factors considerations include changes in single pilot workload in relation to legacy aircraft, use of EFVS and SVS for all-weather ground operations, and changes in visual and manual workload associated with use of data communications controls and displays. Special training may be required to allow a single-person flight crew to participate in certain NextGen operations.

NextGen HUMAN FACTORS AND ATC

Automation and technology must work in concert with the humans in the ATC system to meet NextGen efficiency levels. Human factors facilitate the effective and efficient integration of the various NextGen capabilities into a workable solution that intelligently adds the many new technologies, decision support tools, and other automation to the different workstations to achieve the desired performance outcome.

A core human factors issue is to ensure that safety is maintained. Information on intent as well as positive information on delegation of authority for separation responsibility must be clear and unambiguous. Analysis of new types of human error modes is essential to manage safety risk in the changing environment. Interoperability of air and ground decision support tools necessitates synchronization of conflict probe look-ahead times, trajectory intent information, and alerting functions such as for closely spaced parallel runways in use of cockpit display of traffic information and precision runway monitor.

A parameter representing the complexity of the human factors challenge with NextGen is shown through a preliminary analysis of roadmaps in the enterprise architecture. This analysis indicated that some 12 or more tools will converge into the workstation of the TRACON controller and 10 into the en route workstation. Human factors considerations are important to ensuring the accrual of benefits intended by each of these tools in relation to situation awareness, workload, potential for human error, and so forth as workstations undergo incremental transitions.

Additional human factors considerations include flow of collaborative air traffic management information and communications ensuring shared situation awareness and effective decision-making.

The screening and selecting of applicants for controller jobs and training controllers to use NextGen automation and procedures can be determined as design is completed and new systems transition for implementation.

FINAL THOUGHTS

Design solutions to these challenges and opportunities can be brought about by developing a taxonomy describing the relationships between pilots and controllers and their associated automation systems. Research will develop guidance to support certification personnel in assessing the suitability of design methods to support human error detection and mitigation. Research will also assess legacy flightdeck avionics for compatibility in use of these systems in NextGen flight procedures.

NextGen necessitates close collaboration with stakeholders, customers, sponsors, and researchers in government, industry, and academia. There are many opportunities to leverage expertise and laboratory resources while also avoiding duplication of efforts. For academia, NextGen affords new opportunities to educate and produce the future generation of aviation human factors experts as a national capacity for cutting-edge R&D.

Acumen with human factors necessitates coming up with the right analysis or design, using the right resources, at the right time, to assure the human contribution for delivering NextGen benefits. In doing so, we must be vigilant to ensure situation awareness, manage workload, and mitigate human error. This is a truly a special time to be working in aviation human factors making a contribution to NextGen.

DISCLAIMER

The views expressed are those of the authors and do not represent the Federal Aviation Administration.

References

Federal Aviation Administration. (2009a, January). NAS Architecture 6. Retrieved July 24, 2009, from <http://www.nas-architecture.faa.gov/nas/>.

Federal Aviation Administration. (2009b, February). NextGen Implementation Plan, Revised october 2, 2009, from <http://www.faa.gov/about/initiatives/nextgen/media/NGIP_0130.pdf>.

Kraus, T. L. (2008). *The Federal Aviation Administration: A historical perspective, 1903–2008.* Washington, D.C.: U.S. Department of Transportation, Federal Aviation Administration.

Index